现代农业科技专著大系

中国枣品种资源图鉴

THE ILLUSTRATED GERMPLASM RESOURCES OF CHINESE JUJUBE

李登科　牛西午　田建保　主编

Edited by Dengke Li　Xiwu Niu　Jianbao Tian

中国农业出版社

China Agricultural Press

图书在版编目（CIP）数据

中国枣品种资源图鉴 / 李登科，牛西午，田建保主编．—北京：中国农业出版社，2010.9
ISBN 978-7-109-14752-2

Ⅰ．①中… Ⅱ．①李…②牛…③田… Ⅲ．①枣－品种资源－中国－图谱 Ⅳ．①S665.102.4-64

中国版本图书馆CIP数据核字（2010）第128543号

中国农业出版社出版
（北京市朝阳区农展馆北路2号）
（邮政编码 100125）
责任编辑　舒　薇　杨金妹　贺志清

北京中科印刷有限公司印刷　新华书店北京发行所发行
2013年9月第1版　2013年9月北京第1次印刷

开本：889mm×1194mm　1/16　印张：34.5
字数：1 042千字　印数：1~3 000册
定价：300.00元
（凡本版图书出现印刷、装订错误，请向出版社发行部调换）

《中国枣品种资源图鉴》编委会

主　编　李登科　牛西午　田建保
副主编　王永康　武　威　李　捷
编委会（按姓氏音序排列）
　　　　　毕　平　杜学梅　韩　凤　康振英
　　　　　来发茂　李　捷　李登科　李田丁
　　　　　牛西午　任海燕　隋串玲　田建保
　　　　　王　莉　王永康　武　威　张志善
　　　　　赵爱玲
审　稿　刘孟军

序　一

　　枣是原产我国的特有果树，栽培历史悠久，品种资源丰富。山西省农业科学院果树研究所国家枣种质资源圃是农业部批准建立的国家资源圃之一。建圃30多年来，一直从事枣品种资源的收集保存、鉴定评价和创新利用等工作，积累了大量珍贵的数据资料，取得了丰硕的成果。《中国枣品种资源图鉴》一书正是该圃几代人长期科研积累的结晶，是我国枣种质资源研究领域的代表性著作之一。

　　该书以品种资源图说为主线，采用中英文对照形式，重点详述了250个枣地方品种的原产地、栽培分布及主要性状特性，并补充列出了其他具有优异性状或潜在价值的品种资源图像信息。主题突出、层次分明、内容丰富、图片逼真、数据翔实，具有重要的学术研究和生产利用价值，必将对我国枣种质资源的科研、教学、生产和国际交流合作起到重要的参考和促进作用。

刘旭

中国工程院院士
中国农业科学院研究员
2013年6月30日

Preface 1

    Chinese jujube is a native and characteristic fruit to China with a long cultivation history and rather abundant resources. The National Jujube Germplasm Repository established in Pomology Institute, Shanxi Academy of Agricultural Sciences is one of national-level germplasm repository supported by the Ministry of Agriculture. This germplasm repository has been involved in collection, preservation, identification and evaluation of jujube germplasm for over 30 years from its establishment. A lot of valuable data and achievements have been acquired. *The Illustrated Germplasm Resources of Chinese Jujube* is the very fruit of its long-term research work and could be identified as one of representative works about research field in Chinese jujube germplasm.

    In this book, it is the main line that jujube varieties are illustrated by Chinese and English and pictures. It is described in detail that the original birthplace, cultivation status and main characteristics of 250 cultivated jujube varieties in China and pictures of varieties with special characters or potential research value have been listed supplementarily. The book has prominent subject, rich content, clear layout, detailed data, fine texts and living pictures. Consequently, it is promising that this book play an important reference and promotion role for scientific research, teaching, production and academic exchange.

Xu Liu
Academician of the Chinese Academy of Engineering
Senior Research Fellow of the Chinese Academy of Agricultural Sciences
Jun.30, 2013

序 二

枣是我国原产重要果树和第一大干果经济林树种，栽培历史悠久，品种资源非常丰富。枣品种资源是开展枣树科研和生产的物质基础，对枣产业可持续健康发展具有重要的支撑和保障作用。

农业部国家枣种质资源圃依托山西省农业科学院果树研究所建立，长期致力于枣种质资源的收集保存和鉴定评价研究工作。历经几代人的艰辛努力，收集保存了全国24个省、直辖市、自治区的枣种质资源600多个，并系统鉴定评价了其植物学特征、生物学特性、果实经济性状及适应性、抗逆性等。《中国枣品种资源图鉴》一书，是农业部国家枣种质资源圃科研团队长期科研积累的结晶，是继《中国果树志·枣卷》和《中国枣种质资源》之后我国枣种质资源研究领域的又一代表性著作。

《中国枣品种资源图鉴》一书详述了我国主栽和优良品种资源的原产地、栽培现状和主要性状特性，并概述了具有特异性状或潜在研究价值品种资源的信息。国家枣种质资源圃位于我国枣树栽培的中心地带，该书中的数据是在国家枣种质资源圃多年系统观察和鉴定评价的基础上经过科学分析形成的，可信度高，代表性和可比性强。

该书以品种资源图说为主线，按鲜食、制干、兼用、蜜枣和观赏品种资源等不同用途分类介绍；每类按品种资源的重要性、规模面积等依次排列；每个品种资源从来源分布、植物学特征、物候期、生物学特性、果实经济性状及总体评价等六个方面加以介绍，具有较强的条理性和阅读性，而且还配以树体、开花结果、果实特写和剖面特征等典型彩色图片，直观形象地反映了各品种资源的特性。

该书主题突出、内容丰富、层次分明、数据翔实、图文并茂，理论与实践兼顾，对枣品种资源的生产利用具有很强的指导意义，对于深入开展品种资源和育种研究亦具有重要的参考借鉴价值，同时有助于开展相关的国际交流与合作。该书的出版必将对推动枣品种资源的深入研究和创新利用产生广泛而深远的影响。

国际园艺学会枣属植物工作组主席
中国园艺学会干果分会理事长
河北农业大学教授
2013年6月30日

Preface 2

Jujube is an important fruit tree species and the largest species of economic dry fruits forest in China. It is native to China with a long cultivation history and rather abundant resources. Jujube variety resources are the fundamental materials for scientific research and production of jujube trees and fruits, and play an important supporting and security role for healthy and sustainable development of jujube industry.

The National Jujube Germplasm Repository of the Ministry of Agriculture is established in Pomology Institute, Shanxi Academy of Agricultural Sciences. The research personnel of the Repository have been involved in collection, preservation, identification and evaluation of jujube germplasm for a long time. Under hard work of several generations, more than 600 jujube germplasms from 24 provinces of China have been collected and preserved here, whose botanical and biological characteristics, economic traits of jujube fruit, adaptability and stress tolerance have been systematically identified and evaluated. *The Illustrated Germplasm Resources of Chinese Jujube* is the fruit of long-term hard work of the research team at the National Jujube Germplasm Repository.

The book describes in detail the original birthplace, cultivation status and main characteristics of the mainly-cultivated, elite and rare jujube varieties in China. Varieties with special characters or potential research value are also briefly introduced. The National Jujube Germplasm Repository is located in the central area of jujube cultivation in China. The data in this book are formed as a result of scientific analysis based on long-term systematic observation, identification and evaluation made in the Repository, so they are highly reliable with strong representativeness and comparability.

In this book, it is the main line that jujube varieties are illustrated by words and pictures. The varieties are divided into different categories according to different uses of the fruits, such as table varieties, drying varieties, multipurpose varieties, candied varieties and ornamental varieties. Varieties in each category are ordered in accordance with the importance and cultivation scale, and each variety is introduced in terms of origin, distribution, botanical characters, biological characteristics, economic traits of fruit and general evaluation. Matched by typical color pictures of tree body, flowering and bearing, close-up of jujube fruit and features of fruit section, the characteristics of each variety are reflected visually and vividly.

The book has prominent subject, rich content, clear layout, detailed data, fine texts and pictures, and gives attention to both theory and practice, which makes it a useful guide for production and utilization of jujube variety resources and a valuable reference for further research on germplasm and breeding. It is also helpful for developing related international communication and cooperation. In a word, the publication of this book will surely have a deep and wide effect on promotion of further research and creative utilization of jujube variety resources.

Mengjun Liu
Chair of the ISHS Jujube Working Group
President of Dry-Fruit Branch, CSHS
Professor of Hebei Agricultural University
Jun. 30, 2013

前　言

　　枣（Ziziphus jujuba Mill.）为鼠李科（Rhamnaceae）枣属（Ziziphus Mill.）植物，是最具中国特色的优势果树。据大量出土文物考证、古文献记载和现存的古枣树群落分布表明，我国是枣树的栽培起源中心。山东临朐解河村出土的枣树叶片化石距今已2 400万年，河南密县莪沟北岗新石器时代遗址发掘出的炭化枣核和干枣有7 240多年的历史；文字记载最早的是见于约3 000年前的《诗经》中，其后的《尔雅》、《史记》、《神农本草经》、《齐民要术》和《本草纲目》等古文献对枣的品种、栽培、加工和药用价值等都作了记载和说明；另外，地处黄河流域的山西、陕西、河北、河南和山东等地发现有几百年到上千年的古老枣树林。

　　枣品种资源是枣树科研和生产的物质基础，对新品种选育和生产利用具有基础性和关键性的支撑作用。我国枣品种资源收集保存和利用研究已有7 000多年的历史，在漫长的自然演化和人为选择过程中形成了极为丰富的品种资源。仅古文献记载的枣品种总数就达500个左右，以《广群芳谱》和《植物名实图考》中记载的品种最多，达到87个。尤其是新中国成立后，我国的枣树科技工作者在全国范围内多次开展了品种资源调查和鉴定评价等研究工作，并于1993年编辑出版了《中国果树志·枣卷》。该书主要以文字描述的形式详细记载介绍了枣品种资源700个，成为全国枣树科研、教学、生产和经营者必备的工具书。

　　山西省农业科学院果树研究所建所之初即开展了枣品种资源调查工作，并于1963年开始建立山西省枣品种资源圃，至1965年收集保存品种资源56个（份）。1979年，承担了农业部国家枣种质资源圃的建设任务，从此在全国范围内开始了大规模的品种资源考察和收集保存工作，之后对圃内保存的品种资源进行了植物学性状、生物学特性、果实经济性状及适应性、抗逆性等基本农艺性状的系统观察和鉴定评价。2000年起又承担了农业部农作物种质资源保护项目、国家科技基础条件平台、国家科技基础性工作专项等科研任务，进一步开展了枣品种资源的收集保存和原产地追溯、主要农艺性状和功能性成分的系统调查、鉴定评价技术规范研究、DNA遗传多样性标记等工作，积累了大量的数据资料，取得了丰硕的科研成果。截至2012年底，该圃占地总面积已达11hm^2，收集保存全国24个省、直辖市、自治区的枣品种资源630个（份），是我国目前规模面积最大、种质数量最多、遗传多样性最丰富、基础设施完备、仪器设备齐全、研究手段先进、管理技术一流的现代化国家枣种质资源圃。现已成为基础科学研究和野外观测基地，科技合作和学术交流基地、试验示范和成果展示基地、人才培养和科普教育基地。

　　为了系统总结国家枣种质资源圃多年的研究成果和最新研究进展，充分展示我国枣品种资源的性状特性和现状概况，满足广大枣树科研、教学和生产者的迫切需求，最大限度地实现枣品种资源的有效共享和进一步提高枣品种资源、育种研究及利用水平，山西省农业科学院果树研究所国家枣种质资源圃科研团队编撰了《中国枣品种资源图鉴》一书。

本书以枣品种资源图说为主线，采用中英文对照的方式，以枣果的不同用途为结构单元，将每个品种资源列为一个板块的编排格式，详实介绍了250个品种资源的来源分布、植物学特征、物候期、生物学特性、果实性状及总体评价等，并配以树体、开花结果、果实特写和剖面特征等彩色图片共1 400余幅。本书对枣品种资源中普遍存在的同名异物和同物异名的问题，尽量做到命名规范化，即在原品种名称前冠以原产地名，但对分布范围广泛、原产地难以确定和习惯公认的品种资源则保留其原名称不变。对难以用文字完整描述的114个品种资源，以果实特写图片的形式来反映。为方便查阅使用，书末还列出了以品种资源汉语拼音排序的检索表。

本书的编写，力求结构层次分明、内容系统丰富、数据翔实可信、图片清晰美观、编排规范新颖，集先进性、科学性、全面性、新颖性、实用性和鉴赏性于一体。旨在为进一步开展枣品种资源研究和新品种选育提供有价值的借鉴和参考，为国内外从事枣树科研、教学、生产和经营管理等工作的单位和人员提供有益的帮助和指导，为枣产业的健康和可持续发展提供可靠的支撑和保障。

本书是在国家农作物种质资源保护项目（NB05-070401-2，NB2012-2130135-02）、国家科技基础条件平台（2005DKA10300，2012-051）、国家科技基础性工作专项（2011FY110200，2012FY110100，2013FY111700）、国家林业公益性行业科研专项（201004041）、国家科技支撑计划（2008BAD92B03-13，2013BAD14B03）等国家和省部级项目专项经费资助下，在山西省农业科学院果树研究所国家枣种质资源圃几代科研和管理人员的辛勤努力下以及全国各地枣产区相关人员的鼎力帮助下，以国家枣种质资源圃多年的自主研究成果积累为主体，参考国内外相关文献资料编写而成的，在此向他们表示最诚挚的谢意。但由于时间和水平所限，本书的不足和疏漏之处在所难免，敬请各位专家、同仁和读者批评指正。

山西省农业科学院果树研究所
《中国枣品种资源图鉴》编委会
2013年6月30日

Introduction

Chinese Jujube (*Ziziphus jujuba* Mill.) is a member of *Ziziphus* Mill. in Rhamnaceae. According to a large number of unearthed cultural relics, record of historical documents and existing ancient jujube-tree community, it is found that China is the origin of jujube cultivation. Leaf fossils unearthed at Xiehe Village of Shandong Province are believed to be 24 million years old, and carbonized jujube stone and dried jujube fruits found at Neolithic sites of Mixian County in Henan Province have a history of more than 7 240 years. The earliest written records on jujube cultivation can be found in the Book of Songs of 3 000 years ago. Besides, hundreds-to-thousand-year-old jujube forests also exsit in provinces of Shanxi, Shaanxi, Hebei, Henan and Shandong located in Yellow River Basin.

Jujube variety resources are the fundamental materials for scientific research and production of jujube. Especially for breeding of new cultivars and utilization on production, the varieties play a fundamental and critical supporting role. Collection, preservation, utilization and research for jujube germplasm in China were started more than 7 000 years ago, and rich variety resources were produced in the long process of natural evolution and artificial selection. After 1949, survey and evaluation on jujube variety resources have been carried out for many times by scientific and technological workers on jujube research across the country. The book of *Fruits of China* (*Vol. Jujube*) was published in 1993, which was used as a reference tool by people engaged in scientific research, teaching, producing and marketing of jujube.

From the time of its foundation on, survey on jujube varieties was started by Pomology Institute, Shanxi Academy of Agricultural Sciences. Jujube Variety Resource Center of Shanxi Province was established in the Institute in 1963, and 56 varieties had been collected and preserved by 1965. In 1979, the Institute undertook building task of the National Jujube Germplasm Repository from the Ministry of Agriculture. From then on, large-scale investigation, collection and preservation of jujube variety resources were carried out across the country. And systematic observation, identification and evaluation were conducted on botanical and biological characteristics, economic traits of jujube fruit, adaptability and stress tolerance of preserved varieties in the Repository. Up to the end of 2012, the Repository has occupied a land area of 11 hectares, in which 630 germplasms from 24 provinces have been preserved. It is a modernized national repository with the largest scale and land area, the largest number of germplasms, the richest genetic diversities, complete infrastructure and equipments, advanced research techniques and first-rate management technology. The Repository has become a base for basic scientific research, field inspection, scientific and technological cooperation, academic exchange, experiment and demonstration, achievement exhibition, personnel training and science education.

In order to make a systematic summary on achievements and latest developments of the repository, fully display characteristics and current status of jujube varieties in China, meet the urgent requirements of people engaged in scientific research, teaching and producing of jujube, maximize effective sharing of jujube varieties and further improve research and utilizing level of jujube germplasm, *The Illustrated*

Germplasm Resources of Chinese Jujube is edited and published by scientific research team of the National Jujube Germplasm Repository in Pomology Institute, Shanxi Academy of Agricultural Sciences.

In this book, illustrated explanation of jujube varieties is the main line. The varieties are divided into different categories according to different uses of the fruits, and each variety occupies one page, introduced in both Chinese and English. The origin, distribution, botanical characters, biological characteristics, fruit traits and general evaluation of 250 varieties are introduced in detail, matched by more than 1 400 color pictures of tree body, flowering and bearing, close-up and section features of fruit. For the common problems of homonym and heteronym in jujube varieties, the book tries to give standardized names, i.e., adding name of origin before name of variety. Yet for those widely-distributed varieties with hardly identified origin, the original names remain unchanged. There are also 114 varieties which cannot be described completely in words, so they are introduced by close-up pictures of their fruits. For convenience of reference, a retrieval table for all the varieties is listed in Chinese Pinyin order at the end of the book.

The book has been tried best to be with a well-organized structure, rich and systematic content, detailed and credible data, clear and nice pictures, standard and original layout, meanwhile integrating advancement, scientificity, comprehensiveness, novelty, practicability and appreciation. It aims at offering valuable reference for further research on jujube germplasm and on breeding of new cultivars, affording helpful assistance and guide for units and persons involved in scientific research, teaching, producing and marketing of jujube, and providing reliable support and ensurance for healthy and sustainable development of jujube industry.

The book is successfully completed under the hard work of scientific researchers of several generations at the National Jujube Germplasm Repository in Pomology Institute, Shanxi Academy of Agricultural Sciences, with kind help from relevant persons of jujube producing areas across the nation, and under the financial aids of the National Project on Protection of Crop Germplasms (NB05-070401-2, NB2012-2130135-02), National Infrastructure Platform of Science and Technology (2005DKA10300, 2012-051), National Special Project on Groundwork of Science and Technology (2011FY110200, 2012FY110100, 2013FY111700), Special Scientific Research Project on Public Welfare Industry of State Forestry (201004041) and National Sci-Tech Support Plan (2008BAD92B03-13, 2013BAD14B03). Sincere appreciation is extended here to all of them. Yet because of limited time and compiling level, some omissions and shortcomings are inevitable. So criticism is welcome from experts, colleagues and readers.

Editorial Board of *The Illustrated Germplasm Resources of Chinese Jujube*
Pomology Institute, Shanxi Academy of Agricultural Sciences
Jun. 30, 2013

中国枣品种资源分布图
Distribution Graph of Chinese Jujube Germplasm Resources

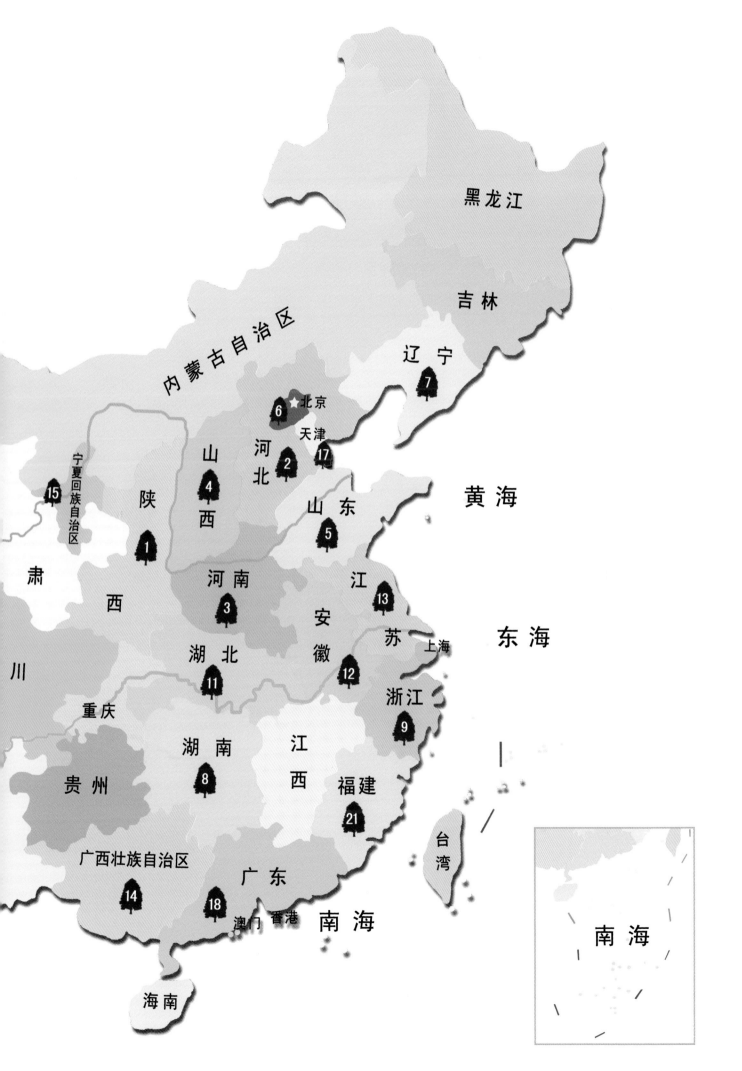

黄河沿岸老枣树（陕西 佳县）
Old Jujube Trees Along the Yellow River (Jiaxian, Shaanxi)

国家枣种质资源圃（山西 太谷）
National Jujube Germplasm Repository (Taigu, Shanxi)

目录 | Contents

序一
Preface 1
序二
Preface 2
前言
Introduction

鲜食品种 Table Varieties

冬枣　Dongzao ··· 2
临猗梨枣　Linyilizao ··· 4
北京白枣　Beijingbaizao ··· 6
湖南鸡蛋枣　Hunanjidanzao ··· 8
大白铃　Dabailing ··· 10
永济蛤蟆枣　Yongjihamazao ·· 12
疙瘩脆　Gedacui ·· 14
北京鸡蛋枣　Beijingjidanzao ·· 16
不落酥　Buluosu ·· 18
宁阳六月鲜　Ningyangliuyuexian ·· 20
孔府酥脆枣　Kongfusucuizao ··· 22
缨络枣　Yingluozao ·· 24
山东梨枣　Shandonglizao ··· 26
临汾蜜枣　Linfenmizao ·· 28
太谷鸡心蜜　Taigujixinmi ·· 30
郎家园枣　Langjiayuanzao ··· 32
平陆尖枣　Pinglujianzao ·· 34
成武冬枣　Chengwudongzao ·· 36
蜂蜜罐　Fengmiguan ·· 38
灵武长枣　Lingwuchangzao ··· 40
马牙白枣　Mayabaizao ··· 42
辣椒枣　Lajiaozao ·· 44
太谷美蜜枣　Taigumeimizao ·· 46
太谷铃铃枣　Taiguglinglingzao ·· 48
夏津妈妈枣　Xiajinmamazao ·· 50
濮阳糖枣　Puyangtangzao ·· 52
林县无头枣　Linxianwutouzao ··· 54
天津尜尜枣　Tianjingagazao ·· 56
大荔鸡蛋枣　Dalijidanzao ·· 58
大瓜枣　Daguazao ·· 60
临潼轱辘枣　Lintongguluzao ·· 62

襄汾圆枣	Xiangfenyuanzao	64
泰安马铃脆	Taianmalingcui	66
大荔马牙枣	Dalimayazao	68
内黄苹果枣	Neihuangpingguozao	70
庆云小梨枣	Qingyunxiaolizao	72
合阳铃铃枣	Heyanglinglingzao	74
溆浦蜜蜂枣	Xupumifengzao	76
连县糖枣	Lianxiantangzao	78
滕州大马枣	Tengzhoudamazao	80
旻枣	Minzao	82
枣强脆枣	Zaoqiangcuizao	84
薛城冬枣	Xuechengdongzao	86
天津快枣	Tianjinkuaizao	88
吴县水团枣	Wuxianshuituanzao	90
交城甜酸枣	Jiaochengtiansuanzao	92
北京坠子白	Beijingzhuizibai	94
溆浦葫芦枣	Xupuhuluzao	96
榆次牙枣	Yuciyazao	98
南京冷枣	Nanjinglengzao	100
天津二秋枣	Tianjinerqiuzao	102
新郑酥枣	Xinzhengsuzao	104
新郑九月青	Xinzhengjiuyueqing	106
永济鸡蛋枣	Yongjijidanzao	108
献县绵枣	Xianxianmianzao	110
清苑大丹枣	Qingyuandadanzao	112
榆次九月青	Yucijiuyueqing	114
濮阳小枣	Puyangxiaozao	116

制干品种 Drying Varieties

吕梁木枣	Lvliangmuzao	120
保德油枣	Baodeyouzao	122
灵宝大枣	Lingbaodazao	124
婆枣	Pozao	126
新乐大枣	Xinledazao	128
官滩枣	Guantanzao	130
相枣	Xiangzao	132
圆铃枣	Yuanlingzao	134
扁核酸	Bianhesuan	136
郎枣	Langzao	138
滕州大马牙	Tengzhoudamaya	140
无核小枣	Wuhexiaozao	142
永城长红	Yongchengchanghong	144
鸡心枣	Jixinzao	146
临泽小枣	Linzexiaozao	148
临泽大枣	Linzedazao	150
稷山圆枣	Jishanyuanzao	152
夏县紫圆枣	Xiaxianziyuanzao	154

品种	拼音	页码
密云小枣	Miyunxiaozao	156
稷山柳罐枣	Jishanliuguanzao	158
大荔圆枣	Daliyuanzao	160
糖枣	Tangzao	162
大荔干尾巴	Daliganweiba	164
俊枣	Junzao	166
蒲城晋枣	Puchengjinzao	168
乐陵长木枣	Lelingchangmuzao	170
稷山长枣	Jishanchangzao	172
大荔小墩墩枣	Dalixiaodundunzao	174
垣曲枣	Yuanquzao	176
洪赵十月红	Hongzhaoshiyuehong	178
大荔疙瘩枣	Daligedazao	180
献县木枣	Xianxianmuzao	182
彬县黑疙瘩	Binxianheigeda	184
溆浦薄皮枣	Xupubopizao	186
溆浦秤砣枣	Xupuchengtuozao	188
新郑尖头灰枣	Xinzhengjiantouhuizao	190
平顺笨枣	Pingshunbenzao	192
平遥大枣	Pingyaodazao	194
献县小大枣	Xianxianxiaodazao	196
溆浦柿饼枣	Xupushibingzao	198
佳县密点脆木枣	Jiaxianmidiancuimuzao	200
新郑长鸡心	Xinzhengchangjixin	202
溆浦岩枣	Xupuyanzao	204
直社疙瘩枣	Zhishegedazao	206
太谷壶瓶酸	Taiguhupingsuan	208
襄汾崖枣	Xiangfenyazao	210
平遥苦端枣	Pingyaokuduanzao	212
溆浦甜酸枣	Xuputiansuanzao	214
遵义甜枣	Zunyitianzao	216
武乡牙枣	Wuxiangyazao	218
新疆小圆枣	Xinjiangxiaoyuanzao	220
衡山长大枣	Hengshanchangdazao	222
溆浦米枣	Xupumizao	224
婆婆枣	Popozao	226
溆浦木枣	Xupumuzao	228
河津条枣	Hejintiaozao	230
临汾木疙瘩	Linfenmugeda	232
万荣翠枣	Wanrongcuizao	234
溆浦秤锤枣	Xupuchengchuizao	236
连县苦楝枣	Lianxiankulianzao	238
临猗笨枣	Linyibenzao	240
临猗鸡蛋枣	Linyijidanzao	242
阿克苏小枣	Akesuxiaozao	244
平陆棒槌枣	Pinglubangchuizao	246
婆枣枝变1号	Pozaozhibian1	248
万荣福枣	Wanrongfuzao	250

滕州落地红	Tengzhouluodihong	252
临猗脖脖枣	Linyibobozao	254
泡泡红	Paopaohong	256
义乌棉絮枣	Yiwumianxuzao	258
中宁小圆枣	Zhongningxiaoyuanzao	260
汝城枣	Ruchengzao	262
离石合钵枣	Lishihebozao	264
西双版纳小枣	Xishuangbannaxiaozao	266
太原圆枣	Taiyuanyuanzao	268
延川白枣	Yanchuanbaizao	270
延川条枣	Yanchuantiaozao	272
北碚小枣	Beibeixiaozao	274
太原驴粪蛋	Taiyuanlvfendan	276
太原长枣	Taiyuanchangzao	278

兼用品种 Multipurpose Varieties

金丝小枣	Jinsixiaozao	282
赞皇大枣	Zanhuangdazao	284
骏枣	Junzao	286
壶瓶枣	Hupingzao	288
灰枣	Huizao	290
板枣	Banzao	292
晋枣	Jinzao	294
中阳团枣	Zhongyangtuanzao	296
临汾团枣	Linfentuanzao	298
榆次团枣	Yucituanzao	300
安阳团枣	Anyangtuanzao	302
延川狗头枣	Yanchuangoutouzao	304
敦煌大枣	Dunhuangdazao	306
定襄星星枣	Dingxiangxingxingzao	308
洪赵小枣	Hongzhaoxiaozao	310
彬县圆枣	Binxianyuanzao	312
鸣山大枣	Mingshandazao	314
赞新大枣	Zanxindazao	316
核桃纹	Hetaowen	318
临汾针葫芦	Linfenzhenhulu	320
太谷墩墩枣	Taigudundunzao	322
彬县耙齿枣	Binxianpachizao	324
玉田小枣	Yutianxiaozao	326
交城端枣	Jiaochengduanzao	328
文水沙枣	Wenshuishazao	330
黎城小枣	Lichengxiaozao	332
黎城大马枣	Lichengdamazao	334
夏县圆脆枣	Xiaxianyuancuizao	336
马连小枣	Malianxiaozao	338
定襄油荷枣	Dingxiangyouhezao	340
八升胡	Bashenghu	342

图鉴

名称	拼音	页码
嵩县大枣	Songxiandazao	344
洪赵葫芦枣	Hongzhaohuluzao	346
民勤小枣	Minqinxiaozao	348
佳县牙枣	Jiaxianyazao	350
广洋枣	Guangyangzao	352
溆浦观音枣	Xupuguanyinzao	354
大荔面枣	Dalimianzao	356
保德小枣	Baodexiaozao	358
溆浦香枣	Xupuxiangzao	360
大荔林檎枣	Dalilinqinzao	362
彬县酸疙瘩	Binxiansuangeda	364
亚腰长红	Yayaochanghong	366
泰安酥圆铃	Taiansuyuanling	368
太谷端子枣	Taiguduanzizao	370
新郑鸡蛋枣	Xinzhengjidanzao	372
清徐圆枣	Qingxuyuanzao	374
大荔墩墩枣	Dalidundunzao	376
献县圆小枣	Xianxianyuanxiaozao	378
沧县傻枣	Cangxianshazao	380
斑枣	Banzao	382
北京笨枣	Beijingbenzao	384
衡阳珍珠枣	Hengyangzhenzhuzao	386
镇平九月寒	Zhenpingjiuyuehan	388
溆浦沙糖枣	Xupushatangzao	390
洪赵脆枣	Hongzhaocuizao	392
河津水枣	Hejinshuizao	394
中宁大红枣	Zhongningdahongzao	396
蒲城绵枣	Puchengmianzao	398
沧县小枣	Cangxianxiaozao	400
蒲城圆梨枣	Puchengyuanlizao	402
献县小小枣	Xianxianxiaoxiaozao	404
沧县屯子枣	Cangxiantunzizao	406
灌阳短枣	Guanyangduanzao	408
大荔知枣	Dalizhizao	410
香山小白枣	Xiangshanxiaobaizao	412
中草笨枣	Zhongcaobenzao	414
沧县长小枣	Cangxianchangxiaozao	416
新郑齐头白	Xinzhengqitoubai	418
定襄小枣	Dingxiangxiaozao	420
南京鸭枣	Nanjingyazao	422
献县酸枣	Xianxiansuanzao	424
赞皇长枣	Zanhuangchangzao	426
大荔小圆枣	Dalixiaoyuanzao	428
新郑大枣	Xinzhengdazao	430
姑苏小枣	Gusuxiaozao	432
延川跌牙枣	Yanchuandieyazao	434
喀什噶尔小枣	Kashigeerxiaozao	436
新郑大马牙	Xinzhengdamaya	438

太谷黑叶枣	Taiguheiyezao	440
曲阜猴头枣	Qufuhoutouzao	442

蜜枣品种 Candied Varieties

灌阳长枣	Guanyangchangzao	446
连县木枣	Lianxianmuzao	448
义乌大枣	Yiwudazao	450
鹅子枣	Ezizao	452
郎溪牛奶枣	Langxiniunaizao	454
兰溪马枣	Lanximazao	456
南京枣	Nanjingzao	458
歙县马枣	Shexianmazao	460
苏南白蒲枣	Sunanbaipuzao	462
定襄山枣	Dingxiangshanzao	464
大荔水枣	Dalishuizao	466
小果算盘枣	Xiaoguosuanpanzao	468
糠头枣	Kangtouzao	470
涪陵鸡蛋枣	Fulingjidanzao	472
宣城尖枣	Xuanchengjianzao	474
阜阳木头枣	Fuyangmutouzao	476
宁阳暄铃枣	Ningyangxuanlingzao	478

观赏品种 Ornamental Varieties

磨盘枣	Mopanzao	482
茶壶枣	Chahuzao	484
三变红	Sanbianhong	486
葫芦长红	Huluchanghong	488
胎里红	Tailihong	490
大荔龙枣	Dalilongzao	492
龙枣	Longzao	494
大柿饼枣	Dashibingzao	496
葫芦枣	Huluzao	498
羊奶枣	Yangnaizao	500
柿顶枣	Shidingzao	502
临猗辣椒枣	Linyilajiaozao	504
大果算盘枣	Daguosuanpanzao	506
大叶无核枣	Dayewuhezao	508

其他品种 Other Varieties

品种索引	Varieties Index	523
主要参考文献	Main References	527

鲜食品种 Table Varieties

冬枣密植园（山西 临猗）
Dongzao Intensive Planting Garden (Linyi, Shanxi)

冬 枣

品种来源及分布 别名冻枣、雁来红、果子枣、苹果枣、水枣、冰糖枣。原产于河北省的黄骅、盐山和山东省的沾化、枣庄等地。因成熟期晚而得名，沾化冬枣、鲁北冬枣、黄骅冬枣等品种均属此类型。该品种栽培历史悠久，河北黄骅市齐家务乡聚馆村现存大面积古枣林，树龄最大的约有500年。

植物学性状 树体较大，树冠自然半圆形，树姿较开张，干性较强，枝条细而密。主干条状皮裂，裂纹深。枣头灰绿色，平均长57.4cm，节间长7.8cm。枣头着生二次枝4～7个，二次枝长32.9cm，5～8节，弯曲度小。针刺退化。枣股较小，平均抽生枣吊4.3个。枣吊平均长19.0cm，着叶11片。叶片中等大，卵圆形，合抱，先端钝尖，叶基偏斜，叶缘锯齿细。花量多，每花序着花3～7朵，花小，花径5.0～5.8mm。

生物学特性 树势强，萌芽率和成枝力强，结果较早，一般定植第二年结果，第三年有一定产量，但产量较低。10年左右进入盛果期，产量中等，一般株产鲜枣15kg。幼龄枝结实能力较差，枣头和2～3年生枝的吊果率分别为44.6%和33.7%。在山西太谷地区，9月中旬果实进入白熟期，10月中旬开始成熟采收，属极晚熟品种，果实生育期120d以上。果实较抗病和抗裂果。

果实性状 果个较小，近圆形，纵径3.10cm，横径2.80cm，单果重11.9g，大小不整齐。果皮薄，红色，果面平滑。果梗细而较长，梗洼中等大，较浅。果顶微凹，柱头遗存，不明显。肉质细嫩酥脆，味甜，汁液多，品质极上，适宜鲜食。鲜枣耐贮藏，普通冷藏条件下可贮藏90d以上。鲜枣可食率94.1%，含可溶性固形物37.80%，单糖14.80%，双糖19.00%，总糖33.80%，100g果肉维生素C含量292.60mg；果皮含黄酮25.59mg/g，cAMP（环磷酸腺苷）含量94.91 μg/g。核较小，椭圆形，纵径1.70cm，横径0.70cm，核重0.70g。核尖短，核纹中度深，含仁率86.7%，种仁较饱满，多为单仁，可作育种亲本。

评价 该品种果实生育期长，成熟晚，适宜年均温11℃以上的地区种植。适宜中密度栽培，每667m²栽80～90株。早期丰产性能较差，为获得较高产量须采取花期开甲等提高坐果率的措施。该品种是目前大面积推广的品质极佳、抗病、耐贮的极晚熟优良鲜食品种。

Dongzao

Source and Distribution The cultivar originated from and spreads in Huanghua, Yanshan in Hebei Province and Zhanhua, Zaozhuang in Shandong Province.

Botanical Characters The tree is spreading with a semi-round crown, a strong central leader trunk and thin and dense branches. The grayish-green 1-year-old shoots are 57.4 cm long with the internodes of 7.8 cm, with almost degraded thorns. There are 4～7 secondary branches on each 1-year-old shoot. The small mother fruiting spurs can germinate 4.3 deciduous fruiting shoots. The medium-sized leaves are oval-shaped. There are many small flowers with a diameter of 5.0～5.8 mm produced.

Biological Characters The small tree has moderate vigor, strong germination and branching ability. The tree enters the full-bearing stage in the 10th year with a medium yield of 15 kg. It is an extremely late-ripening variety.

Fruit Characteristics The small fruit is nearly round with a vertical and cross diameter of 3.10 cm and 2.80 cm, averaging 11.9 g with irregular sizes. It has thin red skin and a smooth surface. The delicate flesh is crisp and juicy, with a sweet taste and a very good quality for fresh-eating. The percentage of edible part of fresh fruit is 94.1%, and the soluble solid content (SSC), monosaccharide, disaccharide, total sugar (TTS) and Vitamin C (Vc) is 37.8%, 14.80%, 19.00%, 33.80% and 292.60 mg per 100 g fresh fruit. The content of flavones and cAMP in mature fruit skin is 25.59 mg/g and 94.91 μg/g. The oval-shaped stone has a vertical and cross diameter of 1.70 cm and 0.70 cm, averaging 0.70 g with a short apex and medium-deep veins. The percentage of containing kernels is 86.7%. Most stones contain a well-developed kernel, which can be used as breeding material.

Evaluation The cultivar can be planted in areas with an average annual temperature of over 11 ℃. In order to get a higher yield, it is necessary to take measures such as girdling in blooming stage to improve fruit set. It is an extremely late-ripening variety with an excellent quality for fresh eating, strong resistance to diseases and good tolerance to storage.

鲜 食
Table Varieties
品 种

临猗梨枣

品种来源及分布　别名梨枣、运城梨枣、山西梨枣。山西十大名枣之一，原产山西省的临猗、运城等地，历史上多为农家庭院栽植，已有3 000多年栽培历史。

植物学性状　树体较小，树姿开张，干性较弱，树冠自然圆头形。主干皮裂较深，呈条状，较易剥落。枣头红褐色，生长势中等，年生长量83.1cm，节间长7.8cm。二次枝长28.3cm，6～8节。针刺不发达。枣股小，圆锥形，抽吊力强，平均抽生枣吊4.4个。枣吊平均长24.3cm，着叶16片。叶片厚，卵圆形，叶长6.1cm，叶宽2.9cm，浓绿色，先端锐尖，叶基圆楔形，叶缘具锐锯齿。花量少，枣吊平均着花28.5朵，每花序平均2.2朵。花中大，零级花花径7.2mm，昼开型。

生物学特性　树体较小，树势中等，萌芽率和成枝力强。开花结果早，定植当年可少量结果，第二年普遍结果，5年左右进入盛果期，早期丰产性极强。幼龄枝结果能力极强，枣头、2～3年生枝和4年生枝的吊果率分别为78.2%、83.0%和31.0%，木质化枣吊最多能结20多个果，枣吊主要坐果部位在枣吊的5～13节。盛果期树株产可达50kg。该品种在山西太谷地区9月上旬果实着色，9月下旬成熟采收，果实生育期110d左右，为晚熟品种类型。存在采前落果、果实易软化及耐贮性差、病害较严重等问题。

果实性状　果实特大，倒卵状，梨形，纵径4.50cm，横径4.00cm，单果重31.6g左右，最大可达100g以上，大小不整齐。果梗细，梗洼窄而较深。果顶平，柱头遗存。果皮薄，浅红色，果面有隆起。果点大，圆形，较明显。果肉厚，肉质松脆，味甜，汁液多，品质上等，适宜鲜食。鲜枣可食率96.6%，含可溶性固形物27.90%，单糖17.00%，双糖5.25%，总糖22.25%，酸0.37%，100g果肉维生素C含量292.25mg，含水量69.8%；果皮含黄酮3.91mg/g，cAMP含量122.17μg/g。核较小，纺锤形，纵径2.67cm，横径0.94cm，核重1.08g，核尖较短，核纹较深，核面粗糙，核内无种仁。

评价　该品种适应性较强，树体小，结果早，特别丰产，果实特大，品质上等，是我国栽培面积最大、分布最广的晚熟鲜食优良品种。但存在采前落果、耐贮性差及病害较严重等问题。

Linyilizao

Source and Distribution　The cultivar originated from Linyi and Yuncheng in Shanxi Province with a cultivation history of over 3 000 years.

Botanical Characters　The small tree is spreading with a weak central leader trunk and a round crown. The trunk bark has deep fissures, easily shelled off. The reddish-brown 1-year-old shoots have strong germination ability and moderate growth potential, averaging 83.1 cm long with 6～8 nodes. The secondary branches are 28.3 cm long with less-developed thorns. The small conical mother fruiting spurs can germinate 4.4 deciduous fruiting shoots. The thick leaves are oval-shaped and dark green. The number of flowers is rather small, averaging 2.2 ones per inflorescence. The medium-sized blossoms have a diameter of 7.2 mm for the zero-level flowers.

Biological Characters　The tree has moderate vigor, blooming and bearing early. It has certain yield in the year of planting and enter the full-bearing stage in the 4th year, with strong early productivity and very strong fruiting ability for young branches. A tree has a yield of 50 kg in its full-bearing stage. In Taigu County of Shanxi Province, it is harvested in late September.

Fruit Characteristics　The extremely large fruit is obovate pear-shaped, with a vertical and cross diameter of 4.50 cm and 4.00 cm, averaging 31.6 g with irregular sizes. It has light-red thin skin. The rough flesh is loose-textured, crisp, juicy and sweet. It has a very good quality for fresh eating. The percentage of edible part of fresh fruit is 96.6%, and SSC, monosaccharide, disaccharide, TTS, titratable acid (TA) and Vc is 27.90%, 17.00%, 5.25%, 22.25%, 0.37% and 292.25 mg per 100 g fresh fruit. The content of flavones and cAMP in mature fruit skin is 3.91 mg/g and 122.17 μg/g. The small spindle-shaped stone weighs 1.08 g.

Evaluation　The small plant of Linyilizao cultivar has strong adaptability, bearing early with a high yield, suitable for intensive planting. The large fruit has a very good quality. This cultivar has the largest cultivation area and widest distribution among the middle-late-ripening table varieties.

北京白枣

品种来源及分布 别名长辛店白枣、白枣、脆枣。原产北京,分布较广,数量较多,以丰台区长辛店乡的朱家坟、张家坟一带较集中,多庭院栽培。有数百年栽培历史。

植物学性状 树体中等大,树姿较直立,干性较强,枝条稀疏,树冠自然圆头形。主干条状不规则纵裂,易剥落。枣头黄褐色,粗壮,年生长量74.0cm,节间长7.8cm。二次枝粗直,弯曲度小,平均长27.1cm,7节左右。针刺多退化。皮目大而密,多呈菱形,凸起,开裂。枣股一般为圆锥形,粗大,多年生枣股有分歧现象,一般抽生枣吊4个,枣吊较粗,长25.9cm,着生叶片17片。叶片中等大,卵状披针形,较薄,绿色。叶尖锐尖,叶基圆楔形,叶缘锯齿较细。花量中等,花序平均着花7朵。

生物学特性 树势强健,中央领导干优势明显,发枝力较弱。结果较早,早期丰产,幼龄枝结实性能较强,吊果率40.3%。盛果期树,一般株产20kg左右。在山西太谷地区,9月中旬果实成熟采收,果实生育期100d左右,为中熟品种类型。成熟期遇雨易裂果,抗病性较差。

果实性状 果实中等大,长卵圆形或长椭圆形,纵径3.70cm,横径2.50cm,单果重12.0g,大小整齐一致。果肩圆或广圆,平斜。梗洼窄,浅平。果梗粗,长4mm左右。果顶尖圆,柱头遗存。果面平滑光亮。果皮薄而脆,暗红色。果肉绿白色,质地致密细脆,汁液多,味酸甜,口感极佳,适宜鲜食。鲜枣可食率96.6%,脆熟期可溶性固形物含量33.00%,总糖32.00%,酸0.60%,100g果肉维生素C含量408.77mg;果皮含黄酮6.47mg/g,cAMP含量210.60μg/g。果核小,纺锤形,纵径1.93cm,横径0.83cm,核重0.41g。核蒂渐尖,先端圆,核尖细长。核纹宽,中等深,呈人字形斜纹。核内多有种子,含仁率85.7%。

评价 该品种树体中等大,树势强健,早期丰产,结果稳定,产量较高,适宜中密度栽培,每667m²栽80~90株。果实中等大,外形光洁美观,肉质细脆多汁,酸甜可口,品质优异,为优质中熟鲜食品种,适宜降雨量较少的城郊、工矿区栽培。应注意采取刻芽、重短截等促进发枝的技术措施。

Beijingbaizao

Source and Distribution The cultivar originated from and widely spreads in Beijing with a large quantity. It is mostly planted in family gardens with a history of hundreds of years.

Botanical Characters The medium-sized tree is vertical with a strong central leader trunk, sparse branches and a natural-round crown. The yellowish-brown 1-year-old shoots are thick and strong, averaging 74.0 cm long with the internodes of 7.8 cm. The secondary branches are 27.1 cm long with 7 nodes and with rare thorns. The conical-shaped mother fruiting spurs can germinate 4 thick deciduous fruiting shoots. The medium-sized leaves are ovate-lanceolate. The number of flowers is medium large, averaging 7 ones per inflorescence.

Biological Characters The tree has strong vigor, a distinct dominance for the central leader trunk, weak branching ability, bearing early with strong early productivity. Young branches have strong fruiting ability. A mature tree in its full-bearing stage has a yield of 20 kg. In Taigu County of Shanxi Province, it matures in mid-September .It is a mid-ripening variety with weak resistance to disease and serious fruit-cracking if it rains in the maturing stage.

Fruit Characteristics The medium-sized obovate fruit has a vertical and cross diameter of 3.70 cm and 2.50 cm, averaging 12.0 g with a regular size. It has a smooth and bright surface and thin, brittle and dark-red skin with large and sparse dots. The greenish-white flesh is tight-textured, crisp, juicy and sweet, with a little sour taste and an excellent quality for fresh eating. The percentage of edible part of fresh fruit is 96.6%, and SSC, TTS, TA and Vc in fruits of crisp-maturing stage is 33.00%, 32.00%, 0.60% and 408.77 mg per 100 g fresh fruit. The content of flavones and cAMP in mature fruit skin is 6.47 mg/g and 210.60 μg/g. The small spindle-shaped stone has a vertical and cross diameter of 1.93 cm and 0.83 cm, averaging 0.41 g.

Evaluation The cultivar has strong vigor, strong early productivity with a high and stable yield. The medium-sized fruit has a bright and attractive appearance, crisp and juicy flesh with an excellent quality. It is an excellent mid-ripening table variety which can be developed in light rainfall areas.

鲜 食
Table Varieties
品 种

湖南鸡蛋枣

品种来源及分布 又名溆浦鸡蛋枣。原产湖南省的溆浦、麻阳、衡山、祁阳等地，为当地主栽品种，栽培历史200年以上。

植物学性状 树体中等偏小，树姿开张，干性较弱，枝条较稀疏，树冠半圆形。主干皮裂条状较深，易剥落。枣头红褐色，平均生长量95.9cm，粗0.97cm，节间长7.9cm，着生二次枝7～11个，二次枝长30.1cm，平均7节。针刺较发达。枣股大，圆柱形，抽吊力中等，一般抽生枣吊3～4个。枣吊较短，平均长25.6cm，着叶20片。叶片中等大，叶长5.0cm，叶宽2.7cm，椭圆形，浓绿色，先端急尖，叶基圆楔形，叶缘锯齿粗钝。花量少，花序平均着花3朵。花较大，零级花花径7.5mm。

生物学特性 树势中庸，成枝力中等。开花结果极早，坐果率高，早期丰产性能极强，盛果期产量高。一般定植第二年普遍结果，3、4年生进入初果期，株产3～5kg，5年生后大量结果，盛果期株产可达20kg左右。幼龄枝结实能力强，枣头、2～3年生枝和4年生以上枝的吊果率分别为70.0%、79.2%和82.8%。在山西太谷地区，9月中旬果实成熟，果实生育期100d左右，为早中熟品种类型。成熟期枣果有软化现象，抗黑斑病能力较差。

果实性状 果实大，近圆形，纵径4.20cm，横径3.70cm，单果重37.8g，最大可达60.0g，大小较整齐。果皮薄，紫红色，果面欠光滑。果点大而明显，圆形，分布密。果梗短，中等粗，梗洼窄而深。果顶凹，柱头遗存。果肉厚，绿白色或乳白色，肉质疏松较脆，味甜，汁液中等多，品质上等，适宜鲜食。鲜枣可食率98.5%，可溶性固形物含量33.00%，酸0.14%，100g果肉维生素C含量333.50mg；果皮含黄酮12.43mg/g，cAMP含量260.92μg/g。核较小，纺锤形，纵径2.00～2.50cm，横径0.90～1.10cm，核重0.55g，核尖较短，核纹较深，核面粗糙，种仁不饱满，含仁率15.0%。

评价 该品种树体较小，结果早，丰产稳产，适宜密植栽培，每667m²栽90～110株。为大果、优质、鲜食品种，可在水肥条件较好、交通便利的城郊地区发展。

Hunanjidanzao

Source and Distribution The cultivar originated from Xupu, Mayang, Hengshan and Qiyang in Hunan Province. It is the dominant variety there with a history of over 200 years.

Botanical Characters The small tree is spreading with a weak central leader trunk, thin branches and a semi-round crown. The trunk bark has deep striped fissures, easily shelled off. The reddish-brown 1-year-old shoots are 95.9 cm long with the internodes of 7.9 cm long. There are generally 7～11 secondary branches which are 30.1 cm long with 7 nodes and developed thorns. The large column-shaped mother fruiting spurs can germinate 3～4 short deciduous fruiting shoots which are 25.6 cm long with 20 leaves. The small leaves are oval-shaped and dark green. The number of flowers is small, averaging 3 per inflorescence. The large blossoms have a diameter of 7.5 mm.

Biological Characters The tree has medium vigor, medium branching ability, blooming and bearing early with high fruit set, strong early productivity and a high yield. It bears heavily after 5 years and the yield in the full-bearing stage reaches 20 kg per tree. In Taigu County of Shanxi Province, it matures in mid-September.

Fruit Characteristics The large fruit is nearly round with a vertical and cross diameter of 4.20 cm and 3.70 cm, averaging 37.8 g with a regular size. It has purplish-red thin skin, an unsmooth surface, a narrow and deep stalk cavity, a protuberant fruit apex and a remnant stigma. The thick flesh is greenish white or ivory white, loose-textured, crisp, sweet and juicy, with a good quality for fresh eating and processing candied fruits. The percentage of edible part of fresh fruit is 98.5%, and SSC, TA and Vc is 33.00%, 0.14% and 333.50 mg per 100 g fresh fruit. The content of flavones and cAMP in mature fruit skin is 12.43 mg/g and 260.92 μg/g. The small spindle-shaped stone has a vertical and cross diameter of 2.00～2.50 cm and 0.90～1.10 cm respectively, averaging 0.55 g.

Evaluation The small plant of Hunanjidanzao cultivar bears early with a high and stable yield, suitable for close planting. It is a good table variety with a large fruit size, which can be developed in suburbs with good conditions of irrigation, fertilization and convenient transportation.

鲜 食
Table Varieties
品 种

大 白 铃

品种来源及分布　别名鸭蛋枣、梨枣、鸭枣青、馒头枣。原产山东省夏津县，分布于山东省的临清、武城、阳谷和河北省的献县等地。多为零星栽植。

植物学性状　树体较小，树姿较开张，干性强，枝条中密，树冠伞形。主干皮裂浅，呈条状，易剥落。枣头红褐色，平均生长量81.0cm，粗1.06cm，节间长8.0cm。二次枝长29.1cm，平均6节。针刺不发达。皮目大，圆形或长圆形，凸起，开裂。枣股中等大，抽吊力中等，抽生枣吊3~4个。枣吊平均长23.5cm，着叶16片。叶片中等大，叶长5.9cm，叶宽3.1cm，椭圆形，浓绿色，先端钝尖，叶基圆楔形，叶缘锯齿较粗，密度中等。花量多，每花序着花10~11朵。花小，花径6.0mm。

生物学特性　树势中庸健壮，成枝力较强。开花结果早，坐果率高，早期丰产性能强，盛果期产量高。一般定植第二年普遍结果，4年生左右进入初果期，株产3~5kg，6年生后大量结果，盛果期株产可达20kg左右。幼龄枝结实能力强，枣头、2~3年生枝和4年生以上枝的吊果率分别为60.0%、100.6%和85.2%。在山西太谷地区，9月中旬开始果实进入脆熟期，9月下旬全红完熟，果实生育期110d左右，为中晚熟品种类型。进入脆熟期后枣果有软化现象，抗黑斑病中等。

果实性状　果实大，近圆形，纵径4.20cm，横径4.00cm，单果重31.9g，最大42.0g，大小较整齐。果梗短而粗，梗洼窄而深。果顶平圆或微凹，柱头遗存。果点小而密，圆形，浅黄色。果皮较薄，紫红色，果面欠平滑。果肉厚，绿白色，肉质松脆，味甜，汁液中多，品质上等，适宜鲜食。鲜枣可食率97.2%，含可溶性固形物33.00%，总糖24.50%，酸0.28%，100g果肉维生素C含量280.08mg；果皮含黄酮55.54mg/g，cAMP含量127.48μg/g。核小，纺锤形，核尖短，纵径2.2~2.4cm，横径1.0~1.2cm，核重0.90g，核内几乎无种仁。

评价　该品种树体中庸，结果早，早期丰产性能强，适宜密植栽培，每667m²栽90~110株。果实大，外观艳丽，品质优异，为优良的中晚熟鲜食品种。目前已在山西、山东、河北、河南、陕西及南方枣区规模发展，表现丰产、质优及适应性较强的特性，可望代替临猗梨枣品种。

Dabailing

Source and Distribution　The cultivar originated from Xiajin County in Shandong Province, and spreads in Linqing, Wucheng, Yanggu in Shandong and Xianxian County in Hebei Province. It is dispersedly planted.

Botanical Characters　The small tree has moderate vigor, spreading with a strong central leader trunk, medium-dense branches and an umbrella-shaped crown. The reddish-brown 1-year-old shoots have moderate growth vigor, 81.0 cm long with the internodes of 8.0 cm. The secondary branches are 29.1 cm long with 6 nodes and less-developed thorns. The medium-sized mother fruiting spurs can germinate 3~4 deciduous fruiting shoot, which are 23.5 cm long with 16 leaves. The medium-sized leaves are oval-shaped and dark green. There are many small flowers with a diameter of 6 mm produced, averaging 10~11 ones per inflorescence.

Biological Characters　The tree has moderate vigor and strong branching ability, blooming and bearing early with high fruit set, strong early productivity and a high yield in full-bearing stage. In the 6th year after being planted, the tree bears heavily, and the yield in the full-bearing stage reaches 20 kg. In Taigu County of Shanxi Province, it enters the crisp-maturing stage in late September.

Fruit Characteristics　The large fruit is nearly round, averaging 31.9 g with a regular size. It has a purplish-red thin skin and an unsmooth surface. The greenish-white thick flesh is loose-textured, crisp and sweet, with medium juice and a good quality for fresh eating. The percentage of edible part of fresh fruit is 97.2%, and SSC, TTS, TA and Vc is 33.00%, 24.50%, 0.28% and 280.08 mg per 100 g fresh fruit. The small stone is spindle-shaped, averaging 0.9 g almost without kernels.

Evaluation　The cultivar has strong vigor, bearing early with strong early productivity, suitable for close planting. The large fruit has an attractive appearance and an excellent quality.

永济蛤蟆枣

品种来源及分布 别名蛤蟆枣。原产于山西省永济市的仁阳、太宁等村，为当地主栽品种。栽培历史不详。

植物学性状 树体高大，树姿较直立，中心干顶端优势较强，枝叶较稀疏。树皮裂纹呈条状且裂纹较深，较易脱落。枣头红褐色，萌发力中等，生长势强，平均生长量75.9cm，节间长7～9cm，二次枝长29.4cm，7～9节。针刺不发达。枣股较大，抽吊力中等，平均抽生枣吊3.3个。枣吊平均长21.8cm，着叶16片。叶片大，叶长6.2cm，叶宽3.4cm，长卵圆形，浓绿色，先端锐尖，叶基圆楔形，叶缘锯齿较细。花量中多，枣吊平均着花57.2朵，每花序着花4朵。花大，零级花花径8.7mm，昼开型。

生物学特性 树势强健，成枝力较弱。一般定植第二年开始结果，10年后进入盛果期。坐果率中等。枣头、2～3年生枝和4年生枝吊果率分别为49.0%、48.9%和29.6%，主要坐果部位在枣吊的5～10节，占坐果总数的66.9%。在山西太谷地区，9月上旬果实着色，9月下旬进入脆熟期，果实生育期110d左右，为晚熟品种类型。成熟期遇雨易裂果，采前落果极轻。

果实性状 果实大，扁柱形，纵径4.93cm，横径3.61cm，单果重25.4g，大小较整齐。果皮薄，紫红色，果面不平滑，有明显小块瘤状隆起和紫黑色斑点，似癞蛤蟆瘤状，故称"蛤蟆枣"。果点较小而不明显，但密度较大。果顶平或微凹，柱头遗存。果梗中等粗，较长，梗洼窄而深。果肉厚，绿白色，肉质细而松脆，味甜，汁液较多，品质上等，适宜鲜食。鲜枣耐贮藏，普通冷库条件可贮藏3个月以上。鲜枣可食率96.3%，含可溶性固形物32.40%，总糖28.10%，酸0.50%，100g果肉维生素C含量420.82mg；果皮含黄酮59.05mg/g，cAMP含量79.49μg/g。果核小，纺锤形，纵径2.87cm，横径0.99cm，核重0.94g。核尖中长，核纹较深，核面粗糙，含仁率仅3.3%。

评价 该品种适应性差，要求良好的水肥条件，在较精细的栽培管理条件下才能获得较高产量。不适宜密植栽培，整形修剪时注意采取刻芽、重短截等促进发枝的措施。该品种果实大，鲜食品质优良，较耐贮藏，为优质、耐贮的晚熟鲜食品种，可适度发展。

Yongjihamazao

Source and Distribution The cultivar originated from Renyang and Taining Village of Yongji County in Shanxi Province. It is the dominant variety there.

Botanical Characters The large tree is vertical with a strong apical dominance for the central leader trunk and sparse branches and leaves. The reddish-brown 1-year-old shoots have moderate germination ability and strong growth vigor, averaging 75.9 cm long with 7～9 nodes. The internodes are 7～9 cm long. The secondary branches are 29.4 cm long with fewer thorns. The large mother fruiting spurs can germinate 3.3 deciduous fruiting shoots which are 21.8 cm long with 16 leaves. The large leaves are long oval-shaped and dark green. The number of flowers is medium large, averaging 57.2 per deciduous fruiting shoot and 4.11 ones per inflorescence. The large blossoms have a diameter of 8.7 mm for the zero-level flowers.

Biological Characters The tree has strong vigor and weak branching ability. It generally bears in the 2nd year after planting and enters the full-bearing stage in the 10th year with medium fruit set. In Taigu County of Shanxi Province, it enters the crisp-maturing stage in late September.

Fruit Characteristics The large fruit is flat column-shaped, averaging 25.4 g with a regular size. It has thin purple skin, an unsmooth surface with obvious knob-like protuberances and purplish-black spots, which makes it like a toad, that is why it is called "Hamazao" (hama means toad in Chinese). The greenish-white flesh is thick, delicate, crisp and juicy, with a sweet taste and a good quality for fresh eating. Fresh fruits have a good storage character. The percentage of edible part of fresh fruit is 96.3%, and SSC, TTS, TA and Vc 32.4%, 28.1%, 0.50% and 420.82 mg per 100 g fresh fruit. The small stone is spindle-shaped, averaging 0.94 g.

Evaluation The cultivar has poor adaptability. It is a late-ripening table varity. The extremely large fruit has high edibility and a good quality for fresh eating and storage.

鲜 食
Table Varieties
品 种

疙 瘩 脆

品种来源及分布 别名大铃枣、大脆枣。原产山东省的泰安、长清、宁阳、济宁、曲阜、滕州等地。多为庭院栽培,部分小片集中栽培,为山东中南部原产的重要鲜食优良品种。栽培历史悠久。

植物学性状 树体较大,树姿开张,干性较强,枝叶中密,树冠乱头形。树皮粗糙,块状皮裂,易剥落,枣头紫褐色,平均生长量83.8cm,粗1.02cm,节间长9.6cm。二次枝长36.0cm,平均6节。针刺不发达或无针刺。枣股粗大,最长3cm左右,抽吊力弱,抽生枣吊2～3个。枣吊粗而较长,平均长23.8cm,着叶14片,常有二次生长。叶片大,叶长7.2cm,叶宽3.3cm,卵圆形,两侧反卷,叶厚,浓绿色,先端钝尖,叶基圆楔形,叶缘锯齿钝。花量多,一般每花序着花7～10朵,最多20朵。花中大,花径7mm左右,昼开型。

生物学特性 树势强健,发枝力中等,枣头生长势较强。开花结果较早,早期丰产性中等。一般栽后2～3年开始结果,10年左右进入盛果期,成龄树株产鲜枣15kg左右。枣头、2～3年生和3年生以上枝的吊果率分别为6.7%、89.4%和27.4%。在山西太谷地区,9月中旬果实成熟,果实生育期100d左右,为早中熟品种类型。果实抗病性较强,一般年份裂果较少,采前落果轻。

果实性状 果实性状与孔府酥脆和宁阳六月鲜相近。果实中等大,短椭圆形或倒卵圆形,纵径4.30cm,横径3.20cm,单果重13.8g,大小较整齐。果梗短而粗,梗洼窄而较深。果顶平,柱头遗存。果皮较厚,棕红色,光亮美观,果面不平滑,有块状隆起,呈疙瘩状。果点小而密,不明显。果肉厚,白色,肉质松脆,甜味浓,略酸,汁液中多,品质上等,适宜鲜食。鲜枣可食率95.0%,含可溶性固形物32.00%,总糖22.92%,酸0.24%,100g果肉维生素C含量355.45mg;果皮含黄酮16.80mg/g,cAMP含量193.73μg/g。果核中大,纺锤形,平均重0.70g,含仁率70.0%。

评价 该品种适应性强,较耐旱、耐瘠薄土壤条件。结果较早,产量较高而稳定。果个中大,色泽艳丽,果肉疏松,甜味浓,为质优、抗病、较抗裂果的早中熟鲜食良种。适宜城镇附近和观光果园生产栽培。

Gedacui

Source and Distribution The cultivar originated from Taian, Changqing, Ningyang, Jining, Qufu and Tengzhou in Shandong Province. It is an important table variety in Shandong Province with a long history.

Botanical Characters The large tree is spreading with a strong central leader trunk, medium-dense branches and an irregular crown. The trunk bark has rough massive fissures. The purplish-brown 1-year-old shoots have strong growth vigor, 83.8 cm long and 1.02 cm thick with the internodes of 9.6 cm. The secondary branches are 36.0 cm long with 6 nodes and under-developed or without thorns. The mother fruiting spurs are large and thick, which can germinate 2～3 deciduous fruiting shoots which are 23.8 cm long with 14 leaves. The large thick leaves are oval-shaped and dark green. There are many medium-sized flowers with a diameter of 7 mm, generally 7～10 ones per inflorescence.

Biological Characters The tree has strong vigor and medium branching ability with strong growth vigor for the 1-year-old shoots. It blooms and bears early with moderate early productivity. In Taigu County of Shanxi Province, it ripens in early September. It has a strong resistance to diseases, light fruit-cracking and light premature fruit-dropping.

Fruit Characteristics The medium-sized fruit is short oval-shaped or obovate, averaging 13.8 g with a regular size. It has a thick, bright and attractive skin and an unsmooth surface. The thick white flesh is crisp and strongly sweet, a little bit sour, with medium juice and an excellent quality for fresh eating. The percentage of edible part of fresh fruit is 95.0%, and SSC, TTS, TA and Vc 32.00%, 22.92%, 0.24% and 355.45 mg per 100 g fresh fruit. The content of flavones and cAMP in mature fruit skin is 16.80 mg/g and 193.73 μg/g. The small stone is short spindle-shaped, averaging 0.70 g. The percentage of containing kernels is 70.0%.

Evaluation The cultivar is a good early-mid-ripening variety with a good quality, strong resistance to diseases and fruit-cracking. It can be developed moderately.

鲜 食
Table Varieties
品 种

·15·

北京鸡蛋枣

品种来源及分布　原产北京。零星分布于古老宅院。因果实大，形似鸡蛋而得名。栽培历史不详。

植物学性状　树体中大，树姿较开张，树冠圆锥形，枝叶较多。树皮呈碎块状皮裂，不易剥落。枣头棕褐色，有扭曲现象，平均长76.0cm，节间长8.2cm，二次枝平均长38.7cm，8节左右。针刺不发达。枣股较大，圆柱形，一般着生枣吊2~4个。枣吊中等长，较粗，平均长23.9cm，着叶13片。叶片大，叶长6.7cm，叶宽3.3cm，椭圆形，中厚，绿色。叶尖锐尖，叶基广楔形，叶缘锯齿粗。花量多。

生物学特性　树体健壮，发枝力强。开花结果较早，早期丰产性能中等，盛果期产量不稳定。一般定植第二年普遍结果，4~5年生进入初果期，株产3~5kg，8年生左右开始大量结果，株产15kg左右。在山西太谷地区，9月中旬果实成熟采收，果实生育期100d左右，为中熟品种类型。该品种进入落叶期较早，一般年份10月10日左右落叶。果实成熟期遇雨易裂果，抗病性较强，但抗早期落叶病能力较差。

果实性状　果实大，卵圆形，纵径4.50cm，横径3.30cm，单果重23.4g，大小较整齐。果肩宽，广圆。梗洼浅广。果顶平圆，先端略凹陷。柱头遗存，不明显。果点小，黄褐色，分布稀疏，不明显。果面略有隆起，果皮较厚，白熟期至脆熟期有黄灰色过渡现象，完熟期为暗红色，富有光泽。果肉绿白色，质地较致密，汁液较多，味甜，品质上等，适宜鲜食。鲜枣可食率98.0%，含可溶性固形物29.40%，总糖26.98%，酸0.39%，100g果肉维生素C含量271.23mg；果皮含黄酮8.29mg/g，cAMP含量441.48μg/g。果核较小，纺锤形，纵径2.40cm，横径0.74cm，核重0.48g，核蒂短，稍尖，核尖较长，渐尖，核纹细而深。核内多有种子，含仁率71.5%。

评价　树体中等大，树势强健，易发枝，早期丰产性能中等，盛果期产量不稳，适宜中密度栽培。该品种为大果、优质的中熟鲜食良种，可在降雨量少的地区发展栽培。须注意防护枣头枝的风折和早期落叶病的防治。

Beijingjidanzao

Source and Distribution　The cultivar originated from Beijing and sparsely spreads in old houses with an unknown history.

Botanical Characters　The small tree has weak vigor, spreading with a conical crown and dense branches and leaves. The trunk bark has vertical striped fissures, uneasily shelled off. The curly reddish-brown 1-year-old shoots have moderate growth vigor, 76 cm long with the internodes of 8.2 cm. The secondary branches are 38.7 cm long with 8 nodes and under-developed thorns. The large column-shaped mother fruiting spurs can germinate 2~4 deciduous fruiting shoots of medium-length and medium-width, which are 23.9 cm long with 13 leaves. The large oval-shaped leaves are medium thick and light green. There are a great number of flowers.

Biological Characters　The tree has strong vigor and strong branching ability, blooming and bearing early with low fruit set, medium early productivity and an unstable yield in full-bearing stage. The tree enters the full-bearing stage in the 8th year with a yield of 15 kg. In Taigu County of Shanxi Province, it matures in mid-September. It is a mid-ripening variety. Fruit cracking easily occurs. It has strong resistance to diseases, yet the leaves have poor resistance to early defoliation disease.

Fruit Characteristics　The large oval-shaped fruit has a vertical and cross diameter of 4.50 cm and 3.30 cm, averaging 23.4 g with a regular size. It has a slightly protuberant surface and thick skin. The skin usually presents a yellowish-gray color from white-maturing stage to crisp-maturing stage, and turns brightly red in complete-maturing stage. The greenish-white flesh is delicate, juicy and sweet, with a higher than normal quality, suitable for fresh-eating. The percentage of edible part of fresh fruit is 98.0%, and SSC, TTS, TA and Vc is 29.40%, 26.98%, 0.39% and 271.23 mg per 100 g fresh fruit. The small stone is spindle-shaped, averaging 0.48 g.

Evaluation　The medium-sized plant of Beijingjidanzao cultivar has strong vigor, strong branching ability with strong early productivity and a medium yield in full-bearing stage. It is a mid-ripening table variety with a large fruit size and a good quality, which can be developed in areas with light rainfall.

鲜 食
Table Varieties
品 种

· 17 ·

不 落 酥

品种来源及分布　原产和集中分布于山西省平遥县辛村乡的赵家庄一带，栽培数量少。

植物学性状　树体较小，树姿开张，干性弱，枝条细而较密，树冠乱头形，树皮裂纹条状，中等深。枣头红褐色，平均生长量62.7cm，粗度0.96cm，节间长6.6cm。二次枝着生部位较高，枣头着生二次枝5～6个，二次枝长27.8cm，6节左右，弯曲度小。针刺不发达，基本退化。皮目小，圆形，灰白色，中密。枣股小，圆锥形，抽吊力较强，平均抽生枣吊4个。枣吊细而长，平均长22.0cm，着叶17片。叶片中等大，叶长5.1cm，叶宽2.6cm，椭圆形，先端锐尖，叶基偏圆形，叶缘锯齿中密，较细。花量较少，枣吊平均着花44朵，每花序3朵左右。花中等大，花柄长，花径6.5mm左右。

生物学特性　树体较小，树势和干性较弱，萌芽率高，成枝力较强。结果较早，定植第二年开始结果，坐果率中等偏低，枣头、2～3年生枝和4年生枝的吊果率分别为19.0%、16.5%和12.1%，主要坐果部位在枣吊的3～8节。产量中等，较稳定。在山西太谷地区，8月下旬果实着色，9月中旬进入脆熟期，果实生育期105d左右，为中早熟品种类型。枣果白熟期后遇雨易裂果，采前落果轻。

果实性状　果个大，倒卵圆形，纵径4.21cm，横径3.32cm，单果重20.6g，大小较整齐。果肩平，果梗细长，梗洼窄深。果顶微凹，柱头遗存。果皮薄，紫红色，果面粗糙而不平整。果点中大，稀疏而不明显。果肉厚，绿白色，肉质细而酥脆，甜味浓，汁液较多，口感极佳，适宜鲜食。鲜枣可食率95.2%，含可溶性固形物34.80%，总糖27.92%，酸0.35%，100g果肉维生素C含量284.12mg；果皮含黄酮31.03mg/g，cAMP含量101.58μg/g。核较小，纺锤形，纵径2.66cm，横径0.95cm，核重0.98g。核尖较长，核面粗糙，部分核内含有种子，种仁不饱满，含仁率45.0%。

评价　该品种适应性较强，适宜密植栽培，但要求较高的肥水条件。果实大，品质优良，为优良的中早熟鲜食品种。但产量一般，果面粗糙而不平整，易裂果，不宜大面积栽培。

Buluosu

Source and Distribution　The cultivar originated from Zhaojiazhuang in Xincun Village of Pingyao County in Shanxi Province with a small quantity.

Botanical Characters　The small tree is spreading with a weak central leading trunk, an irregular crown and thin and dense branches. The trunk bark has medium-deep striped fissures. The reddish-brown 1-year-old shoots have strong germination ability and weak growth vigor, averaging 62.7 cm long with the internodes of 6.6 cm. There are generally 5～6 secondary branches which always grow in a higher position, averaging 27.8 cm long with 6 nodes and under-developed thorns. The small lenticels are round and gray, medium-densely distributed. The small conical-shaped mother fruiting spurs can germinate 4 thin deciduous fruiting shoots, averaging 22.0 cm long with 17 leaves. The medium-sized leaves are oval-shaped. The number of flowers is small, averaging 44.2 ones per deciduous fruiting shoot and 3 ones per inflorescence. The medium-sized blossoms have a diameter of 6.5 mm.

Biological Characters　The small tree has weak vigor, strong germination and branching ability, bearing early (generally in the 2nd year after planting) with low fruit set and a medium and stable yield. In Taigu County of Shanxi Province, it enters the crisp-maturing stage in mid-September. Fruit-cracking easily occurs.

Fruit Characteristics　The large obovate fruit has a vertical, averaging 20.6 g with a regular size. It has light-red thin skin and a rough surface. The greenish-white thick flesh is delicate, crisp, juicy and sweet, suitable for fresh-eating. The percentage of edible part of fresh fruit is 95.2%, and the content of water, SSC, TTS, TA and Vc is 60.2%, 34.80%, 27.92%, 0.35% and 284.12 mg per 100 g fresh fruit. The content of flavones and cAMP in mature fruit skin is 31.03 mg/g and 101.58 μg/g. The small stone is spindle-shaped, averaging 0.98 g. Some stones contain a shriveled kernel, the percentage of which is 45.0%.

Evaluation　The cultivar has strong adaptability and is suitable for close planting with a high demand on fertilization and irrigation. The large fruit has high edibility and a good quality. It is a fine mid-ripening table variety with normal productivity and serious fruit-cracking. It is not suitable for large-scale cultivation.

宁阳六月鲜

品种来源及分布　分布于山东省的宁阳、兖州、济宁等地，为当地原产品种，数量极少。

植物学性状　树体较大，树姿开张，干性强，枝条中密，树冠自然圆头形。树干灰褐色，树皮条状纵裂。枣头紫褐色，生长量大，平均长86.9cm，节间长9.5cm。二次枝生长量大，平均长41.0cm，着生7节左右，弯曲度大。针刺不发达，弱枝常无针刺。皮孔中大，圆形，凸起，开裂。枣股圆柱形，一般抽生3～5个枣吊。枣吊长24.2cm，着叶13片。叶长7.0cm，叶宽3.5cm，叶片卵状披针形，两侧向叶背反卷，中厚，绿色，富有光泽。先端锐尖，叶基偏斜形，叶缘锯齿粗大。花量中等，每花序着花3～7朵。花中大，昼开型，花粉较多。

生物学特性　树势和发枝力强，结果较早，早期丰产性能较强，产量稳定。花期要求温度较高，在气温不稳定或日均温低于24℃的年份坐果不良。气温高的年份坐果好，产量高，吊果率22.3%。在山西太谷地区，9月中旬果实着色成熟，果实生育期110d左右，为中熟品种类型。较抗裂果和抗病。

果实性状　果实中等大，果形不一，有长椭圆形、卵圆形、倒卵形等多种形状。果个大小较整齐，纵径3.60cm，横径2.70cm，单果重13.0g。果肩平圆，有数条浅沟棱。梗洼中广，浅或中深。环洼深，中等大。果柄粗短，果顶圆或广圆。果面不平整，果皮较厚，浅红色。果点圆形，中等大，密度大，不显著。果肉白绿色，质地细脆，汁液多，甜味浓，略具酸味，口感极佳，适宜鲜食。鲜枣可食率93.5%，含可溶性固形物37.20%，总糖29.60%，酸0.57%，100g果肉维生素C含量349.29mg；果皮含黄酮13.27mg/g，cAMP含量248.67μg/g。果核中等大，长纺锤形或椭圆形，核重0.85g，核蒂小，略尖，核尖突尖。核纹中深，呈不规则短斜纹。核内多具饱满种子，含仁率96.7%。

评价　该品种为稀有的优良中熟品种，在较好的肥水条件下，表现丰产性能较强。果实较大，果肉松脆，质细，汁液较多，味浓适口，品质优良。适于城郊、工矿区发展栽培。

Ningyangliuyuexian

Source and Distribution　The cultivar originated from and spreads in Ningyang and Jining of Shandong Province with a very small quantity.

Botanical Characters　The large tree is spreading with a strong central leader trunk, medium-dense branches and a natural-round crown. The purplish-brown 1-year-old shoots have strong growth vigor, averaging 86.9 cm long with the internodes of 9.5 cm. The secondary branches also have strong growth vigor, averaging 41.0 cm long with 7 node and less-developed thorns. Thorns are underdeveloped and there are always no thorns on weak branches. The medium-sized round lenticels are protuberant and cracked. The column-shaped mother fruiting spurs can germinate 3～5 deciduous fruiting shoots, which are 24.2 cm long with 13 leaves. The medium-thick green leaves are ovate-lanceolate and glossy. The number of flowers is medium large, averaging 3～7 ones per inflorescence.

Biological Characters　The tree has strong vigor and strong branching ability, bearing early with strong early productivity and a stable yield. Higher temperature is necessary for fruit-set of flowers. If the temperature is inconsistent or the average daily temperature is below 24 ℃ in the blooming stage, fruit set will be very low; yet it would be higher with a large yield if the temperature is higher. In Taigu County of Shanxi Province, it matures in mid-September. It is a mid-ripening variety with strong resistance to fruit-cracking and diseases.

Fruit Characteristics　The medium-sized fruit has different shapes which are long oval-shaped, oval-shaped or obovate. Yet it has a regular size, averaging 13.0 g. It has an unsmooth surface and light-red thick skin. The greenish-white flesh is delicate, crisp, juicy and sweet, with a little sour taste and an excellent quality for fresh eating. The percentage of edible part of fresh fruit is 93.5%, and SSC, TTS, TA and Vc is 37.20%, 29.60%, 0.57% and 349.29 mg per 100 g fresh fruit. The content of flavones and cAMP in mature fruit skin is 13.27 mg/g and 248.67 μg/g. The medium-sized stone is long spindle-shaped or oval-shaped, averaging 0.85 g.

Evaluation　The cultivar is a rare mid-ripening variety. The large fruit has crisp, sweet and juicy flesh with a good quality.

鲜 食
Table Varieties
品 种

孔府酥脆枣

品种来源及分布 别名脆枣、铃枣。来源于山东省曲阜的孔府院内，由此得名。

植物学性状 树体中大，树姿较开张，干性强，枝条中密，树冠伞形。主干皮裂呈块状，较浅，易剥落。枣头黄褐色，平均生长量69.7cm，粗1.04cm，节间长9.6cm。二次枝长43.6cm，平均7节。无针刺。皮目较小，圆形，凸起，开裂，分布较密。枣股中等大，圆锥形，抽吊力中等，抽生枣吊3～4个。枣吊平均长21.6cm，着叶11片。叶片中等大，椭圆形，叶厚，浓绿色，先端急尖，叶基偏斜，叶缘锯齿粗。花量多，花中等大，花径6.5mm左右，昼开型。

生物学特性 树势强健，发枝力中等，开花结果早，早期丰产性能较强。枣头、2～3年生枝和3年生以上枝的吊果率分别为35.0%、37.5%和34.1%。在山西太谷地区，9月中旬果实成熟，果实生育期100d左右，为中熟品种类型。果实抗病性较强，采前落果较轻，较抗裂果，有隔年结果现象。

果实性状 果实中等大，长圆形或圆柱形，单果重12.5g，大小较整齐。果梗短而粗，梗洼中等深。果顶平，柱头遗存。果皮中厚，紫红色，果面不平滑。果点小而密，圆形或椭圆形，浅黄色。果肉乳白色，肉质酥脆，较细，甜味浓，汁液中多，品质优异，适宜鲜食。鲜枣可食率94.0%，含可溶性固形物36.00%，总糖35.17%，酸0.34%，100g果肉维生素C含量360.14mg；果皮含黄酮12.57mg/g，cAMP含量156.58μg/g。核较大，纺锤形，核重0.75g，核尖较长，核纹深，核面粗糙，多数核内含有种子，含仁率90.0%。

评价 该品种为优质中熟鲜食良种。但产量不稳定，果实表面欠美观，大小不整齐，可适度发展栽培。

Kongfusucuizao

Source and Distribution The cultivar, also called Cuizao or Lingzao, originated from the yard of Kongfu in Qubu of Shandong Province, Which is why it is called Kongfusucui..

Botanical Characters The medium-sized tree has strong vigor, spreading with a strong central leader trunk, medium-dense branches and an umbrella-shaped crown. The trunk bark has shallow massive fissures, easily shelled off. The yellowish-brown 1-year-old shoots have strong growth vigor, 69.7 cm long and 1.04 cm thick with the internodes of 9.6 cm. The secondary branches are 43.6 cm long with 7 nodes and without thorns. The small lenticels are round, protuberant and cracked, densely-distributed. The medium-sized conical-shaped mother fruiting spurs can germinate 3～4 deciduous fruiting shoots, which are 21.6 cm long with 11 leaves. The medium-sized thick leaves are oval-shaped and dark green, with a sharply-cuspate apex, an oblique base and a thick saw-tooth pattern on the margin. There are many medium-sized flowers with a diameter of 6.5 mm produced. It blooms in the daytime.

Biological Characters The tree has strong vigor and medium branching ability, blooming and bearing early with strong early productivity. The percentage of fruits to deciduous fruiting shoots of 1-year-old shoots, 2 or 3-year-old branches and over-4-year-old ones is 35.0%, 37.5% and 34.1%. In Taigu County of Shanxi Province, it matures in mid-September with a fruit growth period of 100 d. It is an early-mid-ripening variety with strong resistance to diseases and fruit-cracking, light premature fruit-dropping. Alternate bearing exists.

Fruit Characteristics The medium-sized fruit is oblong or column-shaped, averaging 12.5 g with a regular size. It has a short and thick stalk, a medium-deep stalk cavity, a flat fruit apex, a remnant stigma, an unsmooth surface and medium-thick and purplish-red skin with small, dense, round or oval-shaped and light-yellow dots. The ivory-white flesh is crisp and strongly sweet, with medium juice and an excellent quality for fresh eating. The percentage of edible part of fresh fruit is 94.0%, and SSC, TTS, TA and Vc is 36.00%, 35.17%, 0.34% and 360.14 mg per 100 g fresh fruit. The content of flavones and cAMP in mature fruit skin is 12.57 mg/g and 156.58 μg/g. The large spindle-shaped stone weighs 0.75 g with a long apex, deep veins and a rough surface. Most stones contain a kernel, the percentage of which is 90.0%.

Evaluation The cultivar is a good mid-ripening table variety. Yet it has an unstable yield, and the fruit has an unattractive appearance and a regular size. So it can only be developed moderately.

· 23 ·

缨 络 枣

品种来源及分布 别名莺落枣、硬不落。因果实着色自梗洼起逐渐向外，近成熟时着色达果实中部，如古代官帽的缨络，故得名。原产于北京近郊，为北京地区分布最广、栽培最为普遍的鲜食品种。大兴区榆垡的东、西翁各庄曾是著名产区。栽培历史1 000年以上。

植物学性状 树体中大，树姿开张，干性中等，树冠自然圆头形或自然半圆形。主干灰褐色，树皮纵裂，呈不规则的窄条状，不易剥落。枣头红褐色，平均长85.7cm，粗1.25cm，节间长7.3cm，无蜡层。二次枝长29.9cm，7节左右，弯曲度大。无针刺。枣股圆柱形，抽吊力中等，平均抽生枣吊4.0个。枣吊长25.7cm，着叶16片。叶片中大，长椭圆形，深绿色，光泽差，先端渐尖，先端钝圆，叶基圆形。叶缘平或略有波状起伏，具钝齿。花量多，花序平均着花8朵，花朵较小，夜开型。

生物学特性 风土适应性强，耐土壤瘠薄和花期干旱或偏低气温。树体强健，成枝力较强，坐果稳定。枣吊平均坐果0.6个，丰产性强，成龄树一般株产40kg左右。在山西太谷地区，果实9月下旬开始着色，10月上旬完全成熟，果实生育期约115d，为晚熟品种类型。果实较抗病和抗裂果。

果实性状 果实中大，短柱形或近圆形，纵径3.00cm，横径2.69cm，单果重14.9g，大小整齐。果肩平圆，较宽，梗洼浅、广、平整，无沟棱。果顶圆或广圆，先端略凹，残柱呈点状，不明显。果面平整，着色前果实阳面有红晕。果皮较厚，紫褐色，有光泽。果肉浅绿色或白色，质地细致酥脆，汁液多，味甜，品质优异，主要用于鲜食。鲜枣可食率96.6%，含总糖20.96%，酸0.33%，100g果肉维生素C含量315.02mg；果皮含黄酮27.28mg/g，cAMP含量88.00μg/g。果核中大，椭圆形，核重0.51g，含仁率95.0%。

评价 该品种风土适应性强，树体中大，坐果多而稳定，丰产性强，成熟期晚，裂果少。果实中大，肉质致密硬脆，鲜食品质极佳，为优良晚熟鲜食品种。

Yingluozao

Source and Distribution The cultivar originated from the suburbs of Beijing. It is the most widely cultivated table variety in Beijing, with a cultivation history of over 1 000 years.

Botanical Characters The medium-sized tree is spreading with a medium-strong central leader trunk and a round or semi-round crown. The grayish-brown trunk bark has irregular, vertical narrow-striped fissures. The reddish-brown 1-year-old shoots are 85.7 cm long and 1.25 cm thick. The internodes are 7.3 cm long, without wax and thorns. The secondary branches are 29.9 cm long with 7 nodes of medium curvature. The column-shaped mother fruiting spurs can germinate 4 deciduous fruiting shoots, which are 25.7 cm long with 16 leaves. The medium-sized leaves are long oval-shaped, dark green and less glossy. There are many small flowers produced, averaging 8 ones per inflorescence. It blooms at night.

Biological Characters The tree has strong adaptability, strong vigor and strong branching ability, tolerant to poor soils, drought and low temperatures in the blooming stage, with stable fruit set and high productivity. The deciduous fruiting shoot bears 0.6 fruits on average. A mature tree has a yield of about 40 kg. In Taigu County of Shanxi Province, it matures in early October. It is a late-ripening variety with strong resistance to fruit-racking.

Fruit Characteristics The medium-sized fruit is short column-shaped or nearly round, averaging 14.9 g with a regular size. It has a smooth surface. Before coloring, the sun-side of the fruit is flushed with red. The thick skin is purplish-brown and glossy. The light-green or white flesh is crisp, juicy and sweet. It has an excellent quality for fresh-eating. The percentage of edible part of fresh fruit is 96.6%, and the content of TTS, TA and Vc is 20.96%, 0.33% and 315.02 mg per 100 g fresh fruit. The content of flavones and cAMP in mature fruit skin is 27.28 mg/g and 88.00 μg/g. The medium-sized stone is oval-shaped, averaging 0.51 g. The percentage of containing kernels is 95.0%.

Evaluation The medium-sized plant of ringluozao cultivar has strong adaptability and a compact structure, with high fruit set, a high and stable yield. It matures late with very light fruit-cracking. The medium-sized fruit has tight-textured and crisp flesh, with an excellent quality for fresh-eating. It is a good late-ripening variety for fresh-eating fruits.

山 东 梨 枣

品种来源及分布 别名鲁北梨枣、大铃枣、铃枣、梨枣、脆枣。原产山东和河北省交界处的乐陵、庆云、无棣、盐山、黄骅等地。多为庭院零星栽植。

植物学性状 树体中等大，树姿开张，干性较强，树冠自然圆头形。树皮呈条状纵裂，不易剥落。枣头红褐色，生长势较强，较粗壮，平均生长量73.7cm，粗1.16cm，节间长7.9cm，二次枝长34.2cm，平均7节。无针刺。皮目中等大，圆形，分布密，凸起，开裂。枣股中等大，圆柱形，5～6年生枣股长1.3cm，粗1.1cm，抽吊力中等，抽生枣吊3～4个。枣吊粗而长，一般长20～24cm。叶片中等大，卵状披针形，浓绿色，中等厚，先端渐尖，叶基圆形或广楔形，叶缘锯齿细而较密。花量特多，每花序着花8～10朵，最多达15朵以上。花较大，花径7～8mm，无花粉，为夜开型。

生物学特性 树势中等，萌芽率和成枝力较强，结果早，早期丰产性能较强，产量稳定。一般定植当年即可结果，3、4年生进入初果期，6年生后大量结果，盛果期株产可达20kg左右。幼龄枝结实能力较强，2～3年生枝吊果率73.3%，4年生以上枝较低，仅为18.3%。在山西太谷地区，9月上旬枣果进入脆熟期，果实生育期90d左右，为早熟品种类型。果实抗病性强，较抗裂果。

果实性状 果个大，果形多为梨形或倒卵圆形，纵径3.90cm，横径3.30cm，单果重22.4g，最大可达55.0g，大小不整齐。果皮较薄，红色，果面有隆起。果点小，圆形，浅黄色，不明显。果梗粗而短，梗洼窄而深。果顶微凹，柱头遗存。果肉厚，绿白色，肉质细而酥脆，味甜微酸，汁液多，品质极上，适宜鲜食。鲜枣可食率95.4%，含可溶性固形物29.40%，总糖22.76%，酸0.33%，100g果肉维生素C含量353.10mg；果皮cAMP含量68.72μg/g。核较小，纺锤形，纵径2.60cm，横径0.90cm，核重1.03g。含仁率86.7%，种仁较饱满。

评价 该品种适宜密植栽培，每667m²栽90株左右为宜。为大果、优质、抗病、较抗裂果的早熟优良品种，可规模发展栽培。

Shandonglizao

Source and Distribution The cultivar originated from Leling, Qingyun, Wudi in Shandong Province and Yanshan, Huanghua in Hebei Province. It is planted dispersedly in family garden.

Botanical Characters The medium-sized tree is spreading with a strong central leader trunk and a round crown. The trunk bark has vertical striped fissures, uneasily shelled off. The strong yellowish-brown 1-year-old shoots have strong growth potential, averaging 73.7 cm long and 1.16 cm thick, with the internodes of 7.9 cm. The secondary branches are 34.2 cm long with 7 nodes and without thorns. The medium-sized lenticels are round, dense, protuberant and cracked. The 5～6-year-old mother fruiting spurs are medium-sized and can germinate 3～4 thick deciduous fruiting shoots, which are 20～24 cm long. The medium-sized leaves are ovate-lanceolate, dark green and medium thick. There are many large flowers with a diameter of 7～8 mm, averaging 8～10 ones per inflorescence.

Biological Characters The tree has moderate vigor, strong germination and branching ability, bearing early with strong early productivity and a stable yield. It generally bears in the year of planting and enters the initial full-bearing stage in the 3rd or 4th year. In the 6th year, the tree has a high yield which reaches 20 kg in the full-bearing stage. In Taigu County of Shanxi Province, it enters the crisp-maturing stage in early September. It is an early-ripening variety with strong resistance to diseases and fruit-cracking.

Fruit Characteristics The large fruit is pear-shaped or obovate, averaging 22.4 g with irregular sizes. It has thin red skin. The greenish-white thick flesh is delicate, crisp, sweet and juicy, with an excellent quality for fresh eating. The percentage of edible part of fresh fruit is 95.4%, and SSC, TTS, TA and Vc is 29.40%, 22.76%, 0.33% and 353.10 mg per 100 g fresh fruit. The content of cAMP in mature fruit skin is 68.72 μg/g. The small stone is spindle-shaped, averaging 1.03 g.

Evaluation The cultivar grows well under intensive planting. The large fruit has a good quality with strong resistance to diseases and fruit-cracking. It is a fine early-ripening variety.

·27·

临汾蜜枣

品种来源及分布　原产和分布于山西临汾市尧都区西孔郭一带。

植物学性状　树体较小，树姿开张，干性较弱，树冠自然乱头形。主干皮裂条状。枣头黄褐色，平均长67.2cm，节间长6.5cm。二次枝长26.9cm，着生5～8节。针刺不发达。枣股平均抽生枣吊3.4个。枣吊长23.6cm，着叶13片。叶片大，叶长7.9cm，叶宽4.2cm，椭圆形，两侧略向叶面合抱，先端锐尖，叶基圆形，叶缘具钝锯齿。花量多，花序平均着花8朵，但花小，花径5.8mm。

生物学特性　树势中庸偏弱，萌芽率和成枝力强，结果较早，坐果率高，枣头、2～3年、4～6年和6年以上生枝的吊果率分别为86.0%、160.1%、188.1%和81.0%。一般定植第二年结果，第三年有一定产量，盛果期产量较高。在山西太谷地区，9月中旬果实进入脆熟期，果实生育期103d左右，为早中熟品种类型。成熟期遇雨裂果严重。

果实性状　果个小，卵圆形，纵径2.84cm，横径2.31cm，单果重7.4g，大小不整齐。果皮薄，红色，果面平滑。果点小而圆，分布稀疏。梗洼广，较浅。果顶尖，柱头遗存，不明显。鲜食品质极佳，鲜枣可食率95.0%，含可溶性固形物36.60%，总糖33.41%，100g果肉维生素C含量284.12mg；果皮含黄酮29.68mg/g，cAMP含量24.56μg/g。核小，纺锤形，核重0.37g。种仁较饱满，含仁率43.3%。

评价　该品种树体较小，树势中庸，早果，丰产性能强。果个较小，肉质酥脆，味极甜，汁液多，鲜食品质极上。抗裂果能力较差，成熟期需注意防雨。

Linfenmizao

Source and Distribution　The cultivar originated from and spreads in Yaodu District of Linfen City in Shanxi Province.

Botanical Characters　The small tree is half-spreading with a weak central leader trunk and an irregular crown. The trunk bark has striped fissures. The yellowish-brown 1-year-old shoots are 67.2 cm long with the internodes of 6.5 cm. The secondary branches are 26.9 cm long with 5～8 nodes and less-developed thorns. The mother fruiting spurs can germinate 3.4 deciduous fruiting shoots, which are 23.6 cm long with 13 leaves. The large oval-shaped leaves are 7.9 cm long and 4.2 cm wide, with both sides slightly folding towards the center. The leaves have a sharply-cuspate apex, a round base and a blunt saw-tooth pattern on the margin. There are many small flowers with a diameter of 5.8 mm produced, averaging 8 ones per inflorescence.

Biological Characters　The tree has moderate or weak vigor, high germination rate and strong branching ability, bearing early with high fruit set. The percentage of fruits to deciduous bearing shoots of 1-year-old shoots, 2～3-year-old branches, 4～6-year-old ones and over-6-year-old ones is 86.0%, 160.1%, 188.1% and 81.0%. It generally bears in the 2nd year after planting. There is certain yield in the 3rd year with medium productivity. In Taigu County of Shanxi Province, it enters the crisp-maturing stage in mid-September and is harvested in late September with a fruit growth period of 103 d. It is a mid-late-ripening variety with serious fruit-cracking if it rains in the maturing stage.

Fruit Characteristics　The small oval-shaped fruit has a vertical and cross diameter of 2.84 cm and 2.31 cm, averaging 7.4 g with irregular sizes. It has thin red skin, a smooth surface with small, round and sparse dots, a wide and shallow stalk cavity, a sharp fruit apex and a remnant yet indistinct stigma. It has a superior quality for fresh-eating. The percentage of edible part of fresh fruit is 95.0%, and SSC, TTS and Vc is 36.60%, 33.41% and 284.12 mg per 100 g fresh fruit. The content of flavones and cAMP in mature fruit skin is 29.68 mg/g and 24.56 μg/g. The small spindle-shaped stone weighs 0.37 g with a well-developed kernel. The percentage of containing kernels is 43.3%.

Evaluation　The small plant of Linfenmizao cultivar has moderate vigor. The small fruit has crisp, juicy and sweet flesh with an excellent quality for fresh eating. It has poor resistance to fruit-cracking, so rain protection should be paid much attention to in the maturing stage.

太谷鸡心蜜

品种来源及分布 别名鸡心蜜枣。原产山西省太谷、榆次和清徐一带。分布较广，但数量较少。

植物学性状 树体小，树势较弱，树姿半开张，干性弱，有弯曲现象，枝条较稀疏，多年生枝存在先端回枯现象，树冠乱头形。主干条状皮裂。枣头黄褐色，生长势中等，平均生长量67.2cm，二次枝6～7节。幼龄枝针刺发达，多年生老弱枝大部分针刺脱落。皮目小而中密，圆形，凸起，灰白色。枣股小，圆锥形，抽枝力中等，抽生枣吊2～4个，多为3个。枣吊短，平均长19.9cm，节间长1.7cm。叶片大，叶长7.6cm，叶宽3.2cm，长卵形，先端渐尖，叶基偏圆形，锯齿细而较密。花量中多，枣吊平均着花61.7朵。花小，零级花花径6.6mm。

生物学特性 树体生长势弱，发枝力中等。结果较早，早期丰产性能较强，盛果期树产量中等。坐果率较高，枣头枝无果，2～3年生枝的吊果率57.8%，4年生以上枝为38.0%。在山西太谷地区，9月初果实着色，9月10日前后进入脆熟期，果实生育期100d左右，为早熟品种类型。果实成熟期遇雨易裂果。

果实性状 果个小，鸡心形，纵径2.90cm，横径2.70cm，单果重7.0g，大小不整齐。果梗细而短，梗洼中广、中深。果顶平，柱头遗存。果皮薄，鲜红色，果面光滑。果点中大，较密，圆形，明显。果肉厚，浅绿色，肉质细脆，味甜略酸，汁液多，品质极佳，适宜鲜食。鲜枣可食率90%左右，含总糖28.10%，酸0.63%，100g果肉维生素C含量277.31mg；果皮含黄酮4.98mg/g，cAMP含量62.10μg/g。核较大，纺锤形，纵径2.90cm，横径2.70cm，核重0.70g，核尖较长，核面粗糙，种仁较饱满，含仁率90.0%左右。

评价 该品种结果较早，产量中等，较稳定。果实小，肉细脆，味甜略酸，汁液多，鲜食品质优异。但果实大小不整齐，核大，可食率低，采前落果重，遇雨易裂果。

Taigujixinmi

Source and Distribution The cultivar originated from Taigu, Yuci and Qingxu in Shanxi Province, with a small quantity and an unknown history.

Botanical Characters The small tree is half-spreading with an irregular crown, a weak central leader trunk of small curvature and sparse branches. Perennial branches have back withering in the apex. The yellowish-brown 1-year-old shoots have moderate growth vigor, averaging 67.2 cm long. There are generally 6～7 nodes on each secondary branch with developed thorns. The small conical-shaped mother fruiting spurs can germinate 2～4 short deciduous fruiting shoots which are 19.9 cm long with the internodes of 1.7 cm. The large leaves are long oval-shaped. The number of flowers is medium large, averaging 61.7 per deciduous fruiting shoot. The small blossoms have a diameter of 6.56 mm for the zero-level flowers.

Biological Characters The tree has weak vigor, medium branching ability, bearing early with poor early productivity and a medium yield in full-bearing stage. It has high fruit set, and there is no fruit on the 1-year-old shoots. In Taigu County of Shanxi Province, it enters the crisp-maturing stage in early September. It is an early-ripening variety with serious fruit-cracking if it rains in maturing stage.

Fruit Characteristics The small fruit is chicken-heart shaped, averaging 7.0 g with irregular sizes. It has a brightly-red thin skin and a smooth surface with medium-large, dense, round and distinct dots. The thick flesh is light green, delicate, crisp, juicy, sweet and a little bit sour, with an excellent quality for fresh eating. The percentage of edible part of fresh fruit is 90.0%, and the content of TTS, TA and Vc is 28.10%, 0.63% and 277.31 mg per 100 g fresh fruit. The content of flavones and cAMP in mature fruit skin is 4.98 mg/g and 62.10 μg/g. The medium-sized or large stone is spindle-shaped, averaging 0.70 g.

Evaluation The cultivar bears early with a medium and stable yield. The small fruit has crisp, juicy and delicate flesh with a sweet and a little sour taste and a good flavor for fresh eating. Yet it has irregular sizes, a large stone, low edibility, serious premature fruit-dropping and serious fruit-cracking if it rains.

鲜 食
Table Varieties
品 种

郎 家 园 枣

品种来源及分布　原产北京市朝阳区郎家园一带，故得此名。北京最著名的优质鲜食良种，是当地传统主栽品种。

植物学性状　树体中等大，树姿开张，干性较弱，树冠多呈自然半圆形或伞形。树皮条状纵裂，中深，不易剥落。枣头黄褐色，较细，平均长81.0cm，粗度1.07cm，节间长7.0cm，二次枝长24.0cm，5节左右。针刺不发达。皮目中等大，分布较稀，圆形，凸起，开裂。枣股较小，圆柱形，抽吊力中等，抽生枣吊2～4个。枣吊细长，有分叉现象，平均长26.4cm，着叶16片。叶片中大，叶长6.3cm，叶宽3.0cm，椭圆形，浅绿色，先端钝尖，叶基圆楔形，叶缘锯齿浅而粗，两侧向叶背反卷。花量多，花小，花径6mm左右，夜开型。

生物学特性　树势中庸偏强，易发枝，进入结果期较晚，坐果率低，产量不高。幼龄枝结实能力较差，枣头枝无果，2～3年生枝吊果率也仅5.6%，4年生枝为50.4%。在山西太谷地区，果实9月中旬成熟采收，果实生育期105d，为中熟品种类型。采前落果严重。

果实性状　果个较小，长圆形，纵径3.20cm，横径2.09cm，单果重9.0g，大小整齐。果梗中粗，梗洼中广、中深。果顶平，柱头遗存。果点小，中密，不明显。果皮薄，紫红色，果面平滑。果肉厚，绿白色，肉质细嫩酥脆，甜味浓，汁液多，口感极佳，适宜鲜食。鲜枣可食率97.3%，含可溶性固形物35.00%，总糖22.36%，酸0.37%，100g果肉维生素C含量359.88mg；果皮含黄酮12.56mg/g，cAMP含量8.44μg/g。果核小，纺锤形，纵径1.50～1.80cm，横径0.50～0.60cm，核重0.24g，核纹细而浅，含仁率46.7%。

评价　该品种是北京名产，古时为皇宫贡品。鲜枣皮薄核小，肉质酥脆，多汁甘甜，品质优异，裂果较轻，为优质中熟鲜食良种，可在城郊地区发展。但坐果少，产量较低，需加强肥水管理和采取提高坐果率的措施。

Langjiayuanzao

Source and Distribution　The cultivar originated from Langjiayuan in Chaoyang District of Beijing. It is selected from wild species and was named after the production area. It is the most famous table variety in Beijing, and is the traditional dominant cultivar there.

Botanical Characters　The medium-sized tree is spreading with a weak central leader trunk and a semi-round or umbrella-shaped crown. The trunk bark has medium-deep vertical striped fissures, uneasily shelled off. The yellowish-brown 1-year-old shoots are 81.0 cm long and 1.07 cm thick with the internodes of 7.0 cm long. The secondary branches are 24.0 cm long with 5 nodes and less-developed thorns. The medium-sized round lenticels are sparsely distributed, protuberant and cracked. The small column-shaped mother fruiting spurs can germinate 2～4 thin and long deciduous fruiting shoots which are 26.4 cm long with 16 leaves. The medium-sized leaves are oval-shaped and light green. There are many small flowers with a diameter of 6 mm produced. It blooms at night.

Biological Characters　The tree has moderate or strong vigor, strong branching ability, bearing late with low fruit set and a low yield. Young branches have weak fruiting ability. In Taigu County of Shanxi Province, it matures in mid-September.

Fruit Characteristics　The small oblong fruit has a vertical and cross diameter of 3.2 cm and 2.09 cm, averaging 9.0 g with a regular size. It has a purplish-red thin skin and a smooth surface. The greenish-white thick flesh is delicate, crisp, sweet and juicy, with an excellent taste for fresh eating. The percentage of edible part of fresh fruit is 97.3%, and SSC, TTS, TA and Vc is 35.00%, 22.36%, 0.37% and 359.88 mg per 100 g fresh fruit. The content of flavones and cAMP in mature fruit skin is 12.56 mg/g and 8.44 μg/g. The small stone is spindle-shaped, averaging 0.24 g.

Evaluation　The cultivar is a famous product in Beijing. Fresh fruit has thin skin, a small stone, crisp, juicy and sweet flesh with an excellent quality. It is a very good mid-ripening table variety which can be developed in suburb areas.

鲜 食
Table Varieties
品 种

· 33 ·

平陆尖枣

品种来源及分布 原产山西省平陆县岳村一带,栽培数量不多,历史不详。

植物学性状 树体中大,树姿开张,干性较强,枝条中密,树冠圆锥形。主干皮裂较浅,呈小块状,不易脱落。枣头红褐色,平均生长量62.7cm,粗1.01cm,节间长8.2cm,着生二次枝5个左右。二次枝长34.3cm,5节。针刺不发达。枣股中等大,圆锥形,抽吊力较强,抽生枣吊3~4个。枣吊短,平均长12.2cm,着叶11片。叶片较小,叶长5.3cm,叶宽2.6cm,卵圆形,绿色,先端锐尖,叶基圆形,叶缘锯齿粗浅、圆钝。花量较多,花小,零级花花径6.2mm,昼开型。

生物学特性 树势中庸健壮,萌芽率和成枝力较强。开花结果早,早期丰产性能强。嫁接苗当年即可挂果,第二年普遍结果,4~5年进入初果期,5年后大量结果,盛果期株产可达25kg左右。幼龄枝结实能力强,枣头吊果率可达100%,2~3年生枝也在50%以上。在山西太谷地区,9月中旬果实着色,9月下旬进入脆熟期,果实生育期115d左右,为晚熟品种类型。

果实性状 果个中大,圆锥形,纵径3.74cm,横径2.36cm,单果重10.2g,大小较整齐。果皮薄,紫红色,果面平滑。果点小而密,圆形,较明显。果梗较粗短,梗洼浅、中广。果顶尖,柱头遗存。果肉厚,肉质酥脆,味甜,汁液多,品质上等,适宜鲜食。鲜枣耐贮藏,鲜枣可食率94.5%,可溶性固形物含量28.50%,单糖11.58%,双糖9.60%,总糖21.18%,酸0.52%,100g果肉维生素C含量407.97mg;果皮含黄酮3.91mg/g,cAMP含量122.17μg/g。果核纺锤形,纵径2.28cm,横径0.73cm,核重0.56g,含仁率91.7%,种仁较饱满。

评价 该品种树体中大,开花结果早,早期丰产性能强,可进行密植栽培,但要加强树体和肥水管理。果实中大,果肉细脆,味甜,优质,耐贮,为优良晚熟鲜食品种。

Pinglujianzao

Source and Distribution The cultivar originated from Yuecun Village of Pinglu County in Shanxi Province, with a small quantity.

Botanical Characters The medium-sized tree has weak vigor, spreading with a strong central leader trunk, dense branches and a conical-shaped crown. The trunk bark has shallow massive fissures, uneasily shelled off. The reddish-brown 1-year-old shoots are 62.7 cm long and 1.01 cm thick with the internodes of 8.2 cm. There are generally 5 secondary branches which are 34.3 cm long with 5 nodes and less-developed thorns. The medium-sized conical-shaped mother fruiting spurs can germinate 3~4 short deciduous fruiting shoots which are 12.2 cm long with 11 leaves. The small green leaves are oval-shaped. There are many flowers with a diameter of 6.18 mm for the zero-level flowers.

Biological Characters The tree has moderate vigor, strong germination and branching ability, blooming and bearing early with strong early productivity. The tree enters the initial full-bearing stage in the 4th to 5th year, and it can bear substantially after 5 years. A tree in its full-bearing stage has an average yield of 25 kg. In Taigu County of Shanxi Province, it enters the crisp-maturing stage in late September.

Fruit Characteristics The large conical-shaped fruit has a vertical and cross diameter of 3.74 cm and 2.36 cm, averaging 10.2 g with a regular size. It has purplish-red thin skin and a smooth surface. The thick flesh is crisp, sweet and juicy, with a very good quality for fresh eating. Fresh fruits have a good storage character. The percentage of edible part of fresh fruit is 94.5%, and SSC, monosaccharide, disaccharide, TTS, TA and Vc is 28.50%, 11.58%, 9.60%, 21.18%, 0.52% and 407.97 mg per 100 g fresh fruit. The content of flavones and cAMP in mature fruit skin is 3.91 mg/g and 122.17 μg/g. The small spindle-shaped stone has a vertical and cross diameter of 2.28 cm and 0.73 cm, averaging 0.56 g with a well-developed kernel. The percentage of containing kernels is 91.7%.

Evaluation The small plant of Pinglujianzao cultivar blooms and bears early with strong early productivity, suitable for intensive planting. The fruit has thick flesh, a small stone, crisp and sweet flesh, with a good quality and good storage character. It is a fine mid-late ripening variety for fresh eating and processing alcoholic fruits.

成武冬枣

品种来源及分布 别名冬枣、芒果冬枣、金芒果。原产山东省成武县，分布于成武、菏泽、曹县等地。多为庭院零星栽植。栽培历史不详。

植物学性状 树体高大，树姿较直立，干性较强，枝条较稀，粗壮，树冠圆锥形。主干皮裂较宽，呈条状，易剥落。枣头黄褐色，生长势较强，平均生长量77.9cm，粗1.32cm，节间长8.0cm。二次枝长31.9cm，平均7节。无针刺。皮目中等大，较稀，椭圆形。枣股较大，圆锥形，最长2cm以上，抽吊力较强，抽生枣吊3～5个。枣吊平均长26.3cm，着叶15片。叶片大而厚，叶长6.1cm，叶宽2.9cm，浓绿色，椭圆形，先端急尖，叶基圆楔形，叶缘锯齿粗，中密。花量较多，每花序着花3～7朵，花中等大，花径7mm左右。

生物学特性 树势较强，干性明显，成枝力较强。结果较早，一般定植第二年挂果，10年生后开始大量结果，早期丰产性能和盛果期产量中等。在山西太谷地区，枣果10月上旬进入脆熟期，果实生育期120d左右，为极晚熟品种类型。枣果抗病性强，较抗裂果。

果实性状 果个大，果形长卵圆形，似芒果状，纵径4.99cm，横径3.31cm，单果重20.6g，最大32.1g，大小不整齐。果梗短而较粗，梗洼窄而较深。果顶平，柱头遗存而明显。果皮中等厚，浅红色，果面不平整。果点小而圆，分布密。果肉厚，乳白色，肉质细而松脆，味甜微酸，汁液中等偏多，品质上等，较耐贮藏，适宜鲜食。鲜枣可食率96.7%，含可溶性固形物33.20%，总糖22.36%，酸0.34%，100g果肉维生素C含量410.23mg；果皮含黄酮2.76mg/g，cAMP含量173.65μg/g。核小，纺锤形，纵径2.90cm，横径0.90cm，核重0.69g，核尖长，核面较粗糙，核纹中度深，多数核内含有种仁，含仁率88.3%。

评价 该品种为大果、优质、抗病、极晚熟的鲜食优良品种，可在生育期较长的地区适度规模发展。在生产栽培中应注意采取控制树势和促进早期丰产的技术措施。

Chengwudongzao

Source and Distribution The cultivar originated from Chengwu County in Shandong Province, and spreads in Chengwu, Heze and Caoxian. It is planted dispersedly in family gardens.

Botanical Characters The large tree is vertical with a strong central leader trunk, thin branches and a conical-shaped crown. The trunk bark has wide striped fissures, easily shelled off. The reddish-brown 1-year-old shoots are 77.9 cm long and 1.32 cm thick with a strong growth potential. The secondary branches are 31.9 cm long with 7 nodes. The internodes are 8.0 cm long without thorns. The large conical-shaped mother fruiting spurs are 2.0 cm long at most and can germinate 3～5 deciduous fruiting shoots which are 26.3 cm long with 15 leaves. The large thick leaves are dark green and oval-shaped. There are many medium-sized flowers with a diameter of 7 mm produced, averaging 3～7 ones per inflorescence.

Biological Characters The tree has strong vigor, weak branching ability and moderate early productivity. It bears early, generally in the 2nd year after planting and enters the full-bearing stage in the 10th year with a medium yield. In Taigu County of Shanxi Province, it enters the crisp-maturing stage in early October. It has strong resistance to diseases and fruit-cracking.

Fruit Characteristics The large fruit is long oval-shaped like a mango, averaging 20.6 g with irregular sizes. It has a medium-thick and light-red skin and a rough surface with small, round and dense dots. The ivory-white thick flesh is delicate, crisp, sweet, a little sour and juicy, with a good quality and good storage character, suitable for fresh eating. The percentage of edible part of fresh fruit is 96.7%, and SSC, TTS, TA and Vc is 33.20%, 22.36%, 0.34% and 410.23 mg per 100 g fresh fruit. The content of flavones and cAMP in mature fruit skin is 2.76 mg/g and 173.65 μg/g. The small stone is spindle-shaped, averaging 0.69 g.

Evaluation The cultivar is a late-ripening table variety with a large fruit size, good quality and strong resistance to diseases. It can be developed moderately in areas with a long fruit growth period. Technical measures to control vigor and promote early productivity should be carried out.

鲜 食
Table Varieties
品 种

· 37 ·

蜂 蜜 罐

品种来源及分布 别名大荔蜂蜜罐、蜜罐枣、甜蜜罐。原产陕西省大荔县的官池、北丁、中草一带，栽培数量不多。

植物学性状 树体中等大，树姿半开张，干性较强，枝系稠密，树冠圆柱形。皮裂中度深，条状，不易剥落。枣头黄褐色，较细，平均长75.2cm，粗1.11cm，节间长6.6cm。二次枝长29.1cm，平均生长6节，弯曲度大。针刺欠发达。皮目中大，椭圆形，分布中密，略突起。枣股较粗大，圆柱形，抽吊力较强，一般抽生枣吊3～5个。枣吊平均长21.5cm，着叶14片。叶片小而较厚，叶长6.3cm，叶宽2.7cm，浓绿色，卵圆形，先端急尖，叶基偏斜，叶缘锯齿粗钝、较密。花量多，花较小，夜开型。

生物学特性 树势较强，萌芽率高，发枝力较强，枣头生长较细弱。开花结果早，产量较高而稳定。枣头、2～3年生枝和3年生以上枝的吊果率分别为35.8%、62.0%和14.5%。在山西太谷地区，9月上旬果实进入脆熟期，果实生育期90d左右，为早熟品种类型。成熟期遇雨很少裂果，较抗病。采前落果严重。

果实性状 果实较小，近圆形，纵径2.70cm，横径2.50cm，单果重9.2g，最大11.0g，大小不整齐。果梗短而较细，梗洼深而狭。果顶平或微凹，柱头遗存。果皮薄，鲜红色，果面较平滑。果点小而密，圆形，浅黄色。果肉较厚，绿白色，肉质致密，细脆，味甜，汁液较多，品质上佳，适宜鲜食。鲜枣可食率94.0%，含可溶性固形物31.80%，总糖25.97%，酸0.51%，100g果肉维生素C含量359.05mg；果皮含黄酮5.07mg/g，cAMP含量86.15 μg/g。核较大，椭圆形，纵径1.10cm，横径0.80cm，核重0.55g，核尖短，核纹中深，部分核内含有饱满种仁，含仁率30.0%。

评价 该品种对土质要求不严，适应范围较广，开花结果早，早期丰产，稳产。果实肉质细脆，总糖含量较高，口感极佳，抗裂果，抗病，是优良早熟品种，可在我国南、北方枣区栽培。近年辽宁省朝阳地区大面积栽植，性状优异而稳定。

Fengmiguan

Source and Distribution The cultivar originated from Guanchi, Beiding and Zhongcao of Dali County in Shaanxi Province, with a small quantity.

Botanical Characters The medium-sized tree is half-spreading with a strong central leader trunk and a conical-shaped crown. The trunk bark has medium-deep striped fissures, uneasily shelled off. The reddish-brown 1-year-old shoots have moderate growth vigor, averaging 75.2 cm long and 1.11 cm thick with the internodes of 6.6 cm long. The secondary branches are 29.1 cm long with 6 nodes and less-developed thorns. The thick, large and column-shaped mother fruiting spurs can germinate 3～5 deciduous fruiting shoots which are 21.5 cm long with 14 leaves. The small and thick leaves are dark green and oval-shaped. There are many small flowers produced.

Biological Characters The tree has strong vigor, medium germination and branching ability, blooming and bearing early with a high and stable yield. In Taigu County of Shanxi Province, it enters the crisp-maturing stage in mid-September. It is an early-mid-ripening variety with strong resistance to diseases and light fruit-cracking even if it rains in the maturing stage.

Fruit Characteristics The small nearly-round fruit has a vertical and cross diameter of 2.70 cm and 2.50 cm, averaging 9.2 g with irregular sizes. It has a brightly-red thin skin and a smooth surface with small, round, dense and light-yellow dots. The greenish-white thick flesh is tight-textured, crisp, sweet and juicey with an excellent quality for fresh eating. The percentage of edible part of fresh fruit is 94.0%, and SSC, TTS, TA and Vc is 31.80%, 25.97%, 0.51% and 359.05 mg per 100 g fresh fruit. The content of flavones and cAMP in mature fruit skin is 5.07 mg/g and 86.15 μg/g. The large stone is oval-shaped, averaging 0.55 g.

Evaluation The cultivar has strong adaptability and a low demand on soil, blooming and bearing early with strong early productivity and a stable yield. The fruit has delicate, crisp and sweet flesh with an excellent taste and strong resistance to fruit-cracking and diseases. It is a good early-ripening variety. It has been widely planted in Chaoyang District of Liaoning Province with a very good and stable performance.

鲜 食
Table Varieties
品 种

· 39 ·

灵 武 长 枣

品种来源及分布 别名宁夏长枣、马牙枣。原产和分布于宁夏灵武市，为当地主栽品种之一，栽培历史300年左右。

植物学性状 树体中大，树姿开张，树冠呈偏斜形。主干块状皮裂。枣头紫褐色，平均长91.7cm，节间长7.6cm，无蜡质。二次枝平均长30.4cm，6～8节，弯曲度中等。针刺不发达。枣股平均抽生枣吊4.0个。枣吊长27.1cm，着叶18片。叶片中大，椭圆形，绿色，平展，先端急尖，叶基圆楔形，叶缘具钝锯齿。花量少，花序平均着花3朵。花较小，花径5.9mm。

生物学特性 树势中庸，萌芽率和成枝力较高。结果较晚，定植第三年结果，10年左右进入盛果期，丰产性较差，枣吊平均结果0.40个。在山西太谷地区，9月下旬果实成熟采收，果实生育期111d左右，为晚熟品种类型。成熟期遇雨裂果重。

果实性状 果个大，圆锥形，纵径4.10cm，横径2.54cm，单果重19.9g，大小较整齐。果皮中厚，紫红色，果面不平整，有隆起。梗洼深、中广。果顶尖，柱头宿存。肉质酥脆，汁液多，味极甜，品质极好，宜鲜食。鲜枣可食率95.1%，含总糖21.24%，酸0.40%，100g果肉维生素C含量379.10mg；果皮含黄酮15.36mg/g，cAMP含量81.55μg/g。核大，纺锤形，平均重0.98g。核内不含种子。

评价 该品种树体中等大，树势中庸，结果较晚，产量较低，易裂果。果个大，品质极好，宜鲜食。

Lingwuchangzao

Source and Distribution The cultivar originated from and spreads in Lingwu County of Ningxia Province.

Botanical Characters The medium-sized tree is spreading with a deflective crown. The trunk bark has massive fissures. The purplish-brown 1-year-old shoots are 91.7 cm long, with the internodes of 7.6 cm long, on wax and undeveloped thorns. The secondary branches are 30.4 cm long with 6~8 nodes of medium curvature. The mother fruiting spurs can germinate 4 deciduous fruiting shoots which are 27.1 cm long with 18 leaves. The medium-sized leaves are oval-shaped, green and flat, with a sharply-cuspate apex, a round-cuneiform base and a blunt saw-tooth pattern on the margin. The number of flowers is small, averaging 3 ones per inflorescence. The small blossoms have a diameter of 5.9 mm.

Biological Characters The tree has moderate vigor, strong germination and branching ability. It bears late, generally in the 3rd year after planting and enters the full-bearing stage in the 10th year with a low yield. The deciduous fruiting shoot bears 0.40 fruits on average. In Taigu County of Shanxi Province, it matures in late September with a fruit growth period of 111 d. It is a late-ripening variety with serious fruit-cracking if it rains in the maturing stage.

Fruit Characteristics The large fruit is flat column-shaped, with a vertical and cross diameter of 4.10 cm and 2.54 cm, averaging 19.9 g with a regular size. It has medium-thick and purplish-red skin, a rough surface with some protuberances, a deep and medium-wide stalk cavity, a pointed fruit apex and a remnant stigma. The flesh is crisp, juicy and strongly sweet. It has an excellent quality for fresh-eating. The percentage of edible part of fresh fruit is 95.1%, and SSC, TA and Vc is 21.24%, 0.40% and 379.10 mg per 100 g fresh fruit. The content of flavones and cAMP in mature fruit skin is 15.36 mg/g and 81.55 μg/g. The large spindle-shaped stone weighs 0.98 g without kernel.

Evaluation The medium-sized plant of Lingwuchangzao cultivar has moderate tree vigor, bearing late with a low yield. The fruit has poor resistance to fruit-cracking. It has a large fruit size and an excellent quality and is suitable for fresh-eating.

马牙白枣

品种来源及分布 别名马牙枣。原产北京，分布范围较广，面积较大，较集中的产区位于海淀区北安河一带。因果形似马牙而得名。栽培历史悠久。

植物学性状 树体中大，树姿半开张，树冠呈自然圆头形。主干灰褐色，树皮裂纹条状，裂片大，易剥落。枣头红褐色，平均长84.7cm，粗1.24cm，节间长7.7cm，蜡质少。二次枝平均长31.7cm，着生6节左右。针刺多退化，1、2年内逐渐脱落。枣股圆锥形，抽吊力强，平均抽生枣吊5.0个。枣吊长22.1cm，着叶13片。叶片中大，叶长5.8cm，叶宽2.7cm，椭圆形，先端渐尖，叶基圆形或宽楔形。叶缘锯齿较锐。花序平均着花5朵。

生物学特性 适土性强，对土壤条件要求不严。树势较强，发枝力中等。坐果率较高，2～3年和3年以上枝的吊果率分别为87.0%和50.4%。较丰产，但花期气候干燥或偏凉时坐果不良，有大小年结果现象。在山西太谷地区，4月中旬萌芽，5月下旬始花，9月中旬果实成熟，果实生育期100d左右，为中熟品种类型。成熟期遇雨易裂果。

果实性状 果实中大，圆锥形，胴体两侧不对称，一侧较平直，一侧弯曲呈半月形，似马牙形，纵径4.10cm，横径2.42cm，单果重10.1g，大小不整齐。果肩平，但不对称。梗洼浅而广。果顶尖圆。果面光滑。果皮薄而脆，红色。果肉浅绿色，质地酥脆，细腻，汁液多，味酸甜，鲜食品质极佳。鲜枣含可溶性固形物31.00%，含总糖27.91%，酸0.30%，100g果肉维生素C含量258.78mg；果皮含黄酮12.27mg/g，cAMP含量8.70μg/g。果核较大，呈两端不等长的纺锤形，核重0.45g。含仁率76.9%。

评价 该品种树体中大，适应性强，具有较强的结果性能，但有大小年现象，要求花期温湿度较高的气候条件。果实中大，皮薄，肉质细脆多汁，甜味浓，为品质优异的中熟鲜食品种。

Mayabaizao

Source and Distribution The cultivar originated from Beijing, with a long cultivation history. It is planted in different districts of Beijing with a large quantity.

Botanical Characters The medium-sized tree is half-spreading with a round crown. The grayish-brown trunk bark has wide striped fissures, easily shelled off. The reddish-brown 1-year-old shoots are 84.7 cm long and 1.24 cm thick. The internodes are 7.7 cm long with less wax. Most thorns are degrading, gradually falling off in 1 or 2 years. The secondary branches are 31.7 cm long with 6 nodes. Young mother fruiting spurs are conical-shaped and mature ones are column-shaped. The mother fruiting spurs can germinate 5 deciduous fruiting shoots which are 22.1 cm long with 13 leaves. The medium-sized leaves are oval-shaped with a gradually-cuspate apex, a round or wide-cuneiform base and a sharp saw-tooth pattern on the margin. Each inflorescence has 5 flowers.

Biological Characters The cultivar has strong adaptability, strong vigor and medium branching ability, with medium fruit set and a high yield. The percentage of fruits to deciduous fruiting shoots of 2 or 3-year-old branches and over-3-year-old ones is 87.0% and 50.4%. Fruit set will be low if it is dry or cold in the blooming stage. It has an alternate bearing habit. In Taigu County of Shanxi Province, it germinates in mid-April, begins blooming in late May and matures in mid-September. It is a mid-ripening variety with serious fruit-cracking if it rains in the maturing stage.

Fruit Characteristics The medium-sized conical fruit has a vertical and cross diameter of 10.10 cm and 4.42 cm, averaging 10.1 g with irregular sizes. One side of the fruit is flat while the other is half-moon-shaped. It has a flat and asymmetrical shoulder, a shallow and wide stalk cavity, a pointed-round fruit apex, a smooth surface and thin red skin. The light-green flesh is crisp, juicy, sour and sweet. It has an excellent quality for fresh-eating. The content of SSC, TTS, TA and Vc in fresh fruit is 31%, 27.91%, 0.30% and 258.78 mg per 100 g fresh fruit. The content of flavones and cAMP in mature fruit skin is 12.27 mg/g and 8.70 μg/g. The large spindle-shaped stone weighs 0.45 g.

Evaluation The medium-sized plant of Mayabaizao cultivar has strong adaptability and strong fruiting ability. Yet it has an alternate bearing habit, and requires higher humidity and temperature in the blooming stage. The medium-sized fruit has thin skin, crisp, juicy and sweet flesh. It is a good mid-ripening table variety.

辣 椒 枣

品种来源及分布 别名长脆枣、长枣、奶头枣。原产和分布于山东、河北交界的夏津、临清、冠县、深县、衡水、交河、成安等地,多零星栽培。

植物学性状 树体高大,树姿直立,枝系紧凑,树冠呈伞形。主干灰褐色,裂纹宽条状,较粗糙,裂片易剥落。枣头红褐色,平均长81.6cm,粗1.06cm,节间长8.2cm。针刺不发达,易脱落。枣股圆柱形,略弯,平均抽生枣吊3.0个。枣吊细长,平均长28.8cm,着叶17片。叶片中大,卵状披针形,深绿色,有光泽,较薄,先端钝尖,叶基偏斜形。叶缘具细锯齿。花多,较大,每花序着生7朵。

生物学特性 风土适应性强,抗风、耐旱、耐涝、耐盐碱。树势强健,发枝力强,当年生枣头主芽易萌发是该品种的重要特性。坐果率较低,枣头、2～3年和3年以上枝的吊果率分别为1.0%、31.1%和17.6%。定植后3年开始结果,15年左右进入盛果期。坐果稳定,生理落果轻,产量中等,成龄树一般株产30kg左右,最高株产60kg。在山西太谷地区,4月中旬萌芽,5月下旬始花,9月下旬果实成熟,果实生育期115d左右,为晚熟品种类型。果实成熟期遇雨较易裂果。

果实性状 果实中大,长锥形,纵径3.8～4.9cm,横径2.4～2.6cm,单果重11.2g,最大22g,大小较整齐。果肩凸圆。梗洼中深,较窄。果顶渐细,顶端圆,中心略凹陷,乳头状,柱头遗存。果面平滑光洁。果皮薄,红色,光亮美观。果点大,圆形,不明显。果肉白色,质地较细,酥脆,稍松软,汁液较多,甜酸可口,鲜食品质优异。鲜枣可食率97.1%,100g果肉维生素C含量395.66mg;果皮含黄酮5.26mg/g,cAMP含量97.95μg/g。果核较小,纺锤形,核重0.33g。核内多不含种子。

评价 该品种适应性强,树体强健,产量中等,果实中大,皮薄,脆甜,鲜食品质优良。

Lajiaozao

Source and Distribution The cultivar originated from and spreads in Xiajin, Linqing, Guanxian, Shenxian, Hengshui, Jiaohe and Cheng'an within the juncture of Shandong and Hebei Province.

Botanical Characters The large tree is vertical with an umbrella-shaped crown. The grayish-brown trunk bark has rough wide-striped fissures, easily shelled off. The reddish-brown 1-year-old shoots are 81.6 cm long and 1.06 cm thick. The internodes are 8.2 cm long with less-developed and easily falling-off thorns. The column-shaped mother fruiting spurs can germinate 3 deciduous fruiting shoots which are 28.8 cm long with 17 leaves. The medium-sized leaves are ovate-lanceolate. There are many large flowers produced, averaging 7 ones per inflorescence.

Biological Characters The tree has strong adaptability, tolerant to wind, drought, water-logging and saline-alkaline, with strong vigor, strong branching ability and low yet stable fruit set. Easy germination of the buds on 1-year-old shoots is an important good character of this cultivar. The percentage of fruits to deciduous fruiting shoots of 1-year-old shoots, 2 or 3-year-old branches and over-3-year-old ones is 1.0%, 31.1% and 17.6%. It generally bears in the 3rd year after planting and enters the full-bearing stage in the 15th year, with a medium yield and light physiological fruit-dropping. A mature tree has a yield of 30 kg on average (maximum 60 kg). In Taigu County of Shanxi Province, it germinates in mid-April, begins blooming in late May and matures in late September, with a fruit growth period of about 110 d. It is a late-ripening variety with serious fruit-cracking if it rains in the maturing stage.

Fruit Characteristics The medium-sized fruit is long conical, with a vertical and cross diameter of 3.8～4.9 cm and 2.4～2.6 cm, averaging 11.2 g (maximum 22 g) with a regular size. It has a pointed-round shoulder, a medium-deep and narrow stalk cavity, a gradually-thinner fruit apex, a remnant stigma, a thin, red and smooth surface with some large round dots. The white flesh is delicate, crisp, loose-textured, sour, sweet and juicy. It has an excellent quality for fresh-eating. The percentage of edible part of fresh fruit is 97.1%, and the content of Vc is 395.66 mg per 100 g fresh fruit. The content of flavones and cAMP in mature fruit skin is 5.26 mg/g and 97.95 μg/g. The small spindle-shaped stone weighs 0.33 g. Most stones contain no kernels.

Evaluation The cultivar has strong adaptability and strong vigor with a medium yield. The medium-sized fruit has thin skin, crisp and sweet flesh with an excellent quality for fresh-eating.

· 45 ·

太谷美蜜枣

品种来源及分布 别名美蜜枣。原产山西省太谷县阳邑乡里美庄村，栽培极少。历史不详。

植物学性状 树体较小，树姿较直立，干性强，枝条细而中密，树冠圆锥形。主干皮裂条状，较浅，较易脱落。枣头红褐色，生长势弱，平均长51.2cm，着生永久性二次枝6个左右，节间长7.2cm，二次枝长30.6cm，5~6节。针刺细而长。皮目灰白色，小而密，卵圆形，凸起，开裂。枣股中大，抽吊力中等，平均抽生枣吊3.5个。枣吊细而短，平均长17.7cm，着叶15片。叶片小而薄，叶长4.9cm，叶宽2.1cm，椭圆形，浅绿色，先端钝尖，叶基圆楔形，叶缘锯齿密而较钝。花量多，枣吊着花70~100朵，每花序6~7朵。花小，零级花花径6.5mm。

生物学特性 树势中等偏弱，萌芽率低，成枝力弱。结果较迟，早期丰产性能较差，产量低而不稳。一般定植第三年开始结果，15年后进入盛果期。坐果率较低，枣头、2~3年生枝和4年生枝的吊果率分别为38.0%、65.3%和41.7%，主要坐果部位在枣吊的3~6节。在山西太谷地区，8月底果实着色，9月上旬进入脆熟期，果实生育期90d左右，为早熟品种类型。枣果成熟期遇雨易裂果。

果实性状 果个中大，长卵形，纵径3.40cm，横径2.20cm，单果重10.5g，大小不整齐。果皮薄，红色，果面光滑。果点小而密，浅黄色，不明显。果梗细而长，梗洼中度广、深。果顶尖，柱头遗存。果肉厚，白色，肉质细嫩而脆，味甜微酸，汁液多，口感极佳，适宜鲜食。鲜枣可食率96.2%，含可溶性固形物33.00%，总糖31.00%，酸0.64%，100g果肉维生素C含量350.97mg。核小，纺锤形，核重0.40g，核尖较长，核面较粗糙，种仁较饱满，含仁率41.7%。

评价 该品种为稀有的早熟、优质鲜食良种。但树体生长势弱、结果较迟、产量低，生产中应加强肥水管理和采取提高坐果率的措施。另外，成熟期遇雨极易裂果，应及时采收。

Taigumeimizao

Source and Distribution The cultivar originated Yangyi Town of Taigu County in Shanxi Province with a small quantity.

Botanical Characters The small tree is vertical with a strong central leader trunk, thin and dense branches and a conical crown. The trunk bark has shallow fissures, easily shelled off. The reddish-brown 1-year-old shoots have weak growth vigor, averaging 51.2 cm long with 6 permanent secondary branches. The secondary branches are 30.6 cm long with 5~6 nodes and the internodes of 7.2 cm. Thorns are thin and developed. The grey lenticels are small, dense, oval-shaped, protuberant and cracked. The small mother fruiting spurs can germinate 3.53 thin and short deciduous fruiting shoots which are 17.7 cm long with 15 leaves. The small thin leaves are oval-shaped and light green. There are many mall flowers produced, averaging 70~100 ones per deciduous fruiting shoot and 6~7 ones per inflorescence. The zero-level flowers have a diameter of 6.52 mm.

Biological Characters The tree has moderate or weak vigor, weak germination and branching ability, bearing late with poor early productivity and a low and unstable yield. It generally bears in the 3rd year after planting and enters the full-bearing stage in the 15th year with low fruit set. The percentage of fruits to deciduous fruiting shoots of 1-year-old shoots, 2 or 3-year-old branches and 4-year-old ones is 38.0%, 65.3% and 41.7%. In Taigu County of Shanxi Province, its fruits begin coloring in late August and enter the crisp-maturing stage in early September, with a fruit growth period of 90 d. It is an early-ripening variety with serious fruit-cracking if it rains in the maturing stage.

Fruit Characteristics The small fruit is long oval-shaped with a vertical and cross diameter of 3.40 cm and 2.20 cm, averaging 10.5 g with irregular sizes. It has light-red thin skin and a smooth surface. The white flesh is thick, crisp, juicy, sweet and a little bit sour, with an excellent taste for fresh eating. The percentage of edible part of fresh fruit is 96.2%, and SSC, TTS, TA and Vc is 33.00%, 31.00%, 0.64% and 350.97 mg per 100 g fresh fruit. The small spindle-shaped stone weighs 0.40 g.

Evaluation The cultivar is a rare early-ripening and good-quality table variety. Yet the tree has weak vigor, bearing late with a low yield. So it requires good fertilization and irrigation and technical measures to improve fruit set. Besides, fruit-cracking easily occurs if it rains in the maturing stage, so harvest should be in time.

鲜 食
Table Varieties
品 种

太谷铃铃枣

品种来源及分布 别名铃铃枣。原产山西省太谷、榆次、祁县、清徐及太原郊区等地。分布较广,但多零星栽植。

植物学性状 树体中等,树姿直立,干性强,枝条中密,树冠圆锥形。主干皮裂浅,条状,不易脱落。枣头红褐色,生长势中等,平均长60.6cm,粗1.11cm,节间长9.0cm,着生永久性二次枝4~7个,二次枝长31.2cm。针刺不发达。皮目中大,卵圆形,较密,凸起,开裂。枣股较小,抽吊力较强,平均抽生枣吊3.9个。枣吊生长量17.7cm,着叶13片。叶片小,叶长6.2cm,叶宽2.5cm,椭圆形,浅绿色,先端急尖,叶基圆楔形,叶缘锯齿细锐。花量多,枣吊平均着花60.1朵,花序平均4.9朵。花小,零级花花径7.0mm,昼开型。

生物学特性 树势旺盛,萌芽率高,成枝力强,结果早,产量较高。一般定植第二年开始结果,10年左右进入盛果期。坐果率中等,枣头、2~3年生枝和4年生枝的吊果率分别为5.2%、46.6%和34.8%。在山西太谷地区,8月中旬果实开始着色,9月初进入脆熟期,果实生育期90d左右,为早熟品种类型。果实成熟期遇雨极易裂果。采前落果严重。

果实性状 果个小,圆形,纵径2.50cm,横径2.53cm,单果重8.5g,大小整齐。果皮薄,红色,果面光滑。果梗中长、中粗,梗洼中广,较浅。果顶平,柱头遗存,不明显。果肉较厚,白色,肉质疏松,味酸甜,汁液中多,品质中上,适宜鲜食。鲜枣可食率94.2%,含可溶性固形物29.40%,总糖24.10%,酸0.47%,100g果肉维生素C含量378.83mg。果核中大,核重0.49g,卵圆形,核纹较浅,种仁较饱满,含仁率55.0%。

评价 该品种抗旱,耐瘠薄,对栽培条件要求不高。结果早,坐果稳定,产量较高。果实鲜食品质较好,但个小、成熟期极易裂果且落果严重,不适宜生产栽培。

Taigulinglingzao

Source and Distribution The cultivar originated from and widely spreads in Taigu, Qixian, Qingxu and suburbs of Taiyuan in Shanxi Province. It is mainly dispersedly planted with an unknown history.

Botanical Characters The medium-sized tree is vertical with a strong central leader trunk, dense branches and a conical crown. The trunk bark has shallow striped fissures, uneasily shelled off. The reddish-brown 1-year-old shoots have moderate growth vigor, averaging 60.6 cm long and 1.11cm thick with the internodes of 9.0 cm. There are 4~7 permanent secondary branches, which are 31.2 cm long with less-developed thorns. The small mother fruiting spurs can germinate 3.93 deciduous fruiting shoots which are 17.7 cm long with 13 leaves. The small leaves are light green and oval-shaped. There are many small flowers produced averaging 60.1 ones per deciduous fruiting shoot and 4.9 ones per inflorescence. The daytime-bloomed blossoms have a diameter of 7 mm for the zero-level flowers.

Biological Characters The tree has strong vigor, strong germination and branching ability, bearing early with a high yield. It generally bears in the 2nd year after planting and enters the full-bearing stage in the 15th year with medium fruit set. In Taigu County of Shanxi Province, it enters the crisp-maturing stage in early September, with a fruit growth period of 95 d. It is an early-ripening variety with serious fruit-cracking if it rains in maturing stage.

Fruit Characteristics The small round fruit has a vertical and cross diameter of 2.50 cm and 2.53 cm, averaging 8.5 g with a regular size. It has thin red skin and a smooth surface. The thick flesh is white, delicate, crisp, juicy and sweet with a good quality for fresh eating. The percentage of edible part of fresh fruit is 94.2%, and SSC, TTS, TA and Vc is 29.40%, 24.10%, 0.47% and 378.83 mg per 100 g fresh fruit. The medium-sized stone is oval-shaped, averaging 0.49 g with deep veins and a well-developed kernel. The percentage of containing kernels is 55.0%.

Evaluation The cultivar has strong tolerance to drought and poor soil with a low demand on cultural conditions. It bears early with stable fruit set and a high yield. The fruit has crisp, juicy and sweet flesh with a good quality for fresh eating. It has a small fruit size, low edibility, serious fruit-dropping and fruit-cracking in maturing stage, so it is unsuitable for commercial production.

夏津妈妈枣

品种来源及分布 别名妈妈枣、铃枣、乳头枣、鸡心枣。分布于山东夏津、平原、武城、陵县、禹城、成武等地，多庭院零星栽培。

植物学性状 树体中大，树姿半开张，树冠多自然圆头形。主干灰黑色，宽条裂纹，树皮易剥落。枣头黄褐色，平均长92.2cm，粗1.18cm，节间长9.3cm。二次枝平均长33.4cm，6节左右。针刺不发达，易脱落。枣股圆柱形，平均抽生枣吊3.0个。枣吊较细，平均长21.2cm，着叶12片。叶片卵状披针形，叶面平滑较光亮，绿色，先端渐尖，尖端钝圆，叶基圆楔形。叶缘具浅细的钝锯齿。花量大，花小，花序平均着花7朵。

生物学特性 树势和发枝力中等。坐果率较低，2～3年和3年以上枝的吊果率分别为37.7%和9.2%。结果较早，定植第二年即有一定产量，成龄结果树产量较低。在山西太谷地区，4月中旬萌芽，5月下旬初花，9月中旬果实成熟，果实生育期110d左右，为中熟品种类型。成熟期遇雨裂果较重。

果实性状 果实中大，卵圆形，多向一侧歪斜，胴部和果顶部衔接处有细缢环痕，果顶乳头状。果实纵径3.78cm，横径2.55cm，单果重10.8g，最大12.0g，大小较整齐。果肩小，平圆，向下斜披。梗洼窄，中深。果顶顶点略凹陷或略凸出，柱头宿存，呈点状凸起。果面光滑，有的略有隆起。果皮薄，紫红色。果肉浅绿色，质细酥脆，汁较多，味甜，略酸，鲜食品质优异。鲜枣可食率96.8%，可溶性固形物含量30.60%，总糖22.44%，酸0.27%，100g果肉维生素C含量386.10mg。果核纺锤形，平均重0.35g，核内多含有1粒饱满种子。

评价 该品种适性很强，结果早，果实可食率高，品质极佳，为中熟优良鲜食品种，适于秋雨少的地区及城郊的采摘观光果园适量发展栽培。花期应采取喷施激素、微肥或摘心等措施促进坐果，提高产量。

Xiajinmamazao

Source and Distribution The cultivar, also called Lingzao, Rutouzao or Jixinzao, spreads in Xiajin, Pingyuan, Wucheng, Lingxian, Yucheng and Chengwu of Shandong Province.

Botanical Characters The medium-sized tree is half-spreading with a round crown. The grayish-black trunk bark has wide striped fissures, easily shelled off. The yellowish-brown 1-year-old shoots are 92.2 cm long and 1.18 cm thick. The internodes are 9.3 cm long with less-developed and easily falling-off thorns. The secondary branches are 33.4 cm long with 6 nodes. The column-shaped mother fruiting spurs can germinate 3 thin deciduous fruiting shoots which are 21.2 cm long with 12 leaves. The green leaves are ovate-lanceolate. There are many small flowers produced, averaging 7 ones per inflorescence.

Biological Characters The tree has strong vigor and medium branching ability, with low fruit set, a high and stable yield. It bears early, generally in the 2nd year after planting. A mature tree has a yield of 35 kg on average. In Taigu County of Shanxi Province, it germinates in mid-April, begins blooming in late May and matures in mid-September, with a fruit growth period of about 100 d. It is a mid-ripening variety with serious fruit-cracking if it rains in the maturing stage.

Fruit Characteristics The medium-sized fruit is oval-shaped, with a vertical and cross diameter of 3.78 cm and 2.55 cm, averaging 10.8 g with a regular size. It has a small flat-round shoulder, a narrow and medium-deep stalk cavity, a slightly-sunken or pointed fruit apex, a remnant and protuberant stigma, a smooth surface and purplish-red thin skin. The light-green flesh is crisp, juicy, sweet and a little sour. It has an excellent quality for fresh-eating. The percentage of edible part of fresh fruit is 96.8%, and SSC, TTS, TA and Vc is 30.60%, 22.44%, 0.27% and 386.10 mg per 100 g fresh fruit. The spindle-shaped stone weighs 0.35 g. Most stones contain a well-developed kernel.

Evaluation The cultivar has strong adaptability, bearing early with a high and stable yield. The fruit has an excellent quality for fresh-eating. It is a good mid-ripening variety which can be developed in areas with light rainfall during autumn.

鲜食
Table Varieties
品 种

· 51 ·

濮阳糖枣

品种来源及分布 原产和分布于河南濮阳等地。

植物学性状 树体大，树姿较直立，干性较强，树冠呈圆锥形。主干皮裂条状。枣头黄褐色，平均长81.1cm，节间长7.4cm，蜡质少。二次枝长35.4cm，5～9节，弯曲度中等。针刺发达。枣股平均抽生枣吊4.1个。枣吊长22.0cm，着叶15片。叶片较小，叶长5.5cm，叶宽2.2cm，椭圆形，平展，先端尖凹，叶基圆楔形，叶缘锯齿细。花量多，花序平均着花8朵，花中大，花径6.3mm。

生物学特性 树势强，萌芽率和成枝力强。坐果率低，枣头、2～3年和4～6年生枝的吊果率分别为18.9%、9.5%和0.5%。定植第三年结果，10年左右进入盛果期，产量低而不稳。在山西太谷地区，9月下旬果实成熟采收，果实生育期107d左右，为中晚熟品种类型。

果实性状 果个较大，长椭圆形，单果重19.8g，大小较整齐。果皮中厚，紫红色，果面平滑。果点大，分布稀疏。梗洼浅而广。果顶突出，柱头遗存，不明显。肉质致密，味甜，汁液少，品质上等，适宜鲜食。鲜枣可食率95.7%，含可溶性固形物31.20%，总糖24.29%，酸0.53%，糖酸比45.83∶1，100g果肉维生素C含量538.57mg；果皮含黄酮4.11mg/g，cAMP含量117.03μg/g。干枣含总糖58.22%，酸0.94%。果核较小，纺锤形，核重0.42g。种仁较饱满，含仁率53.7%。

评价 该品种树体较大，树姿直立，树势强，产量低而不稳，但果个较大，可食率高，鲜食品质上等，可适度发展栽培。

Puyangtangzao

Source and Distribution The cultivar originated from and spreads in Puyang of Henan Province.

Botanical Characters The large tree is vertical with a strong central leader trunk and a conical crown. The trunk bark has massive fissures. The yellowish-brown 1-year-old shoots are 81.1 cm long with the internodes of 7.4 cm, less wax and developed thorns. The secondary branches are 35.4 cm long with 5～9 nodes. The mother fruiting spurs can germinate 4.1 deciduous fruiting shoots, which are 22.0 cm long with 15 leaves. The small leaves are oval-shaped and flat, 5.5 cm long and 2.2 cm wide, with a sharply-sunken apex, a round-cuneiform base and a thin saw-tooth pattern on the margin. There are many medium-sized flowers with a diameter of 6.3 mm produced, averaging 8 ones per inflorescence.

Biological Characters The tree has strong vigor, strong germination and branching ability. Yet the fruit set is low. The percentage of fruits to deciduous bearing shoots of 1-year-old shoots, 2～3-year-old branches and 4～6-year-old ones is 18.9%, 9.5% and 0.5%. It generally bears in the 3rd year after planting and enters the full-bearing stage in the 10th year, with a low and unstable yield. In Taigu County of Shanxi Province, it matures in late September with a fruit growth period of 107 d. It is a late-ripening variety.

Fruit Characteristics The small fruit is flat column-shaped, with a vertical and cross diameter of 3.07 cm and 2.58 cm, averaging 9.8 g with a regular size. It has medium-thick and purplish-red skin, a smooth surface with large and sparse dots, a narrow and deep stalk cavity, a slightly-sunken fruit apex and a remnant stigma. The tight-textured flesh is sweet with less juice. It has medium quality for fresh-eating and dried fruits. The percentage of edible part of fresh fruit is 95.7%, and SSC, TTS, TA and Vc is 31.20%, 24.29%, 0.53% and 538.57 mg per 100 g fresh fruit. The SAR is 45.83∶1. The content of flavones and cAMP in mature fruit skin is 4.11 mg/g and 117.03 μg/g. The content of TTS and TA in dried fruit is 58.22% and 0.94%. The small spindle-shaped stone weighs 0.42 g, with a well-developed kernel. The percentage of containing kernels is 53.7%.

Evaluation The large plant of Puyangtangzao cultivar is vertical with strong tree vigor. It has a low and unstable yield. The bigger fruit has medium quality and is suitable for fresh-eating.

鲜 食
Table Varieties
品 种

林县无头枣

品种来源及分布　原产和分布于河南省林州市。

植物学性状　树体较大，树姿半开张，干性强，树冠圆锥形。主干灰褐色，块状皮裂。枣头黄褐色，平均长67.4cm，节间长6.3cm。二次枝长33.0cm，5～9节，弯曲度大。针刺不发达。枣股平均抽生枣吊4.0个。枣吊长23.4cm，着叶16片。叶片中大，叶长5.6cm，叶宽2.5cm，椭圆形，部分叶片反卷，先端钝尖，叶基圆楔形，叶缘锯齿细。花量多，花序平均着花6朵，花较小，花径5.8mm。

生物学特性　树势强，萌芽率高，成枝力强。坐果率较低，枣头、2～3年和4～6年生枝的吊果率分别为26.6%、60.6%和4.3%。一般定植第三年有一定产量，产量中等。在山西太谷地区，9月中旬果实进入脆熟期，9月下旬成熟采收，果实生育期110d以上，为晚熟品种类型。成熟期遇雨裂果轻。

果实性状　果个较大，卵圆形，纵径3.34cm，横径3.08cm，单果重15.2g，大小整齐。果皮薄，红色，果面平滑。果点大，分布稀疏。梗洼窄而深，果顶微凹，柱头遗存，不明显。肉质细脆，味甜，汁液多，品质上等，耐贮藏，适宜鲜食。鲜枣可食率96.5%，含可溶性固形物27.00%，总糖22.76%，100g果肉维生素C含量360.81mg；果皮含黄酮3.85mg/g，cAMP含量131.31μg/g。果核小，椭圆形，核重0.54g。种仁较饱满，含仁率90.0%。

评价　该品种果个较大，肉质酥脆，味甜，汁液多，鲜食品质优良。果实耐贮藏，抗裂果能力较强。

Linxianwutouzao

Source and Distribution　The cultivar originated from and spreads in Linxian County of Henan Province.

Botanical Characters　The large tree is half-spreading with a strong central leader trunk and a conical crown. The trunk bark has massive fissures. The reddish-brown 1-year-old shoots are 67.4 cm long with the internodes of 6.3 cm. The secondary branches are 33.0 cm long with 5～9 nodes of large curvature and less-developed thorns. The mother fruiting spurs can germinate 4.0 deciduous fruiting shoots which are 23.4 cm long with 16 leaves. The medium-sized leaves are oval-shaped, 5.6 cm long and 2.5 cm wide, with a bluntly-cuspate apex, a round-cuneiform base and a thin saw-tooth pattern on the margin. Some leaves fold towards the back. There are many small flowers with a diameter of 5.8 mm produced, averaging 6 ones per inflorescence.

Biological Characters　The tree has strong vigor, high germination rate and strong branching ability. It has low fruit set and moderate productivity. The percentage of fruits to deciduous fruiting shoots of 1-year-old shoots, 2～3-year-old branches and 4～6-year-old ones is 26.6%, 60.6% and 4.3%. There is certain yield in the 3rd year after planting. In Taigu County of Shanxi Province, it enters the crisp-maturing stage in mid-September and is harvested in late September, with a fruit growth period of over 110 d. It is a late-ripening variety with light fruit-cracking even if it rains in the maturing stage.

Fruit Characteristics　The large oval-shaped fruit has a vertical and cross diameter of 3.34 cm and 3.08 cm, averaging 15.2 g with a regular size. It has thin red skin, a smooth surface with large and sparse dots, a narrow and deep stalk cavity, a slightly sunken fruit apex and a remnant yet indistinct stigma. The flesh is delicate, crisp, strongly sweet and juicy. It has an excellent quality for fresh-eating. The percentage of edible part of fresh fruit is 96.5%, and SSC, TTS and Vc is 27.00%, 22.76% and 360.81 mg per 100 g fresh fruit. The content of flavones and cAMP in mature fruit skin is 3.85 mg/g and 131.31 μg/g. The fruit has good tolerance to storage. The small spindle-shaped stone weighs 0.54 g with a well-developed kernel. The percentage of containing kernels is 90.0%.

Evaluation　The cultivar has a large fruit size, crisp, juicy and sweet flesh, with an excellent quality for fresh eating. It has good tolerance to storage and strong resistance to fruit-cracking.

天津尜尜枣

品种来源及分布 分布于天津西郊各县。

植物学性状 树体中大，树姿半开张，干性较强，树冠呈圆柱形。主干块状皮裂。枣头黄褐色，生长量84.7cm，节间长7.8cm。二次枝平均长33.1cm，5～8节。针刺细弱，不发达。枣股平均抽生枣吊4.0个。枣吊长22.7cm，着叶15片。叶片椭圆形，先端急尖，叶基圆楔形，叶缘钝锯齿。花量较多，花序平均着花7朵，花径6.0mm。

生物学特性 树势强，萌芽率和成枝力强，成枝力达90%以上。一般定植第二年结果，第三年有一定产量，但产量较低且不稳定，吊果率27.5%，采前落果严重。在山西太谷地区，9月中旬果实成熟采收，果实生育期100d左右，为中熟品种类型。成熟期遇雨极易裂果。

果实性状 果个中等，卵圆形，纵径3.83cm，横径2.50cm，单果重11.0g，大小较整齐。果皮薄，红色，果面平滑。梗洼窄、中深。果顶尖，柱头宿存。肉质细嫩酥脆，味酸甜，汁液多，品质极上，适宜鲜食。鲜枣可食率94.7%，含总糖24.50%，100g果肉维生素C含量315.52mg；果皮含黄酮8.49mg/g，cAMP含量194.97μg/g。果核纺锤形，核重0.58g，种仁较饱满，含仁率63.3%。

评价 该品种果个中等，鲜食品质极上，但产量低而不稳定，采前落果严重，果实抗裂果能力差。栽培中应加强管理，提高产量和采取防止裂果等措施。

Tianjingagazao

Source and Distribution The cultivar originated from and spreads in counties around west Tianjin.

Botanical Characters The small tree is vertical with a strong central leader trunk and a column-shaped crown. The trunk bark has massive fissures. The yellowish-brown 1-year-old shoots are 84.7 cm long with the internodes of 7.8 cm. The secondary branches are 33.1 cm long with 5～8 nodes and almost no thorns. The mother fruiting spurs can germinate 4 deciduous fruiting shoots, which are 22.7 cm long with 15 leaves. The oval-shaped leaves have a sharply-cuspate apex, a round-cuneiform base and a blunt saw-tooth pattern on the margin. There are many flowers with a diameter of 6.0 mm produced, averaging 7 ones per inflorescence.

Biological Characters The tree has strong vigor, high germination rate and strong branching ability, with a branching rate of 96%. It generally bears in the 2nd year after planting. There is certain yield in the 3rd year, though not stable. The percentage of fruits to deciduous bearing shoots is 27.5% on average. In Taigu County of Shanxi Province, it is harvested in mid-September with a fruit growth period of 100 d. It is a mid-ripening variety with serious premature fruit-dropping and serious fruit-cracking if it rains in the maturing stage.

Fruit Characteristics The medium-sized fruit is oval-shaped, with a vertical and cross diameter of 3.83 cm and 2.50 cm, averaging 11.0 g with a regular size. It has thin red skin, a smooth surface, a narrow and medium-deep stalk cavity, a cuspate fruit apex and a remnant stigma. The flesh is delicate, crisp, sour, sweet and juicy. It has an excellent quality for fresh-eating. The percentage of edible part of fresh fruit is 94.7%, and the content of TTS and Vc is 24.50% and 315.52 mg per 100 g fresh fruit. The content of flavones and cAMP in mature fruit skin is 8.49 mg/g and 194.97 μg/g. The stone is inverted spindle-shaped, with a well-developed kernel. The percentage of containing kernels is 63.3%.

Evaluation The cultivar has a medium fruit size and excellent quality for fresh eating with a low and unstable yield, serious premature fruit-dropping and poor resistance to fruit-cracking. Technical measures and orchard management should be enhanced to improve the yield and to prevent fruit-cracking.

鲜 食
Table Varieties
品 种

大荔鸡蛋枣

品种来源及分布 别名陕西鸡蛋枣。主要分布于陕西省大荔县石槽乡的三教、王马村一带。

植物学性状 树体较小，树势较弱，树姿半开张，外围枝下垂，枝条稀疏，树冠乱头形。树皮裂纹较深，呈宽条状，易剥落。枣头红褐色，平均长91.2cm，节间长7.6cm。二次枝平均长20.7cm，6节左右，弯曲度小。针刺不发达，退化为剑形软片。皮孔中大，线形，不明显。枣股大，圆锥形，短粗，一般抽生枣吊3～5个。枣吊长21.3cm，着叶14片。叶片小而厚，叶长5.6cm，叶宽2.4cm，椭圆形，绿色，有光泽。叶尖锐尖，叶基圆楔形，叶缘锯齿钝。花量中多，花中大，花径6.0mm，昼开型。

生物学特性 树体较小，生长势弱，发枝力中等。结果晚，产量不稳定，平均吊果率57.7%。在山西太谷地区，9月下旬果实着色成熟，果实生长期110d左右，为晚熟品种类型。果实较抗裂果、抗病。

果实性状 果个大，卵圆形，纵径3.97cm，横径3.14cm，单果重17.2g，大小较整齐。果肩广圆，耸起，向一侧偏斜，有多条辐射沟棱。梗洼深，中广。果顶平圆，顶点略凹陷。果面平滑。果皮中厚，着色前黄绿色，着色后紫红色。果点小而密，圆形。果肉绿白色，肉质酥脆，汁液中多，味酸甜，品质中等，宜鲜食。鲜枣可食率95.2%，含总糖23.50%，酸0.43%，100g果肉维生素C含量266.17mg；果皮含黄酮10.80mg/g，cAMP含量142.81μg/g。果核大，倒纺锤形，略扁，纵径2.10cm，横径0.90cm，核重0.83g。核纹粗短，呈不规则纵斜条纹。核蒂较短，钝圆。核先端尖锐，呈短角状。核内极少含有种子。

评价 该品种果实大，果皮厚，较抗裂果，但鲜食品质一般，大小年结果明显，而且树体易衰弱，不宜大面积生产栽培，可作为抗裂品种育种材料。

Dalijidanzao

Source and Distribution The cultivar mainly spreads in Sanjiao and Wangmacun in Shicao Village of Dali County, Shaanxi Province.

Botanical Characters The small tree is half-spreading with pendulous external branches, sparse branches and an irregular crown. The trunk bark has deep striped fissures, easily shelled off. The reddish-brown 1-year-old shoots are 91.2 cm long with the internodes of 7.6 cm. The secondary branches are 20.7 cm long with 6 nodes. The thorns degrade into sword-shaped soft strips. The medium-sized lenticels are line-shaped and indistinct. The short and thick conical-shaped mother fruiting spurs can germinate 3～5 deciduous fruiting shoots which are 21.3 cm long with 14 leaves. The small thick leaves are oval-shaped, green and glossy, 5.6 cm long and 2.4 cm wide, with a sharply-cuspate apex, a round-cuneiform base and a blunt saw-tooth pattern on the margin. The number of flowers are medium large, and the medium-sized blossoms have a diameter of 6.0 mm. It blooms in the daytime.

Biological Characters The small tree has weak vigor and medium branching ability, bearing late with an unstable yield. The percentage of fruits to deciduous bearing shoots is 57.7% on average. In Taigu County of Shanxi Province, it matures in late September with a fruit growth period of 110 d. It is a mid-late-ripening variety with strong resistance to fruit-cracking and diseases.

Fruit Characteristics The large oval-shaped fruit has a vertical and cross diameter of 3.97 cm and 3.14 cm, averaging 17.2 g with a regular size. It has a wide-round shoulder with several radiating ribs, a deep and medium-wide stalk cavity, a flat-round fruit apex with the top-point a little sunken, a smooth surface and medium-thick skin with small, round and dense dots. It is yellowish-green before coloring and purplish-red after coloring. The greenish-white flesh is loose-textured, crisp, juicy, sweet and sour, with a higher than normal quality. The percentage of edible part of fresh fruit is 95.2%. TTS, TA and Vc is 23.50%, 0.43% and 266.17 mg per 100 g fresh fruit. The content of flavones and cAMP in mature fruit skin is 10.80 mg/g and 142.81 μg/g. The large stone is inverted spindle-shaped, a little bit flat, averaging 0.83 g, a short and blunt-round pedicle and a sharp silicle-shaped apex. The stones seldom contain kernels.

Evaluation The cultivar has a large fruit size and thick skin, resistant to fruit-cracking. Yet it has a poor quality for fresh eating, with obvious alternate bearing and easily-weakening tree. It is unsuitable for large-scale commercial production, yet it can be considered as a breeding material for fruit-crack-resistant variety.

大 瓜 枣

品种来源及分布 原产山东省东阿、东明等地，多为农家庭院栽植。

植物学性状 树体较小，树姿较开张，干性中等，树冠乱头形，枝条较稀疏。主干皮裂呈条状，较易剥落。枣头红褐色，生长势强，平均生长量78.7cm，粗0.99cm，节间长7.7cm。无针刺。皮目较小，分布稀疏，圆形，凸起。枣股圆柱形，抽枝力较强，抽生枣吊3～5个。枣吊平均长21.6cm，着叶18片。叶片中大，叶长5.8cm，叶宽3.2cm，椭圆形，浓绿色，先端急尖，叶基圆楔形，叶缘锯齿粗钝。花量多。

生物学特性 树势中庸，发枝力较弱，枣头生长势较强。嫁接苗当年即能结果，早期丰产性能极强，产量较高。2～3年生枝坐果率极高，吊果率100%以上，最高为393.4%，枣头和3年生以上枝也可达34.0%和64.2%。在山西太谷地区，9月中旬果实成熟，果实生育期110d左右，为中熟品种类型。成熟期遇干旱条件有果实软化和落果现象，抗裂果、抗病性一般。

果实性状 果实大，圆形，纵径3.60cm，横径3.85cm，单果重27.2g，大小较整齐。梗洼窄而深，果顶平，柱头遗存。果皮薄，浅红色，果面平滑，果点不明显。果肉厚，乳白色，肉质致密细脆，甜味浓，汁液中多，品质上等，适宜鲜食。鲜枣可食率95.6%，含总糖26.13%，酸0.25%；果皮含黄酮3.19mg/g，cAMP含量96.29μg/g。核大，倒卵形，纵径2.40cm，横径1.40cm，核重1.20g，核尖短，核纹较深，核面粗糙，个别核内有种仁。

评价 该品种树体小，开花结果早，早期丰产性能极强，产量较高，适宜密植栽培。果实大，肉质细脆，甜味浓，品质上等，为优良的中熟鲜食品种，可适度规模发展。

Daguazao

Source and Distribution The cultivar originated from Donge and Dongming in Shandong Province with a small quantity. It is dispersedly planted in family yards with an unknown history.

Botanical Characters The medium-sized tree is spreading with a medium-strong central leader trunk, sparse branches and an irregular crown. The trunk bark has striped fissures, easily shelled off. The reddish-brown 1-year-old shoots have strong growth vigor, averaging 78.7 cm long and 0.99 cm thick with the internodes of 7.7 cm, without thorns. The small lenticels are sparse, round and protuberant. The column-shaped mother fruiting spurs can germinate 3～5 deciduous fruiting shoots which are 21.6 cm long with 18 leaves. The medium-sized leaves are oval-shaped and dark green, 5.8 cm long and 3.2 cm wide, with a sharply-cuspate apex, a round-cuneiform base and a thick blunt saw-tooth pattern on the margin. There are many flowers produced.

Biological Characters The tree has moderate vigor and weak branching ability with strong growth potential for the 1-year-old shoots. The grafted seedlings bear fruit in the year of planting, with strong early productivity and a high yield. The 2 or 3-year-old branches have very high fruit set, the percentage of fruits to deciduous fruiting shoots of which reaches over 100% (maximum 393.4%), and that for 1-year-old shoots and over-4-year-old branches is 34.0% and 64.2%. In Taigu County of Shanxi Province, it matures in mid-September with a fruit growth period of 110 d. It is a mid-ripening variety with strong resistance to fruit-cracking and medium resistance to diseases. Fruit flesh always becomes soft if it is dry in the maturing stage.

Fruit Characteristics The large round fruit has a vertical and cross diameter of 3.60 cm and 3.85 cm, averaging 27.2 g with a regular size. It has a narrow and deep stalk cavity, a flat fruit apex, a remnant stigma, light-red thin skin and a smooth surface with indistinct dots. The ivory-white thick flesh is tight-textured and crisp, strongly sweet, with medium juice and an excellent quality for fresh eating. The percentage of edible part of fresh fruit is 95.6%, and the content of TTS and TA is 26.13% and 0.25%. The content of flavones and cAMP in mature fruit skin is 3.19 mg/g and 96.29 μg/g. The big obovate stone has a vertical and cross diameter of 2.40 cm and 1.40 cm, averaging 1.20 g with a short apex, deep veins and a rough surface. Some stones contain a kernel.

Evaluation The small plant of Daguazao cultivar blooms and bears early, with strong early productivity and a high yield, suitable for close planting. The large fruit has crisp and sweet flesh with an excellent quality. It is a good mid-ripening table variety and can be developed moderately.

鲜 食
Table Varieties
品 种

· 61 ·

临潼轱辘枣

品种来源及分布　原产和分布于陕西临潼的相桥、康桥及阎良一带，多零星栽培。

植物学性状　树体较大，树姿半开张，树冠自然圆头形。主干黑灰色，裂纹窄条状，不易剥落。枣头红褐色，有灰色斑块，平均长83.0cm，粗1.15cm，节间长6.1cm，蜡质少。二次枝长31cm，6节左右，弯曲度小。针刺发达。枣股中大，圆锥形，平均抽生枣吊3.0个。枣吊长22.5cm，着叶16片。叶长4.9cm，叶宽2.6cm，叶片椭圆形，绿色，较光亮，先端钝尖，叶基圆楔形，叶缘有浅钝锯齿。花量中多，每花序着花3朵。

生物学特性　适应性强，耐旱、耐瘠薄，在沙壤和壤土表现最好。树势强健，坐果率高，产量高而稳定，枣头、2～3年和4～6年生枝的吊果率分别为61.9%、69.6%和8.9%。在山西太谷地区，4月下旬萌芽，6月初始花，9月中旬脆熟，果实生育期100d左右，为中熟品种类型。成熟期遇雨易裂果。

果实性状　果实中大，卵圆形或倒卵圆形，侧面略扁，纵径3.38cm，横径2.76cm，单果重12.1g，大小较整齐。果肩平。梗洼中深广。果顶宽，顶点略凹陷，成一字沟纹。果柄较粗，长5.7mm。果面平整，果皮中厚，紫红色。果肉浅绿色，质地较致密，汁液中多，味酸甜，鲜食品质中上。鲜枣可食率94.8%，含可溶性固形物31.20%，总糖24.12%，酸0.89%，100g果肉维生素C含量431.45mg；果皮含黄酮10.61mg/g，cAMP含量111.19μg/g。果核中大，短纺锤形，核重0.63g，含仁率86.7%。

评价　该品种适应性强，树体较高大，长势旺，丰产稳产。果实中大，汁多味浓，鲜食品质中上，为较好的中熟鲜食品种。但成熟期遇雨易裂果，宜在成熟期少雨地区栽培。

Lintongguluzao

Source and Distribution　The cultivar originated from and spreads in Xiangqiao, Kangqiao and Yanliang of Lintong in Shaanxi Province.

Botanical Characters　The large tree is half-spreading with an irregular crown. The blackish-gray trunk bark has narrow striped fissures, uneasily shelled off. The reddish-brown 1-year-old shoots are 83 cm long and 1.15 cm thick, with some gray spots. The internodes are 6.1 cm long with less wax and developed thorns. The secondary branches are 31 cm long with 6 nodes of small curvature. The mother fruiting spurs can germinate 3 deciduous fruiting shoots, which are 22.5 cm long with 16 leaves. The green leaves are oval-shaped and glossy. The number of flowers is medium-large, averaging 3 per inflorescence.

Biological Characters　The tree has strong adaptability, tolerant to drought and poor soils, with strong vigor, weak branching ability, low fruit set, a high and stable yield. It grows well in sandy loam and loam soils. The percentage of fruits to deciduous fruiting shoots of 1-year-old shoots, 2～3-year-old branches and 4～6-year-old ones is 61.9%, 69.6% and 8.9%. In Taigu County of Shanxi Province, it germinates in mid-April, begins blooming in early June and enters the crisp-maturing stage in mid-September with a fruit growth period of about 100 d. It is a mid-ripening variety with serious fruit-cracking if it rains in the maturing stage.

Fruit Characteristics　The medium-sized fruit is oval-shaped or obovate, with a vertical and cross diameter of 3.38 cm and 2.76 cm, averaging 12.1 g with a regular size. It has a flat shoulder, a medium-deep and wide stalk cavity, a wide fruit apex with a slightly-sunken summit, a thick stalk of 5.7 mm long, a smooth surface and medium-thick and purplish-red skin. The light-green flesh is tight-textured, sour and sweet, with medium juice. It has a better than normal quality for fresh-eating. The percentage of edible part of fresh fruit is 94.8%, and SSC, TTS, TA and Vc is 31.20%, 24.12%, 0.89% and 431.45 mg per 100 g fresh fruit. The content of flavones and cAMP in mature fruit skin is 10.61 mg/g and 111.19 μg/g. The content of TTS and TA in dried fruit is 63.37% and 0.96%. The medium-sized stone is short spindle-shaped, averaging 0.63 g. The percentage of containing kernels is 86.7%

Evaluation　The large plant of Lintongguluzao cultivar has strong adaptability and strong tree vigor with a high and stable yield. The medium-sized fruit has juicy and sweet flesh with a better than normal quality for fresh-eating. It is a fine mid-ripening table variety. Yet fruit-cracking easily occurs if it rains in the maturing stage. It should be developed in areas with light rainfall in the maturing stage.

襄汾圆枣

品种来源及分布 原产山西省襄汾县，栽培面积较小，历史不详。

植物学性状 树体中等大，树姿开张，干性中等，枝量较少且较细，树冠半圆形。主干皮裂较浅，呈条状，较易脱落。枣头棕褐色，生长势中等，平均生长量57.7cm，粗度0.85cm，节间长7.5cm。枣头着生二次枝5个左右，长26.5cm，5节。针刺不发达。皮目中等大，分布较密，卵圆形，凸起，开裂，灰白色。枣股小，圆锥形，抽吊力强，抽生枣吊4～6个。枣吊明显下垂且较长，平均长21.6cm，着叶15片。叶片小，叶长6.3cm，宽2.7cm，椭圆形，浓绿色，先端尖凹，叶基圆楔形，叶缘锯齿中密，较粗钝。花量中多，枣吊平均着花62.3朵，每花序4.5朵。花小，零级花花径6.44mm，昼开型。

生物学特性 树势中等，成枝力一般。结果较迟，一般定植第三年开始结果，10年后进入盛果期，产量中等。幼龄枝坐果率较低，随枝龄增大，坐果率提高。枣头、2～3年生枝和4年生枝的吊果率分别为13.4%、22.2%和26.5%，主要坐果部位在枣吊的5～9节。在山西太谷地区，9月中旬果实着色，9月底至10月初成熟，果实生育期115d左右，为晚熟品种类型。

果实性状 果实较大，卵圆形，纵径3.55cm，横径2.80cm，单果重15.4g，大小较整齐。果皮薄，浅红色，果面平滑。果点小而密，圆形，浅黄色。果梗较粗长，梗洼窄而深。果顶微凹，柱头遗存，不明显。果肉厚，浅绿色，肉质细脆，味甜略酸，汁液多，品质上等，适宜鲜食。鲜枣耐贮藏，冷库保鲜可达3个月左右。鲜枣可食率96.0%，含可溶性固形物30.00%，总糖24.50%，酸0.48%，100g果肉维生素C含量341.39mg；果皮含黄酮20.10mg/g，cAMP含量123.07μg/g。果核较小，纺锤形，核重0.62g，多数核内含饱满的种子，含仁率高达85.0%。

评价 该品种果实较大，果个均匀，品质好，耐贮藏，为晚熟鲜食良种，可适度发展。但树体扩冠速度慢，结果较迟，产量中等，生产中须加强整形修剪和肥水管理，注意采取提高发枝力和坐果率的技术措施。

Xiangfenyuanzao

Source and Distribution The cultivar originated from Xiangfen County of Shanxi Province with a small quantity.

Botanical Characters The medium-sized tree has moderate vigor, spreading with a moderate central leader trunk and sparse and thin branches. The trunk bark has shallow massive fissures. The yellowish-brown 1-year-old shoots have moderate growth vigor, averaging 57.7 cm long and 0.85 cm thick. The internodes are 7.5 cm long. There are generally 5 secondary branches which are 26.5 cm long with 5 nodes and developed thorns. The medium-sized lenticels are oval-shaped, dense and protuberant, cracked and gray. The small conical-shaped mother fruiting spurs can germinate 4～6 pendulous deciduous fruiting shoots which are 21.6 cm long with 15 leaves. The small oval-shaped leaves are dark green. The number of flowers is medium large, averaging 62.3 ones per deciduous fruiting shoot and 4.5 ones per inflorescence. The small blossoms have a diameter of 6.44 mm for the zero-level flowers.

Biological Characters The tree has moderate vigor and weak branching ability. It bears late, generally in the 3rd year after planting and enters the full-bearing stage in the 10th year with a medium yield. In Taigu County of Shanxi Province, it matures in late September or early October.

Fruit Characteristics The medium-sized fruit is oval-shaped, averaging 15.4 g with a regular size. It has light-red thin skin, a smooth surface. The light-green thick flesh is delicate and crisp, with a sweet and a little sour taste and much juice. It has a good quality for fresh-eating. Fresh fruits have strong tolerance to storage, which can be kept fresh for 3 months under CA storage condition. The percentage of edible part of fresh fruits is 96.0%, and SSC, TTS, TA and Vc is 30.00%, 24.50%, 0.48% and 341.39 mg per 100 g fresh fruit. The content of flavones and cAMP in mature fruit skin is 20.10 mg/g and 123.07 μg/g. The small spindle-shaped stone weighs 0.62 g. Most stones contain a well-developed kernel, the percentage of which is 85.0%.

Evaluation The cultivar has a large and regular fruit size, a good quality and good storage character. It is a fine late-ripening table variety and can be developed moderately.

鲜 食
Table Varieties
品 种

泰安马铃脆

品种来源及分布 别名马铃脆。原产和分布于山东省泰安地区。

植物学性状 树体中大，树姿较开张，干性弱，树冠自然圆头形。主干块状皮裂。枣头黄褐色，生长量76.2cm，节间长8.8cm。二次枝长34.8cm，5～7节。针刺基本退化。枣吊长23.6cm，着叶14片。叶片卵状披针形，部分叶片反卷，先端钝尖，叶基圆形，叶缘具钝锯齿。花量多，花序平均着花7朵，花中大，花径6.0mm。

生物学特性 树势较强，成枝力强，产量较低。在山西太谷地区，9月上旬果实进入白熟期，9月中旬开始成熟采收，果实生育期100d左右，为中熟品种类型。果实成熟期遇雨极易裂果。

果实性状 果个较小，圆锥形，纵径3.58cm，横径2.35cm，单果重9.3g，大小较整齐。果皮薄，红色，果面平滑，果点大。梗洼浅、中深。果顶尖，柱头宿存。肉质细嫩酥脆，味酸甜，汁液多，品质极上，适宜鲜食。鲜枣含可溶性固形物30.60%，总糖28.45%，酸0.38%，100g果肉维生素C含量409.20mg。含仁率83.3%，偶有双仁。

评价 该品种丰产性较差，产量不稳定，果个较小，但品质优异，适宜鲜食。栽培上应采取环剥、花期喷施微肥等技术措施促进坐果，提高产量。成熟期还应注意防雨。

Taianmalingcui

Source and Distribution The cultivar originated from and spreads in Taian City of Shandong Province.

Botanical Characters The medium-sized tree is spreading with a weak central leader trunk and a natural-round crown. The trunk bark has massive fissures. The yellowish-brown 1-year-old shoots are 76.2 cm long with the internodes of 8.8 cm. The secondary branches are 34.8 cm long with 5～7 nodes and almost no thorns. The deciduous fruiting shoots are 23.6 cm long with 14 leaves. The leaves are ovate-lanceolate with some of them folding towards the back. The leaves have a bluntly-cuspate apex, a round base and a blunt saw-tooth pattern on the margin. There are many medium-sized flowers with a diameter of 6.0 mm produced, averaging 7 ones per inflorescence.

Biological Characters The tree has strong vigor, strong branching ability and a low yield. In Taigu County of Shanxi Province, it enters the white-maturing stage in early September and is harvested in mid-September with a fruit growth period of 100 d. It is a mid-ripening variety with serious fruit-cracking if it rains in the maturing stage.

Fruit Characteristics The small conical fruit has a vertical and cross diameter of 3.58 cm and 2.35 cm, averaging 9.3 g with a regular size. It has thin red skin, a smooth surface with large dots, a shallow stalk cavity, a cuspate fruit apex and a remnant stigma. The flesh is delicate, crisp, sour, sweet and juicy. It has an excellent quality for fresh-eating. SSC, TTS, TA and Vc in fresh fruit are 30.60%, 28.45%, 0.38% and 409.20 mg per 100 g fresh fruit. The percentage of containing kernels is 83.3%. Sometimes the stone contains double kernels.

Evaluation The cultivar has low and unstable productivity. The small fruit has an excellent quality for fresh eating. Technical measures such as girdling and spraying microelement fertilizers in blooming stage should be carried out to improve fruit set and the yield. Rain protection should be paid much attention to in the maturing stage.

大荔马牙枣

品种来源及分布　别名马蔺枣、马脸枣、马头枣。原产和分布于陕西大荔的石槽、官池、苏村、八渔及彬县城关镇等地。

植物学性状　树体较大，树姿开张，干性强，树冠自然圆头形。主干灰褐色，皮裂较浅，裂纹纵条形，不易剥落。枣头红褐色，平均长80.2cm，粗1.17cm，节间长6.3cm，蜡质少。二次枝长33.7cm，6节左右，弯曲度中等。针刺较发达。枣股圆柱形，平均抽生枣吊5.0个。枣吊长19.6cm，着叶14片。叶长5.0cm，叶宽2.3cm，叶片较小，椭圆形，叶薄，绿色，叶面有光泽。先端尖圆，叶基圆楔形，叶缘锯齿较锐，偶有复齿。花量少，花朵中大。

生物学特性　适应性强，在沙土和壤土都能正常生长结果。树势中庸，成龄树丰产性强，坐果稳定，枣头、2～3年和4～6年生枝的吊果率分别为10.8%、56.2%和18.0%。在山西太谷地区，4月中旬萌芽，6月上旬始花，9月下旬果实成熟，果实生育期115d，为晚熟品种类型。成熟期遇雨易裂果，易受食心虫危害。

果实性状　果实大，圆锥形或圆柱形，纵径4.22cm，横径2.55cm，单果重24.0g，大小整齐。果肩平圆。梗洼浅广。果顶尖或平。果面光滑，果皮薄，紫红色。果点小、密，不明显。果肉浅绿色，质地细脆，汁液较多，味酸甜，鲜食品质中上。鲜枣可食率95.8%，含可溶性固性物28.60%，总糖24.76%，酸0.50%，100g果肉维生素C含量245.29mg；果皮含黄酮5.61mg/g，cAMP含量255.20μg/g。果核大，纺锤形，略弯曲，核重1.00g。含仁率25.0%。

评价　该品种适应性强，丰产稳产。果实肉质细脆，汁液较多，品质中等。主要缺点是核大，易裂果。

Dalimayazao

Source and Distribution　The cultivar originated from and spreads in Shicao, Guanchi, Sucun and Bayu in Dali County and Chengguanzhen of Binxian County in Shaanxi Province.

Botanical Characters　The large tree is spreading with a strong central leader trunk and a round crown. The grayish-brown trunk bark has shallow, vertical striped fissures, uneasily shelled off. The reddish-brown 1-year-old shoots are 80.2 cm long and 1.17 cm thick. The internodes are 6.3 cm long with less wax and developed thorns. The secondary branches are 33.7 cm long with 6 nodes of medium curvature. The column-shaped mother fruiting spurs can germinate 5 deciduous fruiting shoots, which are 19.6 cm long with 14 leaves. The small oval-shaped leaves are thin, green and glossy, with a sharply-round apex, a round-cuneiform base and a sharp, blunt saw-tooth pattern on the margin. The number of flowers is small with medium-sized blossoms.

Biological Characters　The tree has strong adaptability, moderate vigor and high productivity for mature trees, with stable fruit set. It grows well in sandy soil and loam soil. The percentage of fruits to deciduous fruiting shoots of 1-year-old shoots, 2～3-year-old branches and 4～6-year-old ones is 10.8%, 56.2% and 18.0%. In Taigu County of Shanxi Province, it germinates in early April, begins blooming in early June and matures in late September with a fruit growth period of 110 d. It is a mid-late-ripening variety with serious fruit-cracking if it rains in the maturing stage. It is also susceptible to frelt borer.

Fruit Characteristics　The large conical fruit has a vertical and cross diameter of 4.22 cm and 2.55 cm, averaging 24.0 g with a regular size. It has a flat-round shoulder, a shallow and wide stalk cavity, a pointed or flat fruit apex, a smooth surface and purplish-red thin skin with small and dense dots. The light-green flesh is delicate, crisp, sour, sweet and juicy. It has a better than normal quality for fresh-eating. The percentage of edible part of fresh fruit is 95.8%, and SSC, TTS, TA and Vc is 28.60%, 24.76%, 0.50% and 245.29 mg per 100 g fresh fruit. The content of flavones and cAMP in mature fruit skin is 5.61 mg/g and 255.20 μg/g. The large spindle-shaped stone weighs 1.0 g with the percentage of containing kernels of 25.0%.

Evaluation　The cultivar has strong adaptability with a high and stable yield. The flesh is delicate, crisp and juicy with a good taste. Yet it has a large stone along with serious fruit-cracking.

内黄苹果枣

品种来源及分布 原产河南省内黄县,数量较少,栽培历史不详。

植物学性状 树体高大,树姿较直立,枝条密,树冠圆头形。主干灰褐色,皮裂深,易剥落。枣头红褐色,萌发力强,生长较粗壮,二次枝粗而长,弯曲度极小。针刺基本退化。皮目圆形,灰白色,分布整齐。枣股大,圆柱形,老龄枣股长3.7cm。枣吊短,一般长8.4~12.7cm,节间长1~1.7cm。叶片中大,叶长7.4cm,叶宽3.3cm,卵圆形,浓绿色,先端钝尖,叶基圆形,叶缘锯齿粗而中密。花量多,花小,花径6.2mm。

生物学特性 树势强健,枣头及二次枝生长量极大,萌芽率高,成枝力强。结果较晚,早期丰产性和盛果期产量中等。在山西太谷地区,果实成熟期为9月中旬,果实生育期100d左右,为中熟品种类型。果实抗裂果和抗病性较强。

果实性状 果个较大,近圆形或扁圆形,纵径3.23cm,横径3.27cm,单果重16.1g,最大33.2g。果梗长,中粗,梗洼窄而深。果顶凹,柱头遗存。果皮薄,红色,果面光滑。果点小而圆,浅黄色,分布密。果肉厚,白色,肉质致密,酥脆,味甜,汁液多,品质上等,适宜鲜食。鲜枣可食率96.0%,含可溶性固形物32.00%,总糖22.73%,酸0.36%,100g果肉维生素C含量232.61mg;果皮含黄酮12.82mg/g,cAMP含量246.45μg/g。核小,纺锤形,纵径2.30cm,横径1.30cm,核重0.64g,核尖较短,核纹中深,核面粗糙,不含种仁。

评价 该品种适应性广,抗逆性强,为优质、中熟鲜食良种,在我国南北枣区均可发展。但生产中应加强控制树势和采取提高坐果率的技术措施。

Neihuangpingguozao

Source and Distribution The cultivar originated from Neihuang county of Henan Province with a small quantity and an unknown history.

Botanical Characters The large tree is vertical with dense branches and a round crown. The grayish-brown trunk bark has deep fissures, easily shelled off. The reddish-brown 1-year-old shoots have strong germination ability. The secondary branches are thick and long, with a very small curvature and almost no thorns. The grayish-white lenticels are round and regularly distributed. The large column-shaped mother fruiting spurs are 3.7 cm. The short deciduous fruiting shoots are 8.4~12.7 cm long with the internodes of 1.0~1.7 cm. The medium-sized leaves are oval-shaped and dark green, 4.8 cm long and 2.6 cm wide, with a bluntly-cuspate apex, a round base and a thick and medium-dense saw-tooth pattern on the margin. There are many small flowers with a diameter of 6.2 mm produced.

Biological Characters The tree has strong vigor, strong germination and branching ability, bearing late with moderate early productivity and a medium yield in full-bearing stage. 1-year-old shoots and secondary branches have great growth increment. In Taigu County of Shanxi Province, it matures in mid-September with a fruit growth period of 100 d. It is a mid-ripening variety with strong resistance to fruit-cracking and diseases.

Fruit Characteristics The medium-sized fruit is nearly round, with a vertical and cross diameter of 3.23 cm and 3.27 cm, averaging 16.1 g (maximum 33.2 g). It has a long and medium-wide stalk, a narrow and deep stalk cavity, a sunken and conical fruit apex, a remnant stigma, thin red skin and a smooth surface with small, round, light-yellow and dense dots. The white flesh is thick, delicate, crisp, juicy and sweet, with a good quality for fresh eating. The percentage of edible part of fresh fruit is 96.0%, and SSC, TTS, TA and Vc is 32.00%, 22.73%, 0.36% and 232.61 mg per 100 g fresh fruit. The content of flavones and cAMP in mature fruit skin is 12.82 mg/g and 246.45 μg/g. The small spindle-shaped stone has a vertical and cross diameter of 2.30 cm and 1.30 cm, averaging 0.64 g with a short apex, medium-deep veins and a rough surface. There are no kernels inside the stones.

Evaluation The cultivar has strong adaptability and strong resistance to adverse conditions. It is a mid-ripening table variety with a good quality and can be developed in jujube production areas in south and north China. Yet technical measures of controlling growth vigor and improving fruit set should be enforced.

鲜 食
Table Varieties
品 种

· 71 ·

庆云小梨枣

品种来源及分布 别名梨枣、小梨枣、铃枣。原产和分布于山东的庆云、乐陵等地，数量极少，以庭院栽培为主。

植物学性状 树体中大，树姿半开张，枝系结构紧凑，树冠自然半圆形。主干灰褐色，皮裂细条状，粗糙，不易剥落。枣头黄褐色，平均长63.4cm，粗1.20cm，节间长7.1cm，枝面粗糙，被少量灰白色蜡质。二次枝平均长33.9cm，7节左右，弯曲度小。针刺细短，不发达，当年生长季脱落。枣股圆柱形，抽吊力强，平均抽生枣吊4.5个。枣吊粗长，平均长24.7cm，着叶13片。叶片中大，卵圆形，较薄，浅绿色。叶面反卷，先端渐尖，较宽，叶基截形，叶缘波状，锯齿尖细，较深，齿距小。花多而大，花序平均着花7朵。

生物学特性 适应性强，树势强健，发枝力较弱，枝条较细。定植后2年即开始结果，结果稳定，较丰产，枣吊平均结果1.3个。在山西太谷地区，9月上旬果实成熟。果实生育期95d左右，为早熟品种类型。较抗裂果和果实病害。

果实性状 果实中大，卵圆形或倒卵圆形，纵径3.32cm，横径2.77cm，单果重13.1g，大小整齐。果肩平圆。梗洼窄，中等深。果顶平圆，端部略凸。果面有隆起。果皮薄，红色。果肉白色，质地细脆，汁液较多，味极甜略酸，鲜食品质上等。鲜枣可食率95.2%，含可溶性固形物28.40%，含总糖26.64%，酸0.62%，100g果肉维生素C含量286.29mg；果皮含黄酮6.67mg/g，cAMP含量23.82μg/g。果核纺锤形，平均重0.63g，含仁率30.0%。

评价 该品种结果早，较丰产。果实中大，肉质细脆多汁，味浓，质优，为优良早熟品种。

Qingyunxiaolizao

Source and Distribution The cultivar, also called Lizao or Lingzao, originated from and spreads in Qingyun and Leling of Shandong Province. It is mainly planted in family gardens with a small quantity.

Botanical Characters The medium-sized tree is half-spreading with compact branches and a semi-round crown. The grayish-brown trunk bark has thin striped fissures, uneasily shelled off. The yellowish-brown 1-year-old shoots are 63.4 cm long and 1.2 cm thick. The internodes are 7.1 cm long with grayish-white wax. The thin and short thorns are less developed, usually falling off in the growing season. The secondary branches are 33.9 cm long with 7 nodes of small curvature. The column-shaped mother fruiting spurs can germinate 4 or 5 thick and long deciduous fruiting shoots, which are 24.7 cm long with 13 leaves. The medium-sized thin leaves are oval-shaped and light green, with a gradually-cuspate apex, a truncate base and a sharp, thin and deep saw-tooth pattern on the wavy margin. There are many large-sized flowers produced, averaging 7 ones per inflorescence.

Biological Characters The tree has strong adaptability, strong vigor and weak branching ability. It generally bears in the 2nd year after planting, with a high and stable yield. The deciduous fruiting shoot bears 1.3 fruits on average. In Taigu County of Shanxi Province, it matures in early September with a fruit growth period of 95 d. It is an early-ripening variety with strong resistance to fruit-cracking and diseases.

Fruit Characteristics The large fruit is oval-shaped or obovate, with a vertical and cross diameter of 3.32 cm and 2.77 cm, averaging 13.1 g with a regular size. It has a flat-round shoulder, a narrow and medium-deep stalk cavity, a flat-round fruit apex and thin red skin with some protuberances. The white flesh is crisp, juicy, strongly sweet and a little sour. It has a good quality for fresh-eating. The percentage of edible part of fresh fruit is 95.2%, and SSC, TTS, TA and Vc is 28.40%, 26.64%, 0.62% and 286.29 mg per 100 g fresh fruit. The content of flavones and cAMP in mature fruit skin is 6.67 mg/g and 23.82 μg/g. The spindle-shaped stone weighs 0.63 g with the percentage of containing kernels of 30.0%.

Evaluation The cultivar bears early with high productivity. The large fruit has crisp, juicy, sweet flesh and a good quality. It is a good early-ripening variety.

鲜食
Table Varieties
品种

· 73 ·

合阳铃铃枣

品种来源及分布 别名蛋蛋枣。原产和分布于陕西合阳的坊镇夏阳村及百良乡黄河沿岸一带，数量较少。

植物学性状 树体高大，树姿半开张，枝叶密度中等，树冠伞形。主干灰褐色，皮裂纹深，宽条状，易剥落。枣头枝粗壮，红褐色，平均长81.9cm，粗1.10cm，节间长7.1cm，蜡质少。二次枝长22.9cm，6节，弯曲度中等。无针刺。枣股圆锥形，粗大，平均抽生枣吊4个。枣吊长22.9cm，着叶16片。叶片中大，叶长5.4cm，叶宽2.6cm，椭圆形，中厚，绿色，有光泽，先端渐尖，先端钝圆，叶基偏斜形，叶缘锯齿较深，齿尖钝。花量多，花朵小，花序平均着花4朵。

生物学特性 适应性较强，树势强健，发枝力较强。生长结果良好，产量较高，枣头、2~3年和3年以上枝的吊果率分别为3.3%、140.0%和47.3%。在山西太谷地区，4月下旬发芽，5月底始花，9月中旬果实成熟，果实生育期100d，为早中熟品种类型。较抗病，抗裂果性一般。

果实性状 果实中大，近圆形，纵径2.72cm，横径2.72cm，单果重12.1g，大小整齐。果肩平。梗洼浅，中广。环洼中宽广。果顶微凹，柱头遗存。果面平整，果皮薄，紫红色。果点小而明显。果肉绿色，质地疏松，汁液中多，味香甜，鲜食品质上等。鲜枣可食率93.1%，含可溶性固形物26.60%，总糖24.55%，酸0.26%，100g果肉维生素C含量 367.44mg；果皮含黄酮11.45mg/g，cAMP含量448.95μg/g。果核较大，倒纺锤形，核重0.83g。核内大多无种子，含仁率9.7%。

评价 该品种适应性较强，树体强健，产量较高。果实中大，鲜食品质优异。

Heyanglinglingzao

Source and Distribution The cultivar, also called Dandanzao, originated from and spreads in Xiayangcun of Fangzhen Village and Bailiang Village in Heyang of Shaanxi Province with a small quantity.

Botanical Characters The large tree is half-spreading with an umbrella-shaped crown. The grayish-brown trunk bark has deep wide-striped fissures, easily shelled off. The brown 1-year-old shoots are 81.9 cm long and 1.10 cm thick. The internodes are 7.1 cm long with less wax and no thorns. The secondary branches are 22.9 cm long with 6 nodes of medium curvature. The conical mother fruiting spurs can germinate 4 deciduous fruiting shoots, which are 22.9 cm long with 16 leaves. The medium-sized leaves are 5.4 cm long and 2.6 cm wide, oval-shaped, medium thick, green and glossy, with a gradually-cuspate apex, a deflective base and a deep, blunt saw-tooth pattern on the margin. There are many small flowers produced, averaging 4 ones per inflorescence.

Biological Characters The tree has strong adaptability, strong vigor, medium branching ability and medium-dense branches, with an unstable yield. The percentage of fruits to deciduous fruiting shoots of 1-year-old shoots, 2~3-year-old branches and over-3-year-old ones is 3.3%, 140.0% and 47.3%. In Taigu County of Shanxi Province, it germinates in late April, begins blooming in late May and matures in mid-September, with a fruit growth period of 100 d. Fruit-cracking easily occurs if it rains in the maturing stage.

Fruit Characteristics The medium-sized fruit is nearly-round with a vertical and cross diameter of 2.72 cm and 2.72 cm, averaging 12.1 g with a regular size. It has a flat shoulder, a shallow and medium-wide stalk cavity, a slightly-sunken fruit apex, a remnant stigma, a thin, purplish-red and smooth surface with some small and distinct dots. The green flesh is loose-textured and sweet with medium juice. It has a good quality for fresh-eating. The percentage of edible part of fresh fruit is 93.1%, and SSC, TTS, TA and Vc is 26.60%, 24.55%, 0.26% and 367.44 mg per 100 g fresh fruit. Dried fruits have glossy and flat skin with loose-textured flesh. The content of flavones and cAMP in mature fruit skin is 11.45 mg/g and 448.95 μg/g. The content of TTS and TA in dried fruit is 80.70% and 0.96%. The large stone is inverted spindle-shaped, averaging 0.83 g. Most stones contain no kernels, and the percentage of containing kernels is 9.7%.

Evaluation The cultivar has strong adaptability and strong tree vigor. The medium-sized fruit has an excellent quality for fresh-eating. Yet it has poor resistance to fruit-cracking with a high yield.

溆浦蜜蜂枣

品种来源及分布 别名蜜蜂枣。原产湖南溆浦县低庄镇的连山、连圹村和双井乡的圹湾、水口、新圹村等地。

植物学性状 树体较大，树姿开张，枝系稀疏，树冠呈乱头形。主干灰黑色，树皮粗糙，裂纹较深，呈不规则条状，易剥落。枣头黄褐色，平均长63.5cm，粗0.96cm，节间长6.8cm。二次枝平均长25.9cm，6节，弯曲度小。针刺不发达。枣股圆锥形，略弯曲，平均抽生枣吊4个。枣吊平均长24.2cm，平均着叶17片。叶片较小，椭圆形，绿色，先端渐尖，叶基圆楔形，叶缘锯齿细小。花量多，花序平均着花5朵。花朵小，花萼长，昼开型。

生物学特性 树势强，发枝力和萌蘖力较弱。坐果率极高，枣头、2～3年和3年以上枝的吊果率分别为50.0%、297.6%和79.5%。定植后2年开始结果，盛果期产量高而稳定。高产树株产150～180kg。自然落果少，成熟期不易裂果。在山西太谷地区，4月中旬萌芽，5月下旬始花，9月下旬果实成熟，果实生育期115d，属晚熟品种类型。

果实性状 果实中大，卵圆形，纵径2.88cm，横径2.34cm，单果重12.2g。果肩平圆，梗洼中深广，果顶平。果皮中厚，紫红色，光滑。果肉白色，质地疏松，汁液多，味甜酸，鲜食品质中等。鲜枣可食率91.3%，含可溶性固形物27.00%，总糖16.45%，酸0.36%，100g果肉维生素C含量322.68mg；果皮含黄酮13.69mg/g，cAMP含量262.92μg/g。果核大，纺锤形，核重1.06g，核内多不含种子，含仁率仅1.7%。

评价 该品种果实中大，核大，丰产稳产。肉质疏松，味甜略酸，品质中等，不易裂果，为较优良的晚熟品种。

Xupumifengzao

Source and Distribution The cultivar originated from Lianshan, Liankuang in Dizhuang Village and Kuangwan, Shuikou and Xinkuang in Shuangjing Village of Xupu County in Hunan Province.

Botanical Characters The large tree is spreading with sparse branches and an irregular crown. The grayish-black trunk bark has irregular deep-striped fissures, easily shelled off. The yellowish-brown 1-year-old shoots are 63.5 cm long and 0.96 cm thick. The internodes are 6.8 cm long with developed thorns. The secondary branches are 25.9 cm long with 6 nodes of small curvature. The conical mother fruiting spurs can germinate 4 deciduous fruiting shoots which are 24.2 cm long with 17 leaves. The small green leaves are oval-shaped, with a gradually-cuspate apex, a round-cuneiform base and a thin saw-tooth pattern on the margin. There are many small flowers produced, averaging 5 ones per inflorescence. The daytime-bloomed blossom has a long calyx.

Biological Characters The tree has strong vigor, weak branching and suckering ability, with very high fruit set. The percentage of fruits to deciduous fruiting shoots of 1-year-old shoots, 2 or 3-year-old branches and over-3-year-old ones is 50.0%, 297.6% and 79.5%. It generally bears in the 2nd year after planting, with a high and stable yield in the full-bearing stage. A productive tree has a yield of 150～180 kg. Natural fruit-dropping and fruit-cracking seldom occur. In Taigu County of Shanxi Province, it germinates in mid-April, begins blooming in late May and matures in late September with a fruit growth period of 115 d. It is a late-ripening variety.

Fruit Characteristics The medium-sized fruit is oval-shaped with a vertical and cross diameter of 2.88 cm and 2.34 cm, averaging 12.2 g. It has a flat-round shoulder, a medium-deep and wide stalk cavity, a flat fruit apex, medium-thick and purplish-red skin and a smooth surface. The white flesh is loose-textured, sour, sweet and juicy. It has medium quality for fresh-eating. The percentage of edible part of fresh fruit is 91.3%, and SSC, TTS, TA and Vc is 27.00%, 16.45%, 0.36% and 322.68 mg per 100 g fresh fruit. The content of flavones and cAMP in mature fruit skin is 13.69 mg/g and 262.92 μg/g. The large spindle-shaped stone weighs 1.06 g.

Evaluation The cultivar has a large fruit size and a large stone with a high and stable yield. The loose-textured flesh has a sour and sweet taste with medium quality. It is a late-ripening variety with light fruit-cracking.

鲜 食
Table Varieties
品 种

· 77 ·

连 县 糖 枣

品种来源及分布 主要分布于广东连县的星子、大路边等乡镇，是当地的乡土品种，栽培数量占当地枣树总数的30%，有400年左右栽培历史。

植物学性状 树体中大，树姿开张，枝叶稠密，树冠自然半圆形。主干灰黑色，粗糙，裂纹不规则，条状。枣头黄褐色，平均长75.7cm，粗0.97cm，节间长7.5cm，蜡质少。二次枝长26.7cm，5节左右，弯曲度小。无针刺。枣股圆柱形，平均抽生枣吊4.0个。枣吊长19.5cm，着生叶片12片。叶片卵圆形，绿色，较光亮，先端尖凹，叶基圆楔形，叶缘具钝锯齿。花较多且较大，花序平均着花5朵。

生物学特性 适应性较强，树势强旺，发枝力强。坐果率中等，枣头、2～3年和3年以上枝的吊果率分别为6.7%、57.1%和60.3%。结果较早，10年后进入盛果期，产量中等。在山西太谷地区，4月中旬萌芽，5月底始花，8月下旬果实进入脆熟期，果实生育期85d左右，为极早熟品种类型。采前落果和裂果较轻。

果实性状 果实中大，圆柱形，纵径3.00cm，横径2.26cm，单果重10.6g，大小较整齐。梗洼中深广，果顶平。果皮中厚，光滑，浅红色。果点小，不明显。果肉浅绿色，质地较致密，汁液少，味甜，鲜食品质上等。鲜枣可食率94.7%，含可溶性固形物32.00%，100g果肉维生素C含量369.85mg；果皮含黄酮5.70mg/g，cAMP含量43.77μg/g。果核中大，纺锤形，平均重0.56g，含仁率16.7%。

评价 该品种适应性较强，产量中等，裂果轻。果实肉质较致密，汁较少，味甜，鲜食品质上等，为优良的极早熟品种。

Lianxiantangzao

Source and Distribution The cultivar originated from and mainly spreads in Xingzi and Dalubian Villages of Lianxian County in Guangdong Province. Its cultivation area occupies 30% of the total jujube area in the native places. It has a cultivation history of about 400 years.

Botanical Characters The medium-sized tree is spreading with dense branches and a semi-round crown. The grayish-black trunk bark has rough irregular striped fissures. The yellowish-brown 1-year-old shoots are 75.7 cm long and 0.97 cm thick. The internodes are 7.5 cm long with less wax and no thorns. The secondary branches are 26.7 cm long with 5 nodes of small curvature. The column-shaped mother fruiting spurs can germinate 4 deciduous fruiting shoots which are 19.5 cm long with 12 leaves. The green leaves are oval-shaped and glossy, with a cuspate-sunken apex, a round-cuneiform base and a blunt saw-tooth pattern on the margin. There are many large flowers produced, averaging 5 ones per inflorescence.

Biological Characters The tree has strong adaptability, strong vigor and strong branching ability, with medium fruit set. The percentage of fruits to deciduous fruiting shoots of 1-year-old shoots, 2～3-year-old branches and over-3-year-old ones is 6.7%, 57.1% and 60.3%. It bears early and enters the full-bearing stage in the 10th year with a medium yield. In Taigu County of Shanxi Province, it germinates in mid-April, begins blooming in late May and enters the crisp-maturing stage in late August, with a fruit growth period of 85 d. It is an extremely early-ripening variety with light premature fruit-dropping and fruit-cracking.

Fruit Characteristics The medium-sized fruit is column-shaped, with a vertical and cross diameter of 3.00 cm and 2.26 cm, averaging 10.6 g with a regular size. It has a medium-deep and wide stalk cavity, a flat fruit apex, a medium-thick, light-red and smooth surface with some small dots. The light-green flesh is tight-textured and sweet, with less juice. It has a good quality for fresh-eating. The percentage of edible part of fresh fruit is 94.7%, and SSC and Vc is 32.00% and 369.85 mg per 100 g fresh fruit. The content of flavones and cAMP in mature fruit skin is 5.70 mg/g and 43.77 μg/g. The medium-sized stone is spindle-shaped, averaging 0.56 g. The percentage of containing kernels is 16.7%.

Evaluation The cultivar has strong adaptability and a medium yield, with light fruit-cracking. The fruit has tight-textured, less juicy and sweet flesh with an excellent quality for fresh-eating. It is a good extremely-early-ripening variety.

鲜 食
Table Varieties
品 种

· 79 ·

滕州大马枣

品种来源及分布 分布在山东滕州等地,数量少,栽培历史不详。

植物学性状 树体较大,树姿直立,干性较强,树冠多呈伞形。主干黑褐色,皮裂纹深、块状,不易剥落。枣头红褐色,平均长69.2cm,粗1.00cm,节间长8.2cm。无针刺。二次枝平均长32.1cm,6节,弯曲度中等。枣股圆锥形,平均抽生枣吊4个。枣吊长26.5cm左右,着叶17片。叶片中大,叶长5.3cm,叶宽2.2cm,椭圆形,叶面平展,浓绿色,较光亮,先端渐尖,叶基圆楔形,叶缘钝锯齿。花量中多,花序平均着花8朵。

生物学特性 适土性强,花期要求气温较高。花朵坐果率因当地花期气温而异,在气温较低且不稳定的年份,坐果较差,2～3年和3年以上枝的吊果率分别为53.3%和16.3%。结果较早,定植后2年开始结果。在山西太谷地区,9月下旬果实成熟,果实生育期115d左右,属晚熟品种类型。成熟期遇雨易裂果。

果实性状 果实小,圆锥形,单果重7.2g,最大14.0g,大小较整齐。果肩较宽,斜圆或平圆,耸起,有多条深浅不等的沟棱。梗洼深,中广。环洼窄,深或中深。果柄细,长3.5mm,果顶较肩部瘦小,端部平圆,顶洼中深、广或浅小。果面较平整,果皮红色。果肉白色,质地酥脆,汁液多,味甜,宜鲜食,品质极上。鲜枣可食率96.2%,含总糖23.39%,酸0.47%,100g果肉维生素C含量363.54mg。果核小,纺锤形,核重0.53g,含仁率91.7%,种仁较饱满。

评价 该品种适应性强,较耐瘠薄土壤条件。果实较小,品质优异,晚熟,适宜鲜食,但抗裂果能力较差。

Tengzhoudamazao

Source and Distribution The cultivar spreads in Tengzhou of Shandong Province with a small quantity and an unknown history.

Botanical Characters The small tree is vertical with a strong central leader trunk and an umbrella-shaped crown. The blackish-brown trunk bark has deep massive fissures, uneasily shelled off. The reddish-brown 1-year-old shoots are 69.2 cm long and 1.0 cm thick. The internodes are 8.2 cm long, without thorns. The secondary branches are 32.1 cm long with 6 nodes of medium curvature. The conical mother fruiting spurs can germinate 4 deciduous fruiting shoots which are 26.5 cm long with 17 leaves. The medium-sized flat leaves are oval-shaped and dark green, 5.3 cm long and 2.2 cm wide, with a sharply-cuspate apex, a round-cuneiform base and a blunt saw-tooth pattern on the margin. The number of flowers is medium large, averaging 8 per inflorescence.

Biological Characters The cultivar has strong adaptability. Yet it requires a higher temperature in the blooming stage, for the fruit set depends on the local temperature during that time. In years with unstable low temperatures, the fruit set is low. The percentage of fruits to deciduous bearing shoots of 2～3-year-old branches and over-3-year-old ones is 53.3% and 16.3%. It bears early, generally in the 2nd year after planting. In Taigu County of Shanxi Province, it matures in late September with a fruit growth period of 115 d. It is a late-ripening variety with serious fruit-cracking if it rains in the maturing stage.

Fruit Characteristics The small conical fruit weighs 7.2 g (maximum 14.0 g) with a regular size. It has a wide shoulder, a deep and medium-wide stalk cavity, a thin stem (3.5 mm long), a smooth surface and red skin. The white flesh is crisp, juicy and sweet. It has a very good quality for fresh eating. The percentage of edible part of fresh fruit is 96.2%, and the content of TTS and TA is 23.39% and 0.47%. The small spindle-shaped stone weighs 0.53 g, with a well-developed kernel. The percentage of containing kernels is 91.7%.

Evaluation The cultivar has strong adaptability, tolerant to poor soils. The small fruit has a very good quality for fresh-eating. It is a late-ripening variety with poor resistance to fruit-cracking.

旻 枣

品种来源及分布 主要分布丁天津西郊的张窝、傅村、木厂一带，为当地主栽品种。栽培历史300~400年。

植物学性状 树体较大，树姿较直立，树冠呈圆柱形。主干灰褐色，树皮裂纹块状，不易剥落。枣头红褐色，平均长73.5cm，粗0.97cm，节间长7.6cm，被灰色蜡质。二次枝平直，平均长22.7cm，5节左右，弯曲度中等。针刺细弱。枣股平均抽生枣吊4.0个。枣吊长15.0cm，着叶12片。叶片中大，卵圆形，绿色，有光泽，先端渐尖，叶基宽楔形，叶缘具细锯齿。花多，花中大，花序平均着花9朵。

生物学特性 适应性较强，耐旱、耐涝、耐盐碱，抗病。树势强健，发枝力中等，结果母枝寿命长，产量稳定。坐果率低，枣头、2~3年和3年以上枝的吊果率分别为2.0%、12.3%和2.9%。定植后2年开始结果，盛果期树株产50kg左右。在山西太谷地区，9月中旬果实成熟，果实生育期100d，为中熟品种类型。成熟期遇雨裂果较重，裂纹小而多。

果实性状 果实较小，长圆形，纵径2.47cm，横径2.15cm，单果重5.3g，大小不整齐。果肩平圆，略斜，梗洼中深。果顶平圆。果皮厚，浅红色。果肉绿色，质地较致密，汁液中多，味甜，品质上等，适宜鲜食和制作醉枣。鲜枣可食率93.0%，含可溶性固形物31.70%，总糖27.37%，酸0.51%。果核较小，纺锤形，核重0.37g。核内多含有不饱满的种子，含仁率16.7%。

评价 该品种适应性强，耐旱涝、盐碱，丰产稳产。果实较小，肉质细，鲜食品质上等。成熟期应注意防治裂果。

Minzao

Source and Distribution The cultivar originated from and spreads in Zhangwo, Fucun and Muchang villages of western Tianjin City. It is a dominant variety there with a cultivation history of 300~400 years.

Botanical Characters The large tree is vertical with a column-shaped crown. The grayish-brown trunk bark has massive fissures, uneasily shelled off. The reddish-brown 1-year-old shoots are 73.5 cm long and 0.97 cm thick. The internodes are 7.6 cm long with gray wax and no thorns. The secondary branches are 22.7 cm long with 5 nodes of medium curvature. The mother fruiting spurs can germinate 4 deciduous fruiting shoots which are 15.0 cm long with 12 leaves. The medium-sized leaves are oval-shaped, green and glossy, with a gradually-cuspate apex, a wide-cuneiform base and a thin saw-tooth pattern on the margin. There are many medium-sized flowers produced, averaging 9 ones per inflorescence.

Biological Characters The tree has strong adaptability, tolerant to drought, water-logging and saline-alkaline, along with strong resistance to diseases, strong vigor, medium branching ability and a long economic life for the mother fruiting spurs. It has stable productivity and low fruit set. The percentage of fruits to deciduous fruiting shoots of 1-year-old shoots, 2 or 3-year-old branches and over-3-year-old ones is 2.0%, 12.3% and 2.9%. It generally bears in the 2nd year after planting. A tree in its full-bearing stage has a yield of 50 kg on average. In Taigu County of Shanxi Province, it matures in mid-September with a fruit growth period of 100 d. It is a mid-ripening variety with serious fruit-cracking if it rains in the maturing stage.

Fruit Characteristics The small oblong fruit has a vertical and cross diameter of 2.47 cm and 2.15 cm, averaging 5.3 g with irregular sizes. It has a flat-round shoulder, a medium-deep stalk cavity, a flat-round fruit apex and light-red thick skin. The green flesh is tight-textured and sweet, with medium juice. It has a good quality for fresh-eating, also used for processing alcoholic jujubes. The percentage of edible part of fresh fruit is 93.0%, and SSC, TTS and TA is 31.70%, 27.37% and 0.51%. The small spindle-shaped stone weighs 0.37 g. Most kernels are shriveling and the percentage of containing kernels is 16.7%.

Evaluation The cultivar has strong adaptability, tolerant to drought, water-logging and saline-alkaline along with a high and stable yield. The small fruit has delicate flesh with a good quality for fresh-eating. Prevention of fruit-cracking should be paid much attention to in the maturing stage.

枣 强 脆 枣

品种来源及分布 原产和分布于河北枣强县。

植物学性状 树体较大,树姿开张,干性较弱,树冠呈圆头形。主干灰黑色,块状皮裂。枣头红褐色,平均长79.1cm,节间长7.5cm,无蜡质。二次枝长35.8cm,着生5～7节,弯曲度大。针刺不发达。枣股小,圆锥形,平均抽生枣吊3.6个。枣吊长21.0cm,着叶14片。叶片中等大,叶长5.2cm,叶宽2.2cm,椭圆形,平展,先端尖,叶基圆楔形,叶缘锯齿锐。花量多,花序平均着花8朵,花中大,花径6.3mm。

生物学特性 树势强。萌芽率和成枝力强,坐果率低,枣头、2～3年和4～6年生枝的吊果率分别为35.5％、10.4％和5.5％。一般定植第三年结果,10年左右进入盛果期,产量较低。在山西太谷地区,9月中旬果实进入脆熟期采收,果实生育期100d左右,为早中熟品种类型。成熟期遇雨裂果严重。

果实性状 果个中大,卵圆形,纵径3.69cm,横径2.49cm,单果重10.8g,大小较整齐。果皮薄,紫红色,果面平滑。果点小而稀疏。梗洼中深、广。果顶尖,柱头遗存。肉质细嫩,品质极上,适宜鲜食。鲜枣可食率94.0％,含可溶性固形物30.00％,总糖24.54％,100g果肉维生素C含量378.38mg;果皮含黄酮3.75mg/g,cAMP含量249.20μg/g。核较小,纺锤形,核重0.65g。种仁不饱满,含仁率85.0％。

评价 该品种树体较大,树势强,产量较低。果个中大,酥脆,味极甜,汁液多,鲜食品质极上。但果实抗裂果能力差,成熟期需注意防雨。生产中需采取花期喷施微肥、环剥的措施促进坐果,提高产量。

Zaoqiangcuizao

Source and Distribution The cultivar originated from and spreads in Zaoqiang County of Hebei Province.

Botanical Characters The large tree is spreading with a weak central leader trunk and a round crown. The trunk bark has massive fissures. The reddish-brown 1-year-old shoots are 79.1 cm long with the internodes of 7.5 cm, without wax. The secondary branches are 35.8 cm long with 5～7 nodes of large curvature and less-developed thorns. The mother fruiting spurs can germinate 3.6 deciduous fruiting shoots which are 21.0 cm long with 14 leaves. The medium-sized leaves are oval-shaped and flat, 5.2 cm long and 2.2 cm wide, with a cuspate apex, a round-cuneiform base and a sharp saw-tooth pattern on the margin. There are many medium-sized flowers with a diameter of 6.3 mm produced, averaging 8 ones per inflorescence.

Biological Characters The tree has strong vigor, high germination rate and strong branching ability. It has low fruit set. The percentage of fruits to deciduous fruiting shoots of 1-year-old shoots, 2～3-year-old branches and 4～6-year-old ones is 35.5％, 10.4％ and 5.5％. It has certain yield in the 3rd year after planting and enters the full-bearing stage in the 10th year with poor productivity. In Taigu County of Shanxi Province, it enters the crisp-maturing stage and is harvested in mid-September with a fruit growth period of 100 d. It is an early-mid-ripening variety with serious fruit-cracking if it rains in the maturing stage.

Fruit Characteristics The medium-sized fruit is oval-shaped, with a vertical and cross diameter of 3.69 cm and 2.49 cm, averaging 10.8 g with a regular size. It has purplish-red thin skin, a smooth surface with small and sparse dots, a medium-deep and wide stalk cavity, a pointed fruit apex and a remnant stigma. The flesh is delicate and crisp, with an excellent quality for fresh-eating. The percentage of edible part of fresh fruit is 94.0％, and SSC, TTS and Vc is 30.00％, 24.54％ and 378.38 mg per 100 g fresh fruit. The content of flavones and cAMP in mature fruit skin is 3.75 mg/g and 249.20 μg/g. The small spindle-shaped stone weighs 0.65 g, with a shriveled kernel. The percentage of containing kernels is 85.0％.

Evaluation The large plant of Zaoqiangcuizao cultivar has strong vigor and a low yield. The medium-sized fruit has crisp, juicy and strongly sweet flesh, with an excellent quality for fresh eating. It has poor resistance to fruit-cracking, so rain protection should be paid much attention to in the maturing stage. Girdling and spraying microelement fertilizers should be carried out in the growing season to promote fruit set and to improve the yield.

鲜 食
Table Varieties
品 种

· 85 ·

薛 城 冬 枣

品种来源及分布 别名大雪枣、沂水大雪枣、沂蒙大雪枣、苹果枣。原产和分布于山东沂蒙山区的沂水、蒙阴县和枣庄市薛城区。

植物学性状 树体中大，树姿半开张，干性较弱，树冠呈乱头形。主干皮裂条状。枣头黄褐色，生长量100.2cm，节间长10.7cm，蜡质少。二次枝平均长33.8cm，3～5节，弯曲度中等，无针刺。枣股平均抽生枣吊4.0个，枣吊长24.4cm，着叶14片。叶片卵状披针形，先端锐尖，叶基圆形，叶缘具钝锯齿。花量多，花序平均着花9朵，花中大，花径6.3mm。

生物学特性 树势中庸偏弱，成枝力强。定植后第二年结果，早期丰产性能强，产量较高。在山西太谷地区，10月上旬果实开始进入脆熟期，10月中旬成熟采收，果实生育期130d左右，11月上旬开始落叶，进入落叶期极晚，为极晚熟品种类型。果实抗病、抗裂果。

果实性状 果实大，圆形或扁圆形，纵径3.33cm，横径3.18cm，单果重17.7g，最大45.0g，大小不整齐。果肩凸。梗洼中深、广。果顶平圆，柱头脱落。果面光滑，果皮厚，浅红色。果肉浅绿色，肉质致密而粗硬，汁液中多，酸甜，鲜食品质较差，可加工蜜枣。鲜枣可食率93.2%，含可溶性固形物25.80%，单糖13.18%，双糖3.69%，总糖16.87%，酸0.40%，100g果肉维生素C含量587.40mg；果皮含黄酮3.97mg/g，cAMP含量81.11μg/g。果核大，纺锤形，平均重1.20g，含仁率20.0%，种仁不饱满。

评价 该品种果实个大，核大，果肉存在木栓化现象，鲜食品质较差，现多用于加工蜜枣。成熟期极晚，可作为极晚熟品种资源研究利用。

Xuechengdongzao

Source and Distribution The cultivar originated from and spreads in Xuecheng County of Shandong Province.

Botanical Characters The medium-sized tree is half-spreading with a weak central leader trunk and an irregular crown. The trunk bark has striped fissures. The yellowish-brown 1-year-old shoots are 100.2 cm long with the internodes of 10.7 cm long and less wax. The secondary branches are 33.8 cm long with 3～5 nodes of medium curvature and without thorns. The mother fruiting spurs can germinate 4 deciduous fruiting shoots which are 24.4 cm long with 14 leaves. The leaves are ovate-lanceolate, with a sharply-cuspate apex, a round base and a blunt saw-tooth pattern on the margin. There are many medium-sized flowers with a diameter of 6.3 mm produced, averaging 9 ones per inflorescence.

Biological Characters The tree has weak vigor and strong branching ability, generally bearing in the 2nd year after planting with a high yield. In Taigu County of Shanxi Province, it enters the crisp-maturing stage in early October and is harvested in mid-October with a fruit growth period of 130 d. Defoliation begins in early November, which is rather late compared with other varieties. It is an extremely late-ripening variety with strong resistance to diseases and fruit-cracking.

Fruit Characteristics The large fruit is round or oblate, with a vertical and cross diameter of 3.33 cm and 3.18 cm, averaging 17.7 g (maximum 45.0 g) with irregular sizes. It has a protruding shoulder, a medium-deep and wide stalk cavity, a flat-round fruit apex, a remnant stigma, a smooth surface and light-red thick skin. The light-green flesh is hard and tight-textured with medium juice and a sour and sweet taste. It has a poor quality for fresh-eating, mainly used for processing candied fruits. The percentage of edible part of fresh fruit is 93.2%, and SSC, monosaccharide, disaccharide, TTS, TA and Vc is 25.80%, 13.18%, 3.69%, 16.87%, 0.40% and 587.40 mg per 100 g fresh fruit. The content of flavones and cAMP in mature fruit skin is 3.97 mg/g and 81.11 μg/g. The large spindle-shaped stone weighs 1.20 g with a shriveled kernel. The percentage of containing kernels is 20.0%.

Evaluation The cultivar has a large fruit size and large stone. The flesh has suberification phenomenon, which gives it a poor quality for fresh-eating. It is mainly used for processing candied fruits. The fruit matures extremely late. It can be used to do some research and utilization.

鲜 食
Table Varieties
品 种

天 津 快 枣

品种来源及分布 主要分布于天津西郊的张家窝、傅家、木厂等乡，为当地古老的优良鲜食品种，栽培历史悠久。

植物学性状 树体中大，树姿较直立，树冠呈圆柱形。主干灰褐色，树皮裂纹条片状，较易剥落。枣头黄褐色，平均长71.7cm，粗1.06cm，节间长7.8cm，蜡质多。二次枝平均长23.1cm，5节左右，弯曲度中等。无针刺。枣股圆柱形，平均抽生枣吊4.0个。枣吊长26.2cm，着叶16片。叶片较小，卵状披针形，绿色，叶面有光泽，先端渐尖，叶基圆楔形，叶缘锯齿钝浅。花多，花序平均着花9朵。昼开型。

生物学特性 适应性强，耐旱、耐涝、耐盐碱。树势中等，树体健壮，经济寿命较长。坐果率较高，2～3年和3年以上枝的吊果率分别为133.3%和22.2%。在山西太谷地区，4月中旬发芽，5月下旬始花，9月下旬果实成熟，果实生育期110d，为晚熟品种类型。

果实性状 果实较小，倒卵形，纵径2.50cm，横径2.13cm，单果重9.3g，大小较整齐。果肩平圆，梗洼中深、广，果顶平。果皮薄，红色。果肉浅绿色，酥脆多汁，极甜，鲜食品质上等。鲜枣可食率95.4%，含可溶性固形物28.00%，含总糖26.29%，100g果肉维生素C含量500.50mg；果皮含黄酮14.44mg/g，cAMP含量63.75μg/g。果核细小，椭圆形，核重0.43g。核内多不含种子，含仁率6.7%。

评价 该品种适应性强，耐旱涝、盐碱。树体强健，经济寿命长。产量高而稳定。果实酥脆多汁，可食率较高，品质上等。

Tianjinkuaizao

Source and Distribution The cultivar originated from and spreads in Zhangjiawo, Fujia and Muchang villages of western Tianjin City. It is an early-ripening table variety with a long cultivation history.

Botanical Characters The medium-sized tree is vertical with a column-shaped crown. The grayish-brown trunk bark has striped fissures, easily shelled off. The yellowish-brown 1-year-old shoots are 71.7 cm long and 1.06 cm thick. The internodes are 7.8 cm long with much wax and no thorns. The secondary branches are 23.1 cm long with 5 nodes of medium curvature. The column-shaped mother fruiting spurs can germinate 4 deciduous fruiting shoots which are 26.2 cm long with 16 leaves. The small leaves are ovate-lanceolate, green and glossy, with a gradually-cuspate apex, a round-cuneiform base and a shallow blunt saw-tooth pattern on the margin. There are many flowers produced, averaging 9 ones per inflorescence. It blooms in the daytime.

Biological Characters The tree has strong adaptability, tolerant to drought, water-logging and saline-alkaline, with moderate vigor, a long economic life and high fruit set. The percentage of fruits to deciduous fruiting shoots of 2～3-year-old branches and over-3-year-old ones is 133.3% and 22.2%. In Taigu County of Shanxi Province, it germinates in mid-April, begins blooming in late May and matures in late September, with a fruit growth period of 110 d. It is a late-ripening variety.

Fruit Characteristics The small obovate fruit has a vertical and cross diameter of 2.50 cm and 2.13 cm, averaging 9.3 g with a regular size. It has a flat-round shoulder, a medium-deep and wide stalk cavity, a flat fruit apex and thin red skin. The light-green flesh is crisp, juicy and strongly sweet. It has a good quality for fresh-eating. The percentage of edible part of fresh fruit is 95.4%, and SSC, TTS and Vc is 28.00%, 26.29% and 500.50 mg per 100 g fresh fruit. The content of flavones and cAMP in mature fruit skin is 14.44 mg/g and 63.75 μg/g. The small oval-shaped stone weighs 0.43 g. Most stones contain no kernels, and the percentage of containing kernels is 6.7%.

Evaluation The cultivar has strong adaptability, tolerant to drought, water-logging and saline-alkaline. The tree has moderate vigor and a long economic life with a high and stable yield. Its fruit has crisp and juicy flesh with high edibility and a good quality.

鲜 食
Table Varieties
品 种

吴县水团枣

品种来源及分布 原产于江苏吴县的洞庭山，为当地原产品种，有小规模经济栽培。

植物学性状 树体较小，树姿半开张，树冠呈圆锥形。主干灰褐色，树皮裂纹粗浅，纵条状。枣头黄褐色，平均长75.6cm，粗0.86cm，节间长8.2cm，蜡质少。针刺不发达，短而少。枣股平均抽生枣吊4.0个。枣吊平均长28.4cm，着叶17片。叶片较小，椭圆形，浅绿色，较光亮。花量多，花序平均着花4朵。花小，花蕾五角形，夜开型。

生物学特性 适应性较强，树势健壮，病虫害少。在原产地，结果较早，丰产稳产。在山西太谷地区，坐果率中等，产量低而不稳，枣头、2～3年和3年以上枝的吊果率分别为2.0%、80.4%和41.6%。4月中旬萌芽，5月下旬始花，9月中旬果实成熟，果实生育期约100d，为中熟品种类型。

果实性状 果实小，倒卵圆形，纵径2.94cm，横径2.37cm，单果重5.5g。果肩较小，平圆。梗洼窄深。果柄细。果顶平圆，顶点略凹。果面光滑。果皮较薄，浅红色。果肉质地较致密，汁液中多，味酸甜，鲜食品质中等。鲜枣可食率93.6%，含可溶性固形物31.20%，总糖23.70%，酸0.41%，100g果肉维生素C含量520.18mg；果皮含黄酮2.71mg/g，cAMP含量276.84μg/g。果核小，椭圆形或倒卵形，核尖突尖，核重0.35g，含仁率83.40%。

评价 该品种在气温较高地区结果早，产量高而稳定。果实较小，肉质较致密，多汁，味甜酸适口，为品质中上的中熟鲜食品种，在南方城郊地区可适当发展。

Wuxianshuituanzao

Source and Distribution The cultivar originated from Dongtingshan of Wuxian in Jiangsu Province. It is commercially cultivated on a small scale.

Botanical Characters The small tree is half-spreading with a conical crown. The grayish-brown trunk bark has thick vertical-striped fissures. The yellowish-brown 1-year-old shoots are 75.6 cm long and 0.86 cm thick. The internodes are 8.2 cm long with less wax and less-developed short thorns. The mother fruiting spurs can germinate 4 deciduous fruiting shoots which are 28.4 cm long with 17 leaves. The small oval-shaped leaves are light green and glossy. There are many small flowers produced averaging 4 ones per inflorescence. The night-bloomed blossoms have five-star-shaped flower buds.

Biological Characters The tree has strong adaptability, strong vigor and light diseases. In the original place, it bears early with a high and stable yield. In Taigu County of Shanxi Province, the tree has medium fruit set, a low and unstable yield. The percentage of fruits to deciduous fruiting shoots of 1-year-old shoots, 2～3-year-old branches and over-3-year-old ones is 2.0%, 80.4% and 41.6%. It germinates in mid-April, begins blooming in late May and ripens in mid-September, with a fruit growth period of 100 d. It is a mid-ripening variety.

Fruit Characteristics The small obovate fruit has a vertical and cross diameter of 2.94 cm and 2.37 cm, averaging 5.5 g. It has a small flat-round shoulder, a narrow and deep stalk cavity, a thin stalk, a flat-round fruit apex, a smooth surface and light-red thin skin. The tight-textured flesh is sour and sweet, with medium juice. It has medium quality for fresh-eating. The percentage of edible part of fresh fruit is 93.6%, and SSC, TTS, TA and Vc is 31.20%, 23.70%, 0.41% and 520.18 mg per 100 g fresh fruit. The content of flavones and cAMP in mature fruit skin is 2.71 mg/g and 276.84 μg/g. The small oval-shaped or obovate stone weighs 0.35 g with a sharply-cuspate apex. The percentage of containing kernels is 83.40%.

Evaluation The cultivar bears early with a high and stable yield in areas with higher temperatures. The small fruit has tight-textured and juicy flesh with a proper SAR. It is a good mid-ripening table variety which can be developed in suburb areas of south China.

鲜 食
Table Varieties
品 种

· 91 ·

交城甜酸枣

品种来源及分布 原产山西省交城县边山的田家山村一带,与交城骏枣品种混栽,为当地主栽品种之一。栽培历史不详。

植物学性状 树体中大,树姿半开张,干性弱,枝条较密,树冠呈半圆形或伞形。主干皮裂条状,较浅,易脱落。枣头红褐色,萌发力较强,平均长46.0cm,粗1.05cm,节间长7.9cm,二次枝长36.2cm,7节左右,弯曲度中等。针刺较发达。皮目大,分布中密,圆形,凸起,开裂,灰白色。枣股较大,抽吊力中等,抽生枣吊2~5个,多为3个,枣吊平均长17.3cm,着叶11片。叶片中大,叶长5.5cm,叶宽2.5cm,绿色,椭圆形,先端急尖,叶基圆楔形,叶缘具锐锯齿。花量中多,枣吊平均着花58.4朵,花序平均4.8朵。花较大,零级花花径7.8mm,昼开型。

生物学特性 树势较强,萌蘖力特强,根蘖苗生长旺。结果较早,较丰产,产量稳定。一般定植第二年开始结果,15年后进入盛果期。坐果率较高,枣头、2~3年生枝和4年生枝的吊果率分别为4.4%、60.4%和32.5%,主要坐果部位在枣吊的5~7节。在山西太谷地区,8月25日前后果实着色,9月中旬成熟,果实生育期100d左右,为中熟品种类型。

果实性状 果个较小,扁圆形,纵径3.08cm,横径2.53cm,单果重9.4g,大小整齐。果皮中厚,紫红色,果面光滑。果顶平,柱头遗存。梗洼中广、中深,果点小,较稀疏,不明显。果肉中厚,白色,肉质较松,味甜酸,汁液中多,品质中等,适宜鲜食。鲜枣可食率93.2%,含可溶性固形物25.80%,总糖25.36%,酸0.87%,100g果肉维生素C含量599.00mg,含水量63.6%;果皮含黄酮5.10mg/g,cAMP含量11.42μg/g。核较大,纺锤形,纵径2.11cm,横径0.87cm,核重0.64g。核面较粗糙,种仁饱满,含仁率58.3%。

评价 该品种适应性较强,较抗枣疯病。萌蘖力特强,易归圃繁殖。结果早,产量较高而稳定。缺点是果个偏小,肉质较粗,品质中等,果核大,可食率偏低。适宜作归圃育苗时抗病砧木使用。

Jiaochengtiansuanzao

Source and Distribution The cultivar originated from Tianjiashancun of Bianshan Village in Jiaocheng County of Shanxi Province. It is a dominant variety there with an unknown history.

Botanical Characters The medium-sized tree is half-spreading with a weak central leading trunk, dense branches and a semi-round or umbrella-shaped crown. The reddish-brown 1-year-old shoots have strong germination ability and medium growth vigor, averaging 46.0 cm long and 1.05 cm thick, with the internodes of 7.9 cm. The secondary branches are 36.2 cm long with 7 nodes and developed thorns. The medium-sized mother fruiting spurs can germinate 2~5 deciduous fruiting shoots which are 17.3 cm long with 11 leaves. The medium-sized leaves are green and oval-shaped. The number of flowers is medium large averaging 58.4 per deciduous fruiting shoot. The large blossoms have a diameter of 7.8 mm for the zero-level flowers.

Biological Characters The tree has strong growth vigor and sprouting ability. The root sprouts grow very strong. It bears early and has a fairly high and stable yield. It generally bears in the 2nd year after planting and enters the full-bearing stage in the 15th year with high fruit set. In Taigu County of Shanxi Province, it enters the crisp-maturing stage in mid-September.

Fruit Characteristics The small fruit is flat round, averaging 9.4 g with a regular size. It has medium-thick and purplish-red skin, a smooth surface. The white flesh is medium-thick and loose-textured, juicy, sweet and sour, with a medium quality for fresh eating. The percentage of edible part of fresh fruit is 93.2%, and SSC, TTS, TA and Vc is 25.80%, 25.36%, 0.87% and 599.00 mg per 100 g fresh fruit. The water content is 63.6%. The large stone is spindle-shaped averaging 0.64 g.

Evaluation The cultivar has strong adaptability, strong resistance to jujube witches' broom and strong suckering ability, bearing early with a high and stable yield. Yet it has a small fruit size, a normal taste and low edibility. It can be used as a disease-resistant rootstock when breeding seedlings in the nursery.

北京坠子白

品种来源及分布 别名小白枣。原产北京，主要分布在海淀区北安河一带。

植物学性状 树体较小，干性强，树姿开张，树冠圆柱形。主干红褐色，树皮裂片小，呈条状。枣头红褐色，平均长76.5cm，粗1.10cm，节间长8.3cm，蜡质少。针刺不发达。枣股圆柱形，平均抽生枣吊4.0个。枣吊平均长18.7cm，着叶10片左右。叶长6.1cm，叶宽3.4cm，叶片椭圆形，浅绿色，光亮，先端渐尖，先端钝尖，叶基截形。叶缘波状，具锐锯齿。花序平均着花5朵。

生物学特性 适应性强，耐旱、耐瘠，在低山阳坡栽培表现良好。树势较强，结果稳定，枣头、2～3年和3年以上枝的吊果率分别为6.7%、21.9%和26.0%。产量一般，对病虫有较强的抗性，受桃小食心虫为害较轻，不易感染枣疯病。在山西太谷地区，9月中旬果实完全成熟，果实生育期90d左右，属早中熟品种类型。

果实性状 果实中大，卵形或倒卵圆形，纵径3.39cm，横径2.71cm，单果重11.7g，大小整齐。果肩平，梗洼中深而广。果梗长2～4mm。果顶平。果皮薄，浅红色。果点小，稀疏。果肉浅绿色，质地较致密，多汁，甜味浓，略酸，鲜食品质中上。100g果肉维生素C含量394.68mg；果皮含黄酮6.01mg/g，cAMP含量16.09μg/g。果核椭圆形或纺锤形。核内大多含有种子，含仁率90.0%。

评价 该品种树势强健，抗逆性强，病虫危害轻。果实中大，皮薄，肉质脆嫩，多汁味甜，为较优良的早中熟鲜食品种。

Beijingzhuizibai

Source and Distribution The cultivar, also called Xiaobaizao, originated from and spreads in Haidian District of Beijing City.

Botanical Characters The large tree is spreading with a strong central leader trunk and a column-shaped crown. The reddish-brown trunk bark has small massive fissures. The yellowish-brown 1-year-old shoots are 76.5 cm long and 1.10 cm thick. The internodes are 8.3 cm long with less wax and no thorns. The column-shaped mother fruiting spurs can germinate 4 deciduous fruiting shoots which are 18.7 cm long with 10 leaves. The oval-shaped leaves are green and glossy, 6.1 cm long and 3.4 cm wide with a gradually-cuspate apex, a truncate base and a sharp saw-tooth pattern on the margin. Each inflorescence has 5 flowers.

Biological Characters The tree has strong adaptability, tolerant to drought and poor soils. It has strong vigor with a medium and stable yield. The percentage of fruits to deciduous fruiting shoots of 1-year-old shoots, 2～3-year-old branches and over-3-year-old ones is 6.7%, 21.9% and 26.0%. It has strong resistance to disease and pests, and it is unsusceptible to jujube witches' broom. In Taigu County of Shanxi Province, it matures in mid-September with a fruit growth period of about 90 d. It is an early-mid-ripening variety.

Fruit Characteristics The medium-sized fruit is oval-shaped or obovate, with a vertical and cross diameter of 3.39 cm and 2.71 cm, averaging 11.7 g with a regular size. It has a flat shoulder, a medium-deep and wide stalk cavity, a stalk of 2～4 mm long, a flat fruit apex and light-red thin skin with small and sparse dots. The light-green flesh is tight-textured, juicy, strongly sweet and a little sour. It has a better than normal quality for fresh eating. The content of Vc in fresh fruit is 394.68 mg per 100 g fresh fruit. The content of flavones and cAMP in mature fruit skin is 6.01 mg/g and 16.09 μg/g. It has an oval-shaped or spindle-shaped stone. Most stones contain kernels, the percentage of which is 90.0%.

Evaluation The cultivar has strong tree vigor and strong resistance to adverse conditions. It is unsusceptible to diseases and pests. The medium-sized fruit has thin skin, crisp, juicy and sweet flesh. It is a good table variety.

溆浦葫芦枣

品种来源及分布 原产于湖南溆浦的双影岩元村。

植物学性状 树体较小,树姿开张,下层主枝平展,干性较强,枝系较稀,树冠呈圆锥形。主干黑褐色,纵条裂,树皮呈小块状剥落。枣头褐色,阳面被有灰色蜡质,平均枝长60cm左右,节间长5.3～10.1cm。二次枝3～5节,最多8节。针刺较发达。枣股圆柱形,抽吊力中等,平均抽生枣吊3.2个。枣吊短小,平均长11.4cm,着叶10片。叶片小,卵圆形,浓绿色,先端渐尖,叶基圆形,叶缘锯齿较细。花量少,花序平均着花2.7朵,昼开型。

生物学特性 树势中庸,发枝力中等,枝叶稀疏。坐果率低,2～3年和3年以上枝的吊果率分别为31.7%和19.1%。生理落果严重,可达坐果数的80%左右,产量低,不稳定。在原产地,9月上旬果实成熟,果实生育期90d左右,为早熟品种类型。成熟期遇雨易裂果。

果实性状 果实小,圆锥形,偏果顶处有明显环形缢痕,呈乳头状,纵径2.40～2.70cm,横径1.90～2.10cm,单果重4.4g,大果重5.6g,大小较整齐。果实两端膨大,中腰凹陷。梗洼窄深,果顶略凹。果皮中厚,紫红色。果肉浅黄色,质地松,稍脆,汁液较多,味甜,鲜食品质上等。鲜枣可食率91.8%。果核较大,纺锤形,核重0.36g,含仁率40%左右。

评价 该品种树势中庸,树冠成形慢,结果晚,产量低。鲜食品质较好,但果实小,落果重,抗裂果能力差。

Xupuhuluzao

Source and Distribution The cultivar originated from Shuangyingyanyuan Village of Xupu in Hunan Province.

Botanical Characters The small tree is spreading with a strong central leader trunk, sparse branches and a conical crown. The blackish-brown trunk bark has vertical massive fissures. The brown 1-year-old shoots are 60.0 cm long with gray wax on the sun-side. The internodes are 5.3～10.1 cm long with developed thorns. Generally there are 3～5 nodes (maximum 8) on each secondary branch. The column-shaped mother fruiting spurs can germinate 3.2 deciduous fruiting shoots which are 11.4 cm long with 10 leaves. The small leaves are oval-shaped and dark green, with a gradually-cuspate apex, a round base and a thin saw-tooth pattern on the margin. The number of flowers is small, averaging 2.7 ones per inflorescence. It blooms in the daytime.

Biological Characters The tree has moderate vigor and medium branching ability with low fruit set. The percentage of fruits to deciduous fruiting shoots of 2～3-year-old branches and over-3-year-old ones is 31.7% and 19.1%. It has serious physiological fruit-dropping, which occupies 80% of the total fruit set, so the yield is low and unstable. In the original place, it matures in early September, with a fruit growth period of 90 d. It is an early-ripening variety with serious fruit-cracking if it rains in the maturing stage.

Fruit Characteristics The small conical fruit has a vertical and cross diameter of 2.40～2.70 cm and 1.90～2.10 cm respectively, averaging 4.4 g (maximum 5.6 g) with a regular size. It is enlarged at both ends and sunken in the middle part. It has a narrow and deep stalk cavity, a slightly-sunken fruit apex and medium-thick and purplish-red skin. The light-yellow flesh is loose-textured, crisp and sweet with less juice. It has a good quality for fresh-eating. The percentage of edible part of fresh fruit is 91.8%. The big spindle-shaped stone weighs 0.36 g with the percentage of containing kernels of 40%.

Evaluation The cultivar has weak vigor. The canopy grows slowly and bears late with a low yield. The fruit has a good quality for fresh-eating. Yet it has a small size, serious fruit-dropping and poor resistance to fruit-cracking.

榆次牙枣

品种来源及分布 原产山西省晋中市榆次区的东赵、西赵、小白、训峪等地。分布广，但数量较少，以东赵村较为集中。栽培历史不详。

植物学性状 树体高大，树姿开张，枝条较密，干性弱，中心干有自然弯曲生长现象，树冠乱头形。主干皮裂呈条状。枣头红褐色，萌发力较强，平均长80.9cm，粗0.83cm，节间长9.5cm，着生永久性二次枝5～6个，二次枝长24.3cm，5～6节。针刺细而较长。枣股中大，圆锥形，抽吊力中等，平均抽生枣吊3.4个。枣吊平均长20.9cm，着叶12片。叶片中大，叶长6.6cm，叶宽3.2cm，椭圆形，浓绿色，先端急尖，叶基偏斜，叶缘具钝齿。花量中多或较少，花较大，枣吊平均着花45.5朵，花序平均4.3朵。

生物学特性 树势强健，萌蘖力特强，产量较高，且稳定。枣头、2～3年和4年生以上枝吊果率分别为为6.3%、44.8%和29.6%。在山西太谷地区，8月下旬果实着色，9月中旬成熟，果实生育期100～110d，属中熟品种类型。采前落果较严重，遇雨易裂果。

果实性状 果实较大，圆柱形，纵径4.18cm，横径3.11cm，单果重16.3g。果梗短，中粗，梗洼窄而深。果顶平，柱头遗存，不明显。果皮薄，红色，果面光滑，果点小而密，不明显。果肉中厚，绿白色，肉质细而松，味甜，汁液中，品质上等，适宜鲜食。鲜枣可食率94.4%，含可溶性固形物31.80%，总糖23.90%，酸0.55%，100g果肉维生素C含量373.26mg；果皮含黄酮4.04mg/g，cAMP含量142.77μg/g。果核较大，纺锤形，纵径2.40cm，横径0.80cm，核重0.92g，核纹较深，核面较粗糙，含仁率8.3%，但种仁不饱满。

评价 该品种适应性强，耐干旱，产量较高。果实品质优良，但果核大，采前落果严重，易裂果。

Yuciyazao

Source and Distribution The cultivar originated from and widely spreads in Dongzhao, Xizhao, Xiaobai and Xunyu in Yuci City of Shanxi Province, with a small quantity, mainly planted in Dongzhao Village with an unknown history.

Botanical Characters The large tree is spreading with dense branches, a weak central leader trunk and an irregular crown. The central leader trunk has a natural curling growth habit. The trunk bark has striped fissures. The reddish-brown 1-year-old shoots have strong germination ability, averaging 80.9 cm long and 0.83 cm thick with the internodes of 9.5 cm. There are generally 5～6 secondary branches on each 1-year-old shoot, which are 24.3 cm long with 5～6 nodes and thin long thorns. The medium-sized conical mother fruiting spurs can germinate 3.37 deciduous fruiting shoots, which are 20.9 cm long with 12 leaves. The medium-sized leaves are oval-shaped and dark green. The number of flowers is medium large, averaging 45.5 ones per deciduous fruiting shoot and 4.3 ones per inflorescence.

Biological Characters The tree has strong vigor and strong suckering ability with a high and stable yield. The percentage of fruits to deciduous fruiting shoots of 1-year-old shoots, 2 or 3-year-old branches and over-3-year-old ones is 6.3%, 44.8% and 29.6%. In Taigu County of Shanxi Province, it matures in mid-September. It is a mid-ripening variety with serious premature fruit-dropping and fruit-cracking if it rains.

Fruit Characteristics The medium-sized fruit is column-shaped with a vertical and cross diameter of 4.18 cm and 3.11 cm, averaging 16.3 g. It has a short and medium-thick stalk, a narrow and deep stalk cavity, a flat fruit apex, a remnant yet indistinct stigma, thin red skin and a smooth surface with small, dense and indistinct dots. The greenish-white flesh is medium thick and loose-textured, delicate, sweet and juicy, with an excellent quality for fresh eating. The percentage of edible part of fresh fruit is 94.4%, and SSC, TTS, TA and Vc is 31.80%, 23.90%, 0.55% and 373.26 mg per 100 g fresh fruit. The content of flavones and cAMP in mature fruit skin is 4.04 mg/g and 142.77 μg/g. The large spindle-shaped stone has a vertical and cross diameter of 2.40 cm and 0.80 cm, averaging 0.92 g with deep veins, a rough surface and a shriveled kernel. The percentage of containing kernels is 8.3%.

Evaluation The cultivar has strong adaptability, strong tolerance to drought and frost with a medium yield and a good quality. Yet it has a large stone with low edibility, serious premature fruit-dropping and fruit-cracking.

鲜 食
Table Varieties
品 种

· 99 ·

南京冷枣

品种来源及分布 别名冷枣。原产江苏南京郊区。为当地原产的地方品种，江南地区的鲜食良种，多零星栽植。栽培历史不详。

植物学性状 树体中等大，树姿开张，枝条细而较软，易弯曲下垂，树冠圆头形。主干皮裂条状，细而深，不易剥落。枣头紫褐色，平均长65.4cm，粗度0.75cm，节间长6.2cm。二次枝长22.0cm，平均生长6节。针刺不发达。皮目较大，微凸，分布较密。枣股较细而长，圆柱形，抽吊力强，抽生枣吊4～5个。枣吊平均长24.2cm，着叶17片，易发生二次生长。叶片较小，叶长5.0cm，宽2.2cm，椭圆形，浓绿色，先端急尖，叶基圆楔形，叶缘锯齿钝。花量多，花序平均着花8.5朵。花小，花径6mm左右，昼开型。

生物学特性 树势较弱，萌芽率和成枝力强。结果早，产量较高而稳定。枣头、2～3年生枝和4年生以上枝的吊果率分别为4.0%、72.5%和21.1%。在山西太谷地区，9月中旬果实成熟，果实生育期100～110d，属中熟品种类型。果实抗裂果和抗病性较强。

果实性状 果个小，圆柱形，纵径2.80cm，横径2.20cm，单果重6.0g，大小较整齐。果梗细而短，梗洼窄而较深。果肩稍宽大，广圆或平圆，有5～6条明显沟棱。果顶微凸，柱头遗存。果面平滑，果点小而密，圆形，浅黄色，不明显。果皮薄，浅红色。果肉厚，绿白色，肉质细而脆嫩，甜味浓，汁液多，口感较好，适宜鲜食。鲜枣可食率93.2%，100g果肉维生素C含量364.80mg。核小，纺锤形，纵径1.80～2.20cm，横径0.60～0.70cm，核重0.41g，核纹浅，含仁率高，可达98.4%。

评价 该品种树势较弱，树姿开张，易发枝，适宜密植栽培。为早果、丰产、优质、抗病和抗裂果的中熟鲜食良种，可在江南高温多雨地区发展栽培。

Nanjinglengzao

Source and Distribution The cultivar originated from the suburban areas of Nanjing in Jiangsu Province. It is the dominant table variety there, and is sparsely cultivated. The cultural history is unknown.

Botanical Characters The medium-sized tree is spreading with a round crown and thin and soft branches, easily bending. The trunk bark has thin, deep and striped fissures, uneasily shelled off. The purplish-brown 1-year-old shoots are 65.4 cm long and 0.75 cm thick, with the internodes of 6.2 cm. The secondary branches are 22.0 cm long with 6 nodes and less-developed thorns. The column-shaped mother fruiting spurs are thin and long, and can germinate 4～5 deciduous fruiting shoots, which are 24.2 cm long with 17 leaves. Secondary growth easily appears on the deciduous fruiting shoots. The small leaves are dark green and oval-shaped, 5.0 cm long and 2.2 cm wide with a sharply-cuspate apex, a round-cuneiform base and a sharp saw-tooth pattern on the margin. There are many small flowers with a diameter of 6 mm produced, averaging 8.5 ones per inflorescence. It blooms in the daytime.

Biological Characters The cultivar has weak vigor, strong germination and branching ability, bearing early with a high and stable yield. In Taigu County of Shanxi Province, it matures in mid-September with a fruit growth period of 100～110 d, It is a mid-ripening variety with strong resistance to fruit-cracking and diseases.

Fruit Characteristics The small column-shaped fruit has a vertical and cross diameter of 2.80 cm and 2.20 cm, averaging 6.0 g with a regular size. It has a thin and short stalk, and a narrow and deep stalk cavity, a wide-round shoulder with 5～6 distinct ribs, a slightly raised fruit apex, a remnant stigma, a smooth surface and light-red thin skin with small, round, light-yellow, dense and indistinct dots. The greenish-white thick flesh is delicate, juicy and sweet, with a good taste for fresh eating. The percentage of edible part of fresh fruit is 93.2%, and the content of Vc is 364.80 mg per 100 g fresh fruit. The small spindle-shaped stone has a vertical and cross diameter of 1.80～2.20 cm and 0.60～0.70 cm, averaging 0.41 g with shallow veins. The percentage of containing kernels is 98.4%.

Evaluation The cultivar has weak vigor, spreading with strong branching ability. Close planting is well fit for this cultivar. It is a good mid-ripening table variety with early-fruiting character, high productivity, strong resistance to diseases and fruit-cracking. It can be planted in areas along the Yangtze River with higher temperature and much rainfall.

鲜 食
Table Varieties
品 种

· 101 ·

天津二秋枣

品种来源及分布 别名二秋枣。原产于天津西郊的张窝、傅村、木厂等地，为当地主栽品种之一。栽培历史悠久。

植物学性状 树体中大，树姿半开张，干性较弱，树冠呈伞形。主干灰褐色，主干皮裂窄条状，较易剥落。枣头黄褐色，粗壮，平均长82.3cm，粗1.06cm，节间长7.5cm，蜡质少。二次枝平直，平均长27.5cm，6节左右。针刺不发达，易脱落。枣股圆柱形，平均抽生枣吊3.8个。枣吊长19.7cm，着叶11片。叶片窄小，卵状披针形，浅绿色，叶面有光泽，先端渐尖，叶基圆楔形，叶缘具浅钝锯齿。花多，枣吊中部花序着花7朵。花为昼开型。

生物学特性 适应性强，耐旱、耐涝、耐盐碱。树势中等，树体健壮，丰产，结果稳定，枣头、2～3年和3年以上枝的吊果率分别为10.0%、101.9%和9.2%。定植后3年开始结果，15年左右进入盛果期，成龄树一般株产60kg左右。在山西太谷地区，4月中旬萌芽，5月底开花，9月20日左右成熟采收，果实生育期110d，为中熟品种类型。成熟期遇雨裂果较严重，裂纹小而密。

果实性状 果实小，圆柱形，纵径2.54cm，横径2.14cm，单果重5.9g，大小较整齐。果肩平，梗洼中深广。果面光滑，皮薄，浅红色。果点小，较稀疏。果肉绿色，质地较致密，汁液多，味甜，香味浓郁，鲜食品质中等。鲜枣可食率95.9%，含可溶性固形物29.40%，含总糖21.80%，酸0.40%，100g果肉维生素C含量473.79mg；果皮含黄酮3.61mg/g，cAMP含量166.90μg/g。果核细小，纺锤形，平均重0.24g。核内多有不饱满种子。

评价 该品种适应性强，耐旱涝和盐碱。树体中大，产量高而稳定。果实小，味甜多汁，核小，品质中等。在成熟期少雨地区栽培。

Tianjinerqiuzao

Source and Distribution The cultivar originated from Zhangwo, Fucun and Muchang Villages of western Tianjin. It is a dominant variety there with a long cultivation history.

Botanical Characters The medium-sized tree is half-spreading with a weak central leader trunk and an umbrella-shaped crown. The grayish-brown trunk bark has narrow striped fissures, easily shelled off. The yellowish-brown 1-year-old shoots are 82.3 cm long and 1.06 cm thick. The internodes are 7.5 cm long, with less wax and less-developed thorns. The straight secondary branches are 27.5 cm long with 6 nodes. The column-shaped mother fruiting spurs can germinate 3.8 deciduous fruiting shoots, which are 19.7 cm long with 11 leaves. The small narrow leaves are ovate-lanceolate, light green and glossy. There are many flowers produced, averaging 7 ones per inflorescence. It blooms in the daytime.

Biological Characters The tree has strong adaptability, tolerant to drought, water-logging and saline-alkaline with moderate vigor, a high and stable yield. It generally bears in the 3rd year after planting and enters the full-bearing stage in the 15th year. A mature tree has a yield of 60 kg on average. In Taigu County of Shanxi Province, it germinates in mid-April, begins blooming in late May, begins coloring in late August and ripens in early September with a fruit growth period of 85 d. It is an extremely early-ripening variety with serious fruit-cracking if it rains in the maturing stage.

Fruit Characteristics The small column-shaped fruit has a vertical and cross diameter of 2.54 cm and 2.14 cm, averaging 5.9 g with a regular size. It has a flat shoulder, a medium-deep and wide stalk cavity, a smooth surface and light-red thin skin with small and sparse dots. The green flesh is tight-textured, strongly sweet and juicy. It has medium quality for fresh-eating. The percentage of edible part of fresh fruit is 95.9%, and SSC, TTS, TA and Vc is 29.40%, 21.80%, 0.40% and 473.79 mg per 100 g fresh fruit. The content of flavones and cAMP in mature fruit skin is 3.61 mg/g and 166.90 μg/g. The thin and small spindle-shaped stone weighs 0.24 g. Most stones contain a shriveled kernel.

Evaluation The medium-sized plant of Tianjinerqiuzao cultivar has strong adaptability, tolerant to drought, water-logging and saline-alkaline with a high and stable yield. The small fruit has sweet, crisp and juicy flesh with a small stone. It has a good quality for fresh-eating.

鲜 食
Table Varieties
品 种

· 103 ·

新 郑 酥 枣

品种来源及分布 原产河南省新郑市的辛店、观音寺等地，多零星栽培，数量较少。为当地原有的古老品种。

植物学性状 树体较高大，树姿半开张，树冠自然圆头形。主干灰褐色，树皮宽条纵裂，裂纹深，不易剥落。枣头红褐色，平均长88.1cm，粗1.28cm，节间长6.1cm。针刺不发达。枣股圆柱形，平均抽生枣吊4.0个。枣吊平均长23.5cm，着叶16片。叶片较小，叶长5.2cm，叶宽2.6cm，卵圆形，深绿色，先端钝尖，叶基圆形，叶缘具不整齐的钝锯齿。花量少，花小，花序平均着花2朵。

生物学特性 树体健壮，树势较强，发枝力中等。结果早，但产量较低，枣头、2~3年生枝的吊果率分别为38.3%和43.3%。在山西太谷地区，10月上旬果实成熟采收，果实生育期120天，为极晚熟品种类型。成熟期遇雨易裂果。

果实性状 果实较小，长圆形，纵径3.31cm，横径2.31cm，单果重8.3g，大小整齐。果肩平，梗洼中深、广，果顶平圆。果面平滑。果皮浅红色。果点大而稀疏。果肉浅绿，质地致密，汁液多，味甜，品质好，适宜鲜食。鲜枣可食率95.7%，含可溶性固形物33.00%，总糖27.00%，酸0.65%。果核纺锤形，核重0.36g，含仁率86.7%。

评价 该品种适应性强，树体健壮，产量较低。果实较小，质脆味甜，品质上等，为优良的鲜食品种。但易裂果，可在成熟期少雨的地区适当发展。

Xinzhengsuzao

Source and Distribution The cultivar originated from Xindian and Guanyinsi of Xinzheng City in Henan Province. It is planted in scattered regions with a small quantity.

Botanical Characters The large tree is half-spreading with a round crown. The grayish-brown trunk bark has deep vertical wide-striped fissures, uneasily shelled off. The reddish-brown 1-year-old shoots are 88.1 cm long and 1.28 cm thick. The internodes are 6.1 cm long with developed thorns. The column-shaped mother fruiting spurs can germinate 4 deciduous fruiting shoots which are 23.5 cm long with 16 leaves. The small leaves are 5.2 cm long and 2.6 cm wide, oval-shaped and dark green, with a bluntly-cuspate apex, a round base and an irregular blunt saw-tooth pattern on the margin. Both the number and size of flowers are small, averaging 2 ones per inflorescence.

Biological Characters The tree has strong vigor and medium branching ability, bearing early with a low yield. The percentage of fruits to deciduous fruiting shoots of 1-year-old shoots and 2~3-year-old branches is 38.3% and 43.3%. In Taigu County of Shanxi Province, it matures in early October with a fruit growth period of 120 d. It is an extremely late-ripening variety with serious fruit-cracking if it rains in the maturing stage.

Fruit Characteristics The small column-shaped fruit has a vertical and cross diameter of 3.31 cm and 2.31 cm, averaging 8.3 g with a regular size. It has a flat shoulder, a medium-deep and wide stalk cavity, a pointed fruit apex, a smooth surface and light-red skin with large and sparse dots. The light-green flesh is tight-textured, sweet and juicy. It has a good quality for fresh-eating. The percentage of edible part of fresh fruit is 95.7%, and SSC, TTS and TA is 33.00%, 27.00% and 0.65%. The spindle-shaped stone weighs 0.36 g, with the percentage of containing kernels of 86.7%.

Evaluation The cultivar has strong adaptability and strong vigor, yet with a low yield. The large fruit has crisp and sweet flesh with a good quality for fresh-eating. Yet it has poor resistance to fruit-cracking, so it can only be developed in areas with light rainfall in the maturing stage.

新郑九月青

品种来源及分布　原产于河南省新郑市。

植物学性状　树体较大，树姿直立，干性强，树冠呈伞形。主干皮裂条状。枣头紫褐色，平均生长量77.0cm，节间长7.1cm，蜡质少。二次枝平均长22.0cm，5～7节，弯曲度中等。无针刺。枣股平均抽生枣吊4个，多者6个。枣吊长16.2cm，着叶12片。叶片椭圆形，平展，先端急尖，叶基圆楔形，叶缘具钝锯齿。花量较多，花序平均着花5朵。花小，花径5.6cm。

生物学特性　树势旺，发枝力强。结果晚，定植后第三年结果，10年后开始大量结果，产量极低，枣吊平均结果0.08个。在山西太谷地区，10月上旬果实成熟采收，果实生育期116d左右，为极晚熟品种类型。

果实性状　果实中大，圆锥形，纵径3.63cm，横径2.67cm，单果重10.0g，大小较整齐。果皮薄，浅红色，果面光滑。梗洼窄、中深。果顶平圆，柱头脱落。果肉浅绿色，质地较致密，汁多，味甜，品质优异，适宜鲜食。鲜枣可食率96.0%。核小，纺锤形，平均重0.40g。核内含有较饱满的种子，含仁率90.0%左右。

评价　该品种树体较大，树势强，结果年龄晚，产量极低，成熟期极晚。果实中大，品质优异，适宜鲜食。花期需开甲环剥、喷施微肥等促进坐果，提高产量。

Xinzhengjiuyueqing

Source and Distribution　The cultivar originated from Xinzheng City of Henan Province.

Botanical Characters　The large tree is vertical with a strong central leader trunk and an umbrella-shaped crown. The trunk bark has striped fissures. The purplish-brown 1-year-old shoots are 77.0 cm long with the internodes of 7.1 cm with less wax and no thorns. The secondary branches are 22.0 cm long with 5～7 nodes of medium curvature. The mother fruiting spurs can germinate 4 deciduous fruiting shoots (6 at most) which are 16.2 cm long with 12 leaves. The flat leaves are oval-shaped, with a sharply-cuspate apex, a round-cuneiform base and a blunt saw-tooth pattern on the margin. There are many small flowers with a diameter of 5.6 mm produced, averaging 5 ones per inflorescence.

Biological Characters　The tree has strong vigor and strong branching ability. It bears late, generally in the 3rd year after planting and enters the full-bearing stage in the 10th year with a very low yield. There are only 0.08 fruits on each deciduous fruiting shoot. In Taigu County of Shanxi Province, it matures in early October with a fruit growth period of 116 d. It is an extremely late-ripening variety.

Fruit Characteristics　The medium-sized fruit is conical, with a vertical and cross diameter of 3.63 cm and 2.67 cm, averaging 10.0 g with a regular size. It has light-red thin skin, a smooth surface, a narrow and medium-deep stalk cavity, a flat-round fruit apex and a falling-off stigma. The light-green flesh is tight-textured, sweet and juicy. It has an excellent quality for fresh-eating. The percentage of edible part of fresh fruit is 96.0%. The small spindle-shaped stone weighs 0.40 g with a well-developed kernel. The percentage of containing kernels is 90.0%.

Evaluation　The large plant of Xinzhengjiuyueqing cultivar has strong vigor, bearing late with a very low yield. The medium-sized fruit has an excellent quality for fresh eating. Girdling and spraying microelement fertilizers should be carried out in blooming period to promote fruit set and to improve the yield.

永济鸡蛋枣

品种来源及分布 原产山西永济的仁阳村，栽培数量不多，零星分布于农家宅院。

植物学性状 树体较大，树姿半开张，树冠圆柱形。主干皮裂条状。枣头长73.7cm，粗0.85cm，节间长6.6cm，蜡质多。二次枝平均长24.4cm，5节，弯曲度中等。针刺发达。枣股平均抽生枣吊4.0个。枣吊平均长28.5cm，着叶20片。叶片椭圆形，深绿，合抱，较光亮，先端钝尖，叶基圆楔形。叶缘具钝锯齿。花序平均着花4朵。

生物学特性 树势中庸，成枝力强，枣头生长势旺盛。坐果率极低，枣头、2~3年和3年以上枝的吊果率分别为1.0%、2.6%和4.9%。丰产稳产性差。在山西太谷地区，9月下旬果实成熟，果实生育期110d，为晚熟品种类型。成熟期遇雨易裂果。

果实性状 果实特大，卵圆形，纵径4.45cm，横径3.39cm，单果重30.8g。果皮薄，紫红色，果面有隆起。果肩平，梗洼深，中广。果点小而密。果顶平圆，柱头宿存。果肉浅绿色，质地酥脆，汁液多，甜味浓，鲜食品质中上。鲜枣可食率95.9%，含可溶性固形物30.80%，总糖28.97%，酸0.30%，100g果肉维生素C含量361.14mg。果核大，纺锤形，核重1.26g。含仁率11.8%。

评价 果个特大，鲜食品质中上，丰产性差，成熟期遇雨易裂果。

Yongjijidanzao

Source and Distribution The cultivar originated from Renyang Village of Yongji County in Shanxi Province. It is mainly planted in family garden.

Botanical Characters The large tree is half-spreading with a column-shaped crown. The trunk bark has striped fissures. The 1-year-old shoots are 73.7 cm long and 0.85 cm thick. The internodes are 6.6 cm long with much wax and developed thorns. The secondary branches are 24.4 cm long with 5 nodes of medium curvature. The mother fruiting spurs can germinate 4 deciduous fruiting shoots, which are 28.5 cm long with 20 leaves. The oval-shaped leaves are dark green and glossy, with a bluntly-cuspate apex, a round-cuneiform base and a blunt saw-tooth pattern on the margin. Each inflorescence has 4 flowers.

Biological Characters The tree has moderate vigor and strong branching ability, with strong growth potential for the 1-year-old shoots. Yet it has very low fruit set and a poor yield. The percentage of fruits to deciduous fruiting shoots of 1-year-old shoots, 2~3-year-old branches and over-3-year-old ones is 1.0%, 2.6% and 4.9%. In Taigu County of Shanxi Province, it matures in late September, with a fruit growth period of 110 d. It is a mid-ripening variety with serious fruit-cracking if it rains in the maturing stage.

Fruit Characteristics The large oval-shaped fruit has a vertical and cross diameter of 4.45 cm and 3.39 cm averaging 30.8 g. It has purplish-red thin skin with some protuberances, a flat shoulder, a deep and medium-wide stalk cavity, small and dense dots, a flat-round fruit apex and a remnant stigma. The light-green flesh is crisp, juicy and strongly sweet. It has a better than normal quality for fresh-eating. The percentage of edible part of fresh fruit is 95.9%, and SSC, TTS, TA and Vc is 30.80%, 28.97%, 0.30% and 361.14 mg per 100 g fresh fruit. The percentage of containing kernels inside the spindle-shaped stones is 11.8%.

Evaluation The cultivar has a very large fruit size and a good quality for fresh-eating. Yet it has poor productivity, and fruit-cracking easily occurs if it rains in the maturing stage.

鲜食
Table Varieties
品种

献县绵枣

品种来源及分布 别名铃枣、脆枣、绵枣。主要分布在河北沧州地区的献县、沧县等地，多与金丝小枣混植。山东乐陵、庆云、无棣和崂山一带也有分布，多零星栽培。

植物学性状 树体中大，树姿半开张，树冠自然圆头形。主干灰褐色，表面粗糙，条状皮裂，不易剥落。枣头红褐色，平均长56.1cm，粗1.07cm，节间长6.4cm。无针刺。二次枝平均长27.9cm，6节，弯曲度小。枣股大，圆柱形，平均抽生枣吊4个。枣吊平均长19.7cm，着叶14片。叶长5.6cm，叶宽2.5cm，叶片椭圆形，绿色或深绿色，较光亮，先端渐尖，叶基圆楔形，叶缘波形，具钝锯齿。花多，花序平均着花8朵。花较小，夜开型。

生物学特性 树势中庸偏弱，发枝力中等。结果较早，定植后2年开始结果。坐果不稳定，产量较低，枣头、2～3年生枝的吊果率分别为13.1%和11.2%。6年生树株产3kg左右，盛果期株产40kg左右。在山西太谷地区，4月中旬萌芽，6月上旬始花，10月上旬果实成熟，果实生育期120d，为极晚熟品种类型。抗裂果能力强。

果实性状 果实小，卵圆形，纵径2.76cm，横径1.92cm，单果重5.5g，大小整齐。果肩平圆。梗洼深，中等广。果顶微凹。果皮厚，平滑光洁，红色，外形美观。果点细小，不明显。果肉浅绿色，质地致密，汁液少，味甜，鲜食品质中等。鲜枣可食率92.7%，含可溶性固形物34.80%，总糖27.61%，酸0.91%，100g果肉维生素C含量522.56mg。核细小，纺锤形，核重0.40g。多数果核含不饱满种子，含仁率45.0%。

评价 该品种适应性强，结果早，抗裂果，但产量较低，果实小，品质一般。

Xianxianmianzao

Source and Distribution The cultivar, also called Lingzao or Cuizao, mainly spreads in Xianxian County of Hebei Province. It is cultivated in mixture with 'Jinsixiaozao'. There are also some scattered distributions in Leling, Qingyun, Wudi and Laoshan in Shandong Province.

Botanical Characters The medium-sized tree is half-spreading with a round crown. The grayish-brown trunk bark has rough striped fissures, uneasily shelled off. The reddish-brown 1-year-old shoots are 56.1 cm long and 1.07 cm thick. The internodes are 6.4 cm long without thorns. The secondary branches are 27.9 cm long with 6 nodes of small curvature. The column-shaped mother fruiting spurs can germinate 4 deciduous fruiting shoots which are 19.7 cm long with 14 leaves. The green or dark-green leaves are oval-shaped and glossy, 5.6 cm long and 2.5 cm wide with a gradually-cuspate apex, a round-cuneiform base and a blunt saw-tooth pattern on the wavy margin. There are many small flowers produced, averaging 8 ones per inflorescence. It blooms at night.

Biological Characters The tree has moderate vigor and medium branching ability. It bears early, generally in the 2nd year after planting, with stable fruit set and a low yield. The percentage of fruits to deciduous bearing shoots of 1-year-old shoots and 2～3-year-old branches is 13.1% and 11.2%. A 6-year-old tree has a yield of 3 kg on average, and a tree in its full-bearing stage can produce 40 kg of fresh fruits. In Taigu County of Shanxi Province, it germinates in mid-April, begins blooming in early June and matures in early October with a fruit growth period of 120 d. It is an extremely late-ripening variety with strong resistance to fruit-cracking.

Fruit Characteristics The small ovoid-shaped fruit has a vertical and cross diameter of 2.76 cm and 1.92 cm, averaging 5.5 g with a regular size. It has a flat-round shoulder, a deep and medium-wide stalk cavity, a slightly-sunken fruit apex and a thick, red, smooth and attractive surface. The light-green flesh is tight-textured and sweet, with less juice. It has medium quality for fresh-eating. The percentage of edible part of fresh fruit is 92.7%, and SSC, TTS, TA and Vc is 34.80%, 27.61%, 0.91% and 522.56 mg per 100 g fresh fruit. The percentage of fresh fruits which can be made into dried ones is 43.3%, and the content of TTS, TA and Vc in dried fruit is 64.44%, 0.90% and 66.44 mg per 100 g fresh fruit. The thin small spindle-shaped stone weighs 0.4 g. Most kernels are shriveling, and the percentage of containing kernels is 45.0%.

Evaluation The cultivar has strong adaptability, bearing early with a low yield. The fruit has strong resistance to fruit-cracking. Yet it has a small size and medium quality.

鲜 食
Table Varieties
品 种

清苑大丹枣

品种来源及分布 分布河北保定市清苑等县（市）。

植物学性状 树体较小，树姿开张，干性较弱，树冠呈圆柱形。主干条状皮裂。枣头黄褐色，生长量71.0cm，节间长7.3cm，二次枝长29.4cm，5～7节。无针刺。枣股平均抽生枣吊4.0个。枣吊长22.3cm，着叶14片。叶片椭圆形，绿色，先端急尖，叶基偏斜形，叶缘具钝锯齿。花量少，花序平均着花4朵。

生物学特性 树势中等，成枝力较强。开花结果早，定植第二年普遍结果，5年生左右进入盛果期。该品种成熟期早，在山西太谷地区，8月下旬果实着色，9月上旬成熟采收，果实生育期95d左右，为早熟品种类型。

果实性状 果实中大，短柱形或近圆形，纵径2.82cm，横径2.55cm，果重10.2g左右，大小整齐。梗洼中深、广。果顶平，柱头遗存。果皮薄，浅红色，果面光滑。肉质酥脆，味甜，汁液多，品质上等，适宜鲜食。鲜枣可食率95.5%，含可溶性固形物29.00%，总糖23.59%，酸0.48%，100g果肉维生素C含量343.04mg；果皮含黄酮14.69mg/g，cAMP含量16.92μg/g。果核椭圆形，核重0.46g，核内种仁较饱满，含仁率86.7%。

评价 该品种树体小，结果早，品质较好，是优良的早熟鲜食品种。

Qingyuandadanzao

Source and Distribution The cultivar spreads in Qingyuan County of Baoding City in Hebei Province.

Botanical Characters The small tree is spreading with a weak central leader trunk and a column-shaped crown. The trunk bark has striped fissures. The yellowish-brown 1-year-old shoots are 71.0 cm long with the internodes of 7.3 cm. The secondary branches are 29.4 cm long with 5～7 nodes and without thorns. The mother fruiting spurs can germinate 4 deciduous fruiting shoots which are 22.3 cm long with 14 leaves. The green oval-shaped leaves have a sharply-cuspate apex, a deflective base and a blunt saw-tooth pattern on the margin. The number of flowers is small, averaging 4 ones per inflorescence.

Biological Characters The tree has strong vigor and strong branching ability, bearing early (generally in the 2nd year after planting) and entering the full-bearing stage in the 5th year. In Taigu County of Shanxi Province, it begins coloring in late August and is harvested in early September with a fruit growth period of 95 d. It is an extremely early-ripening variety.

Fruit Characteristics The medium-sized oblong fruit has a vertical and cross diameter of 2.82 cm and 2.55 cm, averaging 10.2 g with a regular size. It has a medium-deep and wide stalk cavity, a flat fruit apex, a remnant stigma, a smooth surface and light-red thin skin. The crisp flesh has a sweet taste and much juice, with a superior quality for fresh-eating. The percentage of edible part of fresh fruit is 95.5%, and SSC and Vc is 29.00% and 343.04 mg per 100 g fresh fruit. The content of flavones and cAMP in fruit skin is 14.69 mg/g and 16.92 μg/g. The oval-shaped stone weighs 0.46 g with a well-developed stone. The percentage of containing kernels is 86.7%.

Evaluation The small plant of Qingyuandadanzao cultivar bears early. It is a fine early-ripening variety with a superior quality.

鲜食
Table Varieties
品种

榆次九月青

品种来源及分布 原产于山西晋中市榆次区的王湖村，栽培数量不多。

植物学性状 树体中等大小，树姿开张，树冠半圆形。主干皮裂条状。枣头红褐色，平均长53.0cm，粗0.92cm，节间长7.7cm，蜡质多。二次枝平均长25.5cm，6节，弯曲度中等。针刺不发达。枣股抽吊力较弱，平均抽生枣吊3.0个。枣吊平均长17.9cm，着叶11片。叶片较大，叶长7.0cm，叶宽3.8cm，卵圆形，绿色，先端渐尖，叶基截形。叶缘具锐锯齿。花序平均着花6朵。

生物学特性 抗逆性强。坐果率高，枣头、2～3年和3年以上枝的吊果率分别为116.7%、147.2%和28.8%。丰产而稳定，株产50kg左右，最高株产100kg。在山西太谷地区，果实9月下旬成熟，为晚熟品种类型。成熟期遇雨裂果较轻。

果实性状 果实较小，卵圆形，纵径3.40cm，横径2.95cm，单果重10.8g。果肩平，梗洼浅而窄。果面光滑，红色，果顶平圆，果点中大，较密。肉质较松，味甜略酸，略具清香，鲜食品质上等。鲜枣含可溶性固形物26.00%，总糖22.15%，酸0.51%，100g果肉维生素C含量337.66mg；果皮含黄酮3.36mg/g，cAMP含量398.52μg/g。含仁率15.5%，

评价 该品种抗逆性强，丰产稳产，裂果轻，鲜食品质上等，可适度规模栽培。

Yucijiuyueqing

Source and Distribution The cultivar originated from Wanghu Village of Yuci City in Shanxi Province, with a small quantity.

Botanical Characters The medium-sized tree is spreading with a semi-round crown. The trunk bark has striped fissures. The reddish-brown 1-year-old shoots are 53.0 cm long and 0.92 cm thick. The internodes are 7.7 cm long with much wax and less-developed thorns. The secondary branches are 25.5 cm long with 6 nodes of medium curvature. The mother fruiting spurs can germinate 3 deciduous fruiting shoots which are 17.9 cm long with 11 leaves. The large leaves are 7.0 cm long and 3.8 cm wide, oval-shaped and green, with a gradually-cuspate apex, a truncate base and a sharp saw-tooth pattern on the margin. Each inflorescence has 6 flowers.

Biological Characters The tree has strong resistance to adverse conditions, high fruit set, a high and stable yield. The percentage of fruits to deciduous fruiting shoots of 1-year-old shoots, 2～3-year-old branches and over-3-year-old ones is 116.7%, 147.2% and 28.8%. The average yield per tree is about 50 kg (maximum 100 kg). In Taigu County of Shanxi Province, it matures in late September. It is a mid-late-ripening variety with light fruit-cracking even if it rains in the maturing stage.

Fruit Characteristics The small oval-shaped fruit has a vertical and cross diameter of 3.40 cm and 2.95 cm averaging 10.8 g. It has a flat shoulder, a shallow and narrow stalk cavity, a smooth surface, a flat-round fruit apex and medium-large dense dots. The flesh fruit is loose-textured, sweet and sour. It has a good quality for fresh-eating. The content of Vc in fresh fruit is 337.66 mg per 100 g fresh fruit. The content of flavones and cAMP in mature fruit skin is 3.36 mg/g and 398.52 μg/g. The percentage of containing kernels is 15.5%.

Evaluation The cultivar has strong resistance to adverse conditions with a high and stable yield and light fruit-cracking. The fruit has a good quality for fresh-eating. It can be developed widely.

鲜 食
Table Varieties
品 种

濮 阳 小 枣

品种来源及分布 原产和分布于河南省濮阳市等地。

植物学性状 树体较大，树姿半开张，树冠呈伞形。主干皮裂条状。枣头黄褐色，生长量81.3cm，节间长7.7cm，二次枝长25.9cm，5～7节。无针刺。枣股平均抽生枣吊4.1个。枣吊长23.6cm，着叶13片。叶片椭圆形，绿色，先端急尖，叶基偏斜形，叶缘具钝锯齿。花量较多，花序平均着花9朵。花中大，花径6.0mm。

生物学特性 树势较强，较丰产。在山西太谷地区，9月中旬果实着色，10月上旬成熟采收，果实生育期113d左右，为极晚熟品种类型。

果实性状 果实小，长圆形，纵径2.70cm，横径2.10cm，果重7.6g左右，大小较整齐。梗洼较广、中深。果顶平，柱头脱落。果皮中厚，浅红色，果面光滑。肉质较致密，味甜，汁液中多，品质上等，适宜鲜食。鲜枣可食率94.5%，总糖27.36%，酸0.46%，糖酸比59.48∶1，100g果肉维生素C含量394.66mg；果皮含黄酮3.09mg/g，cAMP含量211.16μg/g。果核较大，纺锤形，核重0.42g，大多核内无种仁，含仁率5.0%。

评价 该品种树势较强，较丰产，果实品质上等，适宜鲜食。

Puyangxiaozao

Source and Distribution The cultivar originated from and spreads in Puyang of Henan Province.

Botanical Characters The large tree is half-spreading with an umbrella-shaped crown. The trunk bark has striped fissures. The yellowish-brown 1-year-old shoots are 81.3 cm long with the internodes of 7.7 cm. The secondary branches are 25.9 cm long with 5～7 nodes and without thorns. The mother fruiting spurs can germinate 4.1 deciduous fruiting shoots which are 23.6 cm long with 13 leaves. The green leaves are oval-shaped with a sharply-cuspate apex, a deflective base and a blunt saw-tooth pattern on the margin. There are many medium-sized flowers with a diameter of 6.0 mm produced averaging 9 ones per inflorescence.

Biological Characters The tree has strong vigor and a high yield. In Taigu County of Shanxi Province, it begins coloring in mid-September and is harvested in early October with a fruit growth period of 113 d. It is an extremely late-ripening variety.

Fruit Characteristics The small oblong fruit has a vertical and cross diameter of 2.70 cm and 2.10 cm, averaging 7.6 g with a regular size. It has a medium-deep and wide stalk cavity, a flat fruit apex, a falling-off stigma, a smooth surface and medium-thick and light-red skin. The tight-textured flesh has a sweet taste and medium juice, with a superior quality for fresh-eating. The percentage of edible part of fresh fruit is 94.5%, and TTS, TA and Vc is 27.36%, 0.46% and 394.66 mg per 100 g fresh fruit. The SAR is 59.48∶1. The content of flavones and cAMP in fruit skin is 3.09 mg/g and 211.16 μg/g. The large spindle-shaped stone weighs 0.42 g. Most stones contain no kernels, with the percentage of containing kernels 5.0%.

Evaluation The cultivar has strong vigor and a high yield. The fruit has a superior quality for fresh eating.

鲜食
Table Varieties
品种

制干品种 Drying Varieties

黄河沿岸枣林（山西　临县）
Jujube Forest Along the Yellow River (Linxian, Shanxi)

吕梁木枣

品种来源及分布 别名木枣、中阳木枣、临县木枣、柳林木枣、绥德木枣、木条枣、条枣、长枣。广泛分布于山西省吕梁地区的临县、柳林、石楼、中阳和陕西省榆林地区的绥德、清涧、延川等黄河中游沿岸，是晋、陕两省的主栽品种。该品种栽培历史悠久，陕西省清涧县的王宿村和绥德县的鱼湾村等地至今尚有千年生以上的古老枣树林。

植物学性状 树体较大，树姿开张，干性较强，枝条中密，树冠圆锥形。主干皮裂呈条状。枣头红褐色，生长势较强，平均长50.0cm，粗0.95cm，节间长8.5cm，着生永久性二次枝4～6个，二次枝平均长27.6cm，7节左右。针刺不发达，一般长1.2cm。枣股较大，圆柱形，抽吊力中等，抽生枣吊2～5个，多为3～4个。枣吊平均长20.1cm，着叶12片。叶片中大，叶长7.0cm，叶宽3.05cm，卵圆形，叶片较厚，浓绿色，先端急尖，叶基偏斜，叶缘锯齿粗钝。花量多，枣吊平均着花71.7朵，每花序6朵左右。花较大，花径7.3mm。

生物学特性 树势较强，萌芽率和成枝力中等。结果较早，一般定植第二年开始结果，10年后进入盛果期，产量较高而稳定。坐果率较高，枣头、2～3年生枝和4年生枝的吊果率分别为29.5%、105.1%和52.9%，主要坐果部位在枣吊的3～11节。在山西太谷地区，9月上旬果实开始着色，9月下旬脆熟，10月10日后完熟，果实生育期110d左右，属晚熟品种类型。成熟期遇雨裂果轻。

果实性状 果个中大，圆柱形，纵径3.75cm，横径2.41cm，单果重14.2g，大小较整齐。果皮厚，深红色，果面光滑。梗洼较广，中深。果顶平，柱头遗存。果肉厚，绿白色，肉质较致密，味酸甜，汁液中多，品质中上，适宜制干。鲜枣可食率97.4%，含可溶性固形物25.20%，总糖22.70%，酸0.74%，100g果肉维生素C含量502.90mg；果皮含黄酮19.06mg/g，cAMP含量43.07μg/g。制干率49.0%，干枣含总糖70.88%，酸1.87%，含水量20.35%。核小，纺锤形，核重0.37g。种仁不饱满，含仁率10%左右。

评价 该品种结果早，产量较高而稳定，品质中上，适宜制干。适应性广，抗逆性强，尤其是对土壤和气候条件要求不严格，耐瘠薄土壤和干旱，较抗裂果和抗病，可在黄河流域黄土丘陵区规模发展。

Lvliangmuzao

Source and Distribution As the dominant variety, the cultivar widely spreads in Lvliang District of Shanxi Province and in Yulin District of Shaanxi Province.

Botanical Characters The large tree is spreading with a strong central leader trunk, medium-dense branches and a conical-shaped crown. The reddish-brown 1-year-old shoots have strong growth vigor, averaging 50.0 cm long. The secondary branches are 27.6 cm long with 7 nodes and less-developed thorns. The large column-shaped mother fruiting spurs can germinate 2～5 deciduous fruiting shoots. The medium-sized leaves are oval-shaped.

Biological Characters The tree has moderate vigor, medium germination and branching ability with high fruit set. It enters the full-bearing stage in the 10th year with a high and stable yield. It is a late-ripening variety with light fruit-cracking.

Fruit Characteristics The medium-sized fruit is column-shaped with a vertical and cross diameter of 3.75 cm and 2.41 cm, averaging 14.2 g with a regular size. It has dark-red thick skin, a smooth surface with small. The greenish-white flesh is thick, hard, sweet and sour with medium juice and a higher than normal quality for dried fruits. The percentage of edible part of fresh fruit is 97.4%, and SSC, TTS, TA, Vc and water is 25.20%, 22.70%, 0.74%, 502.90 mg per 100 g fresh fruit and 68.0% respectively. The content of flavones and cAMP in mature fruit skin is 19.06 mg/g and 43.07 μg/g. The percentage of fresh fruits which can be made into dried ones is 49.0%. The content of TTS, TA and water in dried fruit is 70.88%, 1.87% and 20.35%. The small spindle-shaped stone has a vertical and cross diameter of 2.34 cm and 0.61 cm, averaging 0.37 g. The percentage of containing kernels is 10%.

Evaluation The cultivar bears early with a high and stable yield. The fruit has a higher than normal quality and is suitable for making dried fruits. It has strong adaptability and resistance to adverse conditions with a low demand on soils and climates. It has strong tolerance to poor soils and drought and strong resistance to fruit-cracking and diseases. It can be developed on a large scale in the loess hilly region of the Yellow River Valley.

制 干
Drying Varieties
品 种

保 德 油 枣

品种来源及分布 别名佳县油枣、油枣。原产和集中分布于黄河中游沿岸的山西保德、兴县和陕西府谷、佳县等地，为当地主栽品种，约占当地枣树面积的80%以上，是当地的古老品种，栽培历史已有1 000多年。佳县朱家洼乡泥河沟村黄河岸边现存一片唐代老枣树群落，其中树体最大的高约13m，干周3.41m，年产红枣100kg左右。

植物学性状 树体中大，树姿开张，干性较弱，枝条较密，树冠呈乱头形。主干皮裂呈块状。枣头红褐色，生长势较强，平均生长量67.0cm，粗度1.10cm，节间长7.5cm，着生永久性二次枝6～8个。二次枝长32.6cm，平均生长6节。针刺较发达。枣股中大，抽吊力中等，抽生枣吊2～4个，多为3个。枣吊平均长19.9cm，着叶14片。叶片中大，叶长5.8cm，叶宽2.9cm，卵圆形，浓绿色，先端急尖，叶基圆形，锯齿钝。花量较多，枣吊平均着花92.9朵，花序平均3.9朵。花大，花径8.1mm，昼开型。

生物学特性 树势中庸偏强，萌芽率中等，成枝力较强，萌蘖力强。结果较早，根蘖苗一般第二年开始结果，15年后进入盛果期。盛果期和寿命较长，丰产，产量较稳定。坐果率高，枣头枝吊果率83.0%，2～3年生枝为38.3%，4年生枝为6.0%。在山西太谷地区，10月上旬完熟，果实生育期110d左右，为晚熟品种类型。果实成熟期较抗病和抗裂果，落果也较轻。

果实性状 果实中大，扁柱形，纵径3.42cm，横径2.66cm，单果重13.9g，大小整齐。果梗长，较粗，梗洼广，中度深。果顶微凹，柱头遗存。果面平整光亮，紫红色。果肉厚，绿白色，肉质致密，味甜酸，汁液中多，品质中上，制干和加工蜜枣，多以制干为主。鲜枣可食率96.5%，含可溶性固形物33.60%，总糖26.65%，酸0.78%，100g果肉维生素C含量511.44mg；果皮含黄酮8.92mg/g，cAMP含量83.46μg/g。制干率50%左右，干枣含可溶性固形物75.9%，总糖71.29%，酸1.87%。果核小，纺锤形，纵径2.68cm，横径0.69cm，核重0.48g，核尖短，核纹浅，含仁率8.3%，种仁不饱满。

评价 该品种适应性广，抗逆性强。结果早，丰产，采前落果、裂果较轻。果实中大，肉厚、核小，可食率高，品质中上，用于制干。该品种的变异类型多，应注重大果形、制干品质优异品系的筛选。

Baodeyouzao

Source and Distribution The cultivar originated from Shanxi and Shaanxi Province along the middle reaches of Yellow River. It is the dominant variety in those areas. It is an old local variety with a history of over 1 000 years.

Botanical Characters The large tree is spreading with a weak central leader trunk, dense branches and an irregular crown. The reddish-brown 1-year-old shoots have strong growth vigor, averaging 67.0 cm long. There are 6～8 permanent secondary branches, which are 32.6 cm long with developed thorns. The medium-sized mother fruiting spurs can germinate 2～4 deciduous fruiting shoots which are 19.9 cm long with 14 leaves. The medium-sized leaves are oval-shaped and dark green. The large flowers has a diameter of 8.1 mm.

Biological Characters The tree has moderate or strong vigor, medium germination rate, strong branching ability and suckering ability. In Taigu County of Shanxi Province, it begins coloring in early September and enters the crisp-maturing stage around September. 20 and completely matures in early October. It is a late-ripening variety with light fruit-dropping and strong resistance to diseases and fruit-cracking in the maturing stage.

Fruit Characteristics The medium-sized fruit is flat column-shaped, averaging 13.9 g with a regular size. It has a smooth and bright surface. The greenish-white thick flesh is tight-textured, sweet and sour, with medium juice and a better than normal quality for fresh-eating, processing dried fruits and candied fruits. It is mainly used for dried fruits. The percentage of edible part of fresh fruit is 96.5%, and SSC, TTS, TA and Vc is 33.60%, 26.65%, 0.78% and 511.44 mg per 100 g fresh fruit. The content of flavones and cAMP in mature fruit skin is 8.92 mg/g and 83.46 μg/g. The rate of fresh fruits which can be made into dried ones is 50%, and the content of SSC, TTS and TA in dried fruit is 75.9%, 71.29% and 1.87%. The small stone is spindle-shaped, averaging 0.48 g.

Evaluation The cultivar has strong adaptability and strong resistance to adverse conditions, bearing early with a high yield and light premature fruit-dropping and fruit-cracking. The medium-sized fruit has thick flesh, a small stone and high edibility with a better quality. It is mainly used for dried fruits.

制 干
Drying Varieties
品 种

灵 宝 大 枣

品种来源及分布 别名灵宝圆枣、平陆屯屯枣、疙瘩枣。原产山西省南部和河南省西部交界处的黄河两岸，以山西省芮城、平陆，河南省灵宝市、陕县栽培较集中，为当地主栽品种，也是全国主要品种之一。据灵宝县志记载，该品种的栽培历史始于明代前，距今已有600多年，原产地现有大量百年以上的挂果树。

植物学性状 树体高大，树姿直立或半开张，干性较强，枝条粗壮，树冠呈伞形。主干皮裂呈条状。枣头紫褐色，生长势较强，平均生长量63.5cm，粗1.06cm，节间长8.0cm，着生永久性二次枝3～5个，二次枝长28.3cm，平均6节左右。针刺较发达。枣股中大，抽吊力中等，平均抽生枣吊3.4个。枣吊平均长16.8cm，着叶12片。叶长5.2cm，叶宽2.9cm，叶片卵圆形，深绿色，先端尖凹，叶基心形，叶片较厚，叶缘锯齿钝。花量少，枣吊平均着花23.8朵，花序着花1.9朵，零级花花径6.5mm。

生物学特性 生长势较强，发枝力中等，萌蘖力较弱。结果较迟，早期丰产性中等，盛果期较长，产量较高，但存在隔年结果现象。一般定植3～4年开始结果，15年后进入盛果期。坐果率中等，枣头、2～3年生枝和4年生枝的吊果率分别为17.1％、40.5％和17.5％，主要坐果部位在枣吊的1～7节。在山西太谷地区，9月20日果实成熟，果实生育期105d左右，属中熟品种类型。采前落果严重，但裂果较轻。

果实性状 果个较大，扁圆形，纵径2.99cm，横径3.34cm，单果重15.7g，最大34.0g，大小较整齐，成熟较一致。果梗较短，梗洼广而浅。果顶微凹，柱头遗存。果皮中厚，深红色或紫红色，果面呈现五棱突起，并有不规则的黑斑。果点中大，圆形，分布密而明显。果肉厚，绿白色，肉质致密，味甜略酸，汁液较少，品质中上，可制干和加工无核糖枣。鲜枣可食率96.3％，含可溶性固形物32.40％，总糖22.95％，酸0.50％，100g果肉维生素C含量359.47mg。制干率58.0％，干枣含总糖70.17％，酸1.11％。核中大，椭圆形，纵径1.58cm，横径1.00cm，核重0.58g，种仁饱满，含仁率80％左右。

评价 该品种树体高大，生长势强，产量较高，但不稳定。果个较大，肉厚，核中大，可食率高，品质中上，适宜制干和加工蜜枣。不易裂果，可在北方秋雨较多的地区适当发展。

Lingbaodazao

Source and Distribution The cultivar originated from the two banks of the Yellow River in the juncture of south Shanxi Province and west Henan Province. It is the dominant variety in those places, and even in China.

Botanical Characters The large tree is vertical or half-spreading with a strong central leader trunk, strong branches and an umbrella-shaped crown. The purplish-brown 1-year-old shoots have strong growth vigor, 63.5 cm long. There are 3～5 permanent secondary branches on each one-year-old shoot with developed thorns. The mother fruiting spurs can germinate 3.4 deciduous fruiting shoots. The thick leaves are oval-shaped and dark green. The number of flowers is small, averaging 23.8 ones per deciduous fruiting shoot. The blossoms are 6.45 mm in diameter.

Biological Characters The tree has strong vigor, medium branching ability and weak suckering ability, bearing late with moderate early productivity, a long full-bearing stage, a high yield and medium fruit set. In Taigu County of Shanxi Province, it begins coloring in early September and matures around September. 20th. It is a mid-ripening variety with serious premature fruit-dropping and light fruit-cracking.

Fruit Characteristics The large fruit is flat-round, averaging 15.7 g with a regular size. It has a medium-thick and purplish-red or dark-red skin and a surface. The greenish-white thick flesh is tight-textured, sweet and a little sour with little juice and a higher than normal quality and is suitable for processing dried fruits and seedless candied fruits. The percentage of edible part of fresh fruit is 96.3％, and SSC, TTS, TA and Vc is 32.40％, 22.95％, 0.50％ and 359.47 mg per 100 g fresh fruit. The percentage of fresh fruits which can be made into dried ones is 58.0％, and the content of TTS and TA in dried fruit is 70.17％ and 1.11％. The medium stone is oval-shaped, averaging 0.58 g. The percentage of containing kernels is 80％.

Evaluation The large plant of Lingbaodazao cultivar has strong vigor with a high and unstable yield. The large fruit has thick flesh, a small stone and high edibility with light fruit-cracking and higher than normal quality and is suitable for processing dried fruits and candied fruits.

制 干
Drying Varieties
品 种

婆 枣

品种来源及分布 别名阜平大枣、串干、行唐大枣。原产河北西部的阜平、曲阳、唐县、行唐等太行山中段丘陵地带，为当地主栽品种。另外，河北省的衡水、沧州及山东省的夏津、乐陵等地也有小片栽培，是全国的主要制干品种之一。该品种有千年以上栽培历史。

植物学性状 树体较高大，树姿较开张，干性强，枝条中密，树冠圆锥形。主干皮裂浅，小块状，易剥落。枣头紫褐色，生长势较强，平均长85.3cm，粗1.20cm，节间长8.5cm。二次枝长36.5cm，7节左右，弯曲度中等。针刺发达。枣股中大，抽生枣吊3～4个。枣吊平均长18.3cm，着叶15片。叶片中大，叶长4.3cm，叶宽2.7cm，卵圆形，浓绿色，先端急尖，叶基心形，叶缘锯齿浅钝。花量较少，每个枣吊着生5～7个花序，每花序着花3～5朵。花中大，花径7mm左右，昼开型。

生物学特性 树体生长势强，发枝力弱。结果较迟，但盛果期产量高而稳定。一般栽后3～4年开始结果，15年后进入盛果期。坐果稳定，枣头、2～3年生枝和4年生枝吊果率分别为46.6%、94.0%和5.7%。在山西太谷地区，9月下旬果实成熟采收，为晚熟品种类型。成熟期遇雨易裂果。

果实性状 果个较大，长圆形或倒卵圆形，纵径3.40cm，横径3.10cm，单果重14.3g，大小较均匀整齐。果梗较细，中长，梗洼中广而深。果顶平或微凹，柱头遗存，不明显。果皮较薄，红色，果面平滑，着色前阳面有褐色晕斑。果点大而中密，圆形，浅黄色。果肉厚，乳白色，肉质较粗松，味甜，汁液少，品质中等，可制干和加工醉枣。鲜枣可食率96.4%，含可溶性固形物33.00%，总糖25.25%，酸1.02%，100g果肉维生素C含量808.83mg；果皮含黄酮6.01mg/g，cAMP含量51.82μg/g。制干率53.1%，干枣含总糖72.55%，酸1.76%。核小，纺锤形，核重0.52g，种仁不饱满，含仁率26.7%。

评价 该品种适应性很强，耐旱，耐瘠薄土壤条件，花期能适应较低的空气温湿度，产量高而稳定，可在土壤条件较差和成熟期少雨的地区栽植。果形整齐，品质中等，制干率较高，但新建枣园时应注意选择果实品质优良和较抗裂果的变异类型栽植。

Pozao

Source and Distribution The cultivar originated from Fuping in Hebei Province. It is the dominant variety there and one of the main drying varieties in China with a history of over 1 000 years.

Botanical Characters The large tree is vertical with a strong central leader trunk, medium-dense branches and a conical-shaped crown. The purplish-brown 1-year-old shoots mostly grow vertically and extending with strong growth vigor. The secondary branches are 36.5 cm long with developed thorns. The medium-sized column-shaped mother fruiting spurs can germinate 3～4 deciduous fruiting shoots. The medium-sized leaves are oval-shaped and dark green. The number of flowers is small, averaging 5～7 inflorescences per deciduous fruiting shoot. The medium-sized blossoms have a diameter of 7 mm.

Biological Characters The tree has strong vigor and weak branching ability, bearing late with stable fruit set and a high and stable yield in full-bearing stage. It enters the full-bearing stage in the 15th year. In Taigu County of Shanxi Province, it matures in late September with a fruit growth period of 100～110 d. It is a mid-late-ripening variety with serious fruit-cracking if it rains in maturing stage.

Fruit Characteristics The large fruit is oblong or obovate, averaging 14.3 g with a regular size. It has a thin red skin and a smooth surface. The ivory-white flesh is loose-textured, thick and sweet with less juice and a medium quality for processing dried fruits and alcoholic fruits. The percentage of edible part of fresh fruit is 96.4%, and SSC, TTS, TA and Vc is 33.00%, 25.25%, 1.02% and 808.83 mg per 100 g fresh fruit. The content of flavones and cAMP in mature fruit skin is 6.01 mg/g and 51.82 μg/g. The percentage of fresh fruits which can be made into dried ones is 53.1%, and the content of TTS and TA in dried fruit is 72.55% and 1.76%. The small stone is spindle-shaped averaging 0.52 g.

Evaluation The cultivar has strong adaptability, strong tolerance to drought, poor soils and lower temperatures and humidity in blooming stage with a high and stable yield.

新 乐 大 枣

品种来源及分布 分布于河北省新乐县。

植物学性状 树体高大，树姿直立，树冠圆头形。主干碎块状皮裂。枣头紫褐色，平均长76.7cm，节间长6.4cm。二次枝长37.2cm，7～9节。针刺发达。枣股平均抽生枣吊3.4个。枣吊长20.7cm，着叶14片。叶片中等大，卵圆形，绿色，平展，先端锐尖，叶基圆楔形，叶缘具锐锯齿。花量多，花序平均着花5朵，花小，花径5.6mm，萼片黄绿色。

生物学特性 树势强，萌芽率和成枝力高，结果较晚，一般定植第三年结果，10年左右进入盛果期，产量中等，枣吊平均结果0.49个。在山西太谷地区，9月下旬果实成熟采收，果实生育期111d左右，属晚熟品种类型。成熟期遇雨易裂果。

果实性状 果个中等大，卵圆形，纵径3.15cm，横径2.95cm，单果重12.4g，大小整齐。果皮中等厚，红色，果面平滑。梗洼窄深。果顶微凹，柱头宿存。肉质致密，汁液少，味甜，品质中等，适宜制干。鲜枣可食率96.4%，含可溶性固形物28.20%，总糖22.09%，酸0.83%，100g果肉维生素C含量404.59mg；果皮含黄酮4.62mg/g，cAMP含量99.69μg/g。制干率57.0%，干枣含总糖66.45%，酸2.63%。核较小，倒纺锤形，平均重0.45g。核内不含种子。

评价 该品种树体高大，树势强，结果较晚，产量中等，抗裂果能力差。果个中等大，品质中等，适宜制干。

Xinledazao

Source and Distribution The cultivar spreads in Xinle County of Hebei Province.

Botanical Characters The large tree is vertical with a natural-round crown. The trunk bark has massive fissures. The purplish-brown 1-year-old shoots are 76.7 cm long with the internodes of 6.4 cm and developed thorns. The secondary branches are 37.2 cm long with 7～9 nodes. The mother fruiting spurs can germinate 3.4 deciduous fruiting shoots which are 20.7 cm long with 14 leaves. The medium-sized leaves are oval-shaped, green and flat, with a sharply-cuspate apex, a round-cuneiform base and a sharp saw-tooth pattern on the margin. There are many small flowers with a diameter of 5.6 mm produced, averaging 5 ones per inflorescence. The blossoms have yellowish-green sepals.

Biological Characters The tree has strong vigor, strong germination and branching ability. It bears late, generally in the 3rd year after planting and enters the full-bearing stage in the 10th year with a medium yield. The deciduous fruiting shoot bears 0.49 fruits on average. In Taigu County of Shanxi Province, it matures in late September with a fruit growth period of 111 d. Fruit-cracking easily occurs if it rains in the maturing stage.

Fruit Characteristics The medium-sized fruit is column-shaped, with a vertical and cross diameter of 3.15 cm and 2.95 cm, averaging 12.4 g with a regular size. It has medium-thick red skin, a smooth surface, a narrow and deep stalk cavity, a slightly-sunken fruit apex and a remnant stigma. The tight-textured flesh is sweet with less juice. It has medium quality for fresh-eating and dried fruits. The percentage of edible part of fresh fruit is 96.4%, and SSC, TTS, TA and Vc is 28.20%, 22.09%, 0.83% and 404.59 mg per 100 g fresh fruit. The content of flavones and cAMP in mature fruit skin is 4.62 mg/g and 99.69 μg /g. The percentage of fresh fruits which can be made into dried ones is 57.0%, and the content of TTS, TA and Vc in dried fruit is 66.45%, 2.63% and 62.24 mg per 100 g fresh fruit. The small stone is inverted spindle-shaped, averaging 0.45 g, without kernel.

Evaluation The large plant of Xinledazao cultivar has strong tree vigor, bearing late with a medium yield. The medium-sized fruit has medium quality and is suitable for fresh-eating and making dried fruits. Yet it has poor resistance to fruit-cracking.

官 滩 枣

品种来源及分布 山西十大名枣之一。原产山西省襄汾县城关镇的官滩村，由此而得名。现还有百年以上枣树林，为当地主栽品种，约占当地栽培枣树的90%以上，年产鲜枣20多万千克。现已成为襄汾县的主要推广开发品种。

植物学性状 树体较大，树势中等，树姿半开张，干性较弱，枝条细而较密，树冠呈伞形。主干皮裂呈条状。枣头红褐色，平均长60.0cm，粗度0.84cm，节间长7.5cm，着生二次枝5个左右，二次枝平均长26.9cm，弯曲度较小。针刺不发达。枣股较小，抽吊力强，抽生枣吊4~5个，枣吊平均长17.7cm，着叶12片。叶片小，叶长4.8cm，叶宽2.3cm，椭圆形，绿色，先端急尖，叶基圆形，叶缘锯齿粗钝。花量中多，枣吊平均着花50.5朵，每花序4.7朵。花小，零级花花径6.2mm。

生物学特性 生长势中庸，萌芽率较弱，发枝力较强。结果较早，早期丰产，盛果期产量较高而稳定。一般定植第二年开始结果，3~4年普遍结果，10年左右进入盛果期。坐果率高，枣头、2~3年生枝和4年生枝的吊果率分别可达79.1%、69.6%和47.1%，主要坐果部位在枣吊的1~7节，占坐果总数的73.2%。在山西太谷地区，9月下旬脆熟，10月上旬完熟，为晚熟品种类型。成熟期遇雨裂果轻，较抗病。

果实性状 果个中大，扁柱形，纵径3.52cm，横径2.73cm，单果重12.6g，大小不整齐。果梗长3~4mm，采收时不易与果肉分离。梗洼中广而深。果顶平，柱头遗存。果皮中厚，紫红色，果面平滑。果点小而密，浅黄色。果肉厚，绿白色，肉质细而致密，味甜，汁液少，干枣品质上等，适宜制干。鲜枣可食率95.3%，含可溶性固形物30.50%，总糖24.62%，酸0.39%，100g果肉维生素C含量445.90mg；果皮含黄酮13.04mg/g，cAMP含量54.02μg/g。制干率52.0%，干枣含总糖61.14%，酸1.00%。核较大，倒纺锤形，纵径2.03cm，横径0.76cm，核重0.59g，种仁较饱满，含仁率53.3%。

评价 该品种生长势中庸偏弱，结果较早，产量高而稳定。适应性较广，抗逆性强，成熟期遇雨不易裂果，也较抗病，果实肉厚，品质上等，为有商品栽培价值的制干良种。

Guantanzao

Source and Distribution The cultivar originated from Guantancun in Chengguan Town of Xiangfen County, Shanxi Province. It is the dominant variety there which occupies over 90% of the total jujube area.

Botanical Characters The large tree is half-spreading with a weak central leader trunk, thin and dense branches and an irregular crown. The reddish-brown 1-year-old shoots are 60.0 cm long. There are about 5 secondary branches on each one-year-old shoot with small curvature and developed thorns. The small mother fruiting spurs can germinate 4~5 deciduous fruiting shoots. The small green leaves are oval-shaped. There are 50.5 flowers per deciduous fruiting shoot. The small blossoms have a diameter of 6.2 mm for the zero-level flowers.

Biological Characters The tree has moderate vigor, weak germination ability and strong branching ability, bearing early with strong early productivity, high fruit set and a high and stable yield in full-bearing stage. The tree enters the full-bearing stage in the 10th year. In Taigu County of Shanxi Province, it completely matures in early October. It is a late-ripening variety with light fruit-cracking in the maturing stage.

Fruit Characteristics The medium-sized fruit is flat column-shaped, averaging 12.6 g. It has a medium-thick and purplish-red skin and a smooth surface. The greenish-white thick flesh is tight-textured, delicate and sweet with less juice and a good quality for dried fruits. The percentage of edible part of fresh fruit is 95.3%, and SSC, TTS, TA and Vc is 30.50%, 24.62%, 0.39% and 445.90 mg per 100 g fresh fruit. The content of flavones and cAMP in mature fruit skin is 13.04 mg/g and 54.02 μg/g. The percentage of fresh fruits which can be made into dried ones is 52.0%, and the content of TTS, TA and water in dried fruit is 61.14%, 1.00% and 19.2%. The small stone is inverted spindle-shaped averaging 0.59 g.

Evaluation The cultivar has normal vigor, strong adaptability and resistance to aderse conditions bearing early with a high and stable yield. The fruit has thick flesh and high edibility with a good quality, strong resistance to diseases and light fruit-cracking.

制 干
Drying Varieties
品 种

相 枣

品种来源及分布 原产山西运城市北相镇一带。据传说，古时曾作贡品，因而也称"贡枣"，为当地主栽品种，山西省十大名枣之一。据县志记载，该品种已有3 000多年的栽培历史，现尚存许多百年以上大树。

植物学性状 树体中大，树姿半开张，干性较强，枝条较密，树冠呈自然半圆形。树干皮裂呈条状。枣头红褐色，平均长30.0cm，粗0.76cm，节间长9.5cm。二次枝长25.3cm，平均7节。针刺不发达。皮目小，较密，圆形或椭圆形，凸起，开裂，灰白色。枣股中等大，圆柱形，抽生枣吊2～5个，多为3～4个。枣吊平均长18.1cm，着叶13片。叶片小，叶长5.2cm，叶宽2.5cm，卵圆形，浓绿色，先端急尖，叶基偏斜，叶缘具锐齿。花量中等多，枣吊平均着花48.1朵，每花序3.8朵。花小，零级花花径6.8mm，夜开型。

生物学特性 生长势中等或较弱，萌芽率和成枝力中等。结果早，一般定植第二年开始结果，结果株率可达83.3%。10年生后进入盛果期，较丰产，产量较稳定，坐果率高。枣头吊果率36.7%，2～3年生枝55.2%，4年生枝27.7%，主要坐果部位在枣吊的2～7节，占坐果总数的68.7%。在山西太谷地区，9月上旬果实着色，9月下旬成熟。成熟期落果轻、抗裂果。

果实性状 果个大，扁柱形，纵径3.85cm，横径3.52cm，单果重19.1g，大小不整齐。果梗中等长，较粗，梗洼窄而深。果顶凹，柱头遗存。果皮厚，紫红色，果面粗糙。果点较大，分布中密，浅黄色。果肉厚，绿白色，肉质致密，较硬，味甜，汁液少，干枣果肉富有弹性，耐挤压，品质上等，适宜制干。鲜枣可食率96.8%，含可溶性固形物28.50%，总糖25.51%，酸0.34%，100g果肉维生素C含量474.00mg；果皮含黄酮2.95mg/g，cAMP含量101.97μg/g。制干率53.0%，干枣含总糖70.29%，酸0.82%。果核较小，纺锤形，纵径2.55cm，横径0.83cm，核重0.61g。核内多含有不饱满种仁，含仁率66.7%。

评价 该品种果个大，外观亮丽而美观，干枣果肉富弹性，肉厚核小，制干率较高，为优良的制干和蜜枣品种，适宜商品栽培。生产中应注意加强肥水管理，防止树体衰弱。

Xiangzao

Source and Distribution The cultivar originated from Beixiang town in Yuncheng City of Shanxi Province. It is the dominant variety there and has a history of over 3 000 years.

Botanical Characters The medium-sized tree is half-spreading with a strong central leader trunk, dense branches and a semi-round crown. The reddish-brown 1-year-old shoots are 30.0 cm long. The secondary branches are 25.3 cm long with less-developed thorns. The medium-sized column-shaped mother fruiting spurs can germinate 2～5 deciduous fruiting shoots. The small leaves are oval-shaped and dark green. There are 48.1 flowers per deciduous fruiting shoot and 3.8 ones per inflorescence with a diameter of 6.8 mm for the zero-level flowers.

Biological Characters The tree has moderate or weak vigor, medium germination and branching ability. It enters the full-bearing stage in the 10th year with a high and stable yield and high fruit set. In Taigu County of Shanxi Province, it begins coloring in early September and enters the crisp-maturing stage in late September with a fruit growth period of 110 d. It is a mid-late-ripening variety with strong resistance to fruit-cracking and light fruit-dropping in the maturing stage.

Fruit Characteristics The large oval-shaped fruit weighs 19.1 g with irregular sizes. It has purplish-red thick skin and a rough surface. The greenish-white flesh is thick and hard, tight-textured, sweet, with little juice. Dried fruits have whippy flesh, tolerant to extrusion. It has an excellent quality for dried fruits. The percentage of edible part of fresh fruit is 96.8%, and SSC, TTS, TA, Vc is 28.50%, 25.51%, 0.34% and 474.00 mg per 100 g fresh fruit. The content of flavones and cAMP in mature fruit skin is 2.95 mg/g and 101.97 μg/g. The percentage of fresh fruits which can be made into dried ones is 53.0%. And the content of TTS and TA is 70.29% and 0.82%. The small stone is spindle-shaped averaging 0.61 g.

Evaluation The cultivar has a large fruit size and bright and beautiful appearance. Dried fruits have whippy and thick flesh. It is a good variety for processing dried fruits.

制 干
Drying Varieties
品 种

圆 铃 枣

品种来源及分布 别名紫铃、圆红、紫枣、圆果圆铃。原产山东省的聊城、德州等地。以茌平、东阿、聊城、齐河、济阳栽培较集中，泰安、潍坊、济宁、惠民等地也有栽培分布。是山东省的重要制干品种，栽培面积占全省的50%以上。也是全国的著名优良主栽品种之一，产量约为全国的10%左右。

植物学性状 树体较大，树姿开张，干性较强，枝条较密，树冠自然半圆形。树干皮裂细块状，不易剥落。枣头红褐色，生长势较强，平均长74.5cm，粗1.02cm，节间长7.1cm，二次枝长29.1cm，平均7节。针刺较发达。皮目大，黄褐色，分布密。枣股中大，短柱形，抽吊力中等，一般抽生枣吊3～4个。枣吊平均长20.3cm，着叶16片。叶片中大，叶长5.0cm，叶宽2.3cm，卵圆形或宽披针形，绿色，先端钝尖，叶基圆形，叶缘锯齿钝，中密。花量中等，每花序着花3～7朵。花较大，花径7～7.5mm，昼开型。

生物学特性 树势强健，萌芽率和成枝力较强。结果较晚，一般栽后3～4年开始结果。早期丰产性能中等，盛果期产量较高而稳定。坐果率较高，枣头吊果率55.7%，2～3年生枝49.9%，4年生枝9.5%。在山西太谷地区，9月中旬果实成熟，果实生育期105d左右，为中晚熟品种类型。成熟期遇雨不易裂果，抗病性较强。

果实性状 果实较大，近圆柱形或圆柱形，纵径3.89cm，横径3.39cm，单果重19.8g，大小不整齐。果梗细而短，梗洼中等深广。果顶微凹，柱头遗存。果皮较厚，有韧性，紫红色，果面平滑，有紫黑色斑点。果点大而疏，圆形，明显。果肉厚，绿白色，肉质较粗，味甜，汁液少，干枣品质上等，适宜制干。鲜枣可食率96.2%，含可溶性固形物28.20%，总糖24.86%，酸0.18%，100g果肉维生素C含量344.18mg；果皮含黄酮6.96mg/g，cAMP含量118.82μg/g。制干率60.0%～62.0%，干枣含总糖为70.15%，酸1.05%。核小，椭圆形，纵径2.23cm，横径0.91cm，核重0.75g，核尖短，核纹深，多数含饱满种仁，含仁率86.7%。

评价 该品种对土壤、气候的适应性较强，树体强健，产量较高而稳定，干枣品质优良，可在我国北方枣区发展。现已通过株系选优途径从该品种中筛选出圆铃1号和2号等，目前正在区试栽培。

Yuanlingzao

Source and Distribution The cultivar originated from Liaocheng and Dezhou in Shandong Province. It mainly spreads in Chiping, Donge, Liaocheng, Qihe and Jiyang. It is an important drying variety, which occupies over 50% of the total jujube area in Shandong Province.

Botanical Characters The large tree is half-spreading with a strong central leader trunk, dense branches and a natural-round crown. The reddish-brown 1-year-old shoots have strong growth vigor, 74.5 cm long and 1.02 cm thick with the internodes of 7.1 cm. The secondary branches are 29.1 cm long with 7 nodes and developed thorns. The medium-sized short-column-shaped mother fruiting spurs can germinate 3～4 deciduous fruiting shoots which are 20.3 cm long with 16 leaves. The medium-sized green leaves are oval-shaped or wide-lanceolate. The number of flowers is 3～7 per inflorescence. The large blossoms have a diameter of 7～7.5 mm.

Biological Characters The tree has strong vigor, strong germination and branching ability, bearing late with moderate early productivity, a high and stable yield in full-bearing stage and high fruit set. It is a mid-ripening variety with strong resistance to diseases and fruit-cracking even if it rains in the maturing stage.

Fruit Characteristics The large fruit is nearly round or cylinder, averaging 19.8 g with irregular sizes. It has thick, tough and purplish-red skin and a smooth surface. The greenish-white flesh is thick, rough and sweet with little juice and a good quality for dried fruits. The percentage of edible part of fresh fruit is 96.2%, and SSC, TTS, TA and Vc in fresh fruit is 28.2%, 24.86%, 0.18% and 344.18 mg per 100 g fresh fruit. The small stone is oval-shaped, averaging 0.75 g.

Evaluation The cultivar has strong adaptability to different soils and climates with strong vigor and a high and stable yield. Dried fruits have a good quality. It can be developed in jujube production areas of northern China.

制 干
Drying Varieties
品 种

扁 核 酸

品种来源及分布 别名酸铃、铃枣、串干、鞭干，因果核扁而得名。主要原产河南省黄河故道地区的内黄、濮阳、浚县、滑县、清丰、南乐、汤阳等地，河北省的邯郸和山东省的东明等地也有栽培分布，为河南省栽培面积最大、产量最多的品种，也是全国主要制干品种之一。已有2 000多年栽培历史。

植物学性状 树体较大，树姿开张，枝条中密，树冠自然圆头形。主干皮裂较浅，块状，较易剥落。枣头红褐色，生长势较强，平均长103.1cm，粗度1.23cm，节间长7.2cm，二次枝平均长35.1cm，7节。针刺较发达。皮目小而较稀，圆形，凸起，灰白色。枣股小，圆锥形或圆柱形，抽吊力较强，一般抽生枣吊3～4个。枣吊较粗，平均长19.1cm，着叶12片。叶片较大，叶长5.9cm，叶宽3.0cm，浓绿色，先端急尖，叶基圆形或亚心形，叶缘具锐锯齿。花量中等，枣吊着花40～50朵，每花序着花5～7朵。花较小，花径6～7mm，昼开型。

生物学特性 树势较强，枣头粗壮且生长量大，发枝力中等。结果较迟，但进入盛果期后产量较高而稳定。坐果率较高，枣头、2～3年生枝和4年生枝的吊果率分别为78.3%、61.3%和2.0%。在山西太谷地区，9月下旬果实脆熟，果实生育期110d左右，为晚熟品种类型。成熟期遇雨裂果较轻，较抗病。

果实性状 果实中大，卵圆形，侧面略扁，纵径3.20cm，横径2.69cm，单果重10.1g，大小整齐。果梗中长，较粗，梗洼中等深广。果顶凹，柱头遗存，但不明显。果皮中厚，紫红色。果面平滑，果点小，近圆形，分布较稀，不明显。果肉厚，绿白色，肉质粗松，稍脆，味酸甜，汁液少，品质中上，可制干。鲜枣可食率94.2%，含可溶性固形物27.00%，总糖23.77%，酸0.88%，100g果肉维生素C含量351.23mg；果皮含黄酮2.48mg/g，cAMP含量16.67μg/g。制干率56.2%，干枣含总糖75.84%，酸1.58%。核中大，纺锤形，侧面略扁，纵径2.33cm，横径0.90cm，核重0.59g，核尖较短，核纹较深，核内多无种仁，含仁率仅5.0%。

评价 该品种适应性广，抗逆性较强，尤其耐盐碱、抗裂果和抗病，树体强健，丰产性能较强，坐果稳定，果实酸度大，制干品质中等，可在土壤盐碱性较大的地区发展栽培。

Bianhesuan

Source and Distribution The cultivar originated from Neihuang, Puyang, Junxian, Huaxian, Qingfeng and Tangyin in Henan Province. It is a main drying variety in China with a history of over 2 000 years.

Botanical Characters The large tree is spreading with medium-dense branches and a natural-round crown. The reddish-brown 1-year-old shoots have strong growth vigor, 103.1 cm long and 1.23 cm thick, with the internodes of 7.2 cm. The secondary branches are 35.1 cm long with 7 nodes and developed thorns. The small conical or column-shaped mother fruiting spurs can germinate 3～4 thick deciduous fruiting shoots which are 19.1 cm long with 12 leaves. The large leaves are oval-shaped and dark green. The number of flowers is medium large, averaging 40～50 ones per deciduous fruiting shoot. The small daytime-bloomed blossoms have a diameter of 6～7 mm.

Biological Characters The tree has strong vigor with strong 1-year-old shoots and medium branching ability. It bears late, yet with high fruit set and a high and stable yield in full-bearing stage. It is a late-ripening variety with strong resistance to diseases and light fruit-cracking even if it rains in the maturing stage.

Fruit Characteristics The medium-sized fruit is oval-shaped, slightly flat on the lateral sides, averaging 10.1 g with a regular size. It has medium-thick and purplish-red skin and a smooth surface. The greenish-white thick flesh is loose-textured, crisp, sweet and sour, with less juice and a better than normal quality for making dried fruits. The percentage of edible part of fresh fruit is 94.2%, and SSC, TTS, TA and Vc is 27.00%, 23.77%, 0.88% and 351.23 mg per 100 g fresh fruit. The medium-sized spindle-shaped stone has slightly flatened lateral sides, averaging 0.59 g.

Evaluation The cultivar has strong vigor, strong productivity with stable fruit set. Dried fruits have a normal quality and good tolerance to storage and transport. It has strong adaptability and strong resistance to adverse conditions.

制 干
Drying Varieties
品 种

·137·

郎　枣

品种来源及分布　山西十大名枣之一。原产山西晋中的太谷、祁县、平遥等县，为当地主栽品种之一，还可看到很多百年以上的挂果老枣树，为古老的地方栽培品种。

植物学性状　树体高大，树姿半开张，干性较强，枝条较密，树冠自然圆头形。主干皮裂呈条状。枣头红褐色，生长势较强，平均长73.4cm，粗0.96cm，节间长8.5cm。着生6个永久性二次枝，二次枝长27.1cm，6节左右。针刺不发达。皮目中大，分布较密，圆形、凸起，开裂，灰白色。枣股中大，圆柱形，抽吊力中等，一般抽生枣吊3~4个。枣吊平均长18.1cm，着叶11片。叶片中大，叶长6.7cm，叶宽3.4cm，长卵形，浓绿色，先端急尖，叶基圆楔形，叶缘锯齿细而密。花量中多，枣吊平均着花61.6朵，每花序4.9朵。花较大，零级花花径7.6mm，为昼开型。

生物学特性　树势强，萌芽率和发枝力较强，萌蘖力中等，根系不发达。结果较早，早期丰产性能较强，盛果期长，产量较高。一般定植第二年开始结果，10年后进入盛果期。坐果率高，枣头、2~3年生枝和4年生枝的吊果率分别为40.3%、71.0%和34.3%，主要坐果部位在枣吊的8~12节。在山西太谷地区，9月中旬果实脆熟，果实生育期105d左右，为中熟品种类型。成熟期遇雨易裂果，采前落果严重。

果实性状　果实中大，圆柱形或卵圆形，纵径3.34cm，横径2.81cm，单果重12.1g，大小较整齐。梗洼中广、深，果顶平，柱头遗存。果皮较薄，紫红色，果面平滑，果点中等大，较密。果肉厚，绿白色，肉质致密，味酸，汁液中多，品质中上，适宜制干，也可加工蜜枣和酒枣，近年来大量用于加工醺枣。鲜枣可食率95.8%，含可溶性固形物36.90%，总糖32.48%，酸0.85%，100g果肉维生素C含量388.69mg；果皮含黄酮3.20mg/g，cAMP含量49.67μg/g。制干率55.6%，干枣含总糖60.22%，酸1.37%。酒枣含可溶性固形物48.00%，总糖41.20%，酸1.01%。果核纺锤形，纵径1.95cm，横径0.78cm，核重0.51g，核尖较长，核纹较深，核面较粗糙，核内多不含种仁。

评价　该品种树体强健，丰产稳产，果实酸度大，制干率较高，品质中上。成熟期遇雨易裂果，采前落果较重。另外，该品种在长期栽培过程中产生了大果形变异类型，新栽枣树应注意筛选。

Langzao

Source and Distribution　The cultivar originated from Taigu, Qixian and Pingyao in mid-Shanxi Province. It is one of the dominant varieties there, and there are still some fruiting trees of over 100 years old.

Botanical Characters　The large tree is half-spreading with a strong central leader trunk, dense branches and a natural-round crown. The reddish-brown 1-year-old shoots have strong growth vigor with the internodes of 8.5 cm. There are 6 permanent secondary branches on each 1-year-old shoot which are 27.1 cm long with 6 nodes and less-developed thorns. The medium-sized column-shaped mother fruiting spurs can germinate 3~4 deciduous fruiting shoots which are 18.1 cm long with 11 leaves. The medium-sized leaves are long oval-shaped and dark green. The number of flowers is 61.6 per deciduous fruiting shoot. The large blossoms have a diameter of 7.6 mm for the zero-level flowers.

Biological Characters　The tree has strong vigor, strong germination and branching ability, medium suckering ability with a less-developed root. It bears early with strong early productivity and a long full-bearing stage with a high yield.

Fruit Characteristics　The medium-sized fruit is column-or-oval-shaped, averaging 12.1 g with a regular size. It has purplish-red thin skin and a smooth surface. The greenish-white thick flesh is tight-textured, sweet and a little sour, with medium juice and a better than normal quality for processing dried fruits and candied fruits. Recently it is greatly used to process smoked fruits. The percentage of edible part of fresh fruit is 95.8%, and SSC, TTS, TA and Vc is 36.90%, 32.48%, 0.85% and 388.69 mg per 100 g fresh fruit. The stone is spindle-shaped, averaging 0.51 g.

Evaluation　The cultivar has strong vigor with a high and stable yield and high rate for making dried fruits from fresh ones. The fruit has a better than normal quality with serious premature fruit-dropping and fruit-cracking if it rains in the maturing stage.

制 干
Drying Varieties
品 种

滕州大马牙

品种来源及分布 别名长红枣、大马牙。为长红枣品种群的重要品种。原产和分布于山东滕州、邹县、枣庄、益都、临朐等地的丘陵山区，多作为主栽品种大面积栽于梯田地堰或与农作物间作，数量占当地枣树的70%左右。

植物学性状 树体高大，树姿半开张，枝叶中密，层性分明，树冠呈乱头形。主干灰褐色，树皮片状条裂，较浅，容易剥落。枣头紫褐色，细长，平均长73.4cm，粗0.91cm，节间长8.5cm，蜡质少。二次枝长29.7cm，着生6节左右，枝形弯曲。无针刺。枣股圆柱形，抽生枣吊3～4个。枣吊长32.2cm，着叶18片。叶片卵状披针形，深绿色，叶厚，叶面平滑，富光泽，先端长，渐尖，先端尖圆或圆，叶基圆楔形或圆形。锯齿浅小不整齐，齿角尖圆或圆。花量较多，花序平均着花6朵，昼开型。

生物学特性 适应性强，耐旱，耐瘠薄，对枣疯病和枣叶壁虱抗性较强。生长结果要求温热气候，花期凉爽的地区生长结果不良。树势强旺，发枝力中强，在华北地区表现坐果稳定，丰产性好，枣头、2～3年和3年以上枝的吊果率分别为16.7%、79.6%和50.8%，幼树结果较早，成龄结果树株产35kg左右。在山西太谷地区，5月下旬始花，10月初果实成熟，果实生育期120d左右，为极晚熟品种类型。极抗裂果。

果实性状 果实中大，长圆形，侧向略扁，纵径3.69cm，横径2.14cm，侧径2.29cm，单果重10.0g，最大15.8g，大小不整齐。肩部平圆，梗洼浅、中广。果顶凹陷，有一字形缢痕。果柄平均长5.1mm。果面有隆起，富光泽。果皮厚，紫红色，富韧性。果点较大，稀疏。果肉浅绿色，质地较致密，汁液少，味甜酸，可制干，品质中上。鲜枣可食率95.2%，含可溶性固形物29.80%，总糖29.66%，酸0.44%，100g果肉维生素C含量486.39mg；果皮含黄酮5.77mg/g，cAMP含量101.94μg/g。制干率45.0%左右，干制红枣糖分多，总糖含量可达78%以上，略有辣味。果核中大，纺锤形，核重0.48g，核内多无种子。

评价 该品种树势强健，树体高大，耐旱、耐瘠，生长结果要求温度较高，适于花期气候温热的地区栽培。丰产性能强，适宜制干，品质较好。

Tengzhoudamaya

Source and Distribution The cultivar originated from and spreads in hilly areas of Tengzhou, Zouxian and Linqu in Shandong Province. The cultivation area for this cultivar occupies 70% of the total jujube area there.

Botanical Characters The large tree is spreading with medium-dense branches and an irregular crown. The purplish-brown 1-year-old shoots are 73.4 cm long. The internodes are 8.5 cm long with less wax and no thorns. The crooked secondary branches are 29.7 cm long with 6 nodes. The column-shaped mother fruiting spurs can germinate 3～4 deciduous fruiting shoots which are 32.2 cm long with 18 leaves. The dark-green leaves are thick, ovate-lanceolate, smooth and glossy. There are many flowers produced, averaging 6 ones per inflorescence.

Biological Characters The tree has strong adaptability, tolerant to drought and poor soils, with strong vigor, medium branching ability, strong resistance to jujube witches' broom and to mites on the leaves. It requires higher temperature for growth and bearing. In plain areas and hilly areas of north China, it has stable fruit set and high productivity. It is an extremely late-ripening variety with very strong resistance to fruit-cracking.

Fruit Characteristics The medium-sized oblong fruit has a little flattened lateral sides averaging 10.0 g with irregular sizes. It has a glossy surface with some protuberances and purplish-red thick skin. The light-green flesh is tight-textured, sour and sweet with less juice. It has a better than normal quality for dried fruits. The percentage of edible part of fresh fruit is 95.2%, and SSC, TTS, TA and Vc is 29.80%, 29.66%, 0.44% and 486.39 mg per 100 g fresh fruit. Dried fruits have a little hot taste with the TTS content of about 78.00%. The medium-sized stone is spindle-shaped averaging 0.48 g.

Evaluation The large plant of Tengzhoudamaya cultivar has strong tree vigor, tolerant to drought and poor soils with high productivity and a higher demand on temperature. It can only be developed in areas with warm temperatures when blooming. It is suitable for making dried fruits with a good quality.

制 干
Drying Varieties
品 种

· 141 ·

无核小枣

品种来源及分布 别名虚心枣、空心枣。原产山东的乐陵、庆云、无棣及河北的盐山、沧县、交河、献县、青县等地，以乐陵栽培较多。无核小枣栽培历史悠久，在古农书《齐民要术》中便有记载，是古老的地方名优品种。目前乐陵市郭家乡宋文海村还有500多年生的老龄枣树。

植物学性状 树体中大，树姿半开张，枝条中密，干性较强，树冠自然圆头形。树皮粗糙，皮裂呈条状。枣头黄褐色，平均长71.5cm，粗1.08cm，节间长7.5cm，二次枝长32.3cm，平均6节。无针刺。皮目中大，圆形，凸起，分布稀疏。枣股中大，圆柱形，抽吊力较强，抽生枣吊3～5个。枣吊平均长23.3cm，着叶14片。叶片中等大，叶长6.3cm，叶宽3.2cm，卵圆形，浓绿色，先端急尖，叶基心形，叶缘锯齿钝，中密。花量多，枣吊中部花序着花7～11朵。花小，花径5.8～6.2mm，昼开型。

生物学特性 树势中等或较弱，幼树枣头生长势强，发枝力中等，结果迟，产量较低，枣头和2～3年生枝吊果率分别为54.9%和50.4%，4年生枝结果量极少。在山西太谷地区，9月中旬果实着色，9月底成熟，果实生育期115d左右，为晚熟品种类型。成熟期遇雨裂果较少。

果实性状 果实小，圆柱形或长椭圆形，纵径2.81cm，横径2.32cm，单果重7.8g，大小不整齐。果梗细而较长，梗洼中度深广。果顶平或微凹，柱头遗存。果皮薄，鲜红色，果面平滑。果点小而稀，圆形，不明显。果肉厚，白色或乳白色，肉质细而脆，味甜，汁液较少，适宜制干，品质上等。鲜枣可食率96%～100%，含可溶性固形物31.20%，含糖24.30%，酸0.39%，100g果肉维生素C含量408.42mg；果皮含黄酮6.44mg/g，cAMP含量110.95μg/g。制干率53.8%，干枣含总糖64.74%，酸0.70%。中、小果核大部退化为膜状软核或木质化残核，少数大果核发育正常。正常发育核较小，椭圆形，纵径1.65cm，横径0.72cm，核重0.30g，核尖短，核纹浅，含仁率15.0%。

评价 该品种风土适应性较差，结果迟，产量较低。果个小，适宜制干，品质优良，抗裂果。尤其部分果实具有无核的独特性状是其突出特点，具有食用方便、可食率高、经济价值高等优点，也是无核资源研究的重要材料。干枣耐贮运，不易回潮，南方多雨潮湿的地区也可适当发展。

Wuhexiaozao

Source and Distribution The cultivar originated from Leling, Qingyun, Wudi in Shandong Province and Yanshan, Cangxian, Jiaohe, Xianxian and Qingxian in Hebei Province with a long history.

Botanical Characters The medium-sized tree is half-spreading with medium-dense branches, a strong central leader trunk and a natural-round crown. The yellowish-brown 1-year-old shoots are 71.5 cm long and 1.08 cm thick with the internodes of 7.5 cm. The secondary branches are 32.3 cm long with 6 nodes and without thorns. The medium-sized column-shaped mother fruiting spurs can germinate 3～5 deciduous fruiting shoots which are 23.3 cm long with 14 leaves. The medium-sized leaves are oval-shaped and dark green. There are many small flowers with a diameter of 5.8～6.2 mm, generally 7～11 ones per inflorescence.

Biological Characters The tree has moderate or weak vigor and medium branching ability, with strong growth vigor for the 1-year-old shoots of young trees. It bears late with a low yield. It is a late-ripening variety with light fruit-cracking even if it rains in the maturing stage.

Fruit Characteristics The small fruit is column-shaped or long oval-shaped, averaging 7.8 g with a regular size. It has a brightly-red thin skin and a smooth surface. The white or ivory-white flesh is thick, crisp and sweet with less juice and a good quality for dried fruits. The percentage of edible part of fresh fruit is 96.2%, and SSC, TTS, TA and Vc is 31.20%, 24.30%, 0.39% and 408.42 mg per 100 g fresh fruit. The percentage of fresh fruits which can be made into dried ones is 53.8%, and the content of TTS and TA in dried fruit is 64.74% and 0.70%. The small stone is oval-shaped averaging 0.30 g.

Evaluation The cultivar has poor adaptability, bearing late with a low yield. The small fruit has regular sizes with strong resistance to fruit-cracking and is suitable for making dried fruits with excellent quality. Dried fruits have strong tolerance to storage and transport without moisture region.

制 干
Drying Varieties
品 种

· 143 ·

永 城 长 红

品种来源及分布 别名长红枣。主要分布在河南省永城市、夏邑县、虞城县等地，是当地主栽品种。栽培历史500多年。

植物学性状 树体高大，树姿开张，树冠圆柱形。主干灰褐色或紫褐色，皮较粗糙，条裂，裂纹较深，容易剥落。枣头红褐色，生长势弱，平均长52.7cm，粗0.74cm，节间长7.4cm。无针刺。枣股小，圆锥形，粗壮，平均着生枣吊4.0个，最多6个。枣吊平均长20.2cm，着叶13片。叶片中大，椭圆形，绿色，叶尖渐尖，叶基圆楔形，叶缘锯齿圆钝。花量多，花序平均着花5朵，枣吊中部节位的花序着花可达8～12朵。花较大，花径6.8mm，昼开型。

生物学特性 适应性和抗逆性较强，耐旱、耐涝、耐盐碱，较抗枣疯病。树势强健，发枝力中等。枣股寿命长，结果枝系较稳定，树体容易管理。坐果率高，产量高而稳定，枣头、2～3年和3年以上枝的吊果率分别为36.7%、147.1%和60.8%，幼树定植后2年开始结果，盛果期树一般株产60kg左右。在山西太谷地区，9月下旬果实成熟采收，果实生育期110d左右，为晚熟品种类型。成熟期较一致，落果和裂果均轻。

果实性状 果实较小，圆柱形，纵径3.27cm，横径2.11cm，单果重6.9g，最大果重13.5g，大小较整齐。两端稍粗，中腰略细。果肩平圆，微突。梗洼中深。果顶圆，顶点微洼。果面光滑。果皮中厚，橙红色。果肉浅绿色，质地致密，汁液中多，味甜，微酸，品质中上等，适宜制干。鲜枣可食率94.5%，制干率39.6%，干枣含总糖60.90%，酸0.96%。果核较小，呈长纺锤形，核重0.38g。核纹浅，纵条形。含仁率43.3%，但不饱满。

评价 该品种适应性强，耐旱涝和盐碱。树体强健，丰产稳产。果实不易裂果、浆烂，品质中上，但制干率低。

Yongchengchanghong

Source and Distribution The cultivar, also called Changhongzao, mainly spreads in Yongcheng, Xiayi and Yucheng in Henan Province. It is the dominant variety there with a cultivation history of 500 years.

Botanical Characters The large tree is spreading with a column-shaped crown. The reddish-brown 1-year-old shoots are weak, 52.7 cm long and 0.74 cm thick. The internodes are 7.4 cm long without thorns. The column-shaped mother fruiting spurs can germinate 4 deciduous fruiting shoots which are 20.2 cm long with 13 leaves. The green leaves are medium-large and oval-shaped. There are many large flowers with a diameter of 6.8 mm, averaging 5 ones per inflorescence. There are 8～12 flowers on the middle part of deciduous fruiting shoot.

Biological Characters The tree has strong adaptability, tolerant to drought, water-logging and saline-alkaline with strong vigor, medium branching ability, and strong resistance to adverse conditions and to jujube witches' broom. It has high fruit set, a high and stable yield with a long life for mother fruiting spurs. It generally bears in the 2nd year after planting. A tree in its full-bearing stage has a yield of 60 kg. In Taigu County of Shanxi Province, it matures in late September. It is a late-ripening variety with light fruit-cracking and light premature fruit-dropping.

Fruit Characteristics The small column-shaped fruit weighs 6.9 g (maximum 13.5 g) with a regular size. It has thicker ends and thinner middle part. The fruit has a flat-round shoulder, a medium-deep stalk cavity, a round fruit apex, a smooth surface and medium-thick orange skin. The light-green flesh is tight-textured, sweet and a little sour, with medium juice. It has a better than normal quality for dried fruits. The percentage of edible part of fresh fruit is 94.5%. The rate of fresh fruits which can be made into dried ones is 39.6%, and the content of TTS and TA in dried fruit is 60.90% and 0.96%. The small long-spindle-shaped stone weighs 0.38 g with a shriveled kernel and shallow, vertical veins. The percentage of containing kernels is 43.3%.

Evaluation The cultivar has strong adaptability, tolerant to drought, water-logging and saline-alkaline, strong resistance to fruit-cracking with strong tree vigor and a high and stable yield. It has a good quality for dried fruit.

鸡 心 枣

品种来源及分布 别名小枣。原产河南省新郑市、中牟县、西华县和郑州市郊等地，以新郑市栽培较多，为当地次主栽品种，现尚有400多年生的老枣树。

植物学性状 树体中大，树姿直立，树冠圆柱形。主干皮裂条状，较浅。枣头黄褐色，平均长88.2cm，粗1.22cm，节间长6.3cm。二次枝较长，为34.6cm，弯曲度小，平均6节。针刺发达，长1.2cm左右。皮目较大，长圆形，分布密。枣股小，圆柱形，抽吊力中等，枣吊平均长23.5cm，着叶14片。叶片较大，叶长6.1cm，叶宽2.9cm，卵圆形，绿色，先端渐尖，叶基近圆形，叶缘锯齿较细而整齐。花量多，花较大，花径7.6mm，昼开型。

生物学特性 树势中等，萌芽率和发枝力较弱，结果较早，根蘖苗一般第二年开始结果，15年左右进入盛果期，盛果期长，在原产地产量较高而稳定。枣头、2~3年生枝和4年生枝吊果率分别为43.1%、55.0%和6.3%。在山西太谷地区，产量低而不稳，9月下旬果实成熟，果实生育期110d左右，为中晚熟品种类型。成熟期遇雨裂果轻。

果实性状 果实小，鸡心形或卵圆形，纵径2.47cm，横径1.98cm，单果重4.9g，大小较整齐。果梗中长、中粗，梗洼广而较深。果顶尖，柱头遗存。果皮较薄，紫红色，果面平滑，果点小，不明显。果肉中厚，绿白色，肉质致密，略脆，味甘甜，汁液较少，适宜制干，品质上等。鲜枣可食率93.8%，含可溶性固形物37.20%，总糖32.68%，酸0.70%，100g果肉维生素C含量488.90mg；果皮含黄酮13.42mg/g，cAMP含量130.67μg/g。制干率49.9%，干枣含总糖57.03%，酸0.88%。肉质较紧密，富有弹性，耐挤压，耐贮运。核较大，纺锤形，纵径1.64cm，横径0.68cm，核重0.48g，核尖短，核纹较浅，含仁率75%左右，种仁较饱满。

评价 该品种在气温较高地区产量高而稳定，成熟期遇雨裂果轻。果实小，适宜制干，干枣品质优良，果肉富弹性，耐挤压，贮运性好，可在原产地适量发展。

Jixinzao

Source and Distribution The cultivar originated from Xinzheng, Zhongmou, Xihua and the suburbs of Zhengzhou City in Henan Province. It is mainly planted in Xinzheng which is the 2nd dominant variety there. There are still some trees of over 400 years old.

Botanical Characters The medium-sized tree is half-spreading with a round crown. The yellowish-brown 1-year-old shoots are 88.2 cm long and 1.22 cm thick, with the internodes of 6.3 cm. The secondary branches are long, averaging 34.6 cm with 6 nodes of a small curvature and less-developed thorns of 1.2 cm long. The mother fruiting spurs are small and column-shaped. The deciduous fruiting shoots are 23.5 cm long with 14 leaves. The large green leaves are oval-shaped. There are many large flowers with a diameter of 7.6 mm produced. It blooms in the daytime.

Biological Characters The tree has moderate vigor, weak germination and branching ability, bearing early. Generally the suckers bear fruit in the 2nd year and enter the long full-bearing stage in the 15th year with a high and stable yield. It is a mid-late-ripening variety with light fruit-cracking even if it rains in the maturing stage.

Fruit Characteristics The small fruit is chicken-heart-shaped or oval-shaped, averaging 4.9 g with a regular size. It has purplish-red thin skin and a smooth surface. The greenish-white flesh is medium thick and tight-textured, crisp and sweet, with less juice and a good quality for dried fruits. The percentage of edible part of fresh fruit is 93.8%, and SSC, TTS, TA and Vc is 37.20%, 32.68%, 0.70% and 488.90 mg per 100 g fresh fruit. Dried fruits have tight-textured and whippy flesh with strong tolerance to extrusion, storage and transport. The large stone is spindle-shaped, averaging 0.48 g.

Evaluation The cultivar has strong adaptability with a high and stable yield. The small fruit has a regular size with light fruit-cracking even if it rains in the maturing stage and is suitable for making dried fruits with an excellent quality. Dried fruits have whippy flesh with strong tolerance to extrusion and storage and transport.

临泽小枣

品种来源及分布　分布于甘肃的临泽、张掖、高台、酒泉、金塔等地，为当地原有的主栽品种。栽培历史悠久，至今尚有300多年生的老枣树。

植物学性状　树体较大，树姿半开张，干性较强，枝条较密，树冠圆锥形。主干皮裂较浅，不易剥落。枣头黄褐色，平均长73.5cm，粗0.88cm，节间长8.5cm。二次枝长26.7cm，平均6节，无针刺。皮目小而密，圆形，凸起，开裂。枣股较大，圆柱形，抽吊力较强，抽生枣吊3～4个，多的达5个。枣吊长14～17cm，着叶11片。叶片中大，叶长4.9～5.5cm，宽2.6～3.0cm，长卵形，浅绿色，较薄，先端渐尖，叶基圆形或宽楔形，叶缘锯齿中度密，较粗。花量多，枣吊中部每花序着花7～9朵。花中大，花径7mm左右，昼开型。

生物学特性　树势中等，萌芽率和成枝力较强，结果较早，10年生后进入盛果期，产量较高，盛果期长，百年生树仍可正常结果。枣头、2～3年生枝和4年生枝吊果率分别为180.0%、56.9%和6.3%。在山西太谷地区，9月中旬果实成熟，果实生育期100d左右，为中熟品种类型。成熟期遇雨裂果较重。

果实性状　果实小，卵圆形或近圆形，纵径2.50cm，横径2.20cm，单果重6.1g，最大9.5g，大小较整齐。梗洼窄而中深，果顶凹，柱头遗存。果点小，分布中密。果皮较薄，紫红色。果肉较厚，绿白色，肉质致密，较细脆，味甜略酸，汁液中多，适宜制干，品质上等。鲜枣可食率94.9%，含可溶性固形物35%～38%，总糖32.80%，酸0.78%，100g果肉维生素C含量423.08mg；果皮含黄酮4.34mg/g。制干率50%以上，干枣含总糖70.76%，酸1.19%。核较小，纺锤形或倒卵形，纵径1.60cm，横径0.72cm，核重0.31g，核尖短，核纹较深，含仁率10%左右。

评价　该品种适应性广，抗风，极耐干旱。结果较早，丰产性强，果实较小，肉质致密，品质上等，为优良的制干品种，适宜北方干旱地区发展。

Linzexiaozao

Source and Distribution　The cultivar originated from and spreads in Linze, Zhangye and Jiuquan in Gansu Province. It is the dominant variety there with a long history. There are still some trees of over 300 years old.

Botanical Characters　The large tree is half-spreading with a strong central leader trunk, dense branches and a conical-shaped crown. The yellowish-brown 1-year-old shoots are 73.5 cm long and 0.88 cm thick with the internodes of 8.5 cm. The secondary branches are 26.7 cm long with 6 nodes and without thorns. The large column-shaped mother fruiting spurs can germinate 3～4 deciduous fruiting shoots which are 14～17 cm long with 11 leaves. The medium-sized thin leaves are long oval-shaped and light green. There are many medium-sized flowers with a diameter of 7 mm, averaging 7～9 ones per inflorescence in the middle part of deciduous fruiting shoots.

Biological Characters　The tree has moderate vigor, strong germination and branching ability. It bears early and enters the long full-bearing stage in the 10th year with a high yield. Trees of 100 years old can still bear fruit as normal. It is a mid-ripening variety with serious fruit-cracking if it rains in the maturing stage.

Fruit Characteristics　The small oval-shaped fruit weighs 6.1 g with a regular size. It has purplish-red thin skin with small and medium-dense dots. The greenish-white thick flesh is tight-textured, crisp, sweet and a little sour, with medium juice and a good quality for dried fruits. The percentage of edible part of fresh fruit is 94.9%, and SSC, TTS, TA and Vc is 35%～38%, 32.80%, 0.78% and 423.08 mg per 100 g fresh fruit. The percentage of fresh fruits which can be made into dried ones is over 50%, and the content of TTS in dried fruit is 72.8%. The small stone is spindle-shaped or obovate, with a vertical and cross diameter of 1.60 cm and 0.72 cm, averaging 0.31 g with a short apex and deep veins. The percentage of containing kernels is 10%.

Evaluation　The cultivar has strong adaptability, strong tolerance to wind and drought, bearing early with strong productivity. The small fruit has tight-textured flesh and a high quality. Dried fruits have a good quality. It is a good drying variety which can be developed in droughty areas of northern China.

制 干
Drying Varieties
品 种

临泽大枣

品种来源及分布 主要分布在甘肃临泽，为当地的原产品种。此外，张掖、高台等地也有零星栽种。

植物学性状 树体较大，树姿开张，分枝较多，外围枝梢下垂，枝叶稠密，树冠呈乱头形。主干灰褐色，皮裂纹深，呈细条状，不易剥落。枣头红褐色，生长势强，长93.3cm，粗0.89cm，节间长7.2cm。二次枝生长健壮，平均长30.7cm，6节，弯曲度中等。针刺发达，不易脱落。枣股圆柱形，平均抽生枣吊4.0个，最多6个。枣吊长19.9cm，着生13片叶。叶片中大，椭圆形，较薄，绿色，先端渐尖，叶基圆形或偏斜形，叶缘具粗钝锯齿，较整齐。花量较多，花序平均着花5朵，花较大，花径7.0mm，为昼开型。

生物学特性 适应性强，抗风、耐干旱，土质以沙壤土最好。树势强健，发枝力强。结果较晚，根蘖苗定植后3年开始结果，结果较稳定，产量中等，盛果期树株产25kg左右。在山西太谷地区，9月上旬果实开始着色，9月中旬成熟采收，果实生育期100d左右，为中熟品种类型。成熟期遇雨裂果轻。

果实性状 果实大，短柱形或近圆形，纵径3.38cm，横径3.03cm，单果重19.4g，大小整齐。果肩平圆，梗洼窄、中深，柱头遗存。果柄较长，长度4.4mm。果面不平，有较明显的小块起伏。果皮厚，紫红色，果点明显。果肉浅绿色，质地疏松，汁液中多，味甜，略具酸味，制干品质上等。鲜枣可食率93.7%，含可溶性固形物34.00%，100g果肉维生素C含量307.69mg；果皮含黄酮3.77mg/g，cAMP含量505.25μg/g。制干率47.0%，干枣含总糖62.25%，酸0.88%。果核大，纺锤形，核重1.22g，含仁率3.3%，种仁不饱满。

评价 该品种适应性强，树体高大强健，结果稳定，产量中等，抗裂果。果实大，制干品质上等，为较好的制干品种。缺点是果核大，可食率较低。

Linzedazao

Source and Distribution The cultivar originated from and mainly spreads in Linze, Zhangye and Gaotai County of Gansu Province.

Botanical Characters The large tree is spreading with pendulous external branches and an irregular crown. The reddish-brown 1-year-old shoots have strong growth potential, 93.3 cm long and 0.89 cm thick. The internodes are 7.2 cm long with developed and uneasily falling-off thorns. The strong secondary branches are 30.7 cm long with 6 nodes of medium curvature. The column-shaped mother fruiting spurs can germinate 4 (6 at most) deciduous fruiting shoots which are 19.9 cm long with 13 leaves. The thin green leaves are medium-large and oval-shaped. There are many large flowers with a diameter of 7.0 ones mm produced, averaging 5 ones per inflorescence.

Biological Characters The tree has strong adaptability, strong vigor, strong branching ability and dense branches, tolerant to wind and drought. It grows best in sandy loam soil, yet bears late. Root seedlings bear fruit in the 3rd year after planting with a stable and medium yield. A tree in its full-bearing stage has a yield of 25 kg. It is a mid-ripening variety with light fruit-cracking even if it rains in the maturing stage.

Fruit Characteristics The large oblong fruit weighs 19.4 g with a regular size. It has a flat-round shoulder, a medium-deep and narrow stalk cavity, a small fruit apex, a remnant stigma, a stalk of 4.4 mm long, an unsmooth surface with some small massive protuberances and purplish-red thick skin with distinct dots. The light-green flesh is loose-textured, sweet and a little sour with medium juice. It has a good quality for dried fruits. The percentage of edible part of fresh fruit is 93.7%, and SSC and Vc is 34.00% and 307.69 mg per 100 g fresh fruit. The percentage of fresh fruits which can be made into dried ones is 47.0%, and the content of TTS and TA in dried fruit is 78.00% and 0.50%. The large spindle-shaped stone weighs 1.22 g with the percentage of containing kernels of 3.3%. Most kernels are shriveling.

Evaluation The large plant of Linzedazao cultivar has strong adaptability and strong vigor with a stable and medium yield. The large fruit has strong resistance to fruit-cracking with crisp flesh and is suitable for fresh-eating and making dried fruits. Dried fruits have high sugar content and a good quality. Its disadvantage is having a large stones and a low rate of making dried fruits from fresh ones.

制 干
Drying Varieties
品 种

· 151 ·

稷 山 圆 枣

品种来源及分布　原产山西汾河下游稷山县稷峰镇南阳村一带。栽培不多，历史不详。

植物学性状　树体较小，树姿开张，干性弱，枝条细而密，树冠呈乱头形。主干皮裂粗而深，呈宽条状，较易脱落。枣头黄褐色，平均生长量34.0cm，粗度0.74cm，节间长8.1cm，着生永久性二次枝少。二次枝长21.9cm，4～5节。针刺不发达。皮目小而稀，椭圆形。枣股中大，抽吊力较强，平均抽生枣吊3.9个。枣吊一般长16.3cm，着叶12片。叶片较小，叶长5.6cm，叶宽2.6cm，椭圆形，浓绿色，先端急尖，叶基圆楔形，叶缘锯齿中密，粗钝。花量较多，枣吊平均着花70.7朵，花序平均4.6朵。花朵中大，花径7.3mm，昼开型。

生物学特性　树势中庸，萌芽率高，成枝力强，不易生萌蘖。开花结果较迟，一般第三年开始结果，15年后进入盛果期，盛果期长，产量中等。枣头吊果率37.4%，2～3年生枝为44.7%，4年生枝为29.5%，主要坐果部位在枣吊的1～8节，占坐果总数的76.3%。在山西太谷地区，果实9月下旬进入脆熟期，果实生育期110d以上，为晚熟品种类型。较抗裂果，但采前落果严重。

果实性状　果实大，圆形或倒卵圆形，纵径3.63cm，横径3.90cm，单果重23.3g，大小较整齐。果梗较短，梗洼广而浅。果顶微凹，柱头遗存。果皮中厚，紫红色，果面平滑。果肉厚，肉质较致密，味甜，汁液中多，品质中上，可制干和加工酒枣。鲜枣可食率96.1%，含可溶性固形物34.20%，总糖27.96%，酸0.80%，100g果肉维生素C含量347.27mg；果皮含黄酮7.35mg/g，cAMP含量256.96μg/g。制干率51.4%，干枣含总糖52.20%，酸0.70%。果核大，纺锤形，纵径2.17cm，横径1.18cm，核重0.90g，核尖短，核纹较浅，核面较粗糙，含仁率35%左右，种仁不饱满。

评价　该品种树体较小，树势中庸，产量中等且不稳定。但果实较大，外形美观，适宜制干和加工醉枣。

Jishanyuanzao

Source and Distribution　The cultivar originated from Nanyangcun of Jifeng town in Jishan County of Shanxi Province with a small quantity and an unknown history.

Botanical Characters　The small tree is spreading with a weak central leader trunk, thin dense branches and an irregular crown. The yellowish-brown 1-year-old shoots are 34.0 cm long and 0.74 cm thick with the internodes of 8.1 cm. There are only a few secondary branches which are 21.9 cm long with 4～5 nodes and less-developed thorns. The medium-sized mother fruiting spurs can germinate 3.9 deciduous fruiting shoots which are 16.3 cm long with 12 leaves. The small leaves are oval-shaped and dark green. There are many medium-sized flowers with a diameter of 7.34 mm produced, averaging 70.7 ones per deciduous fruiting shoot and 4.6 ones per inflorescence.

Biological Characters　The tree has moderate vigor, high germination rate, strong branching ability and weak suckering ability, blooming and bearing late. It generally bears in the 3rd year after planting and enters the long full-bearing stage in the 15th year with a medium yield. In Taigu County of Shanxi Province, it enters the crisp-maturing stage in late September with a fruit growth period of 110 d. It is a late-ripening variety with strong resistance to fruit-cracking and serious premature fruit-dropping.

Fruit Characteristics　The large globosor obovate fruit weighs 23.3 g with a regular size. It has a short stalk, a wide and shallow stalk cavity, a slightly sunken fruit apex, a remnant stigma, medium-thick and purplish-red skin and a smooth surface. The thick flesh is tight-textured and sweet, with medium juice and a better than normal quality for processing dried fruits and alcoholic jujubes. The percentage of edible part of fresh fruit is 96.1%, and SSC, TTS, TA and Vc is 34.20%, 27.96%, 0.80% and 347.27 mg per 100 g fresh fruit. The percentage of fresh fruits which can be made into dried ones is 51.43%, and the content of TTS and TA in dried fruit is 69.16% and 1.20%. The large spindle-shaped stone has a vertical and cross diameter of 2.17 cm and 1.18 cm averaging 0.90 g.

Evaluation　The small plant of Jishanyuanzao cultivar has moderate or weak vigor with a medium and unstable yield. The large fruit has a beautiful appearance and is suitable for processing dried fruits and alcoholic jujube.

夏县紫圆枣

品种来源及分布 别名紫圆。原产于山西夏县。栽培数量不多，历史不详。

植物学性状 树体较小，树姿直立，干性强，枝条少而粗壮，枝系紧凑，树冠圆锥形，主干皮裂呈条状。枣头红褐色，生长势较强，平均生长量72.8cm，粗1.21cm，节间长7.8cm，二次枝长36.4cm，平均7节。针刺不发达。皮目中等大，分布较稀，圆形。枣股大，圆锥形，抽吊力较强，抽生枣吊3～5个。枣吊平均长23.2cm，着叶17片。叶片中大，叶长5.1cm，叶宽2.5cm，长卵圆形，浓绿色，先端急尖，叶基圆形，叶缘具钝齿。花量中多，枣吊平均着花69朵，花序平均4朵。花朵中大，花径7.2mm，夜开型。

生物学特性 树势较强，萌芽率高，成枝力弱，萌蘖力中等。丰产稳产，坐果率较高，枣头吊果率103.2%，2～3年生枝为51.9%，4年生枝为53.0%，主要坐果部位在枣吊的3～9节，占坐果总数的71%。在山西太谷地区，9月15日果实脆熟。果实生育期105d左右，为中熟品种类型。

果实性状 果实较大，圆形，纵径3.19cm，横径3.32cm，单果重16.2g，大小较整齐。果梗较短，梗洼广而浅。果顶凹，柱头遗存。果皮中厚，紫红色，色泽艳丽，果面光滑。果点中大，分布中密。果肉厚，肉质较松，味甜，汁液少，品质中上，可制干。鲜枣可食率95.8%，含可溶性固形物40.20%，总糖30.25%，酸0.79%，100g果肉维生素C含量371.98mg，含水量60.4%；果皮含黄酮14.69mg/g，cAMP含量258.35μg/g。制干率56.0%，枣果自然晒干后，色泽紫红，果皮光滑不皱，果肉富弹性。干枣含总糖66.54%，酸1.23%。果核中大，椭圆形，纵径1.74cm，横径0.94cm，核重0.68g，核纹浅，含仁率38.3%，种仁较饱满。

评价 该品种树姿直立，树冠紧凑，叶色浓绿，可景观观赏。适应性较强，产量较高而稳定。果实较大，含水量少，适宜制干，制干率较高，品质优良，可适度经济栽培。

Xiaxianziyuanzao

Source and Distribution The cultivar originated from Xiaxian County of Shanxi Province with a small quantity and an unknown history.

Botanical Characters The small tree is vertical with a strong central leader trunk, sparse yet strong branches and a conical-shaped crown. The reddish-brown 1-year-old shoots have strong growth vigor, averaging 72.8 cm long and 1.21 cm thick, with the internodes of 7.8 cm. The secondary branches are 36.4 cm long with 7 nodes and less-developed thorns. The large conical-shaped mother fruiting spurs can germinate 3～5 deciduous fruiting shoots which are 23.2 cm long with 17 leaves. The medium-sized leaves are long oval-shaped and dark green. The number of flowers is medium large, averaging 69 per deciduous fruiting shoot and 4 per inflorescence. The medium-sized blossoms have a diameter of 7.2 mm.

Biological Characters The tree has strong vigor, high germination rate, weak branching ability and medium suckering ability with high fruit set and a high and stable yield. In Taigu County of Shanxi Province, it enters the crisp-maturing stage around September. 15 with a fruit growth period of 105 d. It is a mid-ripening variety.

Fruit Characteristics The medium-sized round fruit has a vertical and cross diameter of 3.19 cm and 3.32 cm, averaging 16.2 g with a regular size. It has a short stalk, a wide and shallow stalk cavity, a sunken fruit apex, a remnant stigma, medium-thick and purplish-red skin and a smooth surface with medium-large and medium-dense dots. The thick flesh is loose-textured and sweet, with medium juice and a medium quality for dried fruits. The percentage of edible part of fresh fruit is 95.8%, and SSC, TTS, TA and Vc is 40.20%, 30.25%, 0.79% and 371.98 mg per 100 g fresh fruit. The water content is 60.4%. The percentage of fresh fruits which can be made into dried ones is 56.0%. The dried fruits have a purplish-red color, smooth skin and whippy flesh. The content of TTS and TA in dried fruit is 66.54% and 1.23%. The medium-sized stone is oval-shaped, averaging 0.68 g.

Evaluation The plant of Xiaxianziyuanzao cultivar is vertical with a compact crown and dark-green leaves and can be used for ornament. It has strong adaptability with a high and stable yield. The large fruit has low water content and high rate for making dried fruits from fresh ones with excellent quality. It is suitable for making dried fruits and can be planted moderately for commercial production.

制 干
Drying Varieties
品 种

· 155 ·

密 云 小 枣

品种来源及分布 别名密云金丝小枣。原产北京密云,主产区分布于白河以西、人沙河以东的冲积平原,以密云西田各庄产品最有名,通州等地也有栽培。栽培历史1 000年以上。

植物学性状 树体中大,树姿较直立,干性强,枝叶较密,树冠呈圆柱形。主干灰褐色,树皮块裂,不易剥落。枣头黄褐色,平均长62.4cm,粗0.92cm,节间长6.9cm,蜡质少。针刺不发达。枣股圆柱形,抽吊能力较强,平均抽生枣吊4.0个。枣吊较长,平均长15.7cm,着叶11片。叶片中大,椭圆形,中厚,绿色,有光泽,先端钝尖,叶基圆楔形。叶缘具锐锯齿。花量多,花序平均着花10朵。

生物学特性 适应性较强,抗寒,耐旱,在沙壤土栽植表现良好。树势中等,发枝力较强,坐果能力中等,产量较低,枣头、2~3年和3年以上枝的吊果率分别为10.0%、67.2%和65.8%。在山西太谷地区,4月下旬萌芽,5月底始花,9月下旬果实成熟,果实生育期约105d,为中晚熟品种类型。

果实性状 果实小,卵圆形,纵径2.45cm,横径1.92cm,单果重4.2g,大小整齐。果肩平圆,梗洼广、中深。果顶平,柱头残存。果面平整光滑,具光泽。果皮红色,中厚,韧性较强。果点中大,不明显,稀疏。果肉浅绿色,质地酥脆,汁液多,味甜,可鲜食,多以制干为主,品质上等。鲜枣可食率95.7%,含可溶性固形物31.64%,酸0.95%,100g果肉维生素C含量190.40mg;果皮含黄酮3.40mg/g,cAMP含量 185.93 μg/g。制干率60.0%,干枣肉厚,皮细,富弹性,耐贮运。果核细小,长纺锤形,核重0.18g,核内多不含种子。

评价 该品种适土性强,耐旱,较耐瘠。果实小,肉质酥脆,糖分多,制干率高,干枣外形好,耐贮运,为品质优良的制干品种。但产量较低,不抗枣疯病。

Miyunxiaozao

Source and Distribution The cultivar, also called Miyunjinsixiaozao, originated from Miyun County of Beijing. It mainly spreads in alluvial plains in west White River and east Dasha River, with a cultivation history of over 1 000 years. The most famous product is in Xitiangezhuang of Miyun County.

Botanical Characters The medium-sized tree is vertical with a strong central leader trunk, dense branches and a column-shaped crown. The yellowish-brown 1-year-old shoots are 62.4 cm long and 0.92 cm thick. The internodes are 6.9 cm long with less wax and no thorns. The column-shaped mother fruiting spurs can germinate 4 deciduous fruiting shoots which are 15.7 cm long with 11 leaves. The medium-sized leaves are medium thick, oval-shaped, green and glossy. There are many flowers produced, averaging 10 ones per inflorescence.

Biological Characters The tree has strong adaptability, tolerant to cold and drought, with moderate vigor, strong branching ability, medium fruit set and a low yield. It grows well in sandy loam soil. In Taigu County of Shanxi Province, it germinates in late April, begins blooming in late May and matures in late September with a fruit growth period of 105 d. It is a mid-late-ripening variety.

Fruit Characteristics The small oval-shaped fruit has a vertical and cross diameter of 2.45 cm and 1.92 cm, averaging 4.2 g with a regular size. It has a flat-round shoulder, a wide and medium-deep stalk cavity, a flat fruit apex, a remnant stigma, a smooth and glossy surface and medium-thick red skin with medium-sized, indistinct and sparse dots. The light-green flesh is crisp, juicy and sweet. It can be used for fresh-eating, yet mainly for making dried fruits with a good quality. The percentage of edible part of fresh fruit is 95.7%, and SSC, TA and Vc is 31.64%, 0.95% and 190.40 mg per 100 g fresh fruit. The content of flavones and cAMP in mature fruit skin is 3.40 mg/g and 185.93 μg/g. The percentage of fresh fruits which can be made into dried ones is 60.0%. Dried fruits have thick, delicate and whippy flesh, tolerant to storage and transport. The small thin stone is long spindle-shaped, averaging 0.18 g. Most stones contain no kernels.

Evaluation The medium-sized plant of Miyunxiaozao cultivar has strong adaptability and is tolerant to drought and poor soils. The small fruit has crisp flesh, high sugar content and high percentage for making dried fruits with good appearance and is tolerant to storage and transport. It is a good drying variety with a low yield and poor resistance to jujube witches' broom.

稷山柳罐枣

品种来源及分布 别名柳罐枣。分布于山西稷山县稷峰镇南阳村一带，栽培数量不多，与板枣为同一产区。栽培历史不详。

植物学性状 树体小，树姿开张，干性弱，枝条细而较密，树冠乱头形。主干皮裂块状。枣头黄褐色，平均生长量45.3cm，粗0.76cm，节间长8.4cm，二次枝长27.8cm，3～4节。针刺不发达。皮目中大，分布较稀，卵圆形，灰白色。枣股中大，抽吊力强，平均抽生枣吊4.1个。枣吊平均长15.3cm，着叶10片。叶片中大，叶长5.6cm，叶宽3.0cm，椭圆形，绿色，先端急尖，叶基圆楔形，叶缘锯齿粗钝。花量中多，枣吊平均着花55.4朵，花序平均5.3朵。花中大，花径6.6mm。

生物学特性 树势较弱，萌芽率低，成枝力较强，开花结果较迟，根蘖苗一般第三年开始结果，15年后进入盛果期。坐果率较低，产量低而不稳，枣头吊果率50.0%，2～3年生枝为38.1%，4年生枝为34.2%，主要坐果部位在枣吊的1～9节，占坐果总数的92.3%。在山西太谷地区，9月下旬果实脆熟，果实生育期110d左右，为晚熟品种。抗裂果，但采前落果严重。

果实性状 果实大，卵圆形，纵径4.34cm，横径3.44cm，单果重21.4g。梗洼较广，中等深，果顶微凹，柱头遗存。果皮厚，深红色，果面不平滑。果肉厚，肉质硬，味甜，汁液较少，品质中上，适宜制干。鲜枣可食率96.2%，含可溶性固形物33.00%，总糖29.48%，酸0.58%，100g果肉维生素C含量397.31mg，含水量61.6%；果皮含黄酮3.23mg/g，cAMP含量101.52μg/g。制干率56.0%，干枣含总糖64.51%，酸0.94%。果核较大，纺锤形，纵径2.46cm，横径0.95cm，核重0.82g，核面粗糙，含仁率55.0%左右，种仁不饱满。

评价 该品种树体小，树势弱，要求较高的肥水栽培条件。果实大，可制干，抗裂果。但结果较晚，产量不稳，采前落果重，不宜生产栽培。

Jishanliuguanzao

Source and Distribution The cultivar spreads in Nanyangcun of Jifeng town in Jishan County of Shanxi Province, the same as Banzao. It has a small quantity with an unknown history.

Botanical Characters The small tree is spreading with a weak central leader trunk, thin dense branches and an irregular crown. The yellowish-brown 1-year-old shoots are 45.3 cm long and 0.76 cm thick with the internodes of 8.4 cm. The secondary branches are 27.8 cm long with 3～4 nodes and less-developed thorns. The medium-sized mother fruiting spurs can germinate 4.1 deciduous fruiting shoots which are 15.3 cm long with 10 leaves. The green leaves are medium-sized and oval-shaped. The number of flowers is medium large, averaging 55.4 ones per deciduous fruiting shoot and 5.3 ones per inflorescence. The medium-sized blossoms have a diameter of 6.6 mm.

Biological Characters The tree has weak vigor, low germination rate and strong branching ability, bearing and blooming late. Generally the suckers bear fruit in the 3rd year after planting and enter the full-bearing stage in the 15th year with low fruit set and a low and unstable yield. It is a late-ripening variety with serious premature fruit-dropping yet strong resistance to fruit-cracking.

Fruit Characteristics The large oval-shaped fruit has a vertical and cross diameter of 4.34 cm and 3.44 cm averaging 21.4 g. It has dark-red thick skin and an unsmooth surface. The flesh is thick, hard, sweet with less juice and a better than normal quality for dried fruits. The percentage of edible part of fresh fruit is 96.2%, and SSC, TTS, TA and Vc is 33.00%, 29.48%, 0.58% and 397.31 mg per 100 g fresh fruit. The water content is 61.6%. The content of flavones and cAMP in mature fruit skin is 3.23 mg/g and 101.52 μg/g. The percentage of fresh fruits which can be made into dried ones is 56.02%, and the content of TTS and TA in dried fruit is 64.51% and 0.94%. The medium-sized stone is spindle-shaped, averaging 0.82 g.

Evaluation The small plant of Jishanliuguanzao cultivar has weak vigor with a high demand on fertilization and irrigation. The large fruit has strong resistance to fruit-cracking and is suitable for dried fruit. Yet it bears late with an unstable yield and serious premature fruit-dropping, which makes it unsuitable for commercial production.

大 荔 圆 枣

品种来源及分布　别名铃铃枣。分布于陕西大荔县的石槽、官池、苏村、八渔等地，为当地主栽品种。

植物学性状　树体中大，干性强，树姿较直立或半开张，树冠自然圆头形，主干皮裂呈条状，较深，不易剥落。枣头红褐色，生长势强，平均生长量76.7cm，粗1.05cm，节间长6.7cm。皮目小，圆形，凸起，开裂、灰白色。二次枝长28.9cm，7节左右。针刺不发达。枣股大，圆柱形，长2cm左右，直径1.5cm，抽吊力中等，抽生枣吊2～4个。枣吊平均长18.9cm，着叶17片。叶片小，叶长4.6cm，叶宽2.2cm，卵圆形，浅绿色，先端钝尖，叶基圆楔形，叶缘锯齿浅钝。花小，花量较少，花径6.0～6.5mm。

生物学特性　树势中等，萌芽率高，成枝力强，结果早，定植后第二年开始结果，盛果期产量高而稳定，枣头、2～3年生枝和4年生枝吊果率分别为25.4%、55.8%和55.5%。在山西太谷地区，9月中旬果实脆熟，果实生育期110d左右，为中晚熟品种类型。成熟期遇雨裂果轻。

果实性状　果个较大，卵圆形，纵径3.92cm，横径3.37cm，单果重20.3g，大小较整齐。果梗中长而细，梗洼中广而较浅。果顶凹，柱头遗存。果皮薄，红色，果面粗糙，果点小而稀，圆形，较明显。果肉厚，绿白色，肉质致密，细脆，味甜，汁液中等多，品质中上，适宜制干。鲜枣可食率95.8%，含可溶性固形物34.20%，总糖27.74%，酸0.72%，100g果肉维生素C含量196.01mg；果皮含黄酮4.17mg/g，cAMP含量386.94μg/g。制干率51.2%，干枣含总糖65.80%，酸0.85%。核小，倒纺锤形，纵径2.26cm，横径0.96cm，核重0.85g，核尖短，核纹浅，含仁率30%左右。

评价　该品种对环境条件要求不严，适应性强，耐瘠薄，在沙质土上表现良好。进入结果期早，产量高而稳定，可制干，品质中上。

Daliyuanzao

Source and Distribution　The cultivar, also called Linglingzao, originated from Shicao, Guanchi, Sucun and Bayu in Dali County of Shaanxi Province. It is local dominant variety with an unknown history.

Botanical Characters　The large tree is vertical or half-spreading with a strong central leader trunk and a natural-round crown. The reddish-brown 1-year-old shoots have strong growth vigor, 76.7 cm long and 1.05 cm thick with the internodes of 6.7 cm. The small grayish-white lenticels are round, protuberant and cracked. The secondary branches are 28.9 cm long with 7 nodes and less-developed thorns. The large column-shaped mother fruiting spurs are 2 cm in length and 1.5 cm in diameter. They can germinate 2～4 deciduous fruiting shoots which are 18.9 cm long with 17 leaves. The small leaves are oval-shaped and light green. The number of flowers is small and few with a diameter of 6.0～6.5 mm.

Biological Characters　The tree has moderate vigor, strong germination and branching ability, bearing early (generally in the 2nd year after planting) with a high and stable yield in the full-bearing stage. In Taigu County of Shanxi Province, it enters the crisp-maturing stage in mid-September with a fruit growth period of 110 d. It is a mid-ripening variety with light fruit-cracking even if it rains in the maturing stage.

Fruit Characteristics　The large oval-shaped fruit weighs 20.3 g with a regular size. It has a medium-long and thin stalk, a medium-wide and shallow stalk cavity, a sunken fruit apex, a remnant stigma, thin red skin and a rough surface with small, round, sparse and distinct dots. The greenish-white thick flesh is tight-textured, crisp and sweet, with medium juice and a better than normal quality for dried fruits. The percentage of edible part of fresh fruit is 95.8%, and SSC, TTS, TA and Vc is 34.20%, 27.74%, 0.72% and 196.01 mg per 100 g fresh fruit. The percentage of fresh fruits which can be made into dried ones is 51.2%, and the content of TTS and TA in dried fruit is 65.80% and 0.85%. The small stone is inverted spindle-shaped, averaging 0.85 g with a short apex and shallow veins. The percentage of containing kernels is 30%.

Evaluation　The cultivar has a low demand on cultivation conditions and strong adaptability and strong tolerance to poor soil. It grows well in sandy soil. It bears early with a high and stable yield. The fruit has a better than normal quality, suitable for processing dried fruits.

制 干
Drying Varieties
品 种

糖　枣

品种来源及分布　原产湖南西部的麻阳、溆浦、花垣和南部的零陵、衡山、桂阳、祁阳等县，为溆浦等产区的主栽品种。

植物学性状　树体较大，树姿开张，外围枝系下垂，枝系密集，树冠多呈乱头形。主干灰褐色，粗糙，皮裂较深，纵条状，易剥落。枣头红褐色，平均长75.3cm，粗1.01cm，节间长5.7cm，阳面被有灰色蜡质。二次枝长29.0cm，7节左右，弯曲度小。针刺发达。枣股圆柱形，平均抽生枣吊4.0个。枣吊长24.3cm，着叶17片。叶片较小，椭圆形，绿色，先端渐尖，叶基圆楔形，叶缘锯齿小，齿角钝圆。花量多，花序平均着花6朵。花中大，昼开型。

生物学特性　适应性广，较耐旱和瘠薄，在肥沃的壤土生长最好。树势强健，发枝力强。结果较早，坐果性能较好，枣头、2~3年和3年以上枝的吊果率分别为3.3%、89.7%和38.6%，产量高而稳定。根蘖苗定植后3年开始结果，15年进入盛果期。成龄树株产可达75kg。在山西太谷地区，4月中旬萌芽，5月底始花，9月下旬果实成熟，果实生育期117d左右，为晚熟品种类型。自然落果轻，抗裂果。

果实性状　果实较小，扁圆柱形，纵径2.54cm，横径2.19cm，单果重8.0g，最大9.9g。果肩平圆，梗洼浅广，果顶平。果皮中厚，红色。果肉浅绿色或白色，质地较致密，汁液中多，味甜，可制干，品质中等。鲜枣可食率93.1%，含总糖25.05%，酸0.44%，100g果肉维生素C含量403.33mg；果皮含黄酮4.78mg/g，cAMP含量97.50μg/g。制干率34.0%，干枣含总糖68.10%，酸0.86%。果核大，倒卵形，核重0.55g，含仁率63.3%。

评价　该品种适应性广，较耐干旱和瘠薄。树体较高大，树势强盛，结实能力强，产量高而稳定，抗裂果，但果个小，制干品质一般，制干率低。

Tangzao

Source and Distribution　The cultivar originated from Mayang, Xupu, Huayuan in west Hunan Province and in Lingling, Hengshan, Guiyang, Qiyang in southern Hunan Province. It is the dominant variety in Xupu County.

Botanical Characters　The large tree is spreading with an irregular crown and pendulous external branches. The reddish-brown 1-year-old shoots are 75.3 cm long and 1.01 cm thick. The internodes are 5.7 cm long with gray wax on the sun-side and developed thorns. The secondary branches are 29 cm long with 7 nodes of small curvature. The column-shaped mother fruiting spurs can germinate 4 deciduous fruiting shoots which are 24.3 cm long with 17 leaves. The small green leaves are oval-shaped. There are many medium-sized flowers produced, averaging 6 ones per inflorescence. It blooms in the daytime.

Biological Characters　The tree has strong adaptability, tolerant to drought and poor soils, with strong vigor, strong branching ability and dense branches. It grows best in fertile loam soil. The tree bears early with high fruit set, a high and stable yield. Root seedlings bear fruit in the 3rd year after planting and enter the full-bearing stage in the 15th year. A mature tree has a yield of about 75 kg. In Taigu County of Shanxi Province, it germinates in mid-April, begins blooming in late May and matures in late September with a fruit growth period of 117 d. It is a late-ripening variety with light natural fruit-dropping and fruit-cracking.

Fruit Characteristics　The small oblate fruit has a vertical and cross diameter of 2.54 cm and 2.19 cm averaging 8.0 g. It has a flat-round shoulder, a shallow and wide stalk cavity, a flat fruit apex and medium-thick red skin. The light-green or white flesh is tight-textured and sweet with medium juice. It can be used for making dried fruits with medium quality. The percentage of edible part of fresh fruit is 93.1%, and the content of TTS, TA and Vc is 25.05%, 0.44% and 403.33 mg per 100 g fresh fruit. The percentage of fresh fruits which can be made into dried ones is 34.0%, and the content of TTS in dried fruit is 63%. The medium-sized obovate stone weighs 0.55 g, and the percentage of containing kernels is 63.3%.

Evaluation　The large plant of Tangzao cultivar has strong adaptability, tolerant to drought and poor soils, with strong tree vigor and a high and stable yield. The fruit has strong resistance to fruit-cracking, yet it has a small size and medium quality for dried fruits, along with a low rate of making dried fruits from fresh ones.

大荔干尾巴

品种来源及分布 别名干尾巴枣。原产和分布于陕西大荔的八渔乡白马村一带。

植物学性状 树体中大，树姿半开张，树冠呈自然圆头形。主干灰褐色，表面粗糙，皮裂较深，呈不规则宽条状，容易剥落。枣头红褐色，长势强，平均长87.5cm，粗1.13cm，节间长6.8cm，蜡质少。二次枝发育良好，6节左右，弯曲度中等。针刺不发达。枣股圆柱形，平均抽生枣吊4.0个。枣吊长20.6cm，着叶17片。叶片较小，叶长4.3cm，叶宽2.1cm，椭圆形，叶尖钝尖，叶基圆形，叶缘具浅钝锯齿，偶有复式锯齿。花量中多，花序平均着花3朵，花较大，花径7.0mm。

生物学特性 树势和发枝力较强，坐果率较高，枣头、2~3年和4~6年生枝的吊果率分别为84.7%、68.8%和2.6%。产量较高而稳定。在山西太谷地区，4月中旬萌芽，5月中下旬始花，9月上旬果实成熟，果实生育期90d左右，为早熟品种类型。成熟期不易裂果。

果实性状 果实较大，柱形，侧面略扁，纵径3.75cm，横径3.12cm，单果重16.4g，大小较整齐。果肩平圆，略耸起。梗洼窄、浅平。果顶平圆，向一侧歪斜，顶点凹陷，柱头遗存。果面粗糙，有小块状隆起。果皮浅红色，着色、干皱自肩部开始，有随着色、随皱皮的特点，故名"干尾巴枣"。果肉白色，近核处浅绿色，质地疏松，汁液中多，味甜，宜制干，品质中等。鲜枣可食率94.6%，含可溶性固形物32.40%，总糖26.35%，酸0.59%，100g果肉维生素C含量209.43mg；果皮含黄酮19.34mg/g，cAMP含量119.45μg/g。干枣含总糖70.00%左右，酸1.00%。果核较大，倒纺锤形，核重0.89g，含仁率33.3%。

评价 该品种适应性强，对土质要求不严。树体中大，产量较高且稳定，成熟早，抗裂果。果实大，主要用于制干，品质中等。

Daliganweiba

Source and Distribution The cultivar originated from and spreads in Baimacun of Bayu Village in Dali County of Shaanxi Province.

Botanical Characters The medium-sized tree is half-spreading with a natural-round crown. The reddish-brown 1-year-old shoots have strong growth potential, 87.5 cm long and 1.13 cm thick. The internodes are 6.8 cm long with less wax and developed thorns. The secondary branches grow well with 6 nodes of medium curvature. The column-shaped mother fruiting spurs can germinate 4 deciduous fruiting shoots which are 20.6 cm long with 17 leaves. The small leaves are oval-shaped. The number of flowers is medium large, averaging 3 ones per inflorescence. The large blossoms have a diameter of 7.0 mm.

Biological Characters The tree has strong adaptability, strong vigor and strong branching ability, with high fruit set, a high and stable yield. In Taigu County of Shanxi Province, it germinates in mid-April, begins blooming in mid-late May and matures in early September with a fruit growth period of 90 d. It is an early-ripening variety with light fruit-cracking in the maturing stage.

Fruit Characteristics The large column-shaped fruit has a vertical and cross diameter of 3.75 cm and 3.12 cm averaging 16.4 g with a regular size. It has light-red skin and a rough surface with some small massive protuberances. Either coloring or shriveling is from the shoulder, and shriveling is in synchrony with coloring, which is why it is called Ganweibazao (Ganweiba means dry tail in Chinese). The white flesh is light-green near the stone, loose-textured with a sweet taste and medium juice. It has medium quality for fresh-eating and dried fruits. The percentage of edible part of fresh fruit is 94.6%, and SSC, TTS, TA and Vc is 32.40%, 26.35%, 0.59% and 209.43 mg per 100 g fresh fruit. The content of flavones and cAMP in mature fruit skin is 19.34 mg/g and 119.45 μg/g, and the content of TTS and TA in dried fruit is 70.00% and 1.00%. The large spindle-shaped stone weighs 0.89 g with the percentage of containing kernels of 33.3%.

Evaluation The large plant of Daliganweibazao cultivar has strong adaptability with a low demand on soils and a high and stable yield. The medium-sized fruit matures early with strong resistance to fruit-cracking. It is mainly used for making dried fruits with a medium quality.

制 干　Drying Varieties
品 种

· 165 ·

俊 枣

品种来源及分布 又名赤壁烟驼枣，山西十大名枣之一。原产和分布于山西太行山区漳河沿岸平顺县北耽车乡海拔500m左右的半山区，以赤壁、烟驼等村最有名且集中分布，为当地主栽品种。现有100余株百年生以上挂果老枣树。

植物学性状 树体高大，树姿较直立，干性较强，枝条细而密，树冠圆锥形。树干皮裂呈块状。枣头黄褐色，平均长85.0cm，粗0.96cm，节间长10.0cm，着生永久性二次枝10～12个。二次枝长33.2cm，5～6节，枣头生长细而较弱。针刺不发达。皮目中大，分布较密，圆形，凸起，开裂，灰白色。枣股中大，抽吊力中等，平均抽生枣吊3.2个。枣吊细长，形似垂柳，平均长20.4cm，着叶12片。叶片较大，叶长7.1cm，叶宽3.1cm，卵圆形，绿色，先端急尖，叶基心形，叶缘锯齿钝。花量少，枣吊平均着花38.6朵，花序平均2.3朵。花朵中大，花径7.6mm，昼开型。

生物学特性 树势较强，萌芽率较高，成枝力强。幼树结果较早，一般第二年开始结果，15年后进入盛果期，产量中等。枣头吊果率20.4%，2～3年生枝为90.8%，4年生枝为47.6%，主要坐果部位在枣吊的2～7节，占坐果总数的90.1%。在山西太谷地区，9月上旬果实着色，10月上旬成熟，果实生育期115d左右，为晚熟品种类型。

果实性状 果实较大，扁柱形，纵径3.82cm，横径2.90cm，单果重15.3g，大小较整齐。果梗细而长，梗洼广，中深。果顶平，柱头遗存。果皮中厚，鲜红色，果面光滑。果肉厚，绿白色，肉质致密，细脆，味甜，汁液较少，品质中上，适宜制干。鲜枣可食率96.4%，含可溶性固形物34.5%，总糖35.13%，酸0.81%，100g果肉维生素C含量463.22mg；果皮含黄酮18.14mg/g，cAMP含量225.29μg/g。制干率55.0%，干枣含总糖76.8%，酸0.79%。核中大，纺锤形，纵径2.24cm，横径0.85cm，核重0.60g，种仁不饱满，含仁率16.7%。

评价 该品种较抗裂果，采前落果轻，干枣果皮皱纹明显，果肉富弹性，制干品质较好，但不抗枣疯病，适应性较差，区域性明显，可在晋东南等秋雨较多的地区适量发展。

Junzao

Source and Distribution The cultivar originated from and spreads in the semi-mountainous region with an altitude of 500 m in Beidanche Village of Pingshun County in Shanxi Province. It is local dominant veriety there.

Botanical Characters The large tree is vertical with a strong central leader trunk, thin and dense branches and a conical-shaped crown. The trunk bark has massive fissures. The yellowish-brown 1-year-old shoots are thin and weak, 85.0 cm long. There are 10～12 permanent secondary branches which are 33.2 cm long with less-developed thorns. The mother fruiting spurs can germinate 3.2 thin and long deciduous fruiting shoots, averaging 20.4 cm long with 12 leaves. The large green leaves are oval-shaped. The number of flowers is small averaging 38.6 per deciduous fruiting shoot. The medium-sized blossoms have a diameter of 7.6 mm.

Biological Characters The tree has strong vigor, high germination rate and strong branching ability. Young trees bear early, generally in the 2nd year after planting and enter the full-bearing stage in the 15th year with a medium yield. It is a late-ripening variety.

Fruit Characteristics The medium-sized fruit is flat column-shaped, averaging 15.3 g with a regular size. It has medium-thick and brightly-red skin and a smooth surface. The greenish-white thick flesh is tight-textured, crisp and sweet with less juice and a better than normal quality for fresh-eating and dried fruits. The percentage of edible part of fresh fruit is 96.4%, and SSC, TTS, TA and Vc is 34.5%, 35.13%, 0.81% and 463.22 mg per 100 g fresh fruit. The rate of fresh fruits which can be made into dried ones is 55.0%. The medium-sized stone is spindle-shaped averaging 0.60 g.

Evaluation The cultivar has strong resistance to fruit-cracking with light premature fruit-dropping. Dried fruits have obviously shriveled skin and whippy flesh with a good quality. Yet it has weak resistance to jujube witches' broom and poor adaptability.

制 干
Drying Varieties
品 种

蒲 城 晋 枣

品种来源及分布 别名夏阳枣。原产陕西蒲城县洛河沿岸的东陈庄、平路庙乡等主要枣区，栽培数量较多，为当地主栽品种。

植物学性状 树体较大，树姿半开张，干性强，枝系较密，树冠呈自然圆头形。主干灰褐色，皮裂纹浅，呈小条块状，不易剥落。枣头红褐色，平均长74.3cm左右，粗1.03cm，节间长7.9cm，最长可达10cm，蜡质多。无刺。二次枝枝长30.3cm，6节左右，弯曲度中等。枣股圆锥形，平均抽生枣吊4.0个。枣吊长26.4cm，着叶15片。叶片小，卵圆形或长圆形，绿色，先端渐尖或突尖，叶基圆楔形，叶缘具钝锯齿。每花序着花7朵左右。花较大，花径7.0mm，昼开型。

生物学特性 适应性较强，树势中等，发枝力强。枣股结果能力以2～3年生枝为主，枣头、2～3年和3年以上枝的吊果率分别为2.0%、166.7%和33.6%，产量高而稳定。在山西太谷地区，10月上旬果实成熟，果实生育期120d左右，为极晚熟品种类型。成熟期遇雨易裂果。

果实性状 果实中大，扁柱形或椭圆形，侧面略扁。纵径3.21cm，横径2.59cm，侧径2.41cm，单果重10.4g，大小整齐。果肩平圆，偏斜。梗洼宽浅，环洼宽深，果顶较瘦，向一侧歪斜，顶点微凹，柱头残存。果面不平整。果皮厚，红色。果肉白绿色，质地较致密，汁液中多，味甜，可制干。鲜枣可食率95.0%，含可溶性固形物26.70%，总糖20.16%，酸0.40%，100g果肉维生素C含量452.82mg；果皮含黄酮32.51mg/g，cAMP含量12.78μg/g。干枣肉厚，富弹性，果皮韧性强，抗揉压。干枣含总糖63.89%，酸1.71%。果核中大，纺锤形或倒卵形，侧面略扁，核重0.52g，含仁率25.0%。

评价 该品种适应性较强，产量高而稳定。果实中大，肉厚，质细，核中大，制干品质中等，抗裂果能力较差。

Puchengjinzao

Source and Distribution The cultivar, also called Xiayangzao, originated from Dongchenzhuang and Pinglumiaoxiang in Pucheng County of Shaanxi Province. It is the dominant variety there with a large quantity.

Botanical Characters The large tree is half-spreading with a strong central leader trunk, dense branches and a natural-round crown. The reddish-brown 1-year-old shoots are 74.3 cm long and 1.03 cm thick. The internodes are 7.9 cm long (maximum 10 cm) with much wax and no thorns. The secondary branches are 30.3 cm long with 6 nodes of medium curvature. The conical mother fruiting spurs can germinate 4 deciduous fruiting shoots which are 26.4 cm long with 15 leaves. The small green leaves are oval-shaped or oblong. Each inflorescence has 7 flowers. The daytime-bloomed large blossoms have a diameter of 7.0 mm.

Biological Characters The tree has strong adaptability, moderate vigor and strong branching ability, with a high and stable yield. 2～3-year-old mother fruiting spurs bear the most fruit. The percentage of fruits to deciduous bearing shoots of 1-year-old shoots, 2～3-year-old branches and over-3-year-old ones is 2.0%, 166.7% and 33.6%. In Taigu County of Shanxi Province, it ripens in early October with a fruit growth period of 120 d. It is an extremely late-ripening variety with serious fruit-cracking in the maturing stage.

Fruit Characteristics The medium-sized fruit is flat-column-shaped or oval-shaped with a vertical, cross and lateral diameter of 3.21cm, 2.59 cm and 2.41 cm respectively averaging 10.4 g with a regular size. It has a flat-round shoulder, a shallow and wide stalk cavity, a slightly-sunken fruit apex, a remnant stigma, an unsmooth surface and thick red skin. The greenish-white flesh is tight-textured and sweet with medium juice. It is suitable for making dried fruits. The percentage of edible part of fresh fruit is 95.0%, and SSC, TTS, TA and Vc is 26.70%, 20.16%, 0.40% and 452.82 mg per 100 g fresh fruit. The content of flavones and cAMP in mature fruit skin is 32.51 mg/g and 12.78 μg/g. Dried fruits have thick flesh, whippy and resistant to pressing. The medium-sized spindle-shaped stone weighs 0.52 g with the percentage of containing kernels of 25.0%.

Evaluation The cultivar has strong adaptability with a high and stable yield. The medium-sized fruit has thick, delicate flesh and a small stone with medium quality for dried fruits. It has poor resistance to fruit-cracking.

制 干
Drying Varieties
品 种

乐陵长木枣

品种来源及分布 别名木枣、大木枣。原产于山东的乐陵、无棣、庆云等地，为当地主栽品种之一。栽培历史悠久。

植物学性状 树体中等大，树姿开张，树冠多呈乱头形。主干灰褐色，皮呈块状浅裂，容易剥落。枣头黄褐色，生长量85.7cm，粗1.07cm，节间较长，平均长8.8cm，蜡质少。二次枝长27.8cm，5节左右，弯曲度中等。针刺不发达。枣股圆锥形，平均抽生枣吊4.0个。枣吊较粗，长20.9cm，着叶12片。叶片大，卵圆形，绿色，反卷，先端渐尖，叶基圆楔形，叶缘锯齿大，钝圆。花量多，每花序着花7朵左右。花中大，花径5.5～6.2mm，昼开型。

生物学特性 适应性较差，要求土壤深厚肥沃。树势中等，结果较晚，栽后3年开始少量结果。坐果率较低，枣头、2～3年和3年以上枝的吊果率分别为10.0%、57.4%和9.2%。落花落果严重，产量中等。在山西太谷地区，10月上旬果实成熟，果实生育期117d，为极晚熟品种类型。

果实性状 果实较大，长圆形，纵径3.55cm，横径2.58cm，单果重13.7g，最大19.4g，大小较整齐。果顶平圆，果肩凹，梗洼深，中广。果柄较粗，长6.3mm。果面光滑，果皮浅红色。果肉厚，白色，质地较致密，汁液中多，甜味浓，可制干，品质中等。鲜枣可食率94.0%，含可溶性固形物32.00%，总糖26.83%，酸0.50%，100g果肉维生素C含量433.97mg；果皮含黄酮0.56mg/g，cAMP含量259.85μg/g。制干率62.8%，干枣富弹性，极耐贮运，含总糖63.89%，酸0.72%，惟稍具苦味。果核大，长纺锤形，核重0.82g，含仁率20.0%。

评价 该品种树体中大，不耐瘠薄土壤条件。结果晚，产量中等，成熟期较晚。果实大，制干率高，品质中等，可在北方平原地区适当发展。

Lelingchangmuzao

Source and Distribution The cultivar, also called Muzao or Damuzao, originated from Leling, Wudi and Qingyun of Shandong Province. It is the dominant variety there, with a long cultivation history.

Botanical Characters The small tree is spreading with an irregular crown. The yellowish-brown 1-year-old shoots are 85.7 cm long and 1.07 cm thick. The internodes are 8.8 cm long, with less wax and less-developed thorns. The secondary branches are 27.8 cm long with 5 nodes of medium curvature. The conical mother fruiting spurs can germinate 4 thick deciduous fruiting shoots which are 20.9 cm long with 12 leaves. The large green leaves are oval-shaped. The leaves bend toward the back. There are many medium-sized flowers with a diameter of 5.5～6.2 mm produced averaging 7 ones per inflorescence.

Biological Characters The tree has poor adaptability and moderate vigor, bearing late (generally in the 3rd year after planting) with low fruit set, serious flower-dropping and a medium yield. It must be planted in deep, fertile soils. In Taigu County of Shanxi Province, it matures in early October with a fruit growth period of 117 d. It is an extremely late-ripening variety.

Fruit Characteristics The large oblong fruit has a vertical and cross diameter of 3.55 cm and 2.58 cm, averaging 13.7 g (maximum 19.4 g) with a regular size. It has a flat-round fruit apex, a sunken shoulder, a deep and medium-wide stalk cavity, a thick stalk of 6.3 mm long, a smooth surface and light-red skin. The thick flesh is white, tight-textured and sweet, with medium juice. It has medium quality for dried fruits. The percentage of edible part of fresh fruit is 94.00%, and SSC, TTS, TA and Vc is 32.00%, 26.83%, 0.50% and 433.97 mg per 100 g fresh fruit. The content of flavones and cAMP in mature fruit skin is 0.56 mg/g and 259.85 μg/g. Dried fruits have whippy flesh, tolerant to storage and transport with a little bitter taste. The percentage of fresh fruits which can be made into dried ones is 62.8%, and the content of TTS and TA is 63.89% and 0.72%. The large stone is long spindle-shaped averaging 0.82 g. The percentage of containing kernels is 20.0%.

Evaluation The small plant of Lelingchangmuzao cultivar has poor tolerance to infertile soils, bearing and maturing late with medium productivity. The large fruit has a high percentage for making dried fruits with medium quality. It can be developed in plain areas of northern China.

稷山长枣

品种来源及分布 别名崖枣。原产和分布于山西稷山的南阳乡和运城的北相镇等地。数量不多，因植株多自然生长在地塄上，故称"崖枣"。

植物学性状 树体较大，树姿半开张，干性较弱，树冠呈自然圆头形。枣头红褐色，粗壮，长50cm左右，粗0.97cm，节间长8.2cm，蜡质少。二次枝长33.7cm，9节左右，弯曲度中等。针刺不发达。枣股粗大，平均抽生枣吊4.0个。枣吊长20.8cm，着叶12片。叶片中大，叶长6.8cm，叶宽3.4cm，卵圆形，绿色，先端渐尖，叶基偏斜形。花量较多，花序平均着花5朵，花较大，为昼开型。

生物学特性 适应性较强，耐旱，对栽培条件要求不严。树势较强，发枝力中等，结果较晚，定植后3年开始结果，15年左右进入盛果期。坐果率较低，枣头、2~3年和3年以上枝的吊果率分别为16.7%、42.6%和10.0%。产量中等，较稳定。在山西太谷地区，4月中旬萌芽，5月底始花，9月下旬进入脆熟期，果实生育期110d左右，为晚熟品种类型。较抗裂果。

果实性状 果实中大，柱形，纵径3.93cm，横径2.65cm，单果重11.7g，大小整齐。果肩平圆，梗洼深窄。果尖，柱头遗存。果皮中厚，紫红色。果肉厚，浅绿色，质地致密，汁液中多，味酸甜，可制干，品质中上。鲜枣可食率95.7%，含可溶性固形物27.90%，总糖24.22%，酸0.76%，100g果肉维生素C含量541.10mg；果皮含黄酮9.51mg/g，cAMP含量46.18μg/g。制干率56.1%，干枣含总糖68.81%，酸2.11%。果核中大，纺锤形，核重0.51g，含仁率16.7%。

评价 该品种树势较强，耐旱，裂果轻。结果较晚，产量中等。制干率较高，品质中上。

Jishanchangzao

Source and Distribution The cultivar, also called Yazao, originated from and spreads in Nanyang Village of Jishan County and Beixiang Village of Yuncheng in Shanxi Province with a small quantity.

Botanical Characters The large tree is half-spreading with a weak central leader trunk and a natural-round crown. The reddish-brown 1-year-old shoots are 50.0 cm long and 0.97 cm thick. The internodes are 8.2 cm long, with less wax and thin developed thorns. The secondary branches are 33.7 cm long with 9 nodes of medium curvature. The thick mother fruiting spurs can germinate 4 deciduous fruiting shoots which are 20.8 cm long with 12 leaves. The medium-sized green leaves are 6.8 cm long and 3.4 cm wide, oval-shaped, with a gradually-cuspate apex and a deflective base. There are many large flowers produced, averaging 5 ones per inflorescence. It blooms in the daytime.

Biological Characters The tree has strong adaptability, tolerant to drought, with a low demand on cultural conditions, strong vigor and medium branching ability. It bears late, generally in the 3rd year after planting and enters the full-bearing stage in the 15th year, with low fruit set, a medium and stable yield. In Taigu County of Shanxi Province, it enters the crisp-maturing stage in late September. It is a late-ripening variety with strong resistance to fruit-cracking.

Fruit Characteristics The medium-sized fruit is column-shaped with a vertical and cross diameter of 3.93 cm and 2.65 cm averaging 11.7 g with a regular size. It has a flat-round shoulder, a deep and narrow stalk cavity, a pointed fruit apex, a remnant stigma, medium-thick and purplish-red skin. The light-green flesh is thick, tight-textured, sour and sweet, with medium juice. It has a better than normal quality for dried fruits. The percentage of edible part of fresh fruit is 95.7%, and SSC, TTS, TA and Vc is 27.90%, 24.22%, 0.76% and 541.10 mg per 100 g fresh fruit. The percentage of fresh fruits which can be made into dried ones is 56.1%, and the content of TTS and TA in dried fruit is 68.81% and 2.11%. The medium-sized stone is spindle-shaped, averaging 0.51 g.

Evaluation The cultivar has strong tree vigor and strong resistance to fruit-cracking and is tolerant to drought. It bears late with medium productivity. Fruits have high percentage for making dried fruits with a better than normal quality.

制 干 品 种
Drying Varieties

大荔小墩墩枣

品种来源及分布 别名小墩墩枣。原产和分布于陕西大荔的西营、北丁、马坊一带，多零星栽培。

植物学性状 树体较大，树姿开张，树冠呈自然圆头形。树干灰褐或黑褐色，皮粗糙，裂纹深，呈条状，不易剥落。枣头红褐色，较细弱，长87.6cm，粗度1.26cm，节间长5.5cm。二次枝长32.6cm，平均7节，弯曲度中等。针刺发达，直刺长1.0～2.5cm。枣股圆柱形或圆锥形，平均抽生枣吊3.0个。枣吊长14.2cm，着叶15片。叶片小，叶长3.9cm、叶宽1.9cm，卵圆形，叶薄，绿色，有光泽，先端渐尖，先端钝圆或微凹。叶基圆形，叶缘锯齿密，浅钝。花量少，每花序着花3朵。花朵小，花径5.0～5.3mm。

生物学特性 适应性中等，耐旱力强。树势中等，发枝力弱。坐果率极高，枣头、2～3年和4～6年生枝的吊果率分别为20.0%、126.4%和129.0%，较丰产。在山西太谷地区，4月中旬萌芽，5月底始花，9月下旬果实成熟，果实生育期107d，为中晚熟品种类型。较抗裂果。

果实性状 果实小，圆形，胴部中腰宽大。纵径2.40cm，横径2.30cm，单果重7.1g，最大9.3g，大小整齐。果肩平，梗洼中深、窄。果顶凹陷，顶洼中深、广。果面平整，果皮中厚，紫红色。果肉浅绿色，质地致密，汁液中多，味甜，宜制干，品质中上。鲜枣可食率90.1%，可溶性固形物含量33.60%，总糖26.73%，酸0.66%，100g果肉维生素C含量334.02mg；果皮含黄酮102.27mg/g，cAMP含量108.57μg/g。干枣含总糖65.76%，酸0.23%。果核较大，椭圆形，核重0.67g。多数核内含有种子，含仁率70.0%。

评价 该品种适应性一般，抗旱性强。产量较高，抗裂果。果实偏小，果核较大，可食率低，制干品质中上。

Dalixiaodundunzao

Source and Distribution The cultivar originated from and spreads in Xiying, Beiding and Mafang in Dali County of Shaanxi Province.

Botanical Characters The large tree is spreading with a round crown. The grayish-brown or blackish-brown trunk bark has rough, deep massive fissures, uneasily shelled off. The reddish-brown 1-year-old shoots are thin and weak, 87.6 cm long and 1.26 cm thick. The internodes are 5.5 cm long with developed thorns of 1.0～2.5 cm long. The secondary branches are 32.6 cm long with 7 nodes of medium curvature. The column-shaped or conical mother fruiting spurs can germinate 3 deciduous fruiting shoots which are 14.2 cm long with 15 leaves. The small oval-shaped leaves are thin, 3.9 cm long and 1.9 cm wide, green and glossy, with a gradually-cuspate apex, a round base and a shallow, blunt saw-tooth pattern on the margin. The number of flowers is small, averaging 3 ones per inflorescence. The small blossoms have a diameter of 5.0～5.3 mm.

Biological Characters The tree has moderate adaptability, strongly tolerant to drought with moderate vigor, weak branching ability, very high fruit set and high productivity. The percentage of fruits to deciduous bearing shoots of 1-year-old shoots, 2～3-year-old branches and over-3-year-old ones is 20.0%, 126.4% and 129.0%. In Taigu County of Shanxi Province, it germinates in mid-April, begins blooming in late May and matures in late September with a fruit growth period of 107 d. It is a mid-late-ripening variety with strong resistance to fruit-cracking.

Fruit Characteristics The small round fruit has a vertical and cross diameter of 2.40 cm and 2.30 cm averaging 7.1 g (maximum 9.3 g) with a regular size. It has a flat shoulder, a medium-deep and narrow stalk cavity, a sunken fruit apex, a smooth surface, medium-thick and purplish-red skin. The light-green flesh is tight-textured and sweet, with medium juice. It has a better than normal quality for dried fruits. The percentage of edible part of fresh fruit is 90.1%, and SSC, TTS, TA and Vc is 33.60%, 26.73%, 0.66% and 334.02 mg per 100 g fresh fruit. The content of flavones and cAMP in mature fruit skin is 102.27 mg/g and 108.57 μg/g. The content of TTS and TA in dried fruit is 65.76% and 0.23%. The larger oval-shaped stone weighs 0.67 g. Most stones contain kernels, the percentage of which is 70.0%.

Evaluation The cultivar has moderate adaptability and strong tolerance to drought and strong resistance to fruit-cracking with high productivity. The small fruit has a large stone with low edibility. It has a better than normal quality.

垣 曲 枣

品种来源及分布 别名苑曲枣。分布于山西稷山的南阳等地。栽培数量不多。

植物学性状 树体较大,树姿开张,干性较弱,枝条细且较密,树冠呈半圆形。主干块状皮裂。枣头红褐色,平均长36.3cm,粗0.84cm,节间长7.0cm,蜡质少。二次枝长23.2cm,6节左右,弯曲度中等。针刺不发达。枣股小,抽吊力强,平均抽生枣吊4.0个。枣吊短小,平均长16.3cm,着叶10片。叶长5.1cm,叶宽3.1cm,叶片卵圆形,浅绿色,叶尖渐尖,叶基圆形,叶缘具钝锯齿,较密。花量多,花序平均着花7朵。蜜盘小,花径5.2～6.4mm。

生物学特性 适应性较强,耐旱,对栽培条件要求不严。树势中等,发枝力较强,萌蘖力弱。坐果能力中等,枣头、2～3年和3年以上枝的吊果率分别为10.0%、38.5%和6.0%。产量中等,较稳定。在山西太谷地区,9月中旬着色,10月上旬成熟采收,果实生育期110d左右,为晚熟品种类型。抗裂果,但采前落果较重。

果实性状 果个小,果实长圆形,纵径2.68cm,横径2.14cm,单果重6.2g,最大果重9.2g,大小整齐。果柄细。果肩小,平圆。梗洼浅而窄。果顶微凹,柱头遗存。果皮较厚,赭红色。果肉中厚,浅绿色,质地较松,汁中多,味酸甜,品质中等,可制干。鲜枣可食率92.4%,含可溶性固形物27.00%,总糖20.09%,酸0.67%,100g果肉维生素C含量444.20mg;果皮含黄酮5.35mg/g,cAMP含量36.84μg/g。制干率56.4%,干枣含总糖70.82%,酸2.27%,品质中等。果核纺锤形,核重0.47g,核面较粗糙,核纹中深,种子较饱满,含仁率70.0%。

评价 该品种适应性较强,产量中等,抗裂果,但采前落果重。果实偏小,可食率低,干枣品质中等。

Yuanquzao

Source and Distribution The cultivar originated from Nanyang of Jishan County in Shanxi Province, with a small quantity.

Botanical Characters The large tree is spreading with a weak central leader trunk, thin and dense branches and an irregular crown. The reddish-brown 1-year-old shoots are 36.3 cm long and 0.84 cm thick. The internodes are 7.0 cm long with less wax. The secondary branches are 23.2 cm long with 6 nodes of medium curvature and less-developed thorns. The small mother fruiting spurs can germinate 4 deciduous fruiting shoots which are 16.3 cm long with 10 leaves. The light-green leaves are oval-shaped, 5.1 cm long and 3.1 cm wide, with a gradually-cuspate apex, a round base and a blunt saw-tooth pattern on the margin. There are many small flowers with a diameter of 5.2～6.4 mm produced, averaging 7 ones per inflorescence.

Biological Characters The tree is strongly adaptable and tolerant to drought, with a low demand on cultural conditions. It has moderate vigor, strong branching ability and weak suckering ability, with medium fruit set, a medium and stable yield. In Taigu County of Shanxi Province, it matures in early October. It is a late-ripening variety with strong resistance to fruit-cracking and serious premature fruit-dropping.

Fruit Characteristics The small oblong fruit has a vertical and cross diameter of 2.68 cm and 2.14 cm averaging 6.2 g (maximum 9.2 g) with a regular size. It has a thin stalk, a small flat-round shoulder, a shallow and narrow stalk cavity, a slightly-sunken fruit apex, a remnant stigma and brownish-red thick skin. The light-green flesh is medium-thick, loose-textured, sour and sweet, with medium juice. It has medium quality for dried fruits. The percentage of edible part of fresh fruit is 92.4%, and SSC, TTS, TA and Vc is 27.00%, 20.09%, 0.67% and 444.20 mg per 100 g fresh fruit. The content of flavones and cAMP in mature fruit skin is 5.35 mg/g and 36.84 μg/g. The percentage of fresh fruits which can be made into dried ones is 56.4%, and the content of TTS and TA in dried fruit is 70.82% and 2.27%. Dried fruits have medium quality. The spindle-shaped stone weighs 0.47 g.

Evaluation The cultivar has strong adaptability and medium productivity. The small fruit has strong resistance to fruit-cracking, yet premature fruit-dropping is serious. It has a low percentage of edible part of fresh fruit with medium quality.

洪赵十月红

品种来源及分布 原产和分布于山西洪洞县赵城镇的稽村东许一带，栽培数量不多。

植物学性状 树体高大，树姿半开张，树冠呈自然圆头形。主干块状皮裂。枣头红褐色，平均长60.0cm左右，节间长8.0cm。二次枝长25.7cm，5节左右。针刺不发达。枣股中大，抽吊力中等，平均抽生枣吊4.0个。枣吊长20.2cm，着叶12片。叶片大，叶长7.0cm，叶宽3.4cm，椭圆形，绿色，先端渐尖，叶基偏斜形，叶缘钝锯齿。花量多，花序平均6朵，花中大。

生物学特性 树势强，发枝力中等，枝叶中密。坐果稳定，产量较高，枣头、2～3年、4～6年和6年生以上枝吊果率分别为39.0%、79.6%、67.4%和53.0%。在山西太谷地区，4月中旬萌芽，5月下旬始花，9月中旬着色，10月上旬完全成熟，果实生育期120d左右，为极晚熟品种类型。采前落果少，且因成熟期晚，裂果较轻。

果实性状 果实中大，卵圆形，纵径3.08cm，横径2.68cm，单果重10.1g，大小整齐。果肩广圆，梗洼中广、较浅。果顶平圆，顶点微凹，柱头遗存。果面光滑，红色，果皮较厚。果肉中厚，浅绿色，质地较致密，汁液少，味酸甜，品质中等，适宜制干。鲜枣可食率92.4%，含可溶性固形物29.40%，总糖22.31%，酸0.44%，100g果肉维生素C含量490.98mg。制干率60.0%，干枣含可溶性固形物71.50%，总糖57.51%，酸1.17%。果核倒纺锤形，核重0.77g。核内多含较饱满的种子，偶有双仁，含仁率78.3%。

评价 该品种树体高大，树势强盛，适应性强。成熟期极晚，坐果稳定，产量较高，采前落果少。果实中大，品质中等，可食率低。抗裂果能力较强。

Hongzhaoshiyuehong

Source and Distribution The cultivar originated from and spreads in Jicun and Dongxu of Zhaocheng Village in Hongtong County of Shanxi Province, with a small quantity.

Botanical Characters The large tree is half-spreading with a natural-round crown. The trunk bark has massive fissures. The reddish-brown 1-year-old shoots are 60.0 cm long with the internodes of 8.0 cm long. The secondary branches are 25.7 cm long with 5 nodes and less-developed thorns. The medium-sized mother fruiting spurs can germinate 4 deciduous fruiting shoots which are 20.2 cm long with 12 leaves. The large green leaves are oval-shaped. There are many flowers produced, averaging 6 per inflorescence. The blossoms are of medium size.

Biological Characters The tree has strong vigor, medium branching ability and medium-dense branches with a high and stable yield. In Taigu County of Shanxi Province, it germinates in mid-April, begins blooming in late May, begins coloring in mid-September and enters the crisp-maturing stage in early October with a fruit growth period of 120 d. It is an extremely late-ripening variety with light premature fruit-dropping. Because it matures very late, fruit-cracking seldom occurs.

Fruit Characteristics The medium-sized fruit is oval-shaped with a vertical and cross diameter of 3.08 cm and 2.68 cm averaging 10.1 g with a regular size. It has a widely-round shoulder, a medium-wide and shallow stalk cavity, a flat-round fruit apex, a remnant stigma, a smooth surface and red thick skin. The light-green flesh is medium-thick and tight-textured, sour and sweet with less juice. It is suitable for making dried fruits. The percentage of edible part of fresh fruit is 92.4%, and SSC, TTS, TA and Vc is 29.40%, 22.31%, 0.44% and 490.98 mg per 100 g fresh fruit. The percentage of fresh fruits which can be made into dried ones with a medium quality is 60.0%, and SSC, TTS and TA in dried fruit is 71.50%, 57.51% and 1.17%. The stone is inverted spindle-shaped averaging 0.77 g. Most stones contain a well-developed kernel, sometimes double kernels. The percentage of containing kernels is 78.3%.

Evaluation The large plant of Hongzhaoshiyuehong cultivar has strong tree vigor and strong adaptability with a high and stable yield, and light premature fruit-dropping. It is a late-ripening variety with strong resistance to fruit-cracking. It has a low percentage of edible part of fresh fruit. Yet the medium-sized fruit has a large stone.

制 干
Drying Varieties
品 种

大荔疙瘩枣

品种来源及分布 别名木枣、十月寒。原产和分布于陕西大荔的八渔乡孝庄、西营等地，数量不多。

植物学性状 树体较大，树姿开张，树冠偏斜形。主干黑褐色，皮裂较浅，呈不规则条状，容易剥落。枣头黄褐色，长势较强，平均长84.3cm，粗1.14cm，节间长6.8cm，无蜡质。二次枝平均长37.5cm，节数8节左右，弯曲度中等。针刺不发达。枣股圆柱形，平均抽生枣吊4.0个。枣吊长24.7cm，着叶15片。叶片较小，叶长5.5cm，叶宽2.7cm，卵圆形，绿色，较光亮，先端渐尖，先端钝尖，叶基偏斜形，叶缘具短锯齿，尖钝不等。花量较多，花序着花7朵。花较大，花径6.5mm。

生物学特性 适应性较强，耐旱、耐涝、耐瘠薄。树势强健，发枝力较强。枣头、2~3年和4~6年生枝的吊果率分别为55.1%、78.5%和23.2%，产量中等。在山西太谷地区，果实10月上旬成熟采收，生长期120d左右，为极晚熟品种类型。抗裂果。

果实性状 果实大，卵圆形，纵径3.62cm，横径3.17cm，单果重18.8g，大小较整齐。果肩圆或平圆，耸起，略偏斜。梗洼浅窄。果顶凹陷成深而窄的顶洼。果面粗糙，有小块状隆起。果皮中厚，紫红色。果肉浅绿色，质地疏松，汁液少，味甜，品质中等，可制干。鲜枣可食率95.0%，含可溶性固形物31.20%，总糖24.31%，酸0.52%，100g果肉维生素C含量322.20mg；果皮含黄酮1.19mg/g，cAMP含量212.63 μg/g。干枣含总糖61.83%，酸1.29%。果核大，椭圆形，核重0.94g，含仁率36.7%。

评价 该品种耐旱、耐涝、耐瘠薄，树势强健，产量中等，不易裂果。果实大，较整齐，可制干，品质中等。

Daligedazao

Source and Distribution The cultivar, also called Muzao or Shiyuehan, originated from and spreads in Xiaozhuang and Xiying of Bayu Village in Dali County of Shaanxi Province, with a small quantity.

Botanical Characters The large tree is half-spreading with a strong central leader trunk and a deflective crown. The blackish-brown trunk bark has irregular shallow massive fissures, easily shelled off. The reddish-brown 1-year-old shoots have strong growth potential, 84.3 cm long and 1.14 cm thick. The internodes are 6.8 cm long without wax and with less-developed thorns. The secondary branches are 37.5 cm long with 8 nodes of medium curvature. The column-shaped mother fruiting spurs can germinate 4 deciduous fruiting shoots which are 24.7 cm long with 15 leaves. The small oval-shaped leaves are 5.5 cm long and 2.7 cm wide, green and glossy, with a gradually-cuspate apex, a deflective base and a short saw-tooth pattern on the margin. There are many large flowers with a diameter of 6.5 mm produced, averaging 7 ones per inflorescence.

Biological Characters The tree has strong adaptability, tolerant to drought, water-logging and poor soils with strong vigor, weak branching ability, strong stretching ability and a medium yield. The percentage of fruits to deciduous bearing shoots of 1-year-old shoots, 2~3-year-old branches and 4~6-year-old ones is 55.1%, 78.5% and 23.2%. In Taigu County of Shanxi Province, it matures in early October with a fruit growth period of 120 d. It is an extremely late-ripening variety with strong resistance to fruit-cracking.

Fruit Characteristics The large oval-shaped fruit has a vertical and cross diameter of 3.62 cm and 3.17 cm, averaging 18.8 g with a regular size. It has a round or flat-round shoulder (a little pointed and deflective), a narrow and shallow stalk cavity, a sunken fruit apex, medium-thick and purplish-red skin and a rough surface with small massive protuberances. The light-green flesh is loose-textured and sweet, with less juice. It has medium quality for dried fruits. The percentage of edible part of fresh fruit is 95.0%, and SSC, TTS, TA and Vc is 31.20%, 24.31%, 0.52% and 322.20 mg per 100 g fresh fruit. The content of flavones and cAMP in mature fruit skin is 1.19 mg/g and 212.63 μg/g, and the content of TTS and TA in dried fruit is 61.83% and 1.29%. The large oval-shaped stone weighs 0.94 g with the percentage of containing kernels of 36.7%.

Evaluation The cultivar has strong tolerance to drought, water-logging and poor soils with strong vigor, a medium yield and light fruit-cracking. The large fruit is suitable for making dried fruits with medium quality and a regular size.

制 干
Drying Varieties
品 种

· 181 ·

献 县 木 枣

品种来源及分布　原产和分布于河北献县。

植物学性状　树体中大，树姿较开张，树冠圆头形。主干皮裂块状。枣头黄褐色，平均长72.7cm，节间长7.8cm。二次枝长32.1cm，5～7节，弯曲度小。针刺发达。枣股平均抽生枣吊3.4个。枣吊长20.7cm，着叶11片。叶片中大，叶长5.9cm，叶宽2.7cm，椭圆形，部分叶片反卷，先端锐尖，叶基圆楔形，叶缘锐锯齿。花量中多，花序平均着花6朵，花较大，花径6.6mm。

生物学特性　树势中庸，萌芽率和成枝力较强。定植第三年结果，坐果率较高，较丰产，枣头、2～3年和4～6年生枝的吊果率分别为87.1%、124.4%和16.2%。在山西太谷地区，9月中旬果实进入脆熟期，9月下旬成熟采收，果实生育期107d左右，为中晚熟品种类型。成熟期遇雨裂果严重。

果实性状　果个中大，卵圆形，纵径2.99cm，横径2.78cm，单果重10.7g，大小整齐。果皮薄，浅红色，果面平滑。梗洼中深、广。果顶平，柱头遗存，不明显。肉质较致密，味酸，汁液中多，品质中等，适宜制干。鲜枣可食率96.2%，含可溶性固形物31.20%，总糖27.22%，酸0.83%，100g果肉维生素C含量440.67mg；果皮含黄酮1.78mg/g，cAMP含量179.55μg/g。制干率59.4%，含总糖60.98%，酸1.48%。果核较小，椭圆形，核重0.41g，种仁饱满，含仁率86.7%。

评价　该品种较丰产，果个中大，品质中等，适宜制干。成熟期应注意防雨。

Xianxianmuzao

Source and Distribution　The cultivar originated from and spreads in Xianxian County of Hebei Province.

Botanical Characters　The medium-sized tree is spreading with a round crown. The trunk bark has massive fissures. The yellowish-brown 1-year-old shoots are 72.7 cm long with the internodes of 7.8 cm long. The secondary branches are 32.1 cm long with 5～7 nodes of small curvature and developed thorns. The mother fruiting spurs can germinate 3.4 deciduous fruiting shoots which are 20.7 cm long with 11 leaves. The medium-sized leaves are oval-shaped, 5.9 cm long and 2.7 cm wide. Some leaves fold towards the back. They have a sharply cuspate apex, a round-cuneiform base and a sharp saw-tooth pattern on the margin. The number of flowers is medium large, averaging 6 ones per inflorescence. The medium-sized blossoms have a diameter of 6.6 mm.

Biological Characters　The tree has moderate vigor, strong germination and branching ability. It generally bears in the 3rd year after planting, with high fruit set and a high yield. The percentage of fruits to deciduous bearing shoots of 1-year-old shoots, 2～3-year-old branches and 4～6-year-old ones is 87.1%, 124.4% and 16.2%. In Taigu County of Shanxi Province, it enters the crisp-maturing stage in mid-September and is harvested in late September with a fruit growth period of 107 d. It is a mid-late-ripening variety with serious fruit-cracking if it rains in the maturing stage.

Fruit Characteristics　The medium-sized ovoid fruit has a vertical and cross diameter of 2.99 cm and 2.78 cm averaging 10.7 g with a regular size. It has light-red thin skin, a smooth surface, a medium-deep and wide stalk cavity, a flat fruit apex and a remnant yet indistinct stigma. The sour flesh is tight-textured with medium juice. It has medium quality for dried fruits. The percentage of edible part of fresh fruit is 96.2%, and SSC, TTS, TA and Vc is 31.20%, 27.22%, 0.83% and 440.67 mg per 100 g fresh fruit. The content of flavones and cAMP in mature fruit skin is 1.78 mg/g and 179.55 μg/g. The small stone is oval-shaped, averaging 0.41 g with a well-developed kernel. The percentage of containing kernels is 86.7%.

Evaluation　The cultivar has a high yield, medium-large fruit size and medium quality and is suitable for making dried fruits. Rain protection should be paid much attention to in the maturing stage.

制 干
Drying Varieties
品 种

· 183 ·

彬县黑疙瘩

品种来源及分布 别名黑月枣、黑疙瘩。主要分布于陕西西部的彬县、长武一带，为当地原产的主栽品种。

植物学性状 树体高大，树姿半开张，枝叶密度适中，树冠呈自然圆头形。主干灰黑色，皮裂较深，小条状，不易剥落。枣头红褐色，生长量87.1cm，粗度1.10cm，节间长6.9cm，蜡质多。二次枝发育较弱，长32.8cm，6节。针刺发达。枣股圆锥形，平均抽生枣吊3.0个。枣吊长20.7cm，着叶16片。叶片中大，叶长4.2cm，叶宽2.1cm，卵圆形，绿色，有光泽，先端钝尖，叶基近圆形，叶缘具钝锯齿。花量少，每花序着花5朵。花小，花径4.0～6.0mm，昼开型。

生物学特性 适应性较强。树势强，发枝力中等，树体容易管理。结果稳定，产量中等，枣头、2～3年和4～6年生枝的吊果率分别为27.3%、40.1%和3.5%。在山西太谷地区，9月中旬果实成熟，果实生育期105d，为中晚熟品种类型。裂果较轻，但采前落果重。

果实性状 果实大，近圆形或短倒卵形，纵径3.78cm，横径3.51cm，单果重20.0g，最大25.8g。果肩圆或窄圆，梗洼窄深。果顶凹陷成一字纹。果面光滑，果皮中厚，赭红色。果肉浅绿色，质地致密，汁液中多，味酸甜，宜制干枣，品质中等。鲜枣可食率95.0%，含可溶性固形物33.00%，总糖23.56%，100g果肉维生素C含量414.74mg；果皮含黄酮15.02mg/g，cAMP含量183.86μg/g。干枣含总糖70.73%，酸1.12%。果核椭圆形，核重0.99g，含仁率21.7%。

评价 该品种适应性较强，产量较高而稳定。果实大，裂果轻，肉质致密，适宜制干，干枣较耐贮运。惟果核较大，品质中等。

Binxianheigeda

Source and Distribution The cultivar, also called Heiyuezao, originated from and mainly spreads in Binxian and Changwu in west Shaanxi Province. It is the dominant variety there.

Botanical Characters The large tree is half-spreading with a natural-round crown. The grayish-black trunk bark has deep striped fissures, uneasily shelled off. The reddish-brown 1-year-old shoots are 87.1 cm long and 1.10 cm thick. The internodes are 6.9 cm long, with much wax and developed thorns. The weak secondary branches are 32.8 cm long with 6 nodes. The conical mother fruiting spurs can germinate 3 deciduous fruiting shoots which are 20.7 cm long with 16 leaves. The medium-sized green leaves are 4.2 cm long and 2.1 cm wide, oval-shaped and glossy, with a bluntly-cuspate apex, a nearly-round base and a blunt saw-tooth pattern on the margin. The number of flowers is small, averaging 5 ones per inflorescence. The daytime-bloomed small blossoms have a diameter of 4.0～6.0 mm.

Biological Characters The tree has strong adaptability, moderate vigor, medium branching ability and medium-dense branches, with a medium and stable yield. The percentage of fruits to deciduous bearing shoots of 1-year-old shoots, 2～3-year-old branches and 4～6-year-old ones is 27.3%, 40.1% and 3.5%. In Taigu County of Shanxi Province, it matures in mid-September with a fruit growth period of 105 d. It is a mid-late-ripening variety with light premature fruit-dropping and light fruit-cracking.

Fruit Characteristics The large fruit is nearly round or obovate, with a vertical and cross diameter of 3.78 cm and 3.51 cm averaging 20.0 g (maximum 25.8 g). It has a narrow-round shoulder, a narrow and deep stalk cavity, a sunken fruit apex, a smooth surface, medium-thick and brownish-red skin. The light-green flesh is tight-textured, sour and sweet with medium juice. It is suitable for making dried fruits with medium quality. The percentage of edible part of fresh fruit is 95.0%, and SSC, TTS and Vc is 33.00%, 23.56% and 414.74 mg per 100 g fresh fruit. The content of flavones and cAMP in mature fruit skin is 15.02 mg/g and 183.86 μg/g. The content of TTS and TA in dried fruit is 70.73% and 1.12%. The oval-shaped stone weighs 0.99 g.

Evaluation The cultivar has strong adaptability with a high and stable yield. The large fruit has soft flesh and is suitable for making dried fruits with strong tolerance to storage and transport. Yet it has a large stone with medium quality.

溆浦薄皮枣

品种来源及分布 别名薄皮枣。原产和分布于湖南溆浦县的低庄、双井、花桥镇和茶陵县的洞头树等地，其中双井乡岩元塘湾村为集中产区。

植物学性状 树体高大，树姿开张，树冠自然圆头形。主干皮裂呈纵向的块状，不规则，易剥落。枣头紫褐色，平均长79.7cm，粗1.05cm，节间长7.8cm，蜡质多。二次枝长29.8cm，6节左右，弯曲度中等。无针刺。枣股圆锥形，平均抽生枣吊4.0个。枣吊长28.5cm，着生16片叶。叶片中大，卵状披针形，绿色，先端渐尖，先端尖圆，叶基圆楔形，叶缘具钝锯齿。花多，花序平均着花4朵，花朵较小，昼开型。

生物学特性 树势强健，发枝力强，萌蘖力弱。结果较晚，丰产性强，自然落果少，坐果稳定，枣头、2~3年和3年以上枝的吊果率分别为30.0%、120.0%和37.6%。定植后3年开始结果，15年进入盛果期，成龄树株产70kg左右。在山西太谷地区，4月中旬萌芽，5月下旬始花，10月上旬果实成熟，果实生育期120d，为极晚熟品种类型。极易裂果。

果实性状 果实较小，长圆柱形，纵径2.54cm，横径2.04cm，单果重8.1g，最大13.1g，大小整齐。果肩平圆，梗洼浅、中广，果顶平。果皮薄，红色，光滑。果点中大，密度中等。果肉浅绿色，质地疏松，汁液中多，味极酸，可制干，但品质差。鲜枣可食率93.2%，含可溶性固形物31.20%，总糖29.42%，酸1.22%，100g果肉维生素C含量504.03mg；果皮含黄酮4.78mg/g，cAMP含量97.50μg/g。制干率59.5%，干枣含总糖60.16%，酸1.58%。果核中大，纺锤形，核重0.55g，含仁率91.7%。

评价 该品种成熟期晚，果实小，品质差，且极易裂果，不宜生产栽培。

Xupubopizao

Source and Distribution The cultivar originated from and spreads in Dizhuang, Shuangjing and Huaqiao villages of Xupu County and in Dongtoushu of Chaling County in Hunan Province. Yanyuantangwancun in Shuangjing Village is the main producing area of this cultivar.

Botanical Characters The large tree is spreading with a natural-round crown. The trunk bark has irregular vertical massive fissures, easily shelled off. The purplish-brown 1-year-old shoots are 79.7 cm long and 1.05 cm thick. The internodes are 7.8 cm long with much wax and no thorns. The secondary branches are 29.8 cm long with 6 nodes of medium curvature. The conical mother fruiting spurs can germinate 4 deciduous fruiting shoots which are 28.5 cm long with 16 leaves. The medium-sized green leaves are ovate-lanceolate, with a gradually-cuspate apex, a round-cuneiform base and a blunt saw-tooth pattern on the margin. The number of flowers is medium large, averaging 4 ones per inflorescence. The daytime-bloomed blossoms have a small size.

Biological Characters The tree has strong vigor, strong branching ability and weak suckering ability bearing late with high productivity, light fruit-drop and stable fruit set. It generally bears in the 3rd year after planting and enters the full-bearing stage in the 15th year. A mature tree has a yield of about 70 kg. In Taigu County of Shanxi Province, it germinates in mid-April, begins blooming in late May and matures in early October with a fruit growth period of 120 d. It is an extremely late-ripening variety with light fruit-cracking.

Fruit Characteristics The small oblong fruit has a vertical and cross diameter of 2.54 cm and 2.04 cm averaging 8.1 g (maximum 13.1 g) with a regular size. It has a flat-round shoulder, a shallow and medium-wide stalk cavity, a flat fruit apex, a smooth surface and thin red skin with medium-sized and medium-dense dots. The light-green flesh is loose-textured with medium juice and a very sour taste. It can be used for making dried fruits with a poor quality. The percentage of edible part of fresh fruit is 93.2%, and SSC, TTS, TA and Vc is 31.20%, 29.42%, 1.22% and 504.03 mg per 100 g fresh fruit. The content of flavones and cAMP in mature fruit skin is 4.78 mg/g and 97.50 μg/g. The medium-sized stone is spindle-shaped averaging 0.55 g. The percentage of containing kernels is 91.7%.

Evaluation The cultivar matures late. The small fruit has a poor quality and poor resistance to fruit-cracking, so it is not suitable for commercial production.

制 干
Drying Varieties
品 种

溆浦秤砣枣

品种来源及分布 别名湖南秤砣枣。主要原产和分布于湖南溆浦县的低庄、花桥等地，麻阳、衡山、石门等县也有栽培。

植物学性状 树体较大，树姿开张，枝叶稀疏，树冠乱头形。主干灰褐色，皮粗糙纵裂，呈条状剥落。枣头黄褐色，枝长73.1cm，粗1.04cm，节间长8.1cm，蜡质多。二次枝平均长25.5cm，6节，最多7节。针刺发达。枣股圆柱形，枣股抽吊力较强，平均抽生枣吊4.0个。枣吊较短，平均长12.5cm，着叶13片。叶片中大，椭圆形，绿色，先端长，急尖，叶基圆楔形，叶缘钝锯齿。花序平均着花6朵。花小，花径5.0～5.5mm，昼开型。

生物学特性 适应性强，较抗枣疯病。树势中等，发枝力较弱，易生根蘖。结果早，产量高而稳定，2～3年和3年以上枝的吊果率分别为112.8%和12.5%。定植后3年开始结果，10年生左右进入盛果期，株产40～50kg。在山西太谷地区，10月上旬果实成熟，果实生育期120d左右，为极晚熟品种类型。成熟期不易裂果。

果实性状 果实中大，近圆形，纵径3.94cm，横径3.09cm，单果重12.7g，最大18.5g。果肩凹，梗洼中深、窄，果顶凹陷。果皮薄，红色。果肉浅绿色，质地致密，汁液中多，味甜，适宜制干，品质中等。鲜枣可食率87.1%，含总糖21.78%，酸0.40%。制干率43.1%，干枣深红色有光泽，果皮韧性较强。果核特大，纺锤形，核重1.64g，含仁率100%。

评价 该品种适应性强，树体较大，发枝力弱。结果早，产量高而稳定。果实中大，肉质细密，汁液中多，果核大，品质中等，抗裂果，为适应南方多雨气候的制干良种。

Xupuchengtuozao

Source and Distribution The cultivar originated from and spreads in Dizhuang, Huaqiao of Xupu County and Mayang, Hengshan and Shimen in Hunan Province.

Botanical Characters The large tree is spreading with sparse branches and an irregular crown. The grayish-brown trunk bark has rough, vertical striped fissures. The yellowish-brown 1-year-old shoots are 73.1 cm long and 1.04 cm thick. The internodes are 8.1 cm long with much wax and developed thorns. The secondary branches are 25.5 cm long with 6 nodes (7 at most). The column-shaped mother fruiting spurs can germinate 4 deciduous fruiting shoots which are 12.5 cm long with 13 leaves. The medium-sized leaves are oval-shaped and green with a sharply-cuspate apex, a round-cuneiform base and a blunt saw-tooth pattern on the margin. Each inflorescence has 6 small flowers with a diameter of 5.0～5.5 mm. It blooms in the daytime.

Biological Characters The tree has strong adaptability, strong resistance to jujube witches' broom, moderate vigor, weak branching ability and strong suckering ability. It bears early with a high and stable yield. The percentage of fruits to deciduous bearing shoots of 2～3-year-old branches and over-3-year-old ones is 112.8% and 12.5%. It generally bears in the 3rd year after planting and enters the full-bearing stage in the 10th year, with a yield of 40～50 kg per tree. In Taigu County of Shanxi Province, it matures in early October with a fruit growth period of 120 d. It is an extremely late-ripening variety, almost no fruit-cracking in the maturing stage.

Fruit Characteristics The medium-sized fruit is nearly round with a vertical and cross diameter of 3.94 cm and 3.09 cm, averaging 12.7 g (maximum 18.5 g). It has a sunken shoulder, a medium-deep and narrow stalk cavity, a slightly-sunken fruit apex and thin red skin. The light-green flesh is tight-textured and sweet, with medium juice. It has medium quality for dried fruits. The percentage of edible part of fresh fruit is 87.1%, and the content of TTS and TA is 21.78% and 0.40%. Dried fruits have dark-red color with glossy and whippy skin. The percentage of fresh fruits which can be made into dried ones is 43.1%. The large spindle-shaped stone weighs 1.64 g. The percentage of containing kernels is 100%.

Evaluation The large plant of Xupuchengtuozao cultivar has strong adaptability and weak branching ability bearing early with a high and stable yield. The medium-sized fruit has less juicy flesh with a medium quality. It has strong resistance to fruit-cracking, so it is a good drying variety for southern China where there is much rainfall.

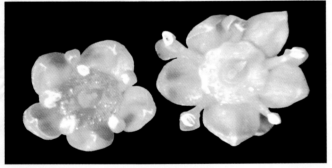

新郑尖头灰枣

品种来源及分布 原产和分布于河南省新郑市。

植物学性状 树体较大,树姿直立,树冠呈圆锥形。主干皮裂条状。枣头黄褐色,平均长81.5cm,节间长7.7cm,无蜡质。二次枝长36.8cm,8～11节,弯曲度中等。针刺不发达。枣股平均抽生枣吊3.8个。枣吊长24.7cm,着叶12片。叶片大,叶长7.4cm,叶宽3.3cm,卵圆形,平展,先端钝尖,叶基圆形,叶缘具钝锯齿。花量多,花序平均着花11朵。花中大,花径6.2mm。

生物学特性 树势强,萌芽率和成枝力强,结果较早,枣头、2～3年和4～6年生枝的吊果率分别为31.8%、107.2%和13.0%。一般定植第二年结果,10年左右进入盛果期,产量一般。在山西太谷地区,10月上旬果实成熟采收,果实生育期115d左右,为极晚熟品种类型。成熟期遇雨裂果较轻。

果实性状 果个极小,圆锥形,纵径2.77cm,横径1.74cm,单果重4.1g,大小整齐。果皮中厚,红色,果面平滑。梗洼窄、深。果顶尖,柱头宿存。肉质致密,汁液中多,味酸甜,品质好,适宜制干。鲜枣可食率92.7%,含可溶性固形物31.20%,总糖25.88%,酸0.51%,100g果肉维生素C含量550.79mg;果皮含黄酮8.92mg/g,cAMP含量172.17μg/g。干枣含总糖57.69%,酸0.90%。果核纺锤形,平均重0.30g,种仁饱满,含仁率63.3%。

评价 该品种树体较大,树势强,结果较早,产量一般,抗裂果能力强。果个极小,品质好,适宜制干。

Xinzhengjiantouhuizao

Source and Distribution The cultivar originated from and spreads in Xinzheng City of Henan Province.

Botanical Characters The large tree is vertical with a conical crown. The trunk bark has striped fissures. The yellowish-brown 1-year-old shoots are 81.5 cm long with the internodes of 7.7 cm, no wax and less-developed thorns. The secondary branches are 36.8 cm long with 8～11 nodes of medium curvature. The mother fruiting spurs can germinate 3.8 deciduous fruiting shoots which are 24.7 cm long with 12 leaves. The large leaves are 7.4 cm long and 3.3 cm wide, oval-shaped and flat, with a bluntly-cuspate apex, a round base and a blunt saw-tooth pattern on the margin. There are many medium-sized flowers with a diameter of 6.2 mm produced, averaging 11 ones per inflorescence.

Biological Characters The tree has strong vigor, strong germination and branching ability. It bears early. The percentage of fruits to deciduous bearing shoots of 1-year-old shoots, 2～3-year-old branches and 4～6-year-old ones is 31.8%, 107.2% and 13.0%. It generally bears in the 2nd year after planting and enters the full-bearing stage in the 10th year with a medium yield. In Taigu County of Shanxi Province, it matures in early October with a fruit growth period of 115 d. It is an extremely late-ripening variety with light fruit-cracking even if it rains in the maturing stage.

Fruit Characteristics The conical fruit has a very small size with a vertical and cross diameter of 2.77 cm and 1.74 cm averaging 4.1 g with a regular size. It has medium-thick red skin, a smooth surface, a narrow and deep stalk cavity, a pointed fruit apex and a remnant stigma. The tight-textured flesh is sour and sweet with medium juice. It has a good quality for dried fruits. The percentage of edible part of fresh fruit is 92.7%, and SSC, TTS, TA and Vc is 31.20%, 25.88%, 0.51% and 550.79 mg per 100 g fresh fruit. The content of flavones and cAMP in mature fruit skin is 8.90 mg/g and 172.17 μg/g. The content of TTS, TA and Vc in dried fruit is 57.69%, 0.90% and 42.54 mg per 100 g fresh fruit. The spindle-shaped stone weighs 0.30 g with a well-developed kernel. The percentage of containing kernels is 63.3%.

Evaluation The large plant of Xinzhengjiantouhuizao cultivar has strong tree vigor bearing early with a medium yield. It has strong resistance to fruit-cracking. The fruit has a very small size and a good quality so as to be suitable for making dried fruits.

平 顺 笨 枣

品种来源及分布　别名笨枣。原产和分布于山西平顺等地。

植物学性状　树体较大，树姿开张，树冠呈乱头形。主干皮裂条状。枣头红褐色，平均长87.0cm，节间长9.0cm，蜡质少。二次枝长26.6cm，4～6节，弯曲度中等。针刺不发达。枣股平均抽生枣吊3.7个。枣吊长22.5cm，着叶12片。叶片大，叶长7.6cm，叶宽3.4cm，卵圆形，两侧略向叶面合抱，先端锐尖，叶基圆楔，叶缘具钝锯齿。花量中多，花序平均着花4朵，花较大，花径6.4mm。

生物学特性　树势中庸，萌芽率和成枝力强。坐果率较高，枣头、2～3年、4～6年和6年以上生枝的吊果率分别为77.7%、80.0%、86.4%和22.5%。定植第三年结果，10年左右进入盛果期，产量较高，一般株产鲜枣63kg。在山西太谷地区，9月中旬果实进入脆熟期，9月下旬成熟采收，果实生育期113d左右，为晚熟品种类型。成熟期遇雨裂果轻。

果实性状　果个中大，长圆形或圆柱形，纵径3.77cm，横径2.58cm，单果重10.8g，大小不整齐。果皮中厚，红色，果面平滑。果点小，分布稀疏。梗洼广、较浅。果顶微凹，柱头遗存，不明显。肉质较致密，味甜，汁液中多，品质中等，可制干。鲜枣可食率96.8%，含可溶性固形物28.20%，总糖23.71%，酸0.76%，100g果肉维生素C含量530.04mg；果皮含黄酮42.50mg/g。制干率57.5%，干枣含总糖63.94%，酸1.55%。果核较小，纺锤形，核重0.35g，种仁不饱满，含仁率31.7%。

评价　该品种树体较大，树势中庸，结果早，产量较高。果个中大，肉质较致密，味甜，汁液中多，品质中等，适宜制干。果实抗裂果能力强。

Pingshunbenzao

Source and Distribution　The cultivar originated from and spreads in Pingshun County of Shanxi Province.

Botanical Characters　The large tree is spreading with an irregular crown. The trunk bark has massive fissures. The reddish-brown 1-year-old shoots are 87.0 cm long, with the internodes of 9.0 cm, with less wax. The secondary branches are 26.6 cm long with 4～6 nodes of medium curvature and less-developed thorns. The mother fruiting spurs can germinate 3.7 deciduous fruiting shoots which are 22.5 cm long with 12 leaves. The large oval-shaped leaves are 7.6 cm long and 3.4 cm wide with both sides slightly folding towards the center. The leaves have a sharply-cuspate apex, a round-cuneiform base and a blunt saw-tooth pattern on the margin. The number of flowers is medium large averaging 4 ones per inflorescence. The medium-sized blossoms have a diameter of 6.4 mm.

Biological Characters　The tree has moderate vigor, strong germination and branching ability, with high fruit set. The percentage of fruits to deciduous bearing shoots of 1-year-old shoots, 2～3-year-old branches, 4～6-year-old ones and over-6-year-old ones is 77.7%, 80.0%, 86.4% and 22.5%. It generally bears in the 3rd year after planting and enters the full-bearing stage in the 10th year with a high yield. The average yield is 63.8 kg per tree. In Taigu County of Shanxi Province, it enters the crisp-maturing stage in mid-September and is harvested in late September with a fruit growth period of 113 d. It is a mid-late-ripening variety with light fruit-cracking even if it rains in the maturing stage.

Fruit Characteristics　The medium-sized fruit is column-shaped, with a vertical and cross diameter of 3.77 cm and 2.58 cm averaging 10.8 g with irregular sizes. It has medium-thick red skin, a smooth surface with small and sparse dots, a wide and shallow stalk cavity, a slightly sunken fruit apex and a remnant yet indistinct stigma. The tight-textured flesh is sweet with medium juice. It has medium quality for dried fruits. The percentage of edible part of fresh fruit is 96.8%, and SSC, TTS, TA and Vc is 28.20%, 23.71%, 0.76% and 530.04 mg per 100 g fresh fruit. The content of flavones in mature fruit skin is 42.50 mg/g. The percentage of fresh fruits which can be made into dried ones is 57.5%. The content of TTS and TA in dried fruit is 63.94% and 1.55%. The small spindle-shaped stone weighs 0.35 g with a shriveled kernel. The percentage of containing kernels is 31.7%.

Evaluation　The medium-sized plant of Pingshunbenzao cultivar has moderate tree vigor bearing early with a high yield. The medium-sized fruit has medium quality and is suitable for making dried fruits. It has strong resistance to fruit-cracking.

平遥大枣

品种来源及分布 原产和分布于山西平遥的梁赵、小胡等村,为当地主栽品种,约占栽培总数的90%。栽培历史不详。

植物学性状 树体较大,树姿半直立,干性较强,枝叶较密,树冠呈圆锥形。主干条状皮裂。枣头黄褐色,平均长63.7cm,粗0.83cm,节间长8.0cm左右,无蜡质。针刺不发达。枣股中大,平均抽生枣吊3.0个。枣吊长22.8cm,着叶12片。叶片大,叶长8.3cm,叶宽4.3cm,卵圆形,浓绿色,先端渐尖,叶基心形或圆形,叶缘钝锯齿。花量多,花序平均着花8朵。蜜盘大,昼开型。

生物学特性 树势强旺,发枝力强。根蘖萌生力中等。结果较早,坐果稳定,产量较高,枣头、2~3年、4~6年和6年以上生枝的吊果率分别为75.5%、111.8%、85.1%和36.3%。在山西太谷地区,9月中旬果实着色,10月上旬成熟采收。果实生育期115d左右,为极晚熟品种类型。采前落果较轻。

果实性状 果实中等大小,长圆柱形,纵径3.53cm,横径2.26cm,单果重7.4g,最大果重12.0g,大小整齐。果肩平圆,果顶平,柱头遗存。果面光滑,红色,果皮中厚。果肉浅绿色,质地致密,汁液中多,味甜略酸,品质中等,可制干。鲜枣含可溶性固形物29.70%,总糖24.20%,酸0.79%,100g果肉维生素C含量447.14mg;果皮含黄酮20.11mg/g,cAMP含量36.10μg/g。制干率55.0%,干枣含总糖66.83%,酸1.93%,品质中等。果核椭圆形,种仁不饱满,含仁率10.0%。

评价 该品种树势强旺,树体较大,适应性强,结果较早,坐果稳定,产量较高。果实品质中等,可制干。

Pingyaodazao

Source and Distribution The cultivar originated from and spreads in Liangzhao and Xiaohu villages of Pingyao County in Shanxi Province. It is the dominant variety there, which occupies 90% of the total jujube area.

Botanical Characters The large tree is vertical with a strong central leader trunk and a conical crown. The trunk bark has massive fissures. The yellowish-brown 1-year-old shoots are 63.7 cm long and 0.83 cm thick. The internodes are 8.0 cm long, without wax and with less-developed thorns. The medium-sized mother fruiting spurs can germinate 3 deciduous fruiting shoots which are 22.8 cm long with 12 leaves. The large leaves are 8.3 cm long and 4.3 cm wide, oval-shaped and dark green, with a gradually-cuspate apex, a heart-shaped or round base and a blunt saw-tooth pattern on the margin. There are many flowers produced, averaging 8 ones per inflorescence. It blooms in the daytime.

Biological Characters The tree has strong vigor, strong branching ability, medium suckering ability and dense branches. It bears early, with stable fruit set and a high yield. The percentage of fruits to deciduous bearing shoots of 1-year-old shoots, 2~3-year-old branches, 4~6-year-old ones and over-6-year-old ones is 75.5%, 111.8%, 85.1% and 36.3%. In Taigu County of Shanxi Province, it begins coloring in mid-September and matures in early October with a fruit growth period of 115 d. It is an extremely late-ripening variety with light premature fruit-dropping.

Fruit Characteristics The medium-sized fruit is column-shaped with a vertical and cross diameter of 3.53 cm and 2.26 cm, averaging 7.4 g (maximum 12.0 g) with a regular size. It has a flat-round shoulder, a flat fruit apex, a remnant stigma, a smooth surface and medium-thick red skin. The light-green flesh is tight-textured, sweet and sour with medium juice. It has medium quality for dried fruits. The content of SSC, TTS, TA and Vc in fresh fruit is 29.70%, 24.20%, 0.79% and 447.14 mg per 100 g fresh fruit. The content of flavones and cAMP in mature fruit skin is 20.11 mg/g and 36.10 μg/g. The percentage of fresh fruits which can be made into dried ones is 55.0%, and the content of TTS and TA in dried fruit is 66.83% and 1.93%. The oval-shaped stone has a shriveled kernel. The percentage of containing kernels is 10.0%.

Evaluation The large plant of Pingyaodazao cultivar has strong tree vigor and strong adaptability bearing early with stable fruit set and high productivity. The fruit has medium quality for dried fruits.

献县小大枣

品种来源及分布 别名大小枣、小大枣。原产河北献县、沧县一带。

植物学性状 树体中大，树姿半开张，枝叶较密，树冠呈自然半圆形。主干灰黑色，树皮裂纹中深，块状，较粗糙，易剥落。枣头黄褐色，枝长81.6cm左右，粗1.24cm，节间长8.3cm，无蜡质。二次枝长31.2cm，6节左右，弯曲度中等。针刺不发达。枣股圆柱形，平均抽生枣吊3.0个。枣吊长19.8cm左右，着生叶片16片。叶片卵圆形，叶长6.8cm，叶宽3.5cm，绿色，较光亮，先端渐尖，叶基圆形，叶缘具钝锯齿。花量较多，每花序着花10朵。花径6.2mm。

生物学特性 树势较强。一般定植后3年开始结果，10年左右进入盛果期，产量较高。枣头、2～3年和4～6年生枝的吊果率分别为24.2%、45.5%和8.5%。在山西太谷地区，10月上旬成熟采收，果实生育期115d左右，为极晚熟品种类型。裂果和落果较轻。

果实性状 果实小，卵圆形，纵径2.76cm，横径2.16cm，单果重6.1g，大小整齐。果肩圆平，梗洼深窄。果顶尖。柱头遗存。果皮中厚，浅红色。果肉浅绿色，质地疏松，汁液中等，味较甜，制干品质上等。鲜枣可食率95.4%，含可溶性固形物34.80%，总糖27.21%，100g果肉维生素C含量612.95mg；果皮含黄酮2.91mg/g，cAMP含量106.95μg/g。干果含总糖70.90%。果核纺锤形，核重0.28g。核内一般不含种子。

评价 该品种适应性较强，丰产性好。果实小，肉厚，适宜制干，有一定经济价值。

Xianxianxiaodazao

Source and Distribution The cultivar, also called Daxiaozao, originated from Xianxian and Cangxian in Hebei Province.

Botanical Characters The medium-sized tree is half-spreading, with dense branches and a semi-round crown. The gray trunk bark has medium-deep massive fissures, easily shelled off. The yellowish-brown 1-year-old shoots are 81.6 cm long and 1.24 cm thick. The internodes are 8.3 cm long, without wax and with less-developed thorns. The secondary branches are 31.2 cm long with 6 nodes of medium curvature. The column-shaped mother fruiting spurs can germinate 3 deciduous fruiting shoots which are 19.8 cm long with 16 leaves. The oval-shaped leaves are green and glossy, 6.8 cm long and 3.5 cm wide, with a gradually-cuspate apex, a round base and a blunt saw-tooth pattern on the margin. There are many flowers with a diameter of 6.2 mm produced, averaging 10 ones per inflorescence.

Biological Characters The tree has strong vigor. It bears in the 3rd year after planting and enters the full-bearing stage in the 10th year. The percentage of fruits to deciduous bearing shoots of 1-year-old shoots, 2～3-year-old branches and over-3-year-old ones is 24.2%, 45.5% and 8.5%. In Taigu County of Shanxi Province, it matures in early October with a fruit growth period of 115 d. It is an extremely late-ripening variety with light fruit-cracking and fruit-dropping.

Fruit Characteristics The small oval-shaped fruit has a vertical and cross diameter of 2.76 cm and 2.16 cm averaging 6.1 g with a regular size. It has a flat-round shoulder, a deep and narrow stalk cavity, a pointed fruit apex, a remnant stigma, and medium-thick and light-red skin. The light-green flesh is loose-textured and sweet with medium juice. It has a good quality for dried fruits. The percentage of edible part of fresh fruit is 95.4%, and SSC, TTS and Vc is 34.80%, 27.21% and 612.95 mg per 100 g fresh fruit. The content of flavones and cAMP in mature fruit skin is 2.91 mg/g and 106.95 μg/g. The content of TTS in dried fruit is 70.90%. The spindle-shaped stone weighs 0.28 g. Most stones contain no kernels.

Evaluation The cultivar has strong adaptability and high productivity. The small fruit has thick flesh and is suitable for making dried fruits with some economic value.

制 干
Drying Varieties
品 种

溆浦柿饼枣

品种来源及分布 主要集中分布于湖南溆浦的观音阁乡铁溪村。

植物学性状 树体较大，树姿开张，干性弱，枝系稀疏，树冠偏斜形。主干黑褐色，粗糙，皮裂条状，易剥落。枣头紫褐色，平均长90.5cm，粗1.20cm，节间长7cm。二次枝长33.5cm，7节左右，弯曲度中等。枣股圆柱形，平均抽生枣吊2.8个。枣吊长23.6cm，着生15片叶。叶片小，卵状披针形，绿色，先端渐尖，叶基圆楔形，叶缘锯齿细锐。花量较大，花序平均着花6朵，昼开型。

生物学特性 适应性较强，树势和发枝力较弱，萌芽率中等。结果较晚，坐果率中等，产量低而不稳，2～3年和3年以上枝的吊果率分别为77.6%和50.4%。定植后3年开始结果，15年后进入盛果期。盛果期树株产25kg左右。在山西太谷地区，4月中旬萌芽，5月下旬始花，9月中旬果实成熟，果实生育期100d左右，为中熟品种类型。落叶迟，11月初开始落叶。

果实性状 果实较小，扁圆形，纵径3.49cm，横径2.29cm，单果重8.3g，大小较整齐。果肩平圆，梗洼窄深，果顶平。果皮厚，浅红色。果肉白色，质地粗，较紧密，汁液中多，味酸甜，可制干，品质中等。鲜枣可食率84.9%，可溶性固形物含量25.20%，总糖18.40%，酸0.64%，100g果肉维生素C含量351.50mg；果皮含黄酮11.41mg/g，cAMP含量193.23μg/g。制干率44.8%。果核大，近圆形，核重1.25g，含仁率10.0%左右。

评价 该品种树体较大，树势较弱，坐果能力差，产量低而不稳。果实小，核特大，可食率极低，肉质粗硬，品质中等，不宜经济栽培。

Xupushibingzao

Source and Distribution The cultivar mainly spreads in Tiexicun of Guanyinge Village in Xupu County of Hunan Province.

Botanical Characters The large tree is spreading with a weak central leader trunk, sparse branches and a deflective crown. The blackish-brown trunk bark has rough striped fissures, easily shelled off. The purplish-brown 1-year-old shoots are 90.5 cm long and 1.20 cm thick. The internodes are 7 cm long. The secondary branches are 33.5 cm long with 7 nodes of medium curvature. The column-shaped mother fruiting spurs can germinate 2.8 deciduous fruiting shoots which are 23.6 cm long with 15 leaves. The small green leaves are ovate-lanceolate, with a gradually-cuspate apex, a round-cuneiform base and a thin, sharp saw-tooth pattern on the margin. There are many flowers produced, averaging 6 ones per inflorescence. It blooms in the daytime.

Biological Characters The tree has strong adaptability, weak vigor, weak branching ability and medium germination ability. It bears late with medium fruit set, a low and unstable yield. The percentage of fruits to deciduous bearing shoots of 2～3-year-old branches and over-3-year-old ones is 77.6% and 50.4%. It generally bears in the 3rd year after planting and enters the full-bearing stage in the 15th year. A tree in its full-bearing stage has a yield of about 25 kg. In Taigu County of Shanxi Province, it germinates in mid-April, begins blooming in late May and matures in mid-September with a fruit growth period of 100 d. It is a mid-ripening variety with late defoliation in early November.

Fruit Characteristics The small oblate-shaped fruit has a vertical and cross diameter of 3.49 cm and 2.29 cm, averaging 8.3 g with a regular size. It has a flat-round shoulder, a narrow and deep stalk cavity, a flat fruit apex and light-red thick skin. The white flesh is rough, tight-textured, sour and sweet, with medium juice. It can be used for making dried fruits with medium quality. The content of Vc in fresh fruit is 351.50 mg per 100 g fresh fruit. The content of flavones and cAMP in mature fruit skin is 11.41 mg/g and 193.23 μg/g. The percentage of fresh fruits which can be made into dried ones is 44.8%. The large nearly-round stone weighs 1.25 g, and the percentage of containing kernels is 10.0%.

Evaluation The large plant of Xupushibingzao cultivar has weak tree vigor, poor fruit set and a low and unstable yield. The small fruit has a large stone, rough and hard flesh with medium quality. It is not suitable for commercial production.

佳县密点脆木枣

品种来源及分布　别名脆木枣。原产于陕西佳县大会坪，数量很少。

植物学性状　树体较大，树姿半开张，呈自然圆头形。树皮裂纹曲折，呈波状条纹，枣头红褐色，被有云斑状灰白色蜡质。针刺较发达。叶片较窄，卵状披针形，叶基呈宽楔形。

生物学特性　在陕北黄河沿岸沙壤土生长良好，丰产稳产，株产30kg左右，枣吊平均结果0.62～0.86个，最多4个。在产地，4月中旬萌芽，6月初始花，9月下旬成熟采收，果实生育期110d左右，为晚熟品种类型。成熟期遇雨易裂果。

果实性状　果实小，圆柱形，纵径3.40cm，横径2.20cm，单果重7.8g，最大11.0g，果个大小整齐。果肩偏斜，平圆。梗洼浅平，环洼大，中深。果顶圆或平圆，顶点微凹，柱头遗存。果柄细，长3.0mm左右。果面平整，果皮中厚，白熟期乳黄色，着色后褐红色，果点密，较明显。果肉白绿色，肉质细，松脆，略松，汁液中多，酸甜，适宜制干，品质较好。鲜枣可食率97.1%，含可溶性固形物27.70%；果皮含黄酮4.30mg/g，cAMP含量110.41μg/g。干枣含总糖70.00%，含酸0.70%，品质中。果核小，长纺锤形，纵径1.70cm，横径0.56cm，核重0.23g。核蒂短，渐尖，近三角形。核尖细长，针状，长0.20～0.40cm，核纹中粗，呈断续相连的纵条纹。含仁率低于10.0%。

评价　该品种适应性强，耐旱，抗寒，结果稳定，丰产性好。果实肉质细，松脆，汁液中多，酸甜适口，为优良的晚熟制干品种。成熟期易裂果。

Jiaxianmidiancuimuzao

Source and Distribution　The cultivar originated from Dahuiping in Jiaxian County of Shaanxi Province, with a small quantity.

Botanical Characters　The large tree is half-spreading with a natural-round crown. The trunk bark has tortuous striped fissures. The reddish-brown 1-year-old shoots have grayish-white cloudy-macula-shaped wax. The thorns are of medium length, generally from 0.5 cm to 1.6 cm long. The narrow leaves are ovate-lanceolate with a wide-cuneiform base.

Biological Characters　The tree grows well in sandy loam soil along Yellow River in north Shaanxi Province. It has a high and stable yield. The average yield per tree is 30 kg. The deciduous fruiting shoot bears 0.62～0.86 fruits (maximum 4). In the main producing areas, it germinates in mid-April, begins blooming in early June and is harvested in late September with a fruit growth period of 105 d. It is a late-ripening variety with serious fruit-cracking if it rains in the maturing stage.

Fruit Characteristics　The small column-shaped fruit has a vertical and cross diameter of 3.40 cm and 2.20 cm averaging 7.8 g (maximum 11.0 g) with a regular size. It has a flat-round deflective shoulder, a shallow and wide stalk cavity, a round or flat-round fruit apex, a remnant stigma, a thin stalk of 3.0 mm long and medium-thick skin with dense and distinct dots. The fruit is milky-yellow in white-maturing stage and turns brownish red after that. The greenish-white flesh is delicate and crisp, loose-textured, with medium juice and a sour and sweet taste. It has medium quality. The percentage of edible part of fresh fruit is 97.1%, and SSC is 27.70%. The content of flavones and cAMP in mature fruit skin is 4.30 mg/g and 110.41 μg/g. Dried fruits with medium quality contain 70.00% of TTS and 0.70% of TA. The small stone is long spindle-shaped. It is round in the middle part with a vertical and cross diameter of 1.70 cm and 0.56 cm, averaging 0.23 g with medium-wide veins, a thin and long needle-shaped apex (0.20～0.40 cm long). The percentage of containing kernels is lower than 10.0%.

Evaluation　The cultivar has strong adaptability, good tolerance to drought and cold with a high and stable yield. The fruit has delicate, crisp and juicy flesh with a proper SAR, all of which makes it a good table variety. Yet it has poor resistance to fruit-cracking.

制干
Drying Varieties
品种

· 201 ·

新郑长鸡心

品种来源及分布　别名长鸡心枣。原产和分布于河南省新郑市的孟庄乡，零星栽种，数量极少。

植物学性状　树体中大，树姿半开张，树冠自然圆头形。主干灰褐色，皮裂纵条形，较浅。枣头红褐色，长63.9cm，粗度0.92cm，节间长7.9cm，蜡质多。二次枝长27.8cm，6节左右，弯曲度中等，节间较短。针刺不发达。枣股圆柱形，平均着生枣吊4个。枣吊平均长22.1cm，着叶15片。叶长6.2cm，叶宽2.5cm，椭圆形，绿色，先端钝尖，叶基圆楔形。花量多，每花序着花6朵，花径6.5mm。

生物学特性　适应性一般，树势中等，发枝力较强。结果早，定植后2年开始结果，10年左右进入盛果期，产量高而稳定，枣头、2~3年和4~6年生枝的吊果率分别为49.2%、36.7%和11.2%。在山西太谷地区，9月下旬果实成熟，果实生育期110d左右，为晚熟品种类型。较抗裂果。

果实性状　果实中大，圆锥形或长鸡心形，纵径4.08cm，横径2.65cm，单果重11.0g，大小整齐。果肩圆平，梗洼浅窄，果顶尖圆。果面平滑，果皮红色。果肉浅绿色，质地较致密，汁液多，味酸甜，可制干，品质上等。鲜枣可食率95.8%，含可溶性固形物34.20%，总糖21.25%，酸0.50%，100g果肉维生素C含量284.44mg。制干率50%以上，干枣含总糖63.77%，酸0.90%。果核纺锤形，核重0.46g，种子较饱满，含仁率73.3%。

评价　该品种树体中大，结果早，产量高而稳定，较抗裂果。制干率较高，干枣质细，味酸甜，饱满，外形美观，为优良的制干品种，可作为主栽品种发展。

Xinzhengchangjixin

Source and Distribution　The cultivar originated from and spreads in Mengzhuang Village of Xinzheng City in Henan Province. It is planted in scattered regions, with a very small quantity.

Botanical Characters　The small tree is half-spreading with a natural-round crown. The grayish-brown trunk bark has shallow, vertical striped fissures. The reddish-brown 1-year-old shoots are 63.9 cm long and 0.92 cm thick. The internodes are 7.9 cm long, with much wax and less-developed thorns. The secondary branches are 27.8 cm long with 6 nodes of medium curvature. The column-shaped mother fruiting spurs can germinate 4 deciduous fruiting shoots which are 22.1 cm long with 15 leaves. The oval-shaped green leaves are 6.2 cm long and 2.5 cm wide with a bluntly-cuspate apex and a round-cuneiform base. There are many flowers with a diameter of 6.5 mm produced, averaging 6 ones per inflorescence.

Biological Characters　The tree has moderate adaptability, moderate vigor and strong branching ability. It bears early, generally in the 2nd year after planting and enters the full-bearing stage in the 10th year, with a high and stable yield. The percentage of fruits to deciduous bearing shoots of 1-year-old shoots, 2~3-year-old branches and 4~6-year-old ones is 49.2%, 36.7% and 11.2%. In Taigu County of Shanxi Province, it matures in late September with a fruit growth period of 110 d. It is a mid-late-ripening variety with strong resistance to fruit-cracking.

Fruit Characteristics　The medium-sized conical fruit has a vertical and cross diameter of 4.08 cm and 2.65 cm averaging 11.0 g with a regular size. It has a flat-round shoulder, a shallow and narrow stalk cavity, a cuspate-round fruit apex, a smooth surface and red skin. The light-green flesh is tight-textured, sour, sweet and juicy. It has a good quality for dried fruits. The percentage of edible part of fresh fruit is 95.8%, and SSC, TTS, TA and Vc is 34.20%, 21.25%, 0.50% and 284.44 mg per 100 g fresh fruit. The percentage of fresh fruits which can be made into dried ones is over 50.0%, and the content of TTS and TA in dried fruit is 63.77% and 0.90%. The spindle-shaped stone weighs 0.46 g, with a well-developed kernel. The percentage of containing kernels is 73.3%.

Evaluation　The small plant of Xinzhengchangjixinzao cultivar bears early with a high and stable yield and strong resistance to fruit-cracking. It has a high percentage for making dried fruits from fresh ones. Dried fruits have delicate, sweet and thick flesh, as well as an attractive appearance. It is a good drying variety which can be developed as a dominant variety.

制 干
Drying Varieties
品 种

· 203 ·

溆浦岩枣

品种来源及分布 别名岩枣。原产和分布于湖南溆浦的祖市殿乡四门村。该省其他枣产区也有零星栽种。

植物学性状 树体小，树姿开张，枝系下垂，较稀疏，树冠偏斜形。树干灰褐色，粗糙，主干皮条状剥落。枣头紫褐色，长78.7cm，节间长8.4cm，蜡质少。无刺。二次枝粗短，弯曲下垂，枝长26.1cm，大多为5节。枣股圆柱形，平均抽生枣吊4.0个。枣吊长25.3cm，着生叶片20片。叶片较小，卵状披针形，绿色，先端渐尖，叶基圆楔形，叶缘粗锯齿。花量少，花序平均着花3朵。花小，昼开型。

生物学特性 适应性广，抗逆性强。树势弱，发枝力强，枣头抽生量多，二次枝生长量小。坐果率高，丰产稳产，枣头、2~3年和3年以上枝的吊果率分别为60.0%、150.0%和132.0%。一般定植后2年开始结果，10年左右进入盛果期，成龄树株产75kg左右。在山西太谷地区，果实10月上旬成熟，果实生育期119d，为极晚熟品种类型。

果实性状 果实中大，卵圆形。纵径3.87cm，横径2.53cm，单果重12.1g，最大21.9g。果肩圆，梗洼窄浅，果顶凹陷，萼片残存。果面光滑，果皮较厚，紫红色。果肉浅绿色，质地致密，汁液少，味甜酸，宜制干。鲜枣可食率93.8%，含可溶性固形物29.00%；果皮含黄酮5.38mg/g，cAMP含量155.76μg/g。制干率43.4%。果核较大，圆形，核重0.75g，含仁率20%左右。

评价 该品种抗逆性强，产量高而稳定。果实肉质粗硬，果核较大，可制干，适口性差，经济价值不高。

Xupuyanzao

Source and Distribution The cultivar originated from Simencun of Zushidian Village in Xupu of Hunan Provincve. There are also some distributions in other jujube production areas of Hunan Province.

Botanical Characters The small tree is spreading with pendulous and sparse branches and a deflective crown. The grayish-brown trunk bark has rough striped fissures. The purplish-brown 1-year-old shoots are 78.7 cm long. The internodes are 8.4 cm long with less wax and no thorns. The pendulous secondary branches are thick and short, 26.1 cm long with 5 nodes. The column-shaped mother fruiting spurs can germinate 4 deciduous fruiting shoots which are 25.3 cm long with 20 leaves. The small green leaves are ovate-lanceolate, with a gradually-cuspate apex, a round-cuneiform base and a thick saw-tooth pattern on the margin. The number of flowers is small, averaging 3 ones per inflorescence. The daytime-bloomed blossoms have a small size.

Biological Characters The tree has strong adaptability, strong resistance to adverse conditions, weak vigor, strong branching ability, high fruit set, a high and stable yield. The 1-year-old shoots have strong growth potential and that for the secondary branches is weak. The percentage of fruits to deciduous bearing shoots of 1-year-old shoots, 2~3-year-old branches and over-3-year-old ones is 60.0%, 150.0% and 132.0%. It generally bears in the 2nd year after planting and enters the full-bearing stage in the 10th year. A mature tree has a yield of about 75 kg. In Taigu County of Shanxi Province, it matures in early October with a fruit growth period of 119 d. It is an extremely late-ripening variety.

Fruit Characteristics The medium-sized fruit is oval-shaped, with a vertical and cross diameter of 3.87 cm and 2.53 cm averaging 12.1 g (maximum 21.9 g). It has a round shoulder, a narrow and shallow stalk cavity, a sunken fruit apex, remnant sepals, a smooth surface and purplish-red thick skin. The light-green flesh is loose-textured, sour and sweet, with less juice. It is suitable for fresh-eating and making dried fruits. The percentage of edible part of fresh fruit is 93.8%, and SSC is 29%. The content of flavones and cAMP in mature fruit skin is 5.38 mg/g and 155.76 μg/g. The percentage of fresh fruits which can be made into dried ones is 43.4%. The larger round stone weighs 0.75 g, with the percentage of containing kernels of 20%.

Evaluation The cultivar has strong resistance to adverse conditions with a high and stable yield. Yet fruit flesh is rough and hard with a poor taste, which makes it low in economic value.

直社疙瘩枣

品种来源及分布 原产和分布于陕西蒲城的直社村一带，是当地的主栽品种。

植物学性状 树体中大，树姿开张，树冠呈圆头形。主干皮裂较深，条状，容易剥落。枣头红褐色，蜡质多，平均长73.1cm，粗0.94cm，节间长7.3cm。针刺不发达。枣股平均抽生枣吊3.0个，最多7个。枣吊长24.6cm，着叶18片。叶片较小，叶长4.9cm，叶宽2.4cm，卵圆形或长卵圆形，叶厚，深绿色，少光泽，先端渐尖，先端钝尖，叶基圆形，叶缘粗锯齿。花量少，花序平均着花4朵。花中大，花径6.0～7.0mm。

生物学特性 适应性和抗逆性均强。树势较强。坐果率较高，丰产性能较强，2～3年和3年以上枝的吊果率分别为119.6%和42.0%。果实着色快，成熟期一致且极少裂果。在山西太谷地区，9月上旬着色，9月中旬成熟采收，果实生育期105d，为中熟品种类型。

果实性状 果实较大，平顶锥形或卵圆形，纵径2.87cm，横径2.76cm，单果重13.2g，大小整齐。果肩平，梗洼浅，狭窄。果柄粗短，长2.3mm。果顶平。果面光滑。果皮厚，赭红色，比一般枣品种果色深。果点中大，圆形或卵圆形，稀疏。果肉浅绿色，质地致密，汁液少，味甘甜，品质中等，适宜制干。鲜枣可食率93.1%，含可溶性固形物32.00%，含总糖21.74%，酸0.24%，100g果肉维生素C含量382.12mg；果皮含黄酮5.79mg/g，cAMP含量298.64μg/g。果核大，椭圆形，侧面略扁，核重0.91g，含仁率20.0%。

评价 该品种适应性和抗逆性较强，成熟期极少裂果，可在秋雨多的地区栽培。

Zhishegedazao

Source and Distribution The cultivar originated from and spreads in Zhishe Village of Pucheng in Shaanxi Province and is local dominant variety there.

Botanical Characters The medium-sized tree is spreading with a round crown. The trunk bark has deep striped fissures, easily shelled off. The reddish-brown 1-year-old shoots are 73.1 cm long and 0.94 cm thick with much wax. The internodes are 7.3 cm long with less-developed thorns. The mother fruiting spurs can germinate 3 (maximum 7) deciduous fruiting shoots which are 24.6 cm long with 18 leaves. The small thick leaves are 4.9 cm long and 2.4 cm wide, oval-shaped, dark green and a little glossy, with a gradually-cuspate apex, a round base and a thick saw-tooth pattern on the margin. The number of flowers is small, averaging 4 ones per inflorescence. The medium-sized blossoms have a diameter of 6.0～7.0 mm.

Biological Characters The tree has strong adaptability, strong resistance to adverse conditions and strong vigor, with high fruit set and a medium yield. The percentage of fruits to deciduous bearing shoots of 2～3-year-old branches and over-3-year-old ones is 119.6% and 42.0%. The fruit begins coloring early and colors rapidly, with uniform maturing stage and light fruit-cracking. In Taigu County of Shanxi Province, it begins coloring in early September and matures in mid-September with a fruit growth period of 105 d. It is a mid-ripening variety.

Fruit Characteristics The large conical or oval-shaped fruit has a vertical and cross diameter of 2.87 cm and 2.76 cm averaging 13.2 g with a regular size. It has a flat shoulder, a shallow and narrow stalk cavity, a thick and short stalk (2.3 mm long), a flat fruit apex, a smooth surface and brownish-red thick skin with medium-large, round and sparse dots. The light-green flesh is tight-textured and sweet with less juice. It has medium quality for dried fruits. The percentage of edible part of fresh fruit is 93.1%, and SSC and Vc is 32.00% and 382.12 mg per 100 g fresh fruit. The content of flavones and cAMP in fruit skin is 5.79 mg/g and 298.64 μg/g. The small oval-shaped stone weighs 0.91 g, with the percentage of containing kernels of 20.0%.

Evaluation The cultivar has strong adaptability and strong resistance to adverse conditions. Fruit-cracking seldom occurs in the maturing stage, so it can be developed in areas with much rainfall.

制干
Drying Varieties
品种

太谷壶瓶酸

品种来源及分布 别名壶瓶酸。主要分布于山西太谷的南张村一带,尚有百年以上结果大树,栽培较少。

植物学性状 树体较大,树姿半开张,干性较强,树冠圆头形。主干块状皮裂。枣头红褐色,平均长54.1cm,粗0.85cm,节间长9.9cm,蜡质多。二次枝长22.7cm,7节左右,弯曲度中等。针刺细,不发达。枣股圆锥形,平均抽生枣吊4.0个。枣吊长22.4cm,着叶16片。叶片椭圆形,叶长6.3cm,叶宽3.2cm,绿色,较光亮,先端渐尖,叶基圆楔形。花量中等,花序平均着花6朵。花较大,昼开型。

生物学特性 适应性较强,树势中庸,发枝力中等,萌蘖力弱。结果早,较丰产,枣头、2～3年、4～6年以及7年以上枝的吊果率分别为72.1%、46.5%、43.6%和4.3%。定植后2年开始结果,15年后进入盛果期。在山西太谷地区,4月下旬萌芽,5月下旬始花,9月20日前后进入脆熟期,果实生育期110d,为中晚熟品种类型。成熟期遇雨裂果较重。

果实性状 果实较大,倒卵圆形,纵径3.87cm,横径2.57cm,单果重19.0g,大小较整齐。梗洼深窄。果顶平。果皮较薄,浅红色,光滑。果肉质地疏松,汁液中多,味酸,品质中等,适宜制干。鲜枣可食率97.0%,含可溶性固形物30.30%,总糖26.73%,酸0.77%,100g果肉维生素C含量336.20mg。干枣含总糖68.29%,酸1.29%。果核中大,倒纺锤形,核重0.58g,含仁率36.7%。

评价 该品种适应性强,结果早,较丰产。果实品质一般,成熟期裂果较重,不宜生产栽培。

Taiguhupingsuan

Source and Distribution The cultivar spreads in Nanzhang Village of Taigu County in Shanxi Province, with a small quantity.

Botanical Characters The large tree is half-spreading with a strong central leader trunk and a round crown. The trunk bark has massive fissures. The reddish-brown 1-year-old shoots are 54.1 cm long and 0.85 cm thick. The internodes are 9.9 cm long with much wax and less-developed thorns. The secondary branches are 22.7 cm long with 7 nodes of medium curvature. The conical mother fruiting spurs can germinate 4 deciduous fruiting shoots which are 22.4 cm long with 16 leaves. The green leaves are oval-shaped and glossy, 6.3 cm long and 3.2 cm wide, with a gradually-cuspate apex and a round-cuneiform base. The number of flowers is medium large, averaging 6 ones per inflorescence. The daytime-bloomed blossoms have a large size.

Biological Characters The tree has strong adaptability, moderate vigor, medium branching ability and weak suckering ability. It bears early with a high yield. The percentage of fruits to deciduous bearing shoots of 1-year-old shoots, 2～3-year-old branches, 4～6-year-old ones and over-7-year-old ones is 72.1%, 46.5%, 43.6% and 4.3%. It generally bears in the 2nd year after planting and enters the full-bearing stage in the 15th year. In Taigu County of Shanxi Province, it germinates in late April, begins blooming in late May and enters the crisp-maturing stage around September 20 with a fruit growth period of 110 d. It is a mid-late-ripening variety with serious fruit-cracking if it rains in the maturing stage.

Fruit Characteristics The large oval-shaped fruit has a vertical and cross diameter of 3.87 cm and 2.57 cm, averaging 19.0 g with a regular size. It has a deep and narrow stalk cavity, a flat fruit apex, a thin, light-red and smooth surface. The flesh is loose-textured, sour and sweet, with medium juice. It has medium quality for dried fruits. The percentage of edible part of fresh fruit is 97.0%, and SSC, TTS, TA and Vc is 30.30%, 26.73%, 0.77% and 336.20 mg per 100 g fresh fruit. The content of TTS, TA and Vc in dried fruit is 68.29%, 1.29% and 68.29 mg per 100 g fresh fruit. The large stone is inverted spindle-shaped averaging 0.58 g. The percentage of containing kernels is 36.7%.

Evaluation The cultivar has strong adaptability, bearing early with a high yield. The fruit has a large stone and low edibility and medium quality for fresh-eating. Fruit-cracking easily occurs in the maturing stage, so it is not suitable for commercial production.

制 干
Drying Varieties
品 种

襄汾崖枣

品种来源及分布 别名崖枣。原产和分布于山西襄汾县，栽培数量不多。

植物学性状 树体中大，树姿半开张，树冠呈圆锥形。主干条状皮裂。枣头黄褐色，平均长71.5cm，粗0.85cm，节间长8.9cm，无蜡质。二次枝长28.5cm，5节左右，弯曲度中等。针刺不发达。枣股抽吊力较强，平均抽生枣吊4.0个，枣吊长22.5cm，着叶12片。叶片卵圆形，绿色，光亮，先端渐尖，先端尖圆。叶基心形或圆形。叶缘钝锯齿。花量中多，花序平均着花7朵。蜜盘大。

生物学特性 适应性强，耐旱，较抗枣疯病。树势和发枝力中等。坐果率中等，产量较高而稳定，枣头、2～3年、4～6年和6年以上枝的吊果率分别为69.3%、71.5%、53.2%和15.0%。在山西太谷地区，9月下旬果实成熟采收，果实生育期109d，为晚熟品种类型。抗裂果且采前落果轻。

果实性状 果实中大，圆柱形，纵径3.27cm，横径2.56cm，单果重10.5g，最大14.5g，大小较整齐。果肩平圆，梗洼广浅，果顶圆，顶点微凹。果面光滑，果皮薄，红色。果肉中厚，浅绿色，质地致密，汁液中多，味酸，品质中等，可制干。鲜枣可食率94.3%，含可溶性固形物27.90%，总糖27.17%，酸0.31%，100g果肉维生素C含量499.75mg；果皮含黄酮0.93mg/g，cAMP含量114.76μg/g。制干率55.8%，干枣含总糖59.10%，酸1.40%。果核纺锤形，核重0.60g，含仁率58.3%。

评价 该品种适应性强，较抗枣疯病。产量较高且稳定。果实中大，品质中上。成熟期遇雨裂果轻。

Xiangfenyazao

Source and Distribution The cultivar originated from and spreads in Xiangfen County of Shanxi Province with a small quantity.

Botanical Characters The medium-sized tree is half-spreading with a round crown. The trunk bark has striped fissures. The yellowish-brown 1-year-old shoots are 71.5 cm long and 0.85 cm thick. The internodes are 8.9 cm long without wax and with less-developed thorns. The secondary branches are 28.5 cm long with 5 nodes of medium curvature. The mother fruiting spurs can germinate 4 deciduous fruiting shoots which are 22.5 cm long with 12 leaves. The oval-shaped leaves are green and glossy, with a gradually-cuspate apex, a heart-shaped or round base and a blunt saw-tooth pattern on the margin. The number of flowers is medium large, averaging 7 ones per inflorescence.

Biological Characters The tree has strong adaptability, tolerant to drought and resistant to jujube witches' broom, with strong vigor, medium branching ability, medium fruit set, and a high and stable yield. The percentage of fruits to deciduous bearing shoots of 1-year-old shoots, 2～3-year-old branches and over-3-year-old ones is 69.3%, 71.5% and 53.2%. In Taigu County of Shanxi Province, it matures in late September with a fruit growth period of 109 d. It is a mid-late-ripening variety with strong resistance to fruit-cracking and light premature fruit-dropping.

Fruit Characteristics The medium-sized fruit is column-shaped, with a vertical and cross diameter of 3.27cm and 2.56 cm averaging 10.5 g (maximum 14.5 g) with a regular size. It has a flat-round shoulder, a shallow and wide stalk cavity, a round fruit apex, a smooth surface and thin red skin. The light-green flesh is sour, medium-thick and tight-textured with medium juice. It has medium quality for dried fruits. The percentage of edible part of fresh fruit is 94.3%, and SSC, TTS, TA and Vc is 27.90%, 23.27%, 0.99% and 499.75 mg per 100 g fresh fruit. The content of flavones and cAMP in mature fruit skin is 0.93 mg/g and 114.76 μg/g. The percentage of fresh fruits which can be made into dried ones is 55.8%, and the content of TTS and TA in dried fruit is 59.10% and 1.40%. The spindle-shaped stone weighs 0.6 g with the percentage of containing kernels of 58.3%.

Evaluation The cultivar has strong adaptability and strong resistance to jujube witches' broom with a high and stable yield. The medium-sized fruit has a good quality. Fruit-cracking is not serious even if it rains in the maturing stage.

平遥苦端枣

品种来源及分布 别名苦端（团）枣。原产和分布于山西平遥县。

植物学性状 树体高大，树姿开张，树冠呈半圆形。主干条状皮裂。枣头红褐色，平均长62.6cm，节间长9.4cm，蜡质多。二次枝平均长31.3cm，4～6节，弯曲度小。针刺不发达。枣股平均抽生枣吊3.6个。枣吊长20.1cm，着叶11片。叶片大，叶长7.1cm，叶宽3.5cm，卵圆形，平展，先端锐尖，叶基心形，叶缘具锐锯齿。花量少，花序平均着花3朵。花大，花径6.7mm。

生物学特性 树势强健，萌芽率和成枝力中等。结果较晚，枣头、2～3年、4～6年和6年以上生枝的吊果率分别为38.8%、40.1%、34.8%和17.2%。定植第三年结果，10年左右进入盛果期，丰产稳产。在山西太谷地区，9月下旬果实成熟采收，果实生育期109d左右，为晚熟品种类型。成熟期遇雨裂果轻。

果实性状 果个较大，近圆柱形，纵径3.51cm，横径3.47cm，单果重18.3g，大小整齐。果皮薄，红色，果面平滑。梗洼中深、较广。果顶凹陷，柱头宿存。肉质疏松，汁液中多，味甜，品质中等，适宜制干。鲜枣可食率96.8%，含可溶性固形物30.00%，总糖24.86%，酸0.61%，100g果肉维生素C含量378.23mg；果皮含黄酮63.49mg/g，cAMP含量241.84μg/g。干枣含总糖57.86%，酸1.14%。核中大，纺锤形，平均重0.59g。核内种仁饱满，含仁率66.7%。

评价 该品种树体高大，树势强，结果晚，丰产稳产，较抗裂果。果个大，品质中等，适宜制干。

Pingyaokuduanzao

Source and Distribution The cultivar originated from and spreads in Pingyao County of Shanxi Province.

Botanical Characters The large tree is spreading with an irregular crown. The trunk bark has striped fissures. The reddish-brown 1-year-old shoots are 62.6 cm long, with the internodes of 9.4 cm, with much wax and less-developed thorns. The secondary branches are 31.3 cm long with 4～6 nodes of small curvature. The mother fruiting spurs can germinate 3.6 deciduous fruiting shoots which are 20.1 cm long with 11 leaves. The large leaves are 7.1 cm long and 3.5 cm wide, oval-shaped and flat with a sharply-cuspate apex, a heart-shaped base and a sharp saw-tooth pattern on the margin. The number of flowers is small, averaging 3 ones per inflorescence. The large blossoms have a diameter of 6.7 mm.

Biological Characters The tree has moderate or strong vigor, medium germination and branching ability. It bears late with low fruit set. The percentage of fruits to deciduous bearing shoots of 1-year-old shoots, 2～3-year-old branches, 4～6-year-old ones and over-6-year-old ones is 38.8%, 40.1%, 34.8% and 17.2%. It generally bears in the 3rd year after planting and enters the full-bearing stage in the 10th year with a high and stable yield. In Taigu County of Shanxi Province, it matures in late September with a fruit growth period of 109 d. It is a late-ripening variety with light fruit-cracking even if it rains in the maturing stage.

Fruit Characteristics The large oval-shaped fruit has a vertical and cross diameter of 3.51 cm and 3.47 cm averaging 18.3 g with a regular size. The fruit has thin red skin, a smooth surface, a medium-deep and wide stalk cavity, a sunken fruit apex and a remnant stigma. The loose-textured flesh is sweet with medium juice. It has medium quality for dried fruits. The percentage of edible part of fresh fruit is 96.8%, and SSC, TTS, TA and Vc is 30.00%, 24.86%, 0.61% and 378.23 mg per 100 g fresh fruit. The content of flavones and cAMP in mature fruit skin is 63.49 mg/g and 241.84 μg/g. Dried fruits have a sugar content of 57.86% and TA content of 1.14%. The medium-sized stone is spindle-shaped, averaging 0.59 g with a well-developed kernel. The percentage of containing kernels is 66.7%.

Evaluation The large plant of Pingyaokuduanzao cultivar has strong tree vigor, bearing late with a high and stable yield. It has strong resistance to fruit-cracking. The large fruit has medium quality and is suitable for making dried fruits.

制 干
Drying Varieties
品 种

溆浦甜酸枣

品种来源及分布 原产于湖南溆浦的双井乡大塘村。

植物学性状 树体高大，树姿半开张，树冠呈自然圆头形。主干灰黑色，皮呈条状剥落。枣头紫褐色，平均长68.9cm，粗0.88cm，节间长7.2cm，蜡质少。二次枝长27.8cm，6节，弯曲度小。枣股圆锥形，平均抽生枣吊4.0个。针刺不发达。枣吊长24.8cm，着叶18片。叶片较小，椭圆形，绿色，先端渐尖，叶基偏斜形，叶缘具钝锯齿。花量较多，花序平均着花5朵，夜开型。

生物学特性 适应性广，树势和发枝力强，始果较晚，坐果稳定，枣吊平均坐果0.46个，定植后3年开始结果，丰产性一般，15年后进入盛果期。在山西太谷地区，4月下旬萌芽，5月底始花，9月下旬果实成熟，果实生育期110d，为中晚熟品种类型。成熟期遇雨易裂果。

果实性状 果实中大，圆柱形或圆形，纵径3.27cm，横径2.47cm，单果重11.2g，最大13.6g，大小不整齐。果肩凸圆，梗洼中深、广，果顶凹陷。果皮厚，紫红色。果肉绿色，质地较致密，汁液中多，味甜酸，可制干，品质中上。鲜枣可食率94.3%，含可溶性固形物28.00%，总糖26.96%，酸0.70%；果皮含黄酮12.18mg/g，cAMP含量56.34μg/g。制干率47.2%。果核中大，椭圆形，核重0.64g，含仁率53.4%。

评价 该品种适应性强，树体强健。果实甜酸适口，适宜制干，品质中上。

Xuputiansuanzao

Source and Distribution The cultivar originated from Datangcun of Shuangjing Village in Xupu County of Hunan Province.

Botanical Characters The large tree is half-spreading with a natural-round crown. The grayish-black trunk bark has striped fissures. The purplish-brown 1-year-old shoots are 68.9 cm long and 0.88 cm thick. The internodes are 7.2 cm long with less wax and less-developed thorns. The secondary branches are 27.8 cm long with 6 nodes of small curvature. The conical mother fruiting spurs can germinate 4 deciduous fruiting shoots which are 24.8 cm long with 18 leaves. The small green leaves are oval-shaped, with a gradually-cuspate apex, a deflective base and a blunt saw-tooth pattern on the margin. There are many flowers produced, averaging 5 ones per inflorescence. It blooms at night.

Biological Characters The tree has strong adaptability, strong vigor and strong branching ability, bearing late with stable fruit set. The deciduous fruiting shoot bears 0.46 fruits on average. It generally bears in the 3rd year after planting with a medium yield, and enters the full-bearing stage in the 15th year. In Taigu County of Shanxi Province, it germinates in late April, begins blooming in late May and matures in late September with a fruit growth period of 110 d. It is a mid-late-ripening variety with serious fruit-cracking if it rains in maturing stage.

Fruit Characteristics The medium-sized fruit is column-shaped, with a vertical and cross diameter of 3.27 cm and 2.47 cm, averaging 11.2 g (maximum 13.6 g) with irregular sizes. It has a pointed-round shoulder, a medium-deep and wide stalk cavity, a sunken fruit apex and purplish-red thick skin. The green flesh is tight-textured, sour and sweet, with medium juice. It has a better than normal quality for dried fruits. The percentage of edible part of fresh fruit is 94.3%, and SSC, TTS and TA is 28%, 26.96% and 0.70%. The content of flavones and cAMP in mature fruit skin is 12.18 mg/g and 56.34 μg/g. The percentage of fresh fruits which can be made into dried ones is 47.2%. The medium-sized stone is oval-shaped, averaging 0.64 g. The percentage of containing kernels is 53.4%.

Evaluation The cultivar has strong adaptability and strong tree vigor. The fruit has a sour and sweet taste and is suitable for making dried fruits with a better than normal quality.

遵 义 甜 枣

品种来源及分布 分布于贵州遵义地区。栽培历史不详。

植物学性状 树体较小，树姿半开张，树冠呈圆锥形。主干皮裂呈条状。枣头紫褐色，平均长70.1cm，节间长6.8cm。二次枝长23.7cm，4～5节，弯曲度小。针刺不发达。枣股平均抽生枣吊3.0个。枣吊长26.0cm，着叶17片。叶片中大，卵状披针形，先端急尖，叶基偏斜形，叶缘钝锯齿。花量多，花序平均着花7朵，花较小，花径6.0mm。

生物学特性 树势中庸，成枝力中等。定植第二年结果，盛果期产量较高，2～3年和4～6生枝坐果率高，吊果率分别达到166.0%和116.5%。在山西太谷地区，9月下旬果实进入脆熟期，10月上旬成熟采收，果实生育期118d左右，属极晚熟品种。成熟期遇雨裂果少。

果实性状 果个较小，长圆形或圆柱形，纵径2.80cm，横径1.90cm，单果重7.6g，大小较整齐。果皮中厚，红色，果面平滑。梗洼窄、中深。果顶平圆，柱头残存。肉质较致密，味甜，汁液少，品质中等，适宜制干。鲜枣可食率93.0%，含可溶性固形物29.00%，总糖21.83%，酸0.47%，糖酸比52.83∶1，100g果肉维生素C含量462.14mg；果皮含黄酮12.81mg/g，cAMP含量4.79μg/g。果核中大，纺锤形，核重0.53g。种仁不饱满，含仁率20.0%。

评价 该品种树势中庸，结果早，坐果多，产量高而稳定。果实小，品质中等，适宜制作干枣。抗裂果能力较强。

Zunyitianzao

Source and Distribution The cultivar mainly spreads in Zunyi of Guizhou Province with an unknown history.

Botanical Characters The small tree is half-spreading with a conical crown. The trunk bark has striped fissures. The purplish-brown 1-year-old shoots are 70.1 cm long with the internodes of 6.8 cm. The secondary branches are 23.7 cm long with 4～5 nodes of small curvature and less-developed thorns. The mother fruiting spurs can germinate 3.0 deciduous fruiting shoots which are 26.0 cm long with 17 leaves. The medium-sized leaves are ovate-lanceolate with a sharply-cuspate apex, a deflective base and a blunt saw-tooth pattern on the margin. There are many small flowers with a diameter of 6.0 mm produced averaging 7 ones per inflorescence.

Biological Characters The tree has moderate vigor and medium branching ability, bearing in the 2nd year after planting with a high yield in the full-bearing stage. The 2～3-year-old branches and 4～6-year-old ones have high fruit set, the percentage of fruits to deciduous bearing shoots of which is 166.0% and 116.5% respectively. In Taigu County of Shanxi Province, it enters the crisp-maturing stage in late September and is harvested in early October with a fruit growth period of 118 d. It is an extremely late-ripening variety with light fruit-cracking even if it rains in the maturing stage.

Fruit Characteristics The small column-shaped fruit has a vertical and cross diameter of 2.8 cm and 1.9 cm, averaging 7.6 g with a regular size. It has medium-thick red skin, a smooth surface, a medium-deep and narrow stalk cavity, a flat and round fruit apex and a remnant stigma. The tight-textured flesh is sweet with less juice. It has medium quality for dried fruits. The percentage of edible part of fresh fruit is 93.0%, and SSC, TTS, TA and Vc is 29.00%, 21.83%, 0.47% and 462.14 mg per 100 g fresh fruit. The SAR is 52.83∶1. The content of flavones and cAMP in mature fruit skin is 12.81 mg/g and 4.79 μg/g. The medium-sized stone is spindle-shaped, averaging 0.53 g with a shriveled kernel. The percentage of containing kernels is 20.0%.

Evaluation The cultivar has moderate vigor, bearing early with high fruit set and a high and stable yield. The small fruit has medium quality and is suitable for making dried fruits. It has strong resistance to fruit-cracking.

制 干
Drying Varieties
品 种

武乡牙枣

品种来源及分布 分布于山西武乡县。

植物学性状 树体较小，树姿开张，树冠偏斜形。主干皮裂条状。枣头红褐色，平均长45.4cm，节间长6.5cm。二次枝长22.8cm，5～7节，弯曲度中等。无针刺。枣股平均抽生枣吊4.0个。枣吊长24.2cm，着叶19片。叶片中大，叶长4.7cm，叶宽2.5cm，椭圆形，先端钝尖，叶基圆形，叶缘锐锯齿。花量中多，花序平均着花5朵，花中大，花径6.1mm。

生物学特性 树势弱，成枝力较强，可达86.7%。定植第二年结果，8年左右进入盛果期，产量中等，2～3年生枝坐果率高，枣头、2～3年生和3年以上枝吊果率分别为6.7%、128.6%和68.2%。在山西太谷地区，9月中旬果实进入脆熟期，9月下旬成熟采收，果实生育期110d左右，为晚熟品种类型。成熟期遇雨裂果严重。

果实性状 果个较大，圆柱形，纵径3.9cm，横径2.7cm，单果重13.1g，大小整齐。果皮中厚，红色，果面平滑。梗洼中深、广。果顶凹，柱头宿存。肉质疏松，味甜酸，汁液中多，品质中等，适宜制干。鲜枣可食率93.4%，含可溶性固形物20.40%，总糖17.80%，酸0.70%，100g果肉维生素C含量380.12mg；果皮含黄酮24.27mg/g，cAMP含量168.03 μg/g。果核较大，纺锤形，核重0.86g。种仁不饱满，含仁率40%左右。

评价 该品种树势弱，产量中等。果实中大，果核较大，品质中等，可制干。抗裂果能力较差。

Wuxiangyazao

Source and Distribution The cultivar mainly spreads in Wuxiang County of Shanxi Province.

Botanical Characters The small tree is spreading with a deflective crown. The trunk bark has striped fissures. The reddish-brown 1-year-old shoots are 45.4 cm long with the internodes of 6.5 cm. The secondary branches are 22.8 cm long with 5～7 nodes of medium curvature and without thorns. The mother fruiting spurs can germinate 4.0 deciduous fruiting shoots which are 24.2 cm long with 19 leaves. The medium-sized leaves are oval-shaped, 4.7 cm long and 2.5 cm wide, with a bluntly-cuspate apex, a round base and a sharp saw-tooth pattern on the margin. The number of flowers is medium large, averaging 5 ones per inflorescence. The medium-sized blossoms have a diameter of 6.1 mm.

Biological Characters The tree has weak vigor and strong branching ability, with a branching rate of 86.7%. It bears in the 2nd year after planting and enters the full-bearing stage in the 8th year with a medium yield. The 2 or 3-year-old branches have high fruit set. The percentage of fruits to deciduous fruiting shoots of 1-year-old shoots, 2～3-year-old branches and over-3-year-old ones is 6.7%, 128.6% and 68.2%. In Taigu County of Shanxi Province, it enters the crisp-maturing stage in mid-September and is harvested in late September with a fruit growth period of 110 d. It is a mid-late-ripening variety with serious fruit-cracking if it rains in the maturing stage.

Fruit Characteristics The large conical fruit has a vertical and cross diameter of 3.9 cm and 2.7 cm, averaging 13.1 g with a regular size. It has medium-thick red skin, a smooth surface, a medium-deep and wide stalk cavity, a sunken fruit apex and a remnant stigma. The loose-textured flesh is sweet and sour with medium juice. It has medium quality for dried fruits. The percentage of edible part of fresh fruit is 93.4%, and SSC, TTS, TA and Vc is 20.40%, 17.80%, 0.70% and 380.12 mg per 100 g fresh fruit. The content of flavones and cAMP in mature fruit skin is 24.27 mg/g and 168.03 μg/g. The large spindle-shaped stone weighs 0.86 g with a shriveled kernel. The percentage of containing kernels is 40%.

Evaluation The cultivar has weak vigor and a medium yield. The medium-sized fruit has medium quality, suitable for making dried fruits. It has weak resistance to fruit-cracking.

新疆小圆枣

品种来源及分布　别名小红枣。分布新疆南部和东部绿洲地带，有200年栽培历史。

植物学性状　树体高大，树姿开张，枝系稠密粗壮，树冠偏斜形。主干灰褐色，皮粗糙，裂纹深，呈块状。枣头红褐色，长40.9cm，节间短。二次枝长17.0cm，5节左右，弯曲度小。针刺发达，不易脱落。枣股圆柱形，抽生枣吊4.0个。枣吊长22.5cm，着叶17片。叶片较小，椭圆形，合抱，绿或深绿色，先端渐尖，先端钝尖，叶基圆形或阔楔形，叶缘具细锯齿，齿尖较长。花量较少，花序平均着花3朵。

生物学特性　适应性强，耐旱、耐涝、较耐盐碱，抗病虫能力较强。萌蘖力强，树势中等，结果早，坐果率低，3年以上枝的吊果率为15.3%。在山西太谷地区，9月下旬果实成熟，果实生育期110d左右，为晚熟品种类型。

果实性状　果实极小，近圆形，纵径1.81cm，横径1.95cm，单果重3.7g，大小整齐。果肩平圆，梗洼浅窄，果顶平，柱头脱落。果面平滑，果皮薄，红色。果点中大，密度中等。果肉质地酥脆，汁液多，味甜微酸，品质较差，可制干。鲜枣可食率87.8%，可溶性固形物含量29.40%。核重0.45g，含仁率100%。

评价　该品种树体高大，树势旺，风土适应性强。果实小，可食率低，品质差，食用价值不大。但其萌蘖力强，含仁率高，种仁饱满，可作为砧木利用。

Xinjiangxiaoyuanzao

Source and Distribution　The cultivar, also called Xiaohongzao, spreads in south and east Xinjiang Vygur Autoromous Region with a cultivation history of 200 years.

Botanical Characters　The medium-sized tree is spreading with dense branches and a deflective crown. The grayish-brown trunk bark has rough, deep massive fissures. The reddish-brown 1-year-old shoots are 4.9 cm long with short internodes and developed thorns. The secondary branches are 17.0 cm long with 5 nodes of small curvature. The column-shaped mother fruiting spurs can germinate 4 deciduous fruiting shoots which are 22.5 cm long with 17 leaves. The small oval-shaped leaves are green or dark green with a gradually-cuspate apex, a round or wide-cuneiform base and a thin long saw-tooth pattern on the margin. The number of flowers is small averaging 3 ones per inflorescence.

Biological Characters　The tree has strong adaptability, strong resistance to pests and diseases, with moderate vigor, strong tolerance to drought, water-logging and saline-alkaline. It bears early with low fruit set. The percentage of fruits to deciduous bearing shoots of over-3-year-old branches is 15.3%. In Taigu County of Shanxi Province, it matures in late September with a fruit growth period of 110 d. It is a late-ripening variety.

Fruit Characteristics　The small fruit is nearly round with a vertical and cross diameter of 1.81 cm and 1.95 cm, averaging 3.7 g with a regular size. It has a flat-round shoulder, a narrow and shallow stalk cavity, a flat fruit apex, a falling-off stigma, a smooth surface and thin red skin with medium-large and medium-dense dots. The flesh is crisp, juicy, sweet and a little sour. It has a poor quality, mainly used for making dried fruits. The percentage of edible part of fresh fruit is 87.8%, and SSC is 29.40%. The stone weighs 0.75 g, with the percentage of containing kernels of 100%.

Evaluation　The large plant of Xinjiangxiaoyuanzao cultivar has strong tree vigor and strong adaptability. The small fruit has a poor quality and low edible value. Yet it has strong suckering ability and high percentage of containing kernels, which make it useful as a rootstock.

衡山长大枣

品种来源及分布 原产和分布于湖南衡阳市。

植物学性状 树体较大，树姿开张，树冠呈伞形。主干皮裂条状。枣头红褐色，平均长67.9cm，节间长8.0cm。二次枝长36.6cm，6～8节，弯曲度中等。针刺发达。枣股平均抽生枣吊3.0个。枣吊长25.4cm，着叶15片。叶片中大，卵圆形，合抱，先端钝尖，叶基圆楔形，叶缘钝锯齿。花量多，花序平均着花6朵，花小，花径6.0mm。

生物学特性 树势强健，易发枝，成枝力96.0%。结果较晚，定植第二至三年结果，10年左右进入盛果期，产量低而不稳，坐果率极低，枣头枝不结果，2～3年生和3年以上枝吊果率分别为3.3%和0.3%。在山西太谷地区，9月中旬果实进入脆熟期，10月上旬成熟采收，果实生育期120d左右，属极晚熟品种类型。成熟期遇雨裂果少。

果实性状 果个大，圆柱形，纵径4.20cm，横径3.30cm，单果重21.3g，大小较整齐。果皮厚，浅紫红色，果面极不平整，有隆起。梗洼中等深广。果顶凹陷，柱头残存。肉质较致密，味酸甜，汁液中多，品质中等，适宜制作干枣。鲜枣可食率96.2%，含可溶性固形物26.40%，总糖23.26%，酸0.31%。果核大，椭圆形，核重0.80g。种仁不饱满，含仁率48.3%。

评价 该品种树势强健，产量低而不稳。果实大，品质中等，适宜制作干枣。抗裂果能力较强。

Hengshanchangdazao

Source and Distribution The cultivar originated from and spreads in Hengyang City of Hunan Province.

Botanical Characters The large tree is spreading with an umbrella-shaped crown. The trunk bark has striped fissures. The reddish-brown 1-year-old shoots are 67.9 cm long with the internodes of 8.0 cm. The secondary branches are 36.6 cm long with 6～8 nodes of medium curvature and developed thorns. The mother fruiting spurs can germinate 3.0 deciduous fruiting shoots which are 25.4 cm long with 15 leaves. The medium-sized leaves are oval-shaped with a bluntly-cuspate apex, a round-cuneiform base and a blunt saw-tooth pattern on the margin. There are many small flowers with a diameter of 6.0 mm produced, averaging 6 ones per inflorescence.

Biological Characters The tree has strong vigor and strong branching ability, with a branching rate of 96.0%. It bears late, generally in the 2nd or 3rd year after planting and enters the full-bearing stage in the 10th year, with a low and unstable yield and low fruit set. The 1-year-old shoots do not bear any fruit. The percentage of fruits to deciduous bearing shoots of 2～3-year-old branches and over-3-year-old ones is 3.3% and 0.3%. In Taigu County of Shanxi Province, it enters the crisp-maturing stage in mid-September and is harvested in early October with a fruit growth period of 120 d. It is an extremely late-ripening variety with light fruit-cracking even if it rains in the maturing stage.

Fruit Characteristics The large oval-shaped fruit has a vertical and cross diameter of 4.20 cm and 3.30 cm averaging 21.3 g with a regular size. It has light-purplish-red thick skin, an unsmooth surface with some protuberances, a medium-deep and medium-wide stalk cavity, a sunken fruit apex and a remnant stigma. The tight-textured flesh is sour and sweet with medium juice. It has medium quality for dried fruits. The percentage of edible part of fresh fruit is 96.2%, and SSC, TTS and TA is 26.40%, 23.26% and 0.31%. The large oval-shaped stone weighs 0.80 g with a shriveled kernel. The percentage of containing kernels is 48.3%.

Evaluation The cultivar has strong vigor with a low and unstable yield. The large fruit has medium quality for dried fruits. It has strong resistance to fruit-cracking.

制 干
Drying Varieties
品 种

溆浦米枣

品种来源及分布　原产和分布于湖南溆浦的低庄镇连山村和双井乡等地。

植物学性状　树体较大，树姿开张，树冠自然圆头形。主干灰褐色，皮裂小条块状，易剥落。枣头紫褐色，平均长75.5cm，粗1.07cm，节间长8.1cm，被少量灰白色蜡质。二次枝长30.7cm，7节。针刺发达。枣股圆锥形，平均抽生枣吊3.0个。枣吊长22.8cm，着叶13片。叶片小，椭圆形，绿色，先端钝尖，叶基圆形，叶缘锯齿中大，钝圆。花量中等，花序平均着花5朵。

生物学特性　适应性强，病虫害较轻。树势中等，定植后2～3年开始结果，成龄树坐果差，产量低，2～3年和3年以上枝的吊果率分别为17.5%和4.2%。在山西太谷地区，9月上旬果实成熟，为早熟品种类型。成熟期遇雨易裂果。

果实性状　果实中大，圆柱形，纵径3.74cm，横径2.65cm，单果重12.3g，大小较整齐。果肩平圆，梗洼中深、广，果顶凹陷。果皮厚，红色。果点小，密度中等。果肉乳白色，质地较致密，汁液较少，味甜，略酸，品质较差，可用于制干。鲜枣可食率97.3%，含可溶性固形物22.00%，总糖17.14%，酸0.22%，100g果肉维生素C含量234.10mg。制干率33.8%，干枣含总糖56.50%。果核较小，倒卵形，核重0.33g。核内一般不含种子。

评价　该品种适应性强，病虫害少。制干率低，品质差，不宜经济栽培。

Xupumizao

Source and Distribution　The cultivar originated from and spreads in Lianshancun of Dizhuang Village and Shuangjing Village of Xupu in Hunan Province.

Botanical Characters　The large tree is spreading with a natural-round crown. The grayish-brown trunk bark has small massive fissures, easily shelled off. The purplish-brown 1-year-old shoots are 75.5 cm long and 1.07 cm thick. The internodes are 8.1 cm long, with grayish-white wax and developed thorns. The secondary branches are 30.7 cm long with 7 nodes. The conical mother fruiting spurs can germinate 3 deciduous fruiting shoots which are 22.8 cm long with 13 leaves. The small green leaves are oval-shaped with a bluntly-cuspate apex, a round base and a medium-sized blunt-round saw-tooth pattern on the margin. The number of flowers is medium large, averaging 5 ones per inflorescence.

Biological Characters　The tree has strong adaptability, fewer pests and diseases and moderate vigor. It generally bears in the 2nd or 3rd year after planting, with low fruit set and a low yield for mature trees. The percentage of fruits to deciduous fruiting shoots of 2～3-year-old branches and over-3-year-old ones is 17.5% and 4.2%. In Taigu County of Shanxi Province, it matures in early September. It is an early-ripening variety with serious fruit-cracking if it rains in the maturing stage.

Fruit Characteristics　The medium-sized fruit is column-shaped, with a vertical and cross diameter of 3.74 cm and 2.65 cm, averaging 12.3 g with a regular size. It has a flat-round shoulder, a medium-deep and wide stalk cavity, a sunken fruit apex and thick red skin with small and medium-dense dots. The ivory-white flesh is tight-textured and less juicy, sweet and a little sour. It has a poor quality, mainly used for making dried fruits. The percentage of edible part of fresh fruit is 97.3%, and the content of Vc is 234.10 mg per 100 g fresh fruit. The percentage of fresh fruits which can be made into dried ones is 33.8%, and the content of TTS in dried fruit is 56.50%. The small stone is inverted oval-shaped averaging 0.33 g. Most stones contain no kernels.

Evaluation　The cultivar has strong adaptability, fewer pests and diseases. Yet it has low percentage for making dried fruits from fresh ones. It is not suitable for commercial production.

制 干
Drying Varieties
品 种

婆 婆 枣

品种来源及分布 分布于山西运城市盐湖区的西曲马、舜帝庙、乔阳等村，为当地的次主栽品种，大多已改接为相枣。起源历史不详，据当地史料推测，是当地较古老的品种。

植物学性状 树体高大，树姿开张，干性较强，枝条中密，树冠自然圆头形，皮裂中等深，条状，不易脱落。枣头红褐色，平均长79.8cm，粗1.08cm，节间长8.5cm，着生永久性二次枝6个左右，二次枝长26.5cm，5~7节。针刺不发达。枣股较大，抽吊力中等，平均抽生枣吊3.5个。枣吊长24.0cm，着叶13片。叶片大，叶长7.1cm，叶宽3.2cm，椭圆形，浓绿色，先端急尖，叶基圆形，叶缘锯齿中密，较钝。花量多，枣吊平均着花69.5朵，花序平均5.6朵。花中等大，花径7.4mm，昼开型。

生物学特性 树势强健，枝系粗壮，萌芽率高，成枝力强，萌蘖力中等，根蘖苗根系发达，生长势强，幼树结果较早，根蘖苗一般第二年开始结果，15年后进入盛果期。盛果期长，坐果率高，枣头、2~3年生枝和4年生枝的吊果率分别为74.2%、81.8%和29.5%，主要坐果部位在枣吊的3~8节，占坐果总数的72.9%。丰产，但产量不稳定。在山西太谷地区，9月中旬果实着色，10月上旬进入完熟期，果实生育期110d左右，为晚熟品种。采前落果、裂果均轻。较抗枣疯病。

果实性状 果实较大，纵径3.51cm，横径3.09cm，单果重14.1g，大小较整齐。果形卵圆形，多数枣果果顶以上的1/3的胴部收缩成乳头状，多呈不规则的葫芦形。果顶平，梗洼广而深，柱头遗存。果皮厚，红色，果面粗糙。果点小而密。果肉厚，浅绿色，肉质硬而较粗，味甜，汁液中多，品质中等，适宜制干。鲜枣可食率95.6%，含可溶性固形物30.60%，总糖29.00%，酸0.37%，100g果肉维生素C含量489.4mg；果皮含黄酮9.69mg/g，cAMP含量77.50μg/g。制干率47.5%，干枣含总糖66.77%，酸1.41%。核较大，纺锤形，纵径2.41cm，横径0.89cm，核重0.62g，种仁较饱满，含仁率93.3%。

评价 该品种适应性和抗逆性均强，抗枣疯病，树势强健，早实，丰产。果实中大，肉质粗，汁液少，品质中，宜制干。成熟晚，采前裂果、落果轻。可作抗枣疯病的育种材料。

Popozao

Source and Distribution The cultivar spreads in Yuncheng City of Shanxi Province. it has a small cultivation area. According to the historical records in the local areas, it is an old variety there.

Botanical Characters The large tree is spreading with a strong central leader trunk, medium-dense branches and a natural-round crown. The reddish-brown 1-year-old shoots are 79.8 cm long. There are generally 6 permanent secondary branches which are 26.5 cm long with less-developed thorns. The large mother fruiting spurs can germinate 3.47 deciduous fruiting shoots. The large leaves are oval-shaped and dark green. There are many medium-sized flowers with a diameter of 7.4 mm, averaging 69.5 per deciduous fruiting shoot.

Biological Characters The tree has strong vigor, strong branches and leaves, strong germination and branching ability and medium suckering ability. It enters the long full-bearing stage in the 15th year after planting with high fruit set. It has a high yet unstable yield. In Taigu County of Shanxi Province, it begins coloring in mid-September and enters the crisp-maturing stage in early October. It is an extremely late-ripening variety with light premature fruit-dropping and fruit-cracking, along with strong resistance to jujube witches' broom.

Fruit Characteristics The medium-sized fruit is oval-shaped, averaging 14.1 g with a regular size. Most fruits have an apex which shrinks into nipple shape from the 1/3 part, and present an irregular calabash shape. It has thick red skin and a rough surface. The light-green thick flesh is hard, rough and sweet with medium juice and a medium quality for dried fruits. The percentage of edible part of fresh fruit is 95.6%, and SSC, TTS, TA and Vc is 30.60%, 29.00%, 0.37% and 489.4 mg per 100 g fresh fruit. The content of flavones and cAMP in mature fruit skin is 9.69 mg/g and 77.50 μg/g. The percentage of fresh fruits which can be made into dried ones is 47.5%, and the content of TTS and TA in dried fruit is 66.77% and 1.41%. The medium-sized stone is spindle-shaped, averaging 0.62 g.

Evaluation The cultivar has strong adaptability, strong resistance to adverse conditions, strong vigor, strong resistance to jujube witches' broom, with early fruiting habit and high productivity.

溆浦木枣

品种来源及分布 别名湖南木枣。主产和分布于湖南溆浦和茶陵等县。

植物学性状 树体中大，树姿半开张，枝叶稠密，树冠呈伞形。主干灰褐色，粗糙，块状，易剥落。枣头红褐色，枝长71.6cm，节间长7.0cm，蜡质少。二次枝平均长31.0cm，6节左右。针刺不发达。枣股圆柱形，平均抽生枣吊5.0个。枣吊长25.0cm，着叶14片。叶片较小，椭圆形，绿色，先端尖圆，叶缘钝锯齿。花量中等，花序平均着花6朵，最多8朵。花中大，花径6.0mm，昼开型。

生物学特性 适应性和抗逆性较强，树势中等。结果早，丰产性能强而稳定。2～3年和3年以上枝的吊果率分别为142.3%和57.1%。定植后2年开始结果，10年左右进入盛果期。在山西太谷地区，10月上旬果实成熟，果实生育期119d左右，为极晚熟品种类型。

果实性状 果实中大，近圆形，纵径3.08cm，横径2.50cm，单果重11.3g，大小较整齐。果肩平圆或圆，梗洼窄、中深。果顶微凹，柱头残存。果皮中厚，红色。果肉厚，浅绿色，质地致密，汁液少，味甜，品质中等，可制干。鲜枣可食率92.3%，含总糖26.95%，酸0.35%，100g果肉维生素C含量326.61mg；果皮含黄酮9.10mg/g，cAMP含量72.32 μg/g。制干率58.1%，干枣含总糖51.60%，酸0.49%。果核椭圆形，核重0.87g，含仁率70.0%。

评价 该品种适应性强，结果早，产量高而稳定。果实中等偏小，制干品质中等。

Xupumuzao

Source and Distribution The cultivar originated from and spreads in Xupu and Chaling of Hunan Province.

Botanical Characters The medium-sized tree is half-spreading with dense branches and an umbrella-shaped crown. The grayish-brown trunk bark has rough massive fissures, easily shelled off. The reddish-brown 1-year-old shoots are 71.6 cm long. The internodes are 7.0 cm long with less wax and less-developed thorns. The secondary branches are 31.0 cm long with 6 nodes. The column-shaped mother fruiting spurs can germinate 5 deciduous fruiting shoots which are 25.0 cm long with 14 leaves. The small oval-shaped green leaves have a cuspate-round apex, a blunt saw-tooth pattern on the margin. The number of flowers is medium large, averaging 6 (maximum 8) ones per inflorescence. The daytime-bloomed blossoms have a diameter of 6.0 mm.

Biological Characters The tree has strong adaptability, strong resistance to adverse conditions and moderate vigor. It bears early, with a high and stable yield. The percentage of fruits to deciduous fruiting shoots of 2～3-year-old branches and over-3-year-old ones is 142.3% and 57.1%. It generally bears in the 2nd year after planting and enters the full-bearing stage in the 10th year. In Taigu County of Shanxi Province, it matures in early October with a fruit growth period of 119 d. It is an extremely late-ripening variety.

Fruit Characteristics The medium-sized fruit is nearly round, with a vertical and cross diameter of 3.08 cm and 2.50 cm, averaging 11.3 g with a regular size. It has a flat-round or round shoulder, a narrow and medium-deep stalk cavity, a slightly-sunken fruit apex, a remnant stigma and medium-thick red skin. The light-green flesh is thick, tight-textured and sweet with less juice. It has medium quality for dried fruits. The percentage of edible part of fresh fruit is 92.3%, and the content of TTS, TA and Vc is 26.95%, 0.35% and 326.61 mg per 100 g fresh fruit. The content of flavones and cAMP in mature fruit skin is 9.10 mg/g and 72.32 μg/g. The oval-shaped stone weighs 0.87 g with the percentage of containing kernels of 70.0%.

Evaluation The cultivar has strong adaptability, bearing early with a high and stable yield. The medium-sized fruit is suitable for making dried fruits.

制 干
Drying Varieties
品 种

·229·

河津条枣

品种来源及分布　原产和分布于山西河津的赵家庄乡北里村，栽培数量较少。

植物学性状　树体较大，树姿直立，枝叶稀疏，树冠自然半圆形。主干条状皮裂。枣头黄褐色，平均长74.2cm，粗0.94cm，节间长8.4cm，蜡质较多。针刺不发达。枣股中大，抽吊力中等，平均抽生枣吊4.0个。枣吊长17.0cm，着叶11片。叶片大，平展，椭圆形，绿色，先端钝尖，叶基圆楔形，叶缘钝锯齿，密度中等。花量较少，每花序着花4朵。

生物学特性　适应性较强，树势中等，发枝力弱。坐果率较高，吊果率35.6%，定植第二年开始结果，较丰产。在山西太谷地区，9月下旬果实成熟，果实生育期110d左右，为晚熟品种类型。抗裂果。

果实性状　果实中大，椭圆形，纵径3.32cm，横径2.25cm，单果重10.8g，大小整齐。果肩圆平，梗洼窄、较深。果顶平，柱头脱落。果面光滑，果皮紫红色，较厚。果肉厚，浅绿色，质地较致密，味酸，汁液少，可制干，品质中等。鲜枣可食率95.1%，含可溶性固形物33.90%，酸0.46%，100g果肉维生素C含量224.40mg；果皮含黄酮3.14mg/g，cAMP含量133.51μg/g。制干率58.1%，干枣含总糖64.50%，酸0.61%。果核中大，纺锤形，核重0.53g，含仁率6.7%。

评价　适应性较强，产量较高。果实中大，大小整齐，品质中等，适宜制干。

Hejintiaozao

Source and Distribution　The cultivar originated from and spreads in Beilicun of Zhaojiazhuang Village in Hejin City of Shanxi Province with a small quantity.

Botanical Characters　The large tree is vertical with sparse branches and a semi-round crown. The trunk bark has striped fissures. The yellowish-brown 1-year-old shoots are 74.2 cm long and 0.94 cm thick. The internodes are 8.4 cm long with much wax and less-developed thorns. The medium-sized mother fruiting spurs can germinate 4 deciduous fruiting shoots which are 17.0 cm long with 11 leaves. The large oval-shaped leaves are flat and green, with a bluntly-cuspate apex, a round-cuneiform base and a medium-dense, blunt saw-tooth pattern on the margin. The number of flowers is small, averaging 4 ones per inflorescence.

Biological Characters　The tree has strong adaptability, moderate vigor and weak branching ability with high fruit set. The percentage of fruits to deciduous bearing shoots is 35.6% on average. It generally bears in the 2nd year after planting, with a high yield. In Taigu County of Shanxi Province, it matures in late September with a fruit growth period of 110 d. It is a mid-late-ripening variety with strong resistance to fruit-cracking.

Fruit Characteristics　The medium-sized fruit is oval-shaped with a vertical and cross diameter of 3.32 cm and 2.25 cm averaging 10.8 g with a regular size. It has a flat-round shoulder, a narrow and deep stalk cavity, a flat fruit apex, a falling-off stigma, a smooth surface and purplish-red thick skin. The light-green flesh is thick, tight-textured and sour, with less juice. It has medium quality for dried fruits. The percentage of edible part of fresh fruit is 95.1%, and SSC, TA and Vc is 33.90%, 0.46% and 224.40 mg per 100 g fresh fruit. The content of flavones and cAMP in mature fruit skin is 3.14 mg/g and 133.51 μg/g, and SSC, TTS and TA in dried fruit is 60.40%, 49.20% and 0.78%. The medium-sized stone is spindle-shaped averaging 0.53 g. The percentage of containing kernels is 6.7%.

Evaluation　The cultivar has strong adaptability and high yield. The medium-sized fruit has medium quality and is suitable for making dried fruit.

制 干
Drying Varieties
品 种

临汾木疙瘩

品种来源及分布 分布于山西临汾市。

植物学性状 树体较大,树姿半开张,树冠呈圆锥形。主干皮裂条状。枣头红褐色,平均长48.0cm,节间长6.8cm。二次枝长23.5cm,5~7节,弯曲度中等。无针刺。枣股平均抽生枣吊3.0个。枣吊长16.0cm,着叶12片。叶片中大,叶长6.3cm,叶宽2.7cm,椭圆形,先端钝尖,叶基圆形,叶缘钝锯齿。花量中等,花序平均着花5朵,花中大,花径6.1mm。

生物学特性 树势较弱,成枝力较弱。定植第二年结果,10年左右进入盛果期,产量中等,坐果率低,枣头、2~3年和4~6年生枝吊果率分别为6.7%、15.0%和36.1%。在山西太谷地区,9月中旬果实进入脆熟期,9月下旬成熟采收,果实生育期110d左右,为晚熟品种类型。成熟期遇雨裂果少。

果实性状 果个中大,椭圆形,纵径3.40cm,横径2.70cm,单果重12.1g,大小整齐。果皮中厚,红色,果面平滑,外观较好。梗洼窄、中等深。果顶平圆,柱头脱落。肉质较致密,味甜,汁液中多,品质中等,适宜制作干枣。鲜枣可食率96.4%,含可溶性固形物24.60%,总糖19.32%,酸0.27%,100g果肉维生素C含量334.17mg;果皮含黄酮4.36mg/g,cAMP含量136.13μg/g。制干率39.6%,干枣含总糖64.81%,酸1.29%。核较小,纺锤形,核重0.44g,一般不含种仁。

评价 该品种树势弱,坐果率低,产量中等。果实中大,品质中等,适宜制作干枣。抗裂果能力较强。

Linfenmugeda

Source and Distribution The cultivar mainly spreads in Linfen City of Shanxi Province.

Botanical Characters The large tree is half-spreading with a conical crown. The trunk bark has striped fissures. The reddish-brown 1-year-old shoots are 48.0 cm long with the internodes of 6.8 cm. The secondary branches are 23.5 cm long with 5~7 nodes of medium curvature and without thorns. The mother fruiting spurs can germinate 3.0 deciduous fruiting shoots which are 16.0 cm long with 12 leaves. The medium-sized leaves are oval-shaped, 6.3 cm long and 2.7 cm wide, with a bluntly-cuspate apex, a round base and a blunt saw-tooth pattern on the margin. The number of flowers is medium large, averaging 5 ones per inflorescence. The medium-sized blossoms have a diameter of 6.1 mm.

Biological Characters The tree has weak vigor and weak branching ability, bearing in the 2nd year after planting. It enters the full-bearing stage in the 10th year with a medium yield and low fruit set. The percentage of fruits to deciduous fruiting shoots of 1-year-old shoots, 2~3-year-old branches and 4~6-year-old ones is 6.7%, 15.0% and 36.1%. In Taigu County of Shanxi Province, it enters the crisp-maturing stage in mid-September and is harvested in late September with a fruit growth period of 110 d. It is a mid-late-ripening variety with light fruit-cracking even if it rains in the maturing stage.

Fruit Characteristics The medium-sized conical fruit has a vertical and cross diameter of 3.40 cm and 2.70 cm, averaging 12.1 g with a regular size. It has medium-thick red skin, a smooth surface, an attractive appearance, a medium-deep and narrow stalk cavity, a flat-round fruit apex and a falling-off stigma. The tight-textured flesh is sweet with medium juice. It has medium quality for dried fruits. The percentage of edible part of fresh fruit is 96.4%, and the content of Vc is 334.17 mg per 100 g fresh fruit. The content of flavones and cAMP in mature fruit skin is 4.36 mg/g and 136.13 μg/g. The small spindle-shaped stone weighs 0.44 g. Most stones contain no kernels.

Evaluation The cultivar has weak vigor, low fruit set and moderate productivity. The medium-sized fruit has strong resistance to fruit-cracking and is suitable for making dried fruits.

万荣翠枣

品种来源及分布 分布于山西省万荣县，栽培历史不详。

植物学性状 树体较小，树姿半开张，树冠呈圆柱形。主干皮裂条状。枣头红褐色，平均长55.5cm，节间长6.3cm。二次枝长24.0cm，5~7节，弯曲度中等。针刺发达。枣股平均抽生枣吊4.0个。枣吊长15.9cm，着叶12片。叶片中大，叶长4.6cm，叶宽2.7cm，卵圆形，浅绿色，叶面不平整，先端钝尖，叶基截形，叶缘锐锯齿。花量少，花序平均着花4朵，花小，花径6.0mm。

生物学特性 树势较弱，成枝力较弱，仅为46.7%。产量较低，坐果率一般，枣头、2~3年生和3年以上枝吊果率分别为2.0%、79.6%和49.2%。在山西太谷地区，9月下旬果实成熟采收，果实生育期106d，属中晚熟品种类型。成熟期遇雨裂果严重。

果实性状 果个较大，圆形，纵径3.60cm，横径3.40cm，单果重16.6g，大小整齐。果皮中厚，红色，果面较粗糙。梗洼浅而窄。果顶凹陷，柱头脱落。肉质疏松，味甜，汁液少，品质中等，适宜制作干枣。鲜枣可食率95.8%，含可溶性固形物33.00%，总糖26.76%，酸0.14%，100g果肉维生素C含量362.64mg；果皮含黄酮9.78mg/g，cAMP含量297.31μg/g。果核中大，纺锤形，核重0.69g。种仁不饱满，含仁率20%左右。

评价 该品种树体较小，树势弱，产量低。果实较大，品质中等，适宜制干。抗裂果能力较差。

Wanrongcuizao

Source and Distribution The cultivar spreads in Wanrong County of Shanxi Province, with an unknown history.

Botanical Characters The small tree is half-spreading with a column-shaped crown. The trunk bark has striped fissures. The reddish-brown 1-year-old shoots are 55.5 cm long with the internodes of 6.3 cm long. The secondary branches are 24.0 cm long with 5~7 nodes of medium curvature and developed thorns. The mother fruiting spurs can germinate 4.0 deciduous fruiting shoots which are 15.9 cm long with 12 leaves. The medium-sized leaves are oval-shaped and light green, 4.6 cm long and 2.7 cm wide, with an unsmooth surface, a bluntly-cuspate apex, a truncate base and a sharp saw-tooth pattern on the margin. The number of flowers is small, averaging 4 ones per inflorescence. The small blossoms have a diameter of 6.0 mm.

Biological Characters The tree has weak vigor and weak branching ability, with a branching rate of 46.7%. It has a low yield and medium fruit set. The percentage of fruits to deciduous fruiting shoots of 1-year-old shoots, 2~3-year-old branches and over-3-year-old ones is 2.0%, 79.6% and 49.2%. In Taigu County of Shanxi Province, it is harvested in late September with a fruit growth period of 106 d. Fruit-cracking easily occurs if it rains in the maturing stage.

Fruit Characteristics The large round fruit has a vertical and cross diameter of 3.6 cm and 3.4 cm, averaging 16.6 g with a regular size. It has medium-thick red skin, a rough surface, a shallow and narrow stalk cavity, a sunken fruit apex and a falling-off stigma. The loose-textured flesh is sweet with less juice. It has medium quality for dried fruits. The percentage of edible part of fresh fruit is 95.8%, and SSC, TTS, TA and Vc is 33.00%, 26.76%, 0.14% and 362.64 mg per 100 g fresh fruit. The content of flavones and cAMP in mature fruit skin is 9.78mg/g and 297.31 μg/g. The medium-sized stone is spindle-shaped, averaging 0.69 g with a shriveled kernel. The percentage of containing kernels is 20%.

Evaluation The small plant of Wanrongcuizao cultivar has weak vigor and a low yield. The large fruit has medium quality for dried fruits. It has weak resistance to fruit-cracking.

制 干
Drying Varieties
品 种

· 235 ·

溆浦秤锤枣

品种来源及分布 原产和分布于湖南溆浦。

植物学性状 树体较大，树姿开张，树冠呈圆锥形。主干皮裂呈条状。枣头红褐色，平均长68.4cm，节间长6.5cm。二次枝长27.0cm，5~7节，弯曲度中等。针刺发达。枣股平均抽生枣吊3个。枣吊长26.8cm，着叶17片。叶片中大，叶长6.2cm，叶宽3.7cm，卵圆形，先端急尖，叶基圆形，叶缘具钝锯齿。花量较多，花序平均着花7朵，花较小，花径5.9mm。

生物学特性 树势较弱，一般定植第二年结果，产量中等，以2~3年生枝结果率最高，吊果率达111.4%。在山西太谷地区，8月下旬果实进入白熟期，9月下旬开始成熟采收，果实生育期110d左右，属晚熟品种类型。果实抗裂果能力中等。

果实性状 果个中大，扁卵圆形，纵径3.20cm，横径3.00cm，侧径2.80cm，单果重13.7g，大小整齐。果皮厚，红色，果面光滑。梗洼窄、中等深。果顶凹，柱头残存。肉质致密，味甜酸，汁液中多，品质中等，适宜制干。鲜枣可食率91.3%，含可溶性固形物29.40%，总糖26.80%，酸0.87%，100g果肉维生素C含量403.66mg；果皮含黄酮13.0mg/g，cAMP含量230.39μg/g。核特大，核纹深，纺锤形，核重1.19g。核大多从缝合线处自然开裂，裂口大。含仁率30.0%，种仁瘪。

评价 该品种生长势弱，产量中等。果个中大，品质中等，核大，可食率较低，适宜制作干枣。

Xupuchengchuizao

Source and Distribution The cultivar originated from and spreads in Xupu County of Hunan Province.

Botanical Characters The large tree is spreading with a conical crown. The trunk bark has striped fissures. The reddish-brown 1-year-old shoots are 68.4 cm long with the internodes of 6.5 cm. The secondary branches are 27.0 cm long with 5~7 nodes of medium curvature and developed thorns. The mother fruiting spurs can germinate 3 deciduous fruiting shoots which are 26.8 cm long with 17 leaves. The medium-sized leaves are oval-shaped, 6.2 cm long and 3.7 cm wide, with a sharply-cuspate apex, a round base and a blunt saw-tooth pattern on the margin. There are many small flowers with a diameter of 5.9 mm produced, averaging 7 per inflorescence.

Biological Characters The tree has weak vigor. It generally bears in the 2nd year after planting with a medium yield. The 2~3-year-old branches have the strongest fruiting ability, the percentage of fruits to deciduous fruiting shoots of which is 111.4%. In Taigu County of Shanxi Province, it enters the white-maturing stage in late August and is harvested in late September, with a fruit growth period of 110 d. It has moderate resistance to fruit-cracking.

Fruit Characteristics The medium-sized oblong fruit has a vertical and cross diameter of 3.20 cm and 3.00 cm, averaging 13.7 g with a regular size. It has thick red skin, a smooth surface, a medium-deep and narrow stalk cavity, a sunken fruit apex and a remnant stigma. The tight-textured flesh is sweet and sour with medium juice. It has medium quality for dried fruits. The percentage of edible part of fresh fruit is 91.3%, and SSC, TTS, TA and Vc is 29.40%, 26.80%, 0.87% and 403.66 mg per 100 g fresh fruit. The content of flavones and cAMP in mature fruit skin is 13.0 mg/g and 230.39 μg/g. The extremely large stone is spindle-shaped, averaging 1.19 g with deep veins and a shriveled kernel. Most stones crack into two parts from the suture line. The percentage of containing kernels is 30.0%.

Evaluation The cultivar has weak vigor and a medium yield. The medium-sized fruit has a large stone and medium quality, with low edibility. It is suitable for making dried fruits.

制 干
Drying Varieties
品 种

· 237 ·

连县苦楝枣

品种来源及分布 别名苦楝枣。原产和分布于广东连县。

植物学性状 树体较小，树姿开张，树冠偏斜形。主干皮裂条状。枣头黄褐色，平均长63.2cm，节间长5.7cm。二次枝长21.2cm，5～7节，弯曲度中等。针刺发达。枣股平均抽生枣吊3.7个。枣吊长28.0cm，着叶23片。叶片小，椭圆形，先端尖凹，叶基偏斜形，叶缘具钝锯齿。花量少，花序平均着花3朵，花中大，花径6.1mm。

生物学特性 树势弱，萌芽率和成枝力较差。结果较早，定植第二年开始结果，10年左右进入盛果期，产量较高，吊果率89.3%。在山西太谷地区，9月中旬果实进入脆熟期，10月初成熟采收，果实生育期119d左右，为极晚熟品种类型。成熟期遇雨裂果少。

果实性状 果个较小，椭圆形，纵径3.17cm，横径2.45cm，单果重9.3g，大小整齐。果皮厚，浅红色，果面平滑。梗洼浅、窄。果顶平圆，柱头残存。肉质较致密，味甜酸，汁液多，品质中等，可制干。鲜枣可食率93.7%，含可溶性固形物22.60%，总糖19.38%，酸0.42%，糖酸比46.14∶1，100g果肉维生素C含量500.00mg；果皮含黄酮5.70mg/g，cAMP含量22.66μg/g。制干率43.6%，干枣含总糖67.41%，酸0.84%。果核中大，纺锤形，核重0.59g，含仁率18.4%，种仁不饱满。

评价 该品种树体较小，生长势弱。果个小，品质中等，可制干。果实抗裂果能力极强。

Lianxiankulianzao

Source and Distribution The cultivar originated from and spreads in Lianxian County of Guangdong Province.

Botanical Characters The small tree is spreading with a deflective crown. The trunk bark has striped fissures. The yellowish-brown 1-year-old shoots are 63.2 cm long, with the internodes of 5.7 cm. The secondary branches are 21.2 cm long with 5～7 nodes of medium curvature and developed thorns. The mother fruiting spurs can germinate 3.7 deciduous fruiting shoots which are 28.0 cm long with 23 leaves. The small oval-shaped leaves have a sharply-sunken apex, a deflective base and a blunt saw-tooth pattern on the margin. The number of flowers is small, averaging 3 ones per inflorescence. The medium-sized blossoms have a diameter of 6.1 mm.

Biological Characters The tree has weak vigor, weak germination and branching ability. It generally bears in the 2nd year after planting and enters the full-bearing stage in the 10th year with a high yield. The percentage of fruits to deciduous fruiting shoots is 89.3% on average. In Taigu County of Shanxi Province, it enters the crisp-maturing stage in mid-September and is harvested in early October with a fruit growth period of 119 d. It is an extremely late-ripening variety with light fruit-cracking even if it rains in the maturing stage.

Fruit Characteristics The small conical fruit has a vertical and cross diameter of 3.17 cm and 2.45 cm, averaging 9.3 g with a regular size. It has light-red thick skin, a smooth surface, a shallow and narrow stalk cavity, a flat-round fruit apex and a remnant stigma. The tight-textured flesh is sour, sweet and juicy. It has medium quality for dried fruits. The percentage of edible part of fresh fruit is 93.7%, and SSC, TTS, TA and Vc is 22.60%, 19.38%, 0.42% and 500.00 mg per 100 g fresh fruit. The SAR is 46.14∶1. The content of flavones and cAMP in mature fruit skin is 5.70 mg/g and 22.66 μg/g. The large spindle-shaped stone weighs 0.59 g with a shriveled kernel. The percentage of containing kernels is 18.4%.

Evaluation The small plant of Lianxiankulianzao cultivar has weak tree vigor. The small fruit has medium quality and is suitable for making dried fruits. It has strong resistance to fruit-cracking.

制 干
Drying Varieties
品 种

· 239 ·

临猗笨枣

品种来源及分布 原产和分布于山西临猗县。

植物学性状 树体中大，树姿开张，树冠呈偏斜形。主干皮裂条状。枣头红褐色，平均长69.2cm，节间长6.9cm，蜡质少。二次枝平均长24.4cm，5～7节，弯曲度中等。针刺不发达。枣股平均抽生枣吊3.7个。枣吊长26.2cm，着叶18片。叶片中大，椭圆形，平展，先端急尖，叶基圆楔形，叶缘具钝锯齿。花量极少，花序平均着花2朵。

生物学特性 树势中庸，萌芽率和成枝力较弱，结果晚，定植第三年结果，10年左右进入盛果期，产量一般，枣吊平均结果0.32个。在山西太谷地区，9月下旬果实成熟采收，果实生育期107d左右，为晚熟品种类型。成熟期遇雨裂果轻。

果实性状 果个大，卵圆形，纵径3.15cm，横径2.83cm，单果重17.0g，大小较整齐。果皮薄，紫红色，果面平滑。梗洼窄、浅。果顶平圆，柱头残存。肉质疏松，汁液中多，味甜，品质中等，适宜制干。鲜枣可食率94.6%，含可溶性固形物24.00%，总糖17.41%，酸0.33%，100g果肉维生素C含量306.69mg；果皮含黄酮12.09mg/g，cAMP含量145.64 μg/g。干枣含总糖59.95%，酸0.55%。果核大，倒纺锤形，平均重0.92g。核内多无种子，含仁率5.0%。

评价 该品种树体中大，树势中庸，结果晚，产量一般，抗裂果能力强。果个大，品质中等，适宜制干。

Linyibenzao

Source and Distribution The cultivar originated from and spreads in Linyi County of Shanxi Province.

Botanical Characters The medium-sized tree is spreading with a deflective crown. The trunk bark has striped fissures. The reddish-brown 1-year-old shoots are 69.2 cm long with the internodes of 6.9 cm long with less wax and less-developed thorns. The secondary branches are 24.4 cm long with 5～7 nodes of medium curvature. The mother fruiting spurs can germinate 3.7 deciduous fruiting shoots which are 26.2 cm long with 18 leaves. The medium-sized leaves are oval-shaped and flat, with a sharply-cuspate apex, a round-cuneiform base and a blunt saw-tooth pattern on the margin. The number of flowers is very small, averaging 2 ones per inflorescence.

Biological Characters The tree has moderate vigor, weak germination and branching ability. It bears late, generally in the 3rd year after planting and enters the full-bearing stage in the 10th year with a medium yield. The deciduous fruiting shoot bears 0.32 fruits on average. In Taigu County of Shanxi Province, it matures in late September with a fruit growth period of 109 d. It is a late-ripening variety with light fruit-cracking even if it rains in the maturing stage.

Fruit Characteristics The large oblong fruit has a vertical and cross diameter of 3.15 cm and 2.83 cm, averaging 17.0 g with a regular size. It has purplish-red thin skin, a smooth surface, a narrow and shallow stalk cavity, a flat-round fruit apex and a remnant stigma. The loose-textured flesh is sweet with medium juice. It has medium quality for dried fruits. The percentage of edible part of fresh fruit is 94.6%, and SSC, TTS, TA and Vc is 24.00%, 17.41%, 0.33% and 306.69 mg per 100 g fresh fruit. The content of flavones and cAMP in mature fruit skin is 12.09 mg/g and 145.64 μg/g. The large stone is inverted spindle-shaped, averaging 0.92 g. Most stones have no kernels, and the percentage of containing kernels is 5.0%.

Evaluation The medium-sized plant of Linyibenzao cultivar has moderate tree vigor, bearing late with a medium yield. It has strong resistance to fruit-cracking. The large fruit has medium quality and is suitable for making dried fruits.

临猗鸡蛋枣

品种来源及分布 分布于山西临猗、运城等地。

植物学性状 树体中大，树姿半开张，树冠呈圆锥形。主干皮裂条状。枣头红褐色，平均长48.0cm，节间长6.8cm。二次枝长23.5cm，5～7节，弯曲度中等。无针刺。枣股平均抽生枣吊3个。枣吊长16.0cm，着叶12片。叶片较大，叶长6.7cm，叶宽3.1cm，椭圆形，先端急尖，叶基截形，叶缘钝锯齿。花量极少，花序平均着花2朵，花中大，花径6.1mm。

生物学特性 树势强，成枝力强，可达90.0%。定植第二年结果，5年左右进入盛果期，产量中等，枣头坐果能力强，枣头、2～3年生和3年以上枝吊果率分别为56.7%、33.3%和21.6%。在山西太谷地区，9月中旬果实进入脆熟期，9月下旬成熟采收，果实生育期110d左右，为晚熟品种类型。抗裂果能力中等。

果实性状 果个中大，近圆形，纵径3.20cm，横径3.10cm，单果重11.8g，大小较整齐。果皮厚，紫红色，果面平滑。梗洼深、中广。果顶凹陷，柱头脱落。肉质疏松，味酸甜，汁液中多，品质中等，适宜制作干枣。鲜枣可食率96.5%，100g果肉维生素C含量420.68mg；果皮含黄酮27.96mg/g，cAMP含量198.98μg/g。极少数核有仁，含仁率仅3.3%。

评价 该品种树体中大，树势强，产量中等。果实中大，品质中等，适宜制干。

Linyijidanzao

Source and Distribution The cultivar mainly spreads in Linyi and Yuncheng of Shanxi Province.

Botanical Characters The medium-sized tree is half-spreading with a conical crown. The trunk bark has striped fissures. The reddish-brown 1-year-old shoots are 48.0 cm long with the internodes of 6.8 cm. The secondary branches are 23.5 cm long with 5～7 nodes of medium curvature and without thorns. The mother fruiting spurs can germinate 3 deciduous fruiting shoots which are 16.0 cm long with 12 leaves. The large leaves are oval-shaped, 6.7 cm long and 3.1 cm wide, with a sharply-cuspate apex, a truncate base and a blunt saw-tooth pattern on the margin. The number of flowers is small, averaging 2 ones per inflorescence. The medium-sized blossoms have a diameter of 6.1 mm.

Biological Characters The tree has strong vigor and strong branching ability, with a branching rate of 90.0%. It bears in the 2nd year after planting and enters the full-bearing stage in the 5th year with a medium yield. The 1-year-old shoots have high fruit set. The percentage of fruits to deciduous fruiting shoots of 1-year-old shoots, 2～3-year-old branches and over-3-year-old ones is 56.7%, 33.3% and 21.6%. In Taigu County of Shanxi Province, it enters the crisp-maturing stage in mid-September and is harvested in late September with a fruit growth period of 110 d. It is a mid-late-ripening variety with moderate resistance to fruit-cracking.

Fruit Characteristics The medium-sized fruit is nearly round with a vertical and cross diameter of 3.20 cm and 3.10 cm, averaging 11.8 g with a regular size. It has purplish-red thick skin, a smooth surface, a deep and medium-wide stalk cavity, a sunken fruit apex and a falling-off stigma. The loose-textured flesh is sour and sweet with medium juice. It has medium quality for dried fruits. The percentage of edible part of fresh fruit is 96.5%, and the content of Vc is 420.68 mg per 100 g fresh fruit. The content of flavones and cAMP in mature fruit skin is 27.96 mg/g and 198.98 μg/g. Most stones contain no kernels with the percentage of containing kernels of only 3.3%.

Evaluation The medium-sized plant of Linyijidanzao cultivar has strong vigor and a medium yield. The medium-sized fruit has medium quality and is suitable for making dried fruits.

阿克苏小枣

品种来源及分布 原产和分布于新疆阿克苏地区。

植物学性状 树体较小，树姿半开张，干性中等，树冠呈圆锥形。主干条状皮裂。枣头黄褐色，年生长量73.6cm，节间长6.6cm，二次枝长21.5cm，5～7节，弯曲度中等。针刺不发达。枣股平均抽生枣吊3.0个。枣吊长32.5cm，着叶22片。叶片椭圆形，绿色，先端尖凹，叶基圆楔形，叶缘具锐锯齿。花量大，花序平均着花9朵。花中大，花径6.1mm。

生物学特性 树势强健。定植第二年普遍结果，10年左右进入盛果期，丰产性一般。在山西太谷地区，9月中旬果实着色，9月下旬成熟采收，果实生育期110d左右，为晚熟品种类型。

果实性状 果实小，圆形，纵径3.00cm，横径2.30cm，果重7.1g左右，最大可达15g以上，大小不整齐。梗洼窄而较深。果顶凹或平，柱头遗存。果皮中厚，红色，果面光滑。肉质较致密，味酸，汁液多，品质差，可制干。鲜枣可食率93.6%，含可溶性固形物24.00%，单糖13.48%，双糖8.54%，总糖22.02%，酸1.05%，糖酸比20.97∶1，100g果肉维生素C含量336.69mg；果皮含黄酮8.35mg/g，cAMP含量10.52μg/g。制干率54.8%，干枣含总糖64.31%，酸2.12%。果核大，纺锤形，核重0.45g，核内具饱满种仁，含仁率100%。

评价 该品种树体小，果实小，品质较差，可制干。含仁率高，可作砧木。

Akesuxiaozao

Source and Distribution The cultivar originated from and spreads in Akesu district of Xinjiang Province.

Botanical Characters The small tree is half-spreading with a medium-strong central leader trunk and a conical crown. The trunk bark has striped fissures. The yellowish-brown 1-year-old shoots are 73.6 cm long with the internodes of 6.6 cm. The secondary branches are 21.5 cm long with 5～7 nodes of medium curvature and undeveloped thorns. The mother fruiting spurs can germinate 3 deciduous fruiting shoots which are 32.5 cm long with 22 leaves. The green leaves are oval-shaped, with a sharply-sunken apex, a round-cuneiform base and a sharp saw-tooth pattern on the margin. There are many medium-sized flowers with a diameter of 6.1 mm produced, averaging 9 ones per inflorescence.

Biological Characters The tree has strong vigor. Generally all the trees bear fruits in the 2nd year after planting and enter the full-bearing stage in the 10th year with a medium yield. In Taigu County of Shanxi Province, it begins coloring in mid-September and is harvested in late September with a fruit growth period of 110 d. It is a late-ripening variety.

Fruit Characteristics The small globose fruit has a vertical and cross diameter of 3.00 cm and 2.30 cm, averaging 7.1 g (maximum 15 g) with irregular sizes. It has a narrow and deep stalk cavity, a sunken or flat fruit apex, a remnant stigma, medium-thick red skin and a smooth surface. The tight-textured flesh is juicy and sour. It has a poor quality, only used for making dried fruits. The percentage of edible part of fresh fruit is 93.6%, and SSC, monosaccharide, disaccharide, TTS, TA and Vc is 24.00%, 13.48%, 8.54%, 22.02%, 1.05% and 336.69 mg per 100 g fresh fruit. The SAR is 20.97∶1. The content of flavones and cAMP in mature fruit skin is 8.35 mg/g and 10.52 μg/g. The large spindle-shaped stone weighs 0.45 g with a well-developed kernel. The percentage of containing kernels is 100%.

Evaluation The small plant of Akesuxiaozao cultivar has a small fruit size and poor quality, mainly used for making dried fruits. It has a high percentage of containing kernels, so it can be used as rootstocks.

制干品种
Drying Varieties

平陆棒棰枣

品种来源及分布 分布于山西平陆县。

植物学性状 树体中大,树姿半开张,树冠呈圆锥形。主干皮裂条状。枣头红褐色,平均长46.2cm,节间长5.5cm,蜡质少。二次枝平均长21.7cm,4～6节,弯曲度中等。针刺发达。枣股平均抽生枣吊3.8个。枣吊长22.5cm,着叶16片。叶片中大,椭圆形,平展,先端急尖,叶基圆形,叶缘具锐锯齿。花量中多,花序平均着花4朵。花中大,花径6.1mm。

生物学特性 树势弱,萌芽率和成枝力中等,结果较早,定植第二年结果,10年左右进入盛果期,产量一般,枣吊平均结果0.29个。在山西太谷地区,9月下旬果实成熟采收,果实生育期110d左右,为中晚熟品种类型。成熟期遇雨裂果较轻。

果实性状 果个较小,长圆柱形,纵径3.62cm,横径2.48cm,单果重9.8g,大小较整齐。果皮中厚,红色,果面平滑。梗洼浅而广。果顶平圆,柱头脱落。肉质致密,汁液少,味酸,鲜食品质差,可制干。鲜枣可食率95.7%,含总糖21.04%,酸0.98%,100g果肉维生素C含量449.55mg;果皮含黄酮7.14mg/g,cAMP含量116.17μg/g。核较小,倒纺锤形,平均重0.42g。大多核内不含种子,含仁率8.3%。

评价 该品种树体中大,树势弱,结果较早,产量一般。果个小,品质差,可制干。

Pinglubangchuizao

Source and Distribution The cultivar originated from and spreads in Pinglu County of Shanxi Province.

Botanical Characters The medium-sized tree is half-spreading with a conical crown. The trunk bark has striped fissures. The reddish-brown 1-year-old shoots are 46.2 cm long with the internodes of 5.5 cm long, less wax and developed thorns. The secondary branches are 21.7 cm long with 4～6 nodes of medium curvature. The mother fruiting spurs can germinate 3.8 deciduous fruiting shoots, which are 22.5 cm long with 16 leaves. The medium-sized leaves are oval-shaped and flat, with a sharply-cuspate apex, a round base and a sharp saw-tooth pattern on the margin. The number of flowers is medium large, averaging 4 ones per inflorescence. The medium-sized blossoms have a diameter of 6.1 mm.

Biological Characters The tree has weak vigor, medium germination and branching ability. It bears early, generally in the 2nd year after planting and enters the full-bearing stage in the 10th year with a medium yield. The deciduous fruiting shoot bears 0.29 fruits on average. In Taigu County of Shanxi Province, it matures in late September with a fruit growth period of 110 d. It is a mid-late-ripening variety with light fruit-cracking even if it rains in the maturing stage.

Fruit Characteristics The small column-shaped fruit has a vertical and cross diameter of 3.62 cm and 2.48 cm, averaging 9.8 g with a regular size. It has medium-thick red skin, a smooth surface, a shallow and wide stalk cavity, a flat-round fruit apex and a falling-off stigma. The tight-textured flesh is sour with less juice. It has a poor quality for fresh-eating, so it is mainly used for dried fruits. The percentage of edible part of fresh fruit is 95.7%, and the content of TTS, TA and Vc is 21.04%, 0.98% and 449.55 mg per 100 g fresh fruit. The content of flavones and cAMP in mature fruit skin is 7.14 mg/g and 116.17 μg/g. The small stone is inverted spindle-shaped, averaging 0.42 g. Most stones have no kernels, and the percentage of containing kernels is 8.3%.

Evaluation The medium-sized plant of Pinglubangchuizao cultivar has weak tree vigor, bearing early with a medium yield. The small fruit has a poor quality, mainly used for making dried fruits.

制干 Drying Varieties
品种

婆枣枝变1号

品种来源及分布　原产河北束鹿的西泽北乡石碑村。

植物学性状　树体高大，树姿开张，树冠呈半圆形。主干灰褐色，裂纹呈块状。枣头粗壮，浅褐色，平均长79.1cm，粗1.07cm，节间长8.5cm，蜡质多。针刺不发达。枣股圆柱形，平均抽生枣吊4.0个。枣吊长16.5cm，着叶11片。叶片卵圆形，深绿色，先端钝尖，叶基心形，叶缘具粗钝锯齿。花量少，花较小，花序平均着花3朵。

生物学特性　树势中等，发枝力弱。枣股寿命长，抽生枣吊数量较多，树体容易管理。丰产稳产，坐果较稳定，枣吊平均坐果0.5个，最多可达2.3个。在山西太谷地区，4月中旬萌芽，5月下旬始花，9月下旬果实成熟，果实生育期115d，为晚熟品种类型。

果实性状　果实较小，椭圆形或卵圆形，纵径3.10cm，横径2.34cm，单果重7.8g。果肩平，梗洼浅、中广。果顶平，微凹。果皮红色，中厚，有光泽。果肉白色，较致密，汁液少，味酸甜，可制干，品质中等。鲜枣可食率93.6%，含可溶性固形物33.00%，总糖31.42%，酸0.30%。制干率58.1%。果核较大，长圆形，核重0.50g，含仁率30.0%。

评价　该品种适土性较好，耐涝，耐旱。丰产，稳产，果实制干率较高，但品质一般。

Pozaozhibian 1

Source and Distribution　The cultivar originated from Shibeicun of Xizebei Village in Shulu, Hebei Province.

Botanical Characters　The medium-sized tree is spreading with a semi-round crown. The grayish-brown trunk bark has massive fissures. The light-brown 1-year-old shoots are 79.1 cm long and 1.07 cm thick. The internodes are 8.5 cm long with much wax and less-developed thorns. The column-shaped mother fruiting spurs can germinate 4 deciduous fruiting shoots which are 16.5 cm long with 11 leaves. The dark-green leaves are oval-shaped, with a bluntly-cuspate apex, a heart-shaped base and a thick-blunt saw-tooth pattern on the margin. Both the number and size of flowers are small, averaging 3 ones per inflorescence.

Biological Characters　The tree has moderate vigor and weak branching ability with a high and stable yield, and stable fruit set. The mother fruiting spurs have a long life and can germinate more deciduous fruiting shoots. The deciduous fruiting shoot bears 0.5 fruits (2.3 at most) on average. In Taigu County of Shanxi Province, it germinates in mid-April, begins blooming in late May and ripens in late September with a fruit growth period of 115 d. It is a late-ripening variety.

Fruit Characteristics　The small column-shaped fruit has a vertical and cross diameter of 3.10 cm and 2.34 cm, averaging 7.8 g. It has a flat shoulder, a shallow and medium-wide stalk cavity, a flat fruit apex (a little sunken) and medium-thick, red and glossy skin. The white flesh is tight-textured, sour and sweet with less juice. It has medium quality for dried fruits. The percentage of edible part of fresh fruit is 93.6%, and SSC, TTS and TA is 33.00%, 31.42% and 0.30%. The percentage of fresh fruits which can be made into dried ones is 58.1%. The large oblong stone weighs 0.50 g, with the percentage of containing kernels of 30.0%.

Evaluation　The cultivar has strong adaptability, tolerant to water-logging and drought with a high and stable yield. It has a high rate of making dried fruits from fresh ones, yet with medium quality.

制干
Drying Varieties
品种

· 249 ·

万荣福枣

品种来源及分布 又名水枣。原产和分布于山西万荣县。

植物学性状 树体较小，树姿开张，树冠乱头形。主干皮裂条状。枣头红褐色，平均长67.9cm，节间长6.3cm。二次枝长23.8cm，4~6节，弯曲度中等。针刺发达。枣股平均抽生枣吊3.6个。枣吊长21.5cm，着叶13片。叶片中大，卵圆形，先端钝尖，叶基圆形，叶缘钝锯齿。花量少，花序平均着花4朵，花中大，花径6.2mm。

生物学特性 树势中庸，萌芽率和成枝力中等。定植第三年结果，产量较低，吊果率29.5%。在山西太谷地区，9月中旬果实进入脆熟期，9月下旬成熟采收，果实生育期108d以上，为中晚熟品种类型。成熟期遇雨裂果少。

果实性状 果个大，圆形或短柱形，纵径3.56cm，横径3.30cm，单果重20.0g，大小较整齐。果皮厚，紫红色，果面平滑。梗洼广、较浅。果顶微凹，柱头脱落。肉质疏松，味甜，汁液少，品质中等，可制干。鲜枣可食率94.9%，含可溶性固形物23.00%，总糖19.12%，酸0.30%，100g果肉维生素C含量369.76mg；果皮含黄酮13.42mg/g，cAMP含量256.09μg/g。核较大，纺锤形，核重1.02g。含仁率6.7%，种仁不饱满。

评价 该品种树体较小，树势中庸。果个大，品质中等，适宜制干。果实抗裂果能力较强。

Wanrongfuzao

Source and Distribution The cultivar originated from and spreads in Wanrong County of Shanxi Province.

Botanical Characters The small tree is spreading with an irregular crown. The trunk bark has striped fissures. The reddish-brown 1-year-old shoots are 67.9 cm long, with the internodes of 6.3 cm long. The secondary branches are 23.8 cm long with 4~6 nodes of medium curvature and developed thorns. The mother fruiting spurs can germinate 3.6 deciduous fruiting shoots which are 21.5 cm long with 13 leaves. The medium-sized leaves are oval-shaped, with a bluntly cuspate apex, a round base and a blunt saw-tooth pattern on the margin. The number of flowers is small, averaging 4 ones per inflorescence. The medium-sized blossoms have a diameter of 6.2 mm.

Biological Characters The tree has moderate vigor, medium germination ability and medium branching ability. It generally bears in the 3rd year after planting with a low yield. The percentage of fruits to deciduous fruiting shoots is 29.5% on average. In Taigu County of Shanxi Province, it enters the crisp-maturing stage in mid-September and is harvested in late September with a fruit growth period of 108 d. It is a mid-late-ripening variety with light fruit-cracking even if it rains in the maturing stage.

Fruit Characteristics The large oval-shaped fruit has a vertical and cross diameter of 3.56 cm and 3.30 cm, averaging 20.0 g (maximum 45 g) with a regular size. It has purplish-red thick skin, a smooth surface, a wide and shallow stalk cavity, a slightly sunken fruit apex and a falling-off stigma. The loose-textured flesh is sweet with less juice. It has medium quality for dried fruits. The percentage of edible part of fresh fruit is 94.9%, and SSC, TTS, TA and Vc is 23.00%, 19.12%, 0.30% and 369.76 mg per 100 g fresh fruit. The content of flavones and cAMP in mature fruit skin is 13.42 mg/g and 256.09 μg/g. The large spindle-shaped stone weighs 1.02 g with a shriveled kernel. The percentage of containing kernels is 6.7%.

Evaluation The small plant of Wanrongfuzao cultivar has moderate tree vigor. The fruit has a large size and medium quality and is suitable for making dried fruits. It has strong resistance to fruit-cracking.

制　干
Drying Varieties
品　种

· 251 ·

滕州落地红

品种来源及分布　原产和分布于山东滕州市。

植物学性状　树体较大，树姿半开张，树冠自然乱头形。枣头灰绿色，平均长57.4cm，节间长7.8cm。二次枝长31.4cm，5～7节。枣股平均抽生枣吊4.0个。针刺基本退化。枣吊长33.5cm，着叶18片。叶片椭圆形，先端急尖，叶基偏斜形，叶缘具钝锯齿。花量多，花序平均着花8朵，花中大，花径7.0mm。

生物学特性　树势较强，一般定植第二年结果，10年左右进入盛果期，极丰产，坐果率较高。在山西太谷地区，9月中旬果实进入脆熟期，9月下旬成熟采收，果实生育期110d左右，为晚熟品种类型。果实抗裂果能力强。

果实性状　果个较小，卵圆形，纵径2.72cm，横径2.54cm，单果重8.7g，大小较整齐。果皮厚，红色，果面平滑，光亮，外形十分美观。梗洼窄、浅，果顶平，柱头脱落。肉质致密，味酸，汁液少，品质较差，可制干。鲜枣可食率93.5%，含总糖24.12%，100g果肉维生素C含量421.33mg；果皮含黄酮7.81mg/g，cAMP含量162.19μg/g。干枣含总糖72.51%，酸1.55%。果核较大，倒纺锤形，核重0.57g，种仁较饱满，含仁率36.7%。

评价　该品种树势强健，丰产稳产。果实大小较均匀，外观艳丽，抗裂果能力强，可制干。但果个较小，品质较差。

Tengzhouluodihong

Source and Distribution　The cultivar originated from and spreads in Tengzhou City of Shandong Province.

Botanical Characters　The large tree is half-spreading with an irregular crown. The grayish-green 1-year-old shoots are 57.4 cm long with the internodes of 7.8 cm long and almost no thorns. The secondary branches are 31.4 cm long with 5～7 nodes. The mother fruiting spurs can germinate 4 deciduous fruiting shoots which are 33.5 cm long with 18 leaves. The oval-shaped leaves have a sharply-cuspate apex, a deflective base and a blunt saw-tooth pattern on the margin. There are many medium-sized flowers with a diameter of 7.0 mm produced, averaging 8 ones per inflorescence.

Biological Characters　The tree has strong vigor. It generally bears in the 2nd year after planting and enters the full-bearing stage in the 10th year with a very high yield and high fruit set. In Taigu County of Shanxi Province, it enters the crisp-maturing stage in mid-September and is harvested in late September with a fruit growth period of 110 d. It is a late-ripening variety with strong resistance to fruit-cracking.

Fruit Characteristics　The small oblong fruit has a vertical and cross diameter of 2.72 cm and 2.54 cm, averaging 8.7 g with a regular size. It has thick red skin, a smooth and bright surface, a narrow and shallow stalk cavity, a flat fruit apex and a falling-off stigma. The tight-textured flesh is sour with less juice. It has a poor quality for fresh-eating, so it is mainly used for making dried fruits. The percentage of edible part of fresh fruit is 93.50%, and the content of TTS and Vc is 24.12% and 421.33 mg per 100 g fresh fruit. The content of flavones and cAMP in mature fruit skin is 7.81 mg/g and 162.19 μg/g. The large stone is inverted spindle-shaped, averaging 0.57 g with a well-developed kernel. The percentage of containing kernels is 36.7%.

Evaluation　The cultivar has strong vigor, a high and stable yield. The fruit has a regular size, an attractive appearance and strong resistance to fruit-cracking. It can be used for making dried fruits. Yet its fruit is too small and the quality is not so good.

临猗脖脖枣

品种来源及分布　原产和分布于山西临猗县。

植物学性状　树体中大，树姿直立，树冠呈圆头形。主干皮裂块状。枣头黄褐色，平均长67.9cm，节间长8.2cm，蜡质少。二次枝平均长23.2cm，4～6节，弯曲度中等。针刺不发达。枣股平均抽生枣吊3.8个。枣吊长30.6cm，着叶18片。叶片中大，椭圆形，平展，先端钝尖，叶基圆楔形，叶缘具钝锯齿。花量少，花序平均着花3朵。

生物学特性　树势中庸，萌芽率和成枝力较弱，结果晚，定植第三年结果，10年左右进入盛果期，产量一般，枣吊平均结果0.21个。在山西太谷地区，9月下旬果实成熟采收，果实生育期107d左右，为晚熟品种类型。

果实性状　果个大，扁圆形，纵径3.12cm，横径3.07cm，单果重22.9g，大小整齐。果皮厚，紫红色，果面平滑。梗洼浅、中广。果顶平圆，柱头残存。肉质疏松，汁液少，味甜，品质中等，适宜制干。鲜枣可食率96.4%，含总糖26.70%，酸0.49%，100g果肉维生素C含量474.65mg；果皮含黄酮2.56mg/g，cAMP含量279.44μg/g。果核较大，纺锤形，平均重0.82g。大多核内无种子，含仁率6.7%。

评价　该品种树体中大，树势中庸，结果晚，产量一般。果个大，品质中等，适宜制干。

Linyibobozao

Source and Distribution　The cultivar originated from and spreads in Linyi County of Shanxi Province.

Botanical Characters　The medium-sized tree is vertical with a round crown. The trunk bark has massive fissures. The yellowish-brown 1-year-old shoots are 67.9 cm long with the internodes of 8.2 cm long, less wax and less-developed thorns. The secondary branches are 23.2 cm long with 4～6 nodes of medium curvature. The mother fruiting spurs can germinate 3.8 deciduous fruiting shoots which are 30.6 cm long with 18 leaves. The medium-sized leaves are oval-shaped and flat, with a bluntly-cuspate apex, a round-cuneiform base and a blunt saw-tooth pattern on the margin. The number of flowers is small, averaging 3 ones per inflorescence.

Biological Characters　The tree has moderate vigor, weak germination and branching ability. It bears late, generally in the 3rd year after planting and enters the full-bearing stage in the 10th year, with a medium yield. The deciduous fruiting shoot bears 0.21 fruits on average. In Taigu County of Shanxi Province, it matures in late September with a fruit growth period of 107 d. It is a late-ripening variety.

Fruit Characteristics　The large oblate fruit has a vertical and cross diameter of 3.12 cm and 3.07 cm, averaging 22.9 g with a regular size. It has purplish-red thick skin, a smooth surface, a shallow and medium-wide stalk cavity, a flat-round fruit apex and a remnant stigma. The loose-textured flesh is sweet with less juice. It has medium quality for dried fruits. The percentage of edible part of fresh fruit is 96.4%, and the content of TTS, TA and Vc is 26.70%, 0.49% and 474.65 mg per 100 g fresh fruit. The content of flavones and cAMP in mature fruit skin is 2.56 mg/g and 279.44 μg/g. The large spindle-shaped stone weighs 0.82 g. Most stones have no kernels, and the percentage of containing kernels is 6.7%.

Evaluation　The medium-sized plant of Linyibobozao cultivar has moderate tree vigor, bearing late with a medium yield. The large fruit has medium quality and is suitable for making dried fruits.

制 干
Drying Varieties
品 种

泡 泡 红

品种来源及分布 别名泡泡枣、泡枣。原产和分布于北京房山区。

植物学性状 树体较大,树姿直立,干性极强,树冠呈伞形。主干条状皮裂。枣头紫褐色,平均枝长72.6cm,粗1.08cm,节间长6.3cm。二次枝平均长29.0cm,8节左右。针刺发达。每枣股平均着生枣吊3个。枣吊平均长18.1cm,着叶13片。叶片较小,叶长5.1cm,叶宽3.1cm,卵圆形,先端渐尖,先端钝尖。叶基圆形或心形。叶缘锯齿粗细深浅不一。每花序着花7朵,花小。

生物学特性 适应性强,树势强健,成枝力较弱。丰产,结果稳定,枣头、2~3年、4~6年和6年以上枝的吊果率分别为10.7%、93.2%、87.0%和50.9%。在山西太谷地区,10月上旬果实成熟采收,为极晚熟品种类型。较抗裂果。

果实性状 果实较大,短柱形或倒卵圆形,纵径3.65cm,横径2.93cm,单果重16.6g。果肩凸圆。梗洼浅广。果顶微凹。果皮薄,红色。果肉质地致密,汁液少,味酸甜,品质中等,多用于制干。鲜枣含总糖22.26%,酸0.61%,100g果肉维生素C含量466.94mg;果皮含黄酮20.24mg/g,cAMP含量85.69 μg/g。干枣含总糖70.00%,酸1.11%。果核纺锤形,种仁饱满,含仁率23.3%。

评价 该品种树体较大,树势强健,丰产稳产。果实大,肉质致密,汁少,为北京地区较优良的制干品种。

Paopaohong

Source and Distribution The cultivar originated from and spreads in Fangshan District of Beijing.

Botanical Characters The large tree is vertical with a strong central leader trunk and an umbrella-shaped crown. The trunk bark has striped fissures. The purplish-brown 1-year-old shoots are 72.6 cm long and 1.08 cm thick. The internodes are 6.3 cm long with developed thorns. The secondary branches are 29.0 cm long with 8 nodes. The mother fruiting spurs can germinate 3 deciduous fruiting shoots which are 18.1 cm long with 13 leaves. The small oval-shaped leaves are 5.1 cm long and 3.1 cm wide with a gradually-cuspate apex, a heart-shaped or round base and an irregular saw-tooth pattern on the margin. Each inflorescence has 7 small flowers.

Biological Characters The cultivar has strong adaptability, strong vigor and weak branching ability, with a high and stable yield. The percentage of fruits to deciduous fruiting shoots of 1-year-old shoots, 2~3-year-old branches, 4~6-year-old ones and over-6-year-old ones is 10.7%, 93.2%, 87.0% and 50.9%. In Taigu County of Shanxi Province, it matures in early October. It is an extremely late-ripening variety with strong resistance to fruit-cracking.

Fruit Characteristics The large oblong fruit has a vertical and cross diameter of 3.65 cm and 2.93 cm, averaging 16.6 g. It has a pointed-round shoulder, a shallow and wide stalk cavity, a slightly-sunken fruit apex and thin red skin. The flesh is tight-textured, sour and sweet, with less juice. It has medium quality for dried fruits. The content of TTS, TA and Vc in fresh fruit is 22.26%, 0.61% and 466.94 mg per 100 g fresh fruit. The content of flavones and cAMP in mature fruit skin is 20.24 mg/g and 85.69 μg/g. The content of TTS and TA in dried fruit is 70.00% and 1.11%. The spindle-shaped stone contains a well-developed kernel. The percentage of containing kernels is 23.3%.

Evaluation The large plant of Paopaohong cultivar has strong tree vigor with a high and stable yield. The large fruit has tight-textured and less juicy flesh and is suitable for making dried fruits.

义乌棉絮枣

品种来源及分布 原产和分布于浙江义乌。

植物学性状 树体中大，树姿开张，树冠呈偏斜形。主干皮裂块状。枣头红褐色，平均长69.4cm，节间长7.2cm，蜡质少。二次枝平均长31.3cm，4~9节，弯曲度中等。针刺发达。枣股平均抽生枣吊3.6个。枣吊长25.5cm，着叶18片。叶片小，卵圆形，平展，先端急尖，叶基偏斜形，叶缘具钝锯齿。花量中多，花序平均着花4朵。

生物学特性 树势弱，萌芽率和成枝力中等，结果晚，定植第三年结果，10年左右进入盛果期，产量较低，枣吊平均结果0.23个。在山西太谷地区，10月初果实成熟采收，果实生育期117d左右，为极晚熟品种类型。成熟期遇雨裂果轻。

果实性状 果个中大，卵圆形，纵径2.94cm，横径2.50cm，单果重10.0g，大小较整齐。果皮中厚，红色，果面平滑。梗洼窄、中深。果顶平圆，柱头残存。肉质较致密，汁液中多，味甜酸，品质中等，可制干。鲜枣可食率91.2%，含总糖23.86%，酸0.48%，100g果肉维生素C含量471.55mg。制干率59.7%，干枣含总糖54.71%，酸2.10%。核较大，纺锤形，平均重0.88g。大多核内含有饱满种子，含仁率93.3%。

评价 该品种树体中大，树势弱，结果晚，产量较低。果个中大，品质中等，适宜制干，抗裂果。

Yiwumianxuzao

Source and Distribution The cultivar originated from and spreads in Yiwu of Zhejiang Province.

Botanical Characters The medium-sized tree is spreading with a deflective crown. The trunk bark has massive fissures. The reddish-brown 1-year-old shoots are 69.4 cm long with the internodes of 7.2 cm long, with less wax and developed thorns. The secondary branches are 31.3 cm long with 4~9 nodes of medium curvature. The mother fruiting spurs can germinate 3.6 deciduous fruiting shoots, which are 25.5 cm long with 18 leaves. The small leaves are oval-shaped and flat, with a sharply-cuspate apex, a deflective base and a blunt saw-tooth pattern on the margin. The number of flowers is medium large, averaging 4 ones per inflorescence.

Biological Characters The tree has weak vigor, medium germination and branching ability. It bears late, generally in the 3rd year after planting and enters the full-bearing stage in the 10th year, with a low yield. The deciduous bearing shoot bears 0.23 fruits on average. In Taigu County of Shanxi Province, it matures in early October with a fruit growth period of 117 d. It is a late-ripening variety with light fruit-cracking even if it rains in the maturing stage.

Fruit Characteristics The medium-sized conical fruit has a vertical and cross diameter of 2.94 cm and 2.50 cm, averaging 10.0 g with a regular size. It has medium-thick red skin, a smooth surface, a narrow and medium-deep stalk cavity, a flat-round fruit apex and a remnant stigma. The tight-textured flesh is sweet and sour with medium juice. It has medium quality for dried fruits. The percentage of edible part of fresh fruit is 91.2%, and the content of TTS and TA is 23.86% and 0.48%. The large spindle-shaped stone weighs 0.88 g. Most stones have a well-developed kernel, and the percentage of containing kernels is 93.3%.

Evaluation The medium-sized plant of Yiwumianxuzao cultivar has weak tree vigor, bearing late with a low yield. The medium-sized fruit has medium quality and is suitable for making dried fruits. It has strong resistance to fruit-cracking.

制 干
Drying Varieties
品 种

中宁小圆枣

品种来源及分布　原产和分布于宁夏中宁。

植物学性状　树体中大，树姿半开张，树冠呈乱头形。枣吊平均长19.3cm，着叶17片。叶片中大，叶长6.2cm，叶宽3.7cm，椭圆形，绿色，平展，先端急尖，叶基圆形，叶缘具钝锯齿。花量中多，花序平均着花5朵。花较大，花径6.4mm。

生物学特性　树势较强，萌芽率和成枝力低，成枝力55.0%，定植第三年结果，10年左右进入盛果期，较丰产，坐果率极高，2~3年和3年以上枝吊果率分别为134.6%和120.9%。在山西太谷地区，9月下旬果实成熟采收，果实生育期110d左右，为晚熟品种类型。成熟期遇雨裂果重。

果实性状　果个中大，扁柱形，纵径3.00cm，横径2.60cm，侧径2.32cm，单果重14.2g，大小较整齐。果皮薄，紫红色，果面平滑。果点黄色，中大，密度大，显明。梗洼中深、广。果顶平圆，柱头脱落。肉质致密，汁液中多，味酸甜，品质中等，适宜制干。鲜枣可食率94.3%，含可溶性固形物35.00%，总糖22.92%，酸0.35%，100g果肉维生素C含量505.55mg；果皮含黄酮4.68mg/g，cAMP含量153.25μg/g。制干率57.8%，干枣含总糖63.59%，酸1.33%。果核较大，倒纺锤形，平均重0.81g，大多无种子，含仁率13.4%。

评价　该品种树体中大，树势较强，结果较晚，较丰产，但易裂果。果个中大，品质中等，主要用于制干。

Zhongningxiaoyuanzao

Source and Distribution　The cultivar originated from and spreads in Zhongning County of Ningxia Province.

Botanical Characters　The medium-sized tree is half-spreading with an irregular crown. The deciduous fruiting shoots are 19.3 cm long with 17 leaves. The medium-sized leaves are oval-shaped, green and flat, 6.2 cm long and 3.7 cm wide with a sharply-cuspate apex, a round base and a blunt saw-tooth pattern on the margin. The number of flowers is medium large, averaging 5 ones per inflorescence. The large blossoms have a diameter of 6.4 mm.

Biological Characters　The tree has moderate vigor, low germination rate and weak branching ability, with a branching rate of 55.0%. It generally bears in the 3rd year after planting and enters the full-bearing stage in the 10th year with a high yield and high fruit set. The percentage of fruits to deciduous fruiting shoots of 2~3-year-old branches and over-3-year-old ones is 134.6% and 120.9%. In Taigu County of Shanxi Province, it is harvested in late September with a fruit growth period of 110 d. It is a late-ripening variety with serious fruit-cracking if it rains in the maturing stage.

Fruit Characteristics　The medium-sized oblong fruit has a vertical and cross diameter of 3.0 cm and 2.6 cm, averaging 14.2 g with a regular size. It has purplish-red thin skin, a smooth surface, a medium-deep and wide stalk cavity, a flat-round fruit apex and a falling-off stigma. The tight-textured flesh is sour and sweet with medium juice. It has medium quality for dried fruits. The percentage of edible part of fresh fruit is 94.3%, and SSC, TTS, TA and Vc is 35.00%, 22.92%, 0.35% and 505.55 mg per 100 g fresh fruit. The content of flavones and cAMP in mature fruit skin is 4.68 mg/g and 153.25 μg/g. The large stone is inverted spindle-shaped, averaging 0.81 g. Most stones contain no kernels with the percentage of containing kernels of 13.4%.

Evaluation　The medium-sized plant of Zhongningxiaoyuanzao cultivar has moderate vigor, bearing late with a high yield. The medium-sized fruit has medium quality, mainly used for making dried fruits. It has poor resistance to fruit-cracking.

制 干
Drying Varieties
品 种

· 261 ·

汝 城 枣

品种来源及分布 分布于湖南汝城等地。

植物学性状 树体中大,树姿开张,树冠偏斜形。主干皮裂条状。枣头黄褐色,平均长55.6cm,节间长5.8cm。二次枝长21.3cm,5~7节,弯曲度中等。针刺发达。枣股平均抽生枣吊4.7个。枣吊长23.6cm,着叶19片。叶片小,叶长4.5cm,叶宽2.4cm,椭圆形,先端急尖,叶基圆楔形,叶缘钝锯齿。花量少,花序平均着花3朵,花小,花径6.0mm。

生物学特性 树势弱,易成枝,成枝力83.3%。产量中等,坐果率一般,枣头枝不结果,2~3年和3年以上枝吊果率分别为78.2%和74.6%。在山西太谷地区,9月下旬果实成熟采收,果实生育期110d以上,为晚熟品种类型。成熟期遇雨裂果少。

果实性状 果个小,长圆形,纵径2.50cm,横径2.10cm,单果重4.4g,大小整齐。果皮中厚,红色,果面粗糙。梗洼窄、中深。果顶平圆,柱头残存。肉质疏松,味甜,汁液中多,品质中等,适宜制作干枣。鲜枣可食率93.0%,含总糖20.91%,酸0.86%,100g果肉维生素C含量375.48mg;果皮含黄酮8.89mg/g,cAMP含量89.57μg/g。果核小,纺锤形,核重0.31g,含仁率98.3%。

评价 该品种树势弱,产量中等。果实小,品质中等,适宜制干。抗裂果能力较强。

Ruchengzao

Source and Distribution The cultivar spreads in Rucheng of Hunan Province.

Botanical Characters The medium-sized tree is spreading with a deflective crown. The trunk bark has striped fissures. The yellowish-brown 1-year-old shoots are 55.6 cm long with the internodes of 5.8 cm long. The secondary branches are 21.3 cm long with 5~7 nodes of medium curvature and developed thorns. The mother fruiting spurs can germinate 4.7 deciduous fruiting shoots which are 23.6 cm long with 19 leaves. The small oval-shaped leaves are 4.5 cm long and 2.4 cm wide with a sharply-cuspate apex, a round-cuneiform base and a blunt saw-tooth pattern on the margin. The number of flowers is small, averaging 3 ones per inflorescence. The small blossoms have a diameter of 6.0 mm.

Biological Characters The tree has weak vigor and strong branching ability, with a branching rate of 83.3%. It has a medium yield and medium fruit set. The 1-year-old shoots bear no fruit. The percentage of fruits to deciduous fruiting shoots of 2 or 3-year-old branches and over-3-year-old ones is 78.2% and 74.6%. In Taigu County of Shanxi Province, it is harvested in late September with a fruit growth period of over 110 d. It is a late-ripening variety with light fruit-cracking even if it rains in the maturing stage.

Fruit Characteristics The small oblong fruit has a vertical and cross diameter of 2.5 cm and 2.1 cm, averaging 4.4 g with a regular size. It has medium-thick red skin, a rough surface, a medium-deep and narrow stalk cavity, a flat-round fruit apex and a remnant stigma. The loose-textured flesh is sweet with medium juice. It has medium quality for dried fruits. The percentage of edible part of fresh fruit is 93.0%, and TTS, TA and Vc is 20.91%, 0.86% and 375.48 mg per 100 g fresh fruit. The content of flavones and cAMP in mature fruit skin is 8.89 mg/g and 89.57 μg/g. The small spindle-shaped stone weighs 0.31 g with the percentage of containing kernels of 98.3%.

Evaluation The cultivar has weak vigor with a medium yield. The small fruit has medium quality and is suitable for making dried fruits. It has strong resistance to fruit-cracking.

制 干
Drying Varieties
品 种

· 263 ·

离石合钵枣

品种来源及分布 分布于山西省离石市。

植物学性状 树体中大，树姿半开张，树冠呈伞形。主干条状皮裂。枣头紫褐色，平均长63.1cm，节间长7.5cm，蜡质少。二次枝长20.9cm，4~6节。针刺不发达。枣股平均抽生枣吊3.9个。枣吊长18.6cm，着叶13片。叶片小，椭圆形，平展，先端钝尖，叶基圆形，叶缘具钝锯齿。花量较少，花序平均着花4朵。

生物学特性 树势强，萌芽率和成枝力强。结果晚，一般定植第三年结果，10年左右进入盛果期，产量中等。在山西太谷地区，9月下旬果实成熟采收，果实生育期117d，为晚熟品种类型。成熟期遇雨裂果较轻。

果实性状 果个较大，扁圆形或近圆形，纵径2.88cm，横径3.01cm，单果重25.2g，大小整齐。果皮中厚，紫红色，果面平滑。梗洼浅、中广。果顶凹，柱头宿存。肉质较致密，汁液中多，味酸甜，品质中等，可制干。鲜枣可食率96.2%，含可溶性固形物32.00%，总糖23.50%，酸0.42%，100g果肉维生素C含量366.63mg；果皮含黄酮4.36mg/g，cAMP含量293.01μg/g。干枣含总糖59.00%，酸1.33%。果核较大，椭圆形，平均重0.96g，种仁不饱满，含仁率46.7%。

评价 该品种树体中大，树势强，结果晚，产量中等，抗裂果能力强。果个和果核较大，品质中等，适宜制干。

Lishihebozao

Source and Distribution The cultivar originated from and spreads in Lishi of Lvliang District in Shanxi Province.

Botanical Characters The medium-sized tree is half-spreading with an umbrella-shaped crown. The trunk bark has striped fissures. The purplish-brown 1-year-old shoots are 63.1 cm long with the internodes of 7.5 cm long less wax and less-developed thorns. The secondary branches are 20.9 cm long with 4~6 nodes. The mother fruiting spurs can germinate 3.9 deciduous fruiting shoots which are 18.6 cm long with 13 leaves. The small flat leaves are oval-shaped, with a bluntly-cuspate apex, a round base and a blunt saw-tooth pattern on the margin. The number of flowers is small, averaging 4 ones per inflorescence.

Biological Characters The tree has strong vigor, strong germination and branching ability. It generally bears in the 3rd year after planting and enters the full-bearing stage in the 10th year with a medium yield. In Taigu County of Shanxi Province, it matures in late September with a fruit growth period of 117 d. It is a late-ripening variety with light fruit-cracking even if it rains in the maturing stage.

Fruit Characteristics The large oblate fruit has a vertical and cross diameter of 2.88 cm and 3.01 cm, averaging 25.2 g with a regular size. It has medium-thick and purplish-red skin, a smooth surface, a shallow and medium-wide stalk cavity, a sunken fruit apex and a remnant stigma. The tight-textured flesh is sour and sweet with medium juice. It has medium quality for dried fruits. The percentage of edible part of fresh fruit is 96.2%, and SSC, TTS, TA and Vc is 32.00%, 23.50%, 0.42% and 366.63 mg per 100 g fresh fruit. The content of flavones and cAMP in mature fruit skin is 4.36 mg/g and 293.01 μg/g. The large oval-shaped stone weighs 0.96 g, with a shriveled kernel. The percentage of containing kernels is 46.7%.

Evaluation The medium-sized plant of Lishihabazao cultivar has strong tree vigor, bearing late with a medium yield. It has strong resistance to fruit-cracking. The large fruit has medium quality and is suitable for making dried fruits.

制 干
Drying Varieties
品 种

西双版纳小枣

品种来源及分布 原产和分布于云南省西双版纳。

植物学性状 树体中大，树姿直立，树冠呈圆柱形。主干皮裂条状。枣头黄褐色，平均长52.6cm，节间长6.1cm，无蜡层。二次枝平均长17.7cm，4～6节，弯曲度中等。针刺不发达。枣股平均抽生枣吊4.0个。枣吊长19.3cm，着叶15片。叶片中大，椭圆形，平展，先端急尖，叶基圆形，叶缘具钝锯齿。花量较多，花序平均着花5朵。花小，花径5.8mm。

生物学特性 树势中庸，萌芽率和成枝力弱，结果较早，一般定植第二年结果，10年左右进入盛果期，丰产性一般。在山西太谷地区，10月上旬果实成熟采收，果实生育期119d左右，为极晚熟品种类型。成熟期遇雨裂果轻。

果实性状 果个中大，长圆柱形，纵径2.60cm，横径2.23cm，单果重10.4g，大小整齐。果皮中厚，红色，果面平滑。梗洼中深、广。果顶平圆，柱头脱落。肉质疏松，汁液多，味甜，品质中等，可制干。鲜枣可食率94.6%。核中大，纺锤形，平均重0.56g，种仁饱满，含仁率98.3%。

评价 该品种树体中大，树势中庸，结果较早，丰产性一般，抗裂果。果个中大，品质中等，适宜制干。

Xishuangbannaxiaozao

Source and Distribution The cultivar originated from and spreads in Xishuangbanna of Yunnan Province.

Botanical Characters The medium-sized tree is vertical with a column-shaped crown. The trunk bark has striped fissures. The yellowish-brown 1-year-old shoots are 52.6 cm long, with the internodes of 6.1 cm, without wax and with less-developed thorns. The secondary branches are 17.7 cm long with 4～6 nodes of medium curvature. The mother fruiting spurs can germinate 4.0 deciduous fruiting shoots which are 19.3 cm long with 15 leaves. The medium-sized leaves are oval-shaped and flat, with a sharply-cuspate apex, a round base and a blunt saw-tooth pattern on the margin. The number of flowers is medium large, averaging 5 ones per inflorescence. The small blossoms have a diameter of 5.8 mm.

Biological Characters The tree has moderate vigor, weak germination ability and strong branching ability. It bears early, generally in the 2nd year after planting and enters the full-bearing stage in the 10th year with a medium yield. In Taigu County of Shanxi Province, it matures in early October with a fruit growth period of 119 d. It is an extremely late-ripening variety with light fruit-cracking even if it rains in the maturing stage.

Fruit Characteristics The small oblong fruit has a vertical and cross diameter of 2.60 cm and 2.23 cm, averaging 10.4 g with a regular size. It has medium-thick red skin, a smooth surface, a medium-deep and wide stalk cavity, a flat-round fruit apex and a falling-off stigma. The loose-textured flesh is sweet and juicy. It has medium quality for dried fruits. The percentage of edible part of fresh fruit is 94.6%. The large spindle-shaped stone weighs 0.56 g with a well-developed kernel. The percentage of containing kernels is 98.3%.

Evaluation The medium-sized plant of Xishuangbannaxiaozao cultivar has moderate tree vigor, bearing early with a medium yield. It has strong resistance to fruit-cracking. The medium sized fruit has medium quality for dried fruits.

制 干
Drying Varieties
品 种

太原圆枣

品种来源及分布　原产和分布于山西太原北郊柴村等地，是当地栽培数量较多的品种。

植物学性状　树体较大，树姿半开张，枝叶较密，树冠呈圆锥形。主干皮裂呈条状。枣头黄褐色，平均长65.4cm，粗0.85cm，节间长7.8cm，蜡质少。二次枝长26.4cm，6节左右。针刺较发达。枣吊长14.8cm，着叶11片。叶片中大，卵圆形，绿色，先端钝尖，叶基心形，叶缘具钝锯齿。花量中等，花序平均着花4朵。

生物学特性　适应性较强、耐干旱，对栽培条件要求不严。树势较强，产量高而稳定。在山西太谷地区，9月上旬果实成熟，果实生育期90d左右，为早熟品种类型。成熟期遇雨易裂果。

果实性状　果实较小，近圆形，纵径3.09cm，横径2.69cm，单果重8.8g，大小较整齐。果肩平，梗洼窄而深。果顶微凹，柱头残存。果皮较薄，紫红色。果肉厚，浅绿色，质地疏松，汁液少，味甜酸，较淡，可制干，品质中下。鲜枣可食率97.7%，果皮含黄酮2.04mg/g，cAMP含量22.66μg/g。果核小，纺锤形，核重0.2g，核内无种子。

评价　该品种适应性较强，耐干旱，产量高，成熟早，但果实品质较差，且易裂果，不宜生产栽培。

Taiyuanyuanzao

Source and Distribution　The cultivar originated from and spreads in Chaicun Village of north Taiyuan City in Shanxi Province with a large quantity there.

Botanical Characters　The large tree is half-spreading with dense branches and a conical crown. The trunk bark has striped fissures. The yellowish-brown 1-year-old shoots are 65.4 cm long and 0.85 cm thick. The internodes are 7.8 cm long with less wax and developed thorns. The secondary branches are 26.4 cm long with 6 nodes. The deciduous fruiting shoots are 14.8 cm long with 11 leaves. The medium-sized leaves are oval-shaped and green, with a bluntly-cuspate apex, a heart-shaped base and a blunt saw-tooth pattern on the margin. The number of flowers is medium large, averaging 4 ones per inflorescence.

Biological Characters　The tree has strong adaptability, tolerant to drought with a low demand on cultural conditions, strong vigor, a high and stable yield. In Taigu County of Shanxi Province, it matures in early September with a fruit growth period of 90 d. It is an early-ripening variety with serious fruit-cracking if it rains in the maturing stage.

Fruit Characteristics　The small round fruit has a vertical and cross diameter of 3.09 cm and 2.69 cm, averaging 8.8 g with a regular size It has a flat shoulder, a narrow and deep stalk cavity, a slightly-sunken fruit apex, a remnant stigma and purplish-red thin skin. The light-green flesh is thick, loose-textured, sour and sweet with less juice. It can be used for making dried fruits, yet with a low quality. The content of flavones and cAMP in mature fruit skin is 2.04 mg/g and 22.66 μg/g. The small spindle-shaped stone weighs 0.2 g, without kernel.

Evaluation　The cultivar has strong adaptability, tolerant to drought, with high productivity. It matures early. Yet the fruit has a poor quality and serious fruit-cracking, so it is not suitable for commercial production.

制 干
Drying Varieties
品 种

延川白枣

品种来源及分布 原产和分布于陕西省延川县。

植物学性状 树体中大，树姿半开张，树冠呈自然半圆形。主干皮裂条状。枣头红褐色，平均长73.7cm，节间长8.6cm。二次枝长25.3cm，4～6节。针刺退化。枣股平均抽生枣吊3.0个。枣吊长22.4cm，着叶13片。叶片大，叶长7.2cm，叶宽3.8cm，椭圆形，先端急尖，叶基偏斜形，叶缘具钝锯齿。花量中多，花序平均着花5朵。

生物学特性 树势较弱，萌芽率和成枝力强，成枝力92.5%。一般定植第三年结果，产量较低，坐果率低，枣头不易坐果，2～3年和3年以上枝吊果率分别为47.5%和5.7%。在山西太谷地区，9月中旬果实进入脆熟期，下旬开始成熟采收，果实生育期110d左右，属晚熟品种类型。

果实性状 果个中大，圆柱形，纵径3.60cm，横径2.90cm，单果重13.2g，大小不整齐。果皮中厚，赭红色，果面光滑。梗洼广而浅。顶部凹陷，柱头宿存。肉质疏松，味甜，汁液少，品质中等，适宜制作干枣。鲜枣可食率93.4%，含可溶性固形物29.40%，总糖29.08%，酸0.29%。核较大，纺锤形，核重0.87g。

评价 该品种树势弱，产量较低。果个中大，品质中等，适宜制干。

Yanchuanbaizao

Source and Distribution The cultivar originated from and spreads in Yanchuan County of Shaanxi Province.

Botanical Characters The medium-sized tree is half-spreading with a natural-round crown. The trunk bark has striped fissures. The reddish-brown 1-year-old shoots are 73.7 cm long with the internodes of 8.6 cm tong. The secondary branches are 25.3 cm long with 4～6 nodes and almost no thorns. The mother fruiting spurs can germinate 3 deciduous fruiting shoots, which are 22.4 cm long with 13 leaves. The large oval-shaped leaves are 7.2 cm long and 3.8 cm wide, with a sharply-cuspate apex, a deflective base and a blunt saw-tooth pattern on the margin. The number of flowers is medium large, averaging 5 ones per inflorescence.

Biological Characters The tree has weak vigor, high germination rate and strong branching ability with a branching rate of 92.5%. It generally bears in the 3rd year after planting with a low yield and low fruit set. The 1-year-old shoots almost bear no fruit. The percentage of fruits to deciduous fruiting shoots of 2 or 3-year-old branches and over-3-year-old ones is 47.5% and 5.7%. In Taigu County of Shanxi Province, it enters the crisp-maturing stage in mid-September and is harvested in late September, with a fruit growth period of 110 d.

Fruit Characteristics The medium-sized fruit is oval-shaped with a vertical and cross diameter of 3.60 cm and 2.90 cm, averaging 13.2 g with irregular sizes. It has medium-thick and brownish-red skin, a smooth surface, a wide and shallow stalk cavity, a sunken fruit apex and a remnant stigma. The loose-textured flesh is sweet with less juice. It has medium quality for dried fruits. The percentage of edible part of fresh fruit is 93.4%, and SSC, TTS and TA is 29.40%, 29.08% and 0.29%. The large spindle-shaped stone weighs 0.87 g.

Evaluation The cultivar has weak vigor, a low yield and low fruit set. The medium-sized fruit has medium quality for dried fruits.

制 干 品 种
Drying Varieties

延 川 条 枣

品种来源及分布 原产和分布于陕西省延川县。

植物学性状 树体较大，树姿直立，树冠呈圆锥形。主干皮裂条状。枣头黄褐色。针刺不发达。枣股平均抽生枣吊3.1个。枣吊长21.0cm，着叶12片。叶片大，椭圆形，平展，先端急尖，叶基偏斜形，叶缘具钝锯齿。花量多，花序平均着花8朵。花中大，花径6.4mm。

生物学特性 树势强，萌芽率和成枝力强，结果较早，一般定植第二年结果，10年左右进入盛果期，丰产稳产。在山西太谷地区，10月上旬果实成熟采收，果实生育期120d左右，为极晚熟品种类型。成熟期遇雨裂果较轻。

果实性状 果个中大，圆柱形或长圆形，纵径4.15cm，横径2.50cm，单果重13.1g，大小整齐。果皮厚，赭红色，果面平滑，外观极好。梗洼浅而广，果顶平，柱头宿存。肉质致密，汁液少，味甜酸，品质中等，适宜制干。鲜枣可食率96.6%，含可溶性固形物30.60%，总糖26.93%，酸0.49%，100g果肉维生素C含量520.33mg。制干率45.6%，干枣含总糖54.71%，酸2.10%。核小，纺锤形，平均重0.44g，核内不含种子。

评价 该品种树体较大，树势强，结果较早，丰产稳产，较抗裂果。果个中大，外观好，品质中等，适宜制干。

Yanchuantiaozao

Source and Distribution The cultivar originated from and spreads in Yanchuan County of Shaanxi Province.

Botanical Characters The large tree is vertical with a conical crown. The trunk bark has striped fissures. The yellowish-brown 1-year-old shoots have less-developed thorns. The mother fruiting spurs can germinate 3.1 deciduous fruiting shoots which are 21.0 cm long with 12 leaves. The large leaves are oval-shaped and flat, with a sharply-cuspate apex, a deflective base and a blunt saw-tooth pattern on the margin. There are many medium-sized flowers with a diameter of 6.4 mm produced, averaging 8 ones per inflorescence.

Biological Characters The tree has strong vigor, strong germination and branching ability. It bears early, generally in the 2nd year after planting and enters the full-bearing stage in the 10th year, with a high and stable yield. In Taigu County of Shanxi Province, it matures in early October, with a fruit growth period of 120 d. It is an extremely late-ripening variety with light fruit-cracking even if it rains in the maturing stage.

Fruit Characteristics The medium-sized fruit is oblong or column-shaped with a vertical and cross diameter of 4.15 cm and 2.50 cm, averaging 13.1 g (maximum 15 g) with a regular size. It has dark-red thick skin, a smooth surface, an attractive appearance, a shallow and wide stalk cavity, a flat fruit apex and a remnant stigma. The tight-textured flesh is sour and sweet with less juice. It has medium quality for dried fruits. The percentage of edible part of fresh fruit is 96.6%. The small spindle-shaped stone weighs 0.44 g without kernel inside.

Evaluation The large plant of Yanchuantiaozao cultivar has strong tree vigor, bearing early with a high and stable yield. It has strong resistance to fruit-cracking. The medium-sized fruit has a good appearance with medium quality for dried fruits.

制 干 品 种
Drying Varieties

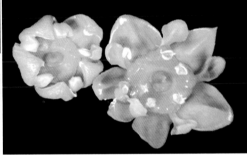

北 碚 小 枣

品种来源及分布　原产和分布于重庆市北碚等地。

植物学性状　树体中大，树姿半开张，干性较强，树冠呈伞形。主干皮裂条状。枣头红褐色，生长量72.5cm，节间长7.6cm，二次枝长28.7cm，4～6节。无针刺。枣股平均抽生枣吊4.4个。枣吊长19.5cm，着叶14片。叶片卵状披针形，深绿色，先端急尖，叶基偏斜形，叶缘具钝锯齿。花量多，花序平均着花8朵。

生物学特性　树势较弱。丰产性中等，结果能力中等。在山西太谷地区，9月上旬果实着色，9月下旬成熟采收，果实生育期110d左右，为晚熟品种类型。

果实性状　果实中大，圆柱形，纵径3.50cm，横径2.60cm，果重13.6g左右，大小较整齐。梗洼窄、中深。果顶凹，柱头残存。果皮厚，浅红色，果面光滑。肉质较致密，味甜，汁液多，品质中等，可制干。鲜枣可食率93.7%，总糖含量21.76%，酸0.57%，糖酸比38.18∶1，100g果肉维生素C含量607.36mg；果皮含黄酮10.26mg/g，cAMP含量224.94μg/g。果核较大，纺锤形，核重0.86g，核内种仁不饱满，含仁率43.3%。

评价　该品种果实中大，维生素C含量极高，品质中等，适宜制干。

Beibeixiaozao

Source and Distribution　The cultivar originated from and spreads in Beibei of Chongqing City.

Botanical Characters　The medium-sized tree is half-spreading with a strong central leader trunk and an umbrella-shaped crown. The trunk bark has striped fissures. The reddish-brown 1-year-old shoots are 72.5 cm long with the internodes of 7.6 cm long. The secondary branches are 28.7 cm long with 4～6 nodes and no thorns. The mother fruiting spurs can germinate 4.4 deciduous fruiting shoots, which are 19.5 cm long with 14 leaves. The leaves are ovate-lanceolate and dark green, with a sharply cuspate apex, a deflective base and a blunt saw-tooth pattern on the margin. There are many flowers produced, averaging 8 ones per inflorescence.

Biological Characters　The tree has weak vigor with a medium yield and medium fruiting ability. In Taigu County of Shanxi Province, it begins coloring in early September and is harvested in late September with a fruit growth period of 110 d. It is a late-ripening variety.

Fruit Characteristics　The medium-sized fruit is column-shaped, with a vertical and cross diameter of 3.50 cm and 2.60 cm, averaging 13.6 g with a regular size. It has a narrow and medium-deep stalk cavity, a sunken fruit apex, a remnant stigma, light-red thick skin and a smooth surface. The flesh is tight-textured and juicy. It has medium quality for dried fruits. The percentage of edible part of fresh fruit is 93.7%, and the content of TTS, TA and Vc is 21.76%, 0.57% and 607.36 mg per 100 g fresh fruit. The SAR is 38.18∶1. The content of flavones and cAMP in mature fruit skin is 10.26 mg/g and 224.94 μg/g. The large spindle-shaped stone weighs 0.86 g, with a shriveled kernel. The percentage of containing kernels is 43.3%.

Evaluation　The cultivar has a large fruit size and high content of Vc. It has medium quality and is suitable for making dried fruits.

太原驴粪蛋

品种来源及分布　原产和分布于山西省太原市小井峪乡下元等村。

植物学性状　树体中大，树姿半开张，树冠自然圆头形。枣头红褐色，年生长量64.3cm，节间长7.7cm。二次枝长22.8cm，4～6节，弯曲度中等，无针刺。枣吊长32.2cm，着叶18片。叶片椭圆形，先端急尖，叶基截形，叶缘具钝锯齿。花量极少，花序平均着花2朵，但花朵大，花径7.0mm。

生物学特性　树势中庸，产量中等。在山西太谷地区，9月中旬果实成熟采收，果实生育期107d左右，为中熟品种类型。成熟期遇雨易裂果。

果实性状　果个大，圆柱形或近圆形，纵径3.55cm，横径3.36cm，单果重16.2g，大小整齐。果皮薄，红色，果面平滑。梗洼浅广。果顶微凹，柱头脱落。肉质疏松，味酸甜，汁液中多，品质中等，适宜制干。鲜枣含可溶性固形物20.40%，总糖17.29%，酸0.34%，100g果肉维生素C含量324.04mg。含仁率86.7%，种仁较饱满，偶有双仁。

评价　该品种果实大，品质中等，适宜制干。果实抗裂果能力差，成熟期应注意防雨。

Taiyuanlvfendan

Source and Distribution　The cultivar originated from and spreads in Xiayuan of Xiaojingyu Village in Taiyuan of Shanxi Province.

Botanical Characters　The medium-sized tree is half-spreading with a natural-round crown. The reddish-brown 1-year-old shoots are 64.3 cm long with the internodes of 7.7 cm. The secondary branches are 22.8 cm long with 4～6 nodes of medium curvature and no thorns. The deciduous fruiting shoots are 32.2 cm long with 18 leaves. The oval-shaped leaves have a sharply-cuspate apex, a truncate base and a blunt saw-tooth pattern on the margin. The number of flowers is very small, averaging 2 ones per inflorescence. The large blossoms have a diameter of 7.0 mm.

Biological Characters　The tree has moderate vigor and a medium yield. In Taigu County of Shanxi Province, it matures in mid-September with a fruit growth period of 107 d. It is a mid-ripening variety with serious fruit-cracking if it rains in the maturing stage.

Fruit Characteristics　The large round fruit has a vertical and cross diameter of 3.55 cm and 3.36 cm, averaging 16.2 g with a regular size. It has thin red skin, a smooth surface, a shallow and wide stalk cavity, a slightly sunken fruit apex and a remnant stigma. The loose-textured flesh is sour and sweet with medium juice. It has medium quality for dried fruits. The content of Vc in fresh fruit is 324.04 mg per 100 g fresh fruit. The percentage of containing kernels is 86.7%. The stones contain a well-developed kernel, sometimes double kernels.

Evaluation　The cultivar has a large fruit size and medium quality and is suitable for making dried fruits. The fruit has poor resistance to fruit-cracking, so rain protection should be paid much attention to in the maturing stage.

制 干
Drying Varieties
品 种

太原长枣

品种来源及分布 原产和分布于山西省太原市及周边地区。

植物学性状 树体较大，树姿开张，树冠偏斜形。主干皮裂条状。枣头黄褐色，年生长量66.4cm，节间长10.2cm，二次枝长19.2cm，4～6节，弯曲度中等。针刺不发达。枣吊长19.2cm，着叶10片。叶片椭圆形，浅绿色，先端锐尖，叶基圆形，叶缘具锐锯齿。花量中等，花序平均着花4朵。花中大，花径6.1mm。

生物学特性 树势强。在山西太谷地区，9月上旬果实着色，9月中旬成熟采收，果实生育期103d左右，为中熟品种类型。

果实性状 果实中大，卵圆形，纵径3.83cm，横径2.86cm，果重12.8g左右，最大可达18.0g，大小不整齐。梗洼窄、中深。果顶平，柱头残存。果皮薄，红色，果面光滑。肉质疏松，味酸甜，汁液少，品质中等，可制干。鲜枣含可溶性固形物23.40%，总糖19.14%，酸0.43%，100g果肉维生素C含量375.62mg；果皮含黄酮4.44mg/g，cAMP含量269.47μg/g。制干率49.3%，干枣含总糖68.38%，酸1.16%。含仁率26.7%。

评价 该品种树势强健，果实中大，品质中等，可制干。

Taiyuanchangzao

Source and Distribution The cultivar originated from and spreads in Taiyuan and its surroundings in Shanxi Province.

Botanical Characters The large tree is spreading with a deflective crown. The trunk bark has striped fissures. The yellowish-brown 1-year-old shoots are 66.4 cm long with the internodes of 10.2 cm long. The secondary branches are 19.2 cm long with 4～6 nodes of medium curvature and undeveloped thorns. The deciduous fruiting shoots are 19.2 cm long with 10 leaves. The leaves are oval-shaped and light green, with a sharply-cuspate apex, a round base and a sharp saw-tooth pattern on the margin. The number of flowers is medium large, averaging 4 ones per inflorescence. The medium-sized blossoms have a diameter of 6.1 mm.

Biological Characters The tree has strong vigor. In Taigu County of Shanxi Province, it begins coloring in early September and is harvested in mid-September with a fruit growth period of 103 d. It is a mid-ripening variety.

Fruit Characteristics The medium-sized fruit is oval-shaped, with a vertical and cross diameter of 3.83 cm and 2.86 cm, averaging 12.8 g (maximum 18.0 g) with irregular sizes. It has a narrow and medium-deep stalk cavity, a flat fruit apex, a remnant stigma, thin red skin and a smooth surface. The loose-textured flesh is sour and sweet with less juice. It has medium quality for dried fruits. The content of Vc is 375.62 mg per 100 g fresh fruit. The content of flavones and cAMP in mature fruit skin is 4.44 mg/g and 269.47 μg/g. The percentage of containing kernels is 26.7%.

Evaluation The cultivar has strong vigor and a medium-large fruit size. It has medium quality and is suitable for making dried fruits.

制干
Drying Varieties
品种

· 279 ·

兼用品种 Multipurpose Varieties

骏枣密植园（新疆　阿克苏）
Junzao Intensive Planting Garden (Akesu, Xinjiang)

金丝小枣

品种来源及分布 原产山东和河北交界地带，主要分布于山东省的乐陵、无棣、庆云、阳信、沾化和河北省的沧县、献县、泊头、南皮、盐山等地，现尚有成片的古老枣林，为当地主栽品种，也是全国栽培面积最大的品种。栽培历史悠久，400年前已有大规模栽培。

植物学性状 树体中大，树姿较开张，干性较弱，枝条中密，树冠圆头形。主干裂纹块状，易剥落。枣头红褐色，平均长63.8cm，粗1.01cm，节间长6.6cm，着生永久性二次枝5个左右。二次枝长33.8cm，平均7节。针刺不发达。枣股中大，圆柱形或圆锥形，抽吊力较强，一般抽生枣吊3～5个。枣吊平均长16.8cm，着叶11片。叶片较大，叶长6.3cm，宽3.1cm，长卵圆形，浓绿色，先端急尖，叶基圆形，叶缘锯齿浅钝。花量多，每花序着花3～9朵。花中大，零级花花径7mm左右。

生物学特性 树势中等，萌芽率较高，成枝力较强。幼树结果较晚，根蘖苗一般第三年开始结果，10年后进入盛果期，较丰产，产量较稳定。坐果率高，枣头、2～3年生枝和4年生枝的吊果率分别为71.6%、95.7%和18.0%。主要坐果部位在枣吊的3～7节，占坐果总数的82.3%。在山西太谷地区，9月中旬果实开始着色，10月上旬完熟，属晚熟品种类型。成熟期不抗裂果。

果实性状 果实小，果形有椭圆形、长圆形、鸡心形、倒卵形等多种，纵径2.60cm，横径2.28cm，单果重6.50g。果梗细，中长，梗洼中深，较窄。果顶平，柱头遗存。果皮薄，红色，果面光滑。果肉厚，乳白色，质地致密，细脆，味甘甜微酸，汁液中多，品质上等，适宜制干和鲜食。鲜枣可食率94.6%，含可溶性固形物36.00%，总糖28.36%，酸0.75%，100g果肉维生素C含量389.13mg。制干率55%～58%，干枣含总糖64.13%，酸1.05%。干枣果形饱满，肉质细，富弹性，耐贮运，味清甜。果核小，纺锤形，核重0.35g，核纹浅，核尖中长，含仁率58.3%，种仁较饱满。

评价 该品种是我国优良的红枣品种之一，果实皮薄肉厚，核小，质地细，糖分高，味甘甜，制干率高，鲜食制干品质兼优。但风土适应性较差，果实成熟期不抗裂果，且结果期较晚。可在花期温热、果实成熟期少雨的地区发展。

Jinsixiaozao

Source and Distribution The cultivar originated from the juncture of Shandong Province and Hebei Province. It is the dominant variety there and also the variety with the largest cultivation area in China with a long history.

Botanical Characters The medium-sized tree is spreading with a weak central leader trunk, medium-dense branches and a round crown. The reddish-brown 1-year-old shoots are 63.8 cm long. There are 5 permanent secondary branches with less-developed thorns. The medium-sized mother fruiting spurs can germinate 3～5 deciduous fruiting shoots. The large leaves are long oval-shaped and dark green. There are 3～9 flowers per inflorescence. The blossoms have a diameter of 7.0 mm for the zero-level flowers.

Biological Characters The tree has moderate vigor, high germination rate and strong branching ability. It enters the full-bearing stage in the 10th year after planting with a high and stable yield and high fruit set. The percentage of fruits to fruiting shoots of 1-year-old shoots, 2～3-year-old branches and 4-year-old ones is 71.6%, 95.7% and 18.0%. In Taigu County of Shanxi Province, it begins coloring in mid-September and completely matures in early October. It is a late-ripening variety with weak resistance to fruit-cracking in the maturing stage.

Fruit Characteristics The small fruit weighs 6.50 g. It may be oval, oblong, chicken-heart or obovate shaped. It has thin red skin and a smooth surface. The ivory-white thick flesh is tight-textured, crisp, sweet and a little sour, with medium juice and a good quality for dried fruits and fresh eating. The percentage of edible part of fresh fruit is 94.6%, and SSC, TTS, TA and Vc is 36.00%, 28.36%, 0.75% and 389.13 mg per 100 g fresh fruit. The percentage of fresh fruits which can be made into dried ones is 55%～58%. Dried fruits have delicate and whippy flesh, good tolerance to storage and transport, with a sweet taste. The small spindle-shaped stone weighs 0.35 g.

Evaluation The Cultivar is one of the best jujubes in China. The fruit has thin skin, thick and delicate flesh, a small stone, high sugar content, a sweet taste and a high rate for making dried fruits from fresh ones.

兼 用
Multipurpose Varieties
品 种

赞 皇 大 枣

品种来源及分布 又名金丝大枣。原产河北省赞皇县，为当地主栽品种，已有400多年的栽培历史，是目前发现的唯一三倍体品种。在山西和新疆地区栽培表现适应性强、丰产、品质更优异。

植物学性状 树体高大，树姿半开张，干性中强，枝条较稀，粗壮，树冠圆锥形。主干皮裂较深条状，不易剥落。枣头黄褐色，平均长83.0cm，粗度1.14cm，节间长8.5cm，着生二次枝7~10个。二次枝长36.2cm，平均8节。针刺发达。枣股较大，圆柱形，抽吊力中等，一般抽生枣吊3~4个。枣吊平均长23.6cm，着叶13片。叶片厚而宽大，叶长6.0cm，宽3.7cm，卵圆形，浓绿色，先端钝圆或急尖，叶基心形或圆形，叶缘锯齿粗钝。花量较多，花序平均着花4朵。花朵大，花径8~9mm。

生物学特性 树势旺，萌芽率中等，成枝力强，枣头生长强旺，节间长。幼树结果较早，坐果率高，枣头、2~3年生枝和4年生枝的吊果率分别为27.5%、97.7%和77.8%。7~8年后进入盛果期，产量高而稳定，在新疆阿克苏地区实验林场盛果期每667m²产可达2 620kg。在山西太谷地区，9月下旬果实进入完熟期，果实生育期100~110d，为晚熟品种类型。果实较抗病和抗裂果。

果实性状 果实较大，圆柱形或近倒卵圆形，纵径3.85cm，横径3.12cm，单果重18.6g，果实大小整齐，果面光滑。果梗中长、中粗，梗洼窄而中深。果顶微凹，柱头遗存不明显。果点小而圆，分布中密，不明显。果皮中厚，红色。果肉厚，近白色，肉质致密细脆，味甜略酸，汁液中多，适宜鲜食、制干和蜜枣加工。鲜枣可食率96.7%，含可溶性固形物33.30%，糖29.32%，酸0.78%，100g果肉维生素C含量324.70mg，鲜食品质中上；果皮含黄酮13.43mg/g，cAMP含量50.58μg/g。制干率47.8%，干枣含总糖66.91%，酸1.93%。干枣果实饱满，富弹性，耐贮运，品质中上。果核较小，纺锤形，核重0.62g，核内无种仁。

评价 该品种适应性较强，耐瘠耐旱，产量较高，坐果稳定。果实品质优良，用途广泛，适宜干制红枣和蜜枣，也可鲜食，适宜北方日照充足、夏季气候温热的地区发展。该品种在新疆和山西栽培表现比原产地优异，为我国北方发展的主要兼用品种之一。另外，近年来河北省选育并审定了综合性状优于赞皇大枣的赞晶、赞玉和赞宝等3个品种，山西省选育出特大果实类型晋赞大枣，正在区试推广。

Zanhuangdazao

Source and Distribution The cultivar originated from Zanhuang County in Hebei Province. It is the dominant variety there with a history of over 400 years and a cell chromosome of $2n=3x=36$.

Botanical Characters The large tree is half-spreading with a medium-strong central leader trunk, sparse and strong branches and a conical-shaped crown. The yellowish-brown 1-year-old shoots are 83.0 cm long. There are 7~10 secondary branches. The large column-shaped mother fruiting spurs can germinate 3~4 deciduous fruiting shoots. The thick wide leaves are oval-shaped and dark green. There are 50 flowers per deciduous fruiting shoot.

Biological Characters The tree has strong vigor, medium germination ability and strong branching ability. Young trees bear with high fruit set. In Taigu County of Shanxi Province, it completely matures in late September. It is a late-ripening variety with strong resistance to diseases and fruit-cracking.

Fruit Characteristics The large fruit is column-shaped or nearly obovate, averaging 18.6 g. It has a smooth surface and medium-thick red skin. The thick flesh is nearly white, tight-textured, crisp, sweet and a little sour, with medium juice and a good quality for fresh-eating, processing dried fruits and candied fruits. The percentage of edible part of fresh fruit is 96.7%, and SSC, TTS, TA and Vc is 33.30%, 29.32%, 0.78% and 324.70 mg per 100 g fresh fruit. The content of flavones and cAMP in mature fruit skin is 13.43 mg/g and 50.58 μg/g. The percentage of fresh fruits which can be made into dried ones is 47.8%. The content of TTS and TA in dried fruit is 66.91% and 1.93%. Dried fruits have plump and whippy flesh, tolerant to storage and transport with a better than normal quality. The small spindle-shaped stone weighs 0.62 g.

Evaluation The cultivar has strong adaptability, strong tolerance to poor soils and drought with a high yield and stable fruit set. The fruit has an excellent quality and wide usage. It can be developed in areas of north China with plentiful sunshine and higher temperature in summer.

兼 用
Multipurpose Varieties
品 种

·285·

骏 枣

品种来源及分布 山西十大名枣之一，原产山西省交城县边山一带，以瓦窑、磁窑、坡底等村栽培较集中，为当地主栽品种。栽培历史1000余年，现尚存百年以上古老枣树林。

植物学性状 树体高大，树姿半开张，干性较强，枝条粗壮，中密，树冠呈自然圆头形。主干皮裂条状。枣头红褐色，平均生长量54.8cm，粗0.95cm，节间长8.5cm，着生永久性二次枝6～7个。二次枝长26.8cm，平均6节。针刺不发达或退化。枣股肥大，抽吊力中等，抽生枣吊3～4个。枣吊平均长15.0cm，着叶10片。叶片大，长卵圆形，浓绿色，先端急尖，叶基圆楔形，叶缘锯齿粗钝。花量中多，花序平均4.5朵。花较大，花径7.6mm，昼开型。

生物学特性 树势强健，萌芽率较高，成枝力强，枣头枝生长粗壮而强旺。易生萌蘖，根蘖苗根系发达，结果较晚，一般第三年开始结果。在新疆阿克苏地区，2年生砧木嫁接苗当年株产可达2.5kg，表现了极强的早期丰产性能。坐果率较高，枣头、2～3年和4年生枝的吊果率可达29.6%、58.5%和44.6%。盛果期较长，但产量不稳定。在山西太谷地区，9月中旬开始成熟，属中熟品种类型。果实成熟期遇雨易裂果且病害严重。

果实性状 果实大，前期果多为柱形，后期果呈长倒卵形，纵径5.07cm，横径3.46cm，单果重26.3g，大小较整齐。果面光滑，果皮薄，深红色。梗洼中广，较深。果顶平，柱头遗存。果肉厚，白色或绿白色，质地细，较松脆，味甜，汁液中多，品质上等，用途广泛，鲜食、制干、加工蜜枣、酒枣均可，是加工酒枣最好的品种之一。鲜枣可食率96.3%，含可溶性固形物33.00%，总糖28.68%，酸0.45%，100g果肉维生素C含量430.20mg；果皮含黄酮1.78mg/g，cAMP含量102.14μg/g。制干率56.8%，干枣含总糖71.77%，酸1.58%。酒枣含可溶性固形物36.30%，总糖30.83%，酸0.83%。果核小，纺锤形，核重0.97g，小果核壁薄而软，有退化现象。含仁率8.3%，种仁不饱满。

评价 该品种适土性强，耐旱涝、盐碱，抗枣疯病。果实品质上等，适宜制作干枣、酒枣和蜜枣。采前易落果，遇雨裂果和病害严重。干枣果肉较松，果皮韧性差，怕挤压，贮运性能较差。

Junzao

Source and Distribution The cultivar originated from remote some mountains of Jiaocheng County in Shanxi Province. It is one of the ten famous jujubes in Shanxi Province with a history of over 1 000 years.

Botanical Characters The large tree is half-spreading with a strong central leader trunk, strong and medium-dense branches and a natural-round crown. The reddish-brown 1-year-old shoots are 54.8 cm long. There are 6～7 permanent secondary branches with less developed or degraded thorns. The large conical-shaped mother fruiting spurs have a long life and can germinate 3～4 deciduous fruiting shoots, which are 15.0 cm long with 10 leaves. The large leaves are long oval-shaped and dark green. The number of flowers is medium large.

Biological Characters The tree has strong vigor, high germination rate, strong branching and suckering ability. The 1-year-old shoots have strong growth vigor. it enters the crisp-maturing stage in mid-September and completely matures in late September. It is a mid-early-ripening variety with serious fruit-cracking.

Fruit Characteristics The large fruit is column-shaped in earlier growth stage and long obovate in later growth stage. It weighs 26.3 g with a regular size. The thick flesh is white or greenish white, loose-textured, crisp and sweet with medium juice and a good quality and wide applications. It can be used for fresh-eating, processing dried fruits, candied fruits and alcoholic fruits. The percentage of edible part of fresh fruit is 96.3%, and SSC, TTS, TA and Vc is 33.00%, 28.68%, 0.45% and 430.20 mg per 100 g fresh fruit. The content of flavones and cAMP in mature fruit skin is 1.78 mg/g and 102.14 μg/g. The percentage of fresh fruits which can be made into dried ones is 56.8%, and the content of TTS and TA in dried fruit is 71.77% and 1.58%. SSC, TTS and TA in alcoholic jujubes is 36.30%, 30.83% and 0.83%.

Evaluation The large plant of Junzao cultivar has strong adaptability, strong tolerance to drought, water-logging and saline-alkaline soil, strong vigor with a long economic life span and a high yield.

兼 用
Multipurpose Varieties
品 种

壶 瓶 枣

品种来源及分布 古老的地方名优品种,山西十大名枣之一。原产和分布于山西省太谷县、清徐县、祁县、榆次区及太原市郊区等地。栽培历史不详,各产区数百年生成片分布老龄结果枣树很多。

植物学性状 树体高大,树姿半开张,干性较强,枝条粗壮,中密,树冠呈自然圆头形。主干皮裂呈块状。枣头红褐色,平均生长量50.0cm,粗0.98cm,节间长8.0cm,二次枝长27.8cm,平均生长7节。针刺不发达或退化。枣股大,抽吊力中等,抽生枣吊2~5个,多为3~4个。枣吊平均长15.7cm,着叶11片。叶片大,叶长6.6cm,叶宽3.2cm,椭圆形,浓绿色,先端急尖,叶基偏圆形,叶缘锯齿中密,较粗钝。花量中多,枣吊平均着花52.1朵,花序平均4.1朵。花较大,花径7.7mm,昼开型。

生物学特性 树势健旺,萌芽率高,成枝力强,枣头枝粗壮且生长强旺。结果较早,萌蘖力较强,根蘖苗根系发达,生长势较强。根蘖苗一般第二年开始结果,15年后进入盛果期,盛果期长,在新疆阿克苏地区表现了极强的早果性和丰产性能。坐果率较高,枣头、2~3年生枝和4年生枝的吊果率分别为22.5%、47.7%和34.1%。丰产,产量较稳定。在山西太谷地区,9月中旬成熟,属中熟品种类型。在原产地,果实成熟期遇雨裂果和病害严重。

果实性状 果实大,倒卵形或圆柱形,纵径4.66cm,横径3.58cm,单果重25.4g,大小较整齐。果梗较短,中粗,梗洼中广、深。果顶平,柱头遗存。果皮薄,紫红色,果面光滑。果点小而密,圆形,浅黄色。果肉厚,绿白色,肉质较松脆,味甜,汁液中多,品质上等,适宜鲜食、制干、加工蜜枣、酒枣。鲜枣可食率96.4%,含可溶性固形物37.80%,总糖30.35%,酸0.57%,100g果肉维生素C含量493.1mg;果皮含黄酮59.84mg/g,cAMP含量167.51μg/g。制干率55.9%,干枣含总糖71.38%,酸1.11%。果核小,纺锤形,核重0.91g,不含种仁,小枣的果核退化成软壁。

评价 该品种树势强,适应性较广,结果较早,产量高而稳定。果实大,品质优良,用途广泛,主要用于干制和加工酒枣。唯进入着色期后,遇雨极易裂果浆烂和病害大发生。另外,新疆南疆地区引入该品种后表现丰产、果个大、抗病和抗裂果,比原产地性状更优异,已成为当地主栽品种之一。

Hupingzao

Source and Distribution The cultivar originated from and spreads in Taigu, Qingxu, Qixian, Yuci and the suburbs of Taiyuan City in Shanxi Province. It is an old famous local variety.

Botanical Characters The large tree is half-spreading with a strong central leader trunk, strong and medium-dense branches and a natural-round crown. The reddish-brown 1-year-old shoots are 50.0 cm long. The secondary branches are 27.8 cm long with less-developed or degraded thorns. The large mother fruiting spurs can germinate 2~5 deciduous fruiting shoots. The large leaves are oval-shaped and dark green. The number of flowers is medium large. The number of flowers is medium large, averaging 52.1 ones per deciduous fruiting shoot and 4.1 ones per inflorescence. The large blossoms have a diameter of 7.7 mm.

Biological Characters The tree has strong vigor, high germination rate, strong branching and suckering ability, bearing early. It generally bears in the 2nd year after planting and enters the long full-bearing stage in the 15th year with high fruit set and a high and stable yield. It is an early-mid-ripening variety. In the original places, fruit-cracking and diseases easily occur.

Fruit Characteristics The large fruit is obovate or column-shaped, averaging 25.4 g with a regular size. It has purplish-red thin skin and a smooth surface. The greenish-white thick flesh is loose-textured, crisp, sweet, with medium juice and a good quality for fresh-eating, processing dried fruits, candied fruits and alcoholic fruits. The content of flavones and cAMP in mature fruit skin is 59.84 mg/g and 167.51 μg/g. The rate of fresh fruits which can be made into dried ones is 55.9%, and the content of TTS and TA in dried fruit is 71.38% and 1.11%. The small stone is spindle-shaped averaging 0.91 g.

Evaluation The cultivar has strong vigor and strong adaptability, bearing early with a high and stable yield. The fruit has an excellent quality with wide applications. Yet fruit-cracking and diseases easily occur.

灰　枣

品种来源及分布　原产和主要分布于河南省新郑市、中牟县和西华县，为当地主栽品种，全国著名的干鲜加工兼用良种。已有2700多年的栽培历史，目前新郑市孟庄镇枣区尚有500多年生的老龄枣树和多处百年以上枣树林。

植物学性状　树体较大，树姿半开张，干性较强，枝系较密，树冠呈自然圆头形或伞形。主干皮裂中深，块状，不易脱落。枣头灰褐色，平均长79.6cm，粗度1.09cm，节间长6.3cm，着生永久性二次枝4~9个。二次枝长40.0cm，平均7节。针刺发达。枣股中大，圆柱形，抽吊力较强，抽生枣吊3~4个。枣吊平均长19.9cm，着叶16片。叶片中大，叶长4.8cm，叶宽2.2cm，椭圆形，浓绿色，较厚，先端钝尖，叶基圆楔形，叶缘锯齿浅钝。花量多，花小，花径7mm左右。

生物学特性　树势中庸偏弱，萌芽率高，成枝力强，枝条细。根蘖苗一般第三年开始结果，酸枣嫁接苗结果较早，3年生结果株率可达100%，15年左右进入盛果期，产量较高。但不同地区产量差异较大，在新疆阿克苏地区表现丰产性极强，成龄树每667m²产可达3 600kg。而在山西中部地区产量较低，经调查，枣头、2~3年生和4年生枝的吊果率仅为36.1%、19.5%和1.6%。在山西太谷地区，9月下旬成熟，为中晚熟品种类型。果实成熟期较抗病和抗裂果。

果实性状　果实较小，倒卵圆形，纵径3.30cm，横径2.23cm，单果重8.3g，大小较整齐。果面较平滑。果梗中粗较长，梗洼窄，中深。果顶平，柱头遗存。果皮中厚，紫红色。果肉厚，绿白色，肉质致密，较脆，味甜，汁液中多，品质上等，适宜制干、鲜食和蜜枣加工。鲜枣可食率94.5%，含可溶性固形物39.00%，总糖34.97%，酸0.49%，100g果肉维生素C含量340.06mg；果皮含黄酮4.06mg/g，cAMP含量106.77μg/g。制干率50%左右，干枣含总糖65.36%，酸1.20%。干枣肉质致密，有弹性，耐贮运，制干品质极佳。果核小，纺锤形，核重0.46g，核尖短，核纹较浅，种仁较饱满，含仁率80.0%。

评价　该品种抗逆性较强，尤其较抗裂果和病害。果实品质优异，用途广泛，鲜食，制干和加工均宜，多以制干为主。目前，已筛选出了大果型的优良变异株系，正在观察区试。

Huizao

Source and Distribution　The cultivar originated from Xinzheng in Henan Province, and mainly spreads in Xinzheng, Zhongmou and Zhengzhou City. It is the dominant variety there with a history of over 2 700 years.

Botanical Characters　The large tree is half-spreading with a strong central leader trunk, dense branches and a natural-round or umbrella-shaped crown. The grayish-brown 1-year-old shoots are 79.6 cm long. There are 4~9 permanent secondary branches with developed thorns. The medium-sized column-shaped mother fruiting spurs can germinate 3~4 deciduous fruiting shoots. The medium-sized thick leaves are oval-shaped and dark green. There are many small flowers with a diameter of 7 mm produced.

Biological Characters　The tree has moderate or weak vigor, high germination rate and strong branching ability with thin branches, blooming and bearing late. The seedlings grafted on wild jujubes enter the full-bearing stage in the 15th year with a high yield. It is a mid-late-ripening variety with strong resistance to diseases and fruit-cracking.

Fruit Characteristics　The small obovate fruit weighs 8.3 g. It has a smooth surface, medium-thick and purplish-red skin. The greenish-white thick flesh is tight-textured, crisp and sweet, with medium juice and a good quality for fresh-eating, processing dried fruits and candied fruits. The percentage of edible part of fresh fruit is 94.5%, and SSC, TTS, TA and Vc is 39.00%, 34.97%, 0.49% and 340.06 mg per 100 g fresh fruit. The rate of fresh fruits which can be made into dried ones is 50%. The small stone is spindle-shaped, averaging 0.46 g.

Evaluation　The cultivar has strong resistance to adverse conditions, especially to fruit-cracking and diseases. The fruit has an excellent quality and is suitable for fresh-eating, dried fruits and processing.

板 枣

品种来源及分布 主要分布于山西稷山县稷峰镇的姚村、陶梁、南阳、下迪等村,为当地主栽品种,是山西十大名枣之一。据《稷山县志》记载,栽培历史始于明代之前,约400年历史,姚村现还有百年以上成片挂果老枣树。

植物学性状 树体较小,树姿半开张,干性较弱,枝条较密,树冠自然圆头形。主干皮裂块状。枣头红褐色,平均长40.0cm,粗度0.93cm,节间长8.5cm,着生永久性二次枝4~5个。二次枝长31.6cm,平均7节。针刺较发达。枣股中大,抽吊力强,一般抽生枣吊4~5个。枣吊一般长16.6cm,着叶11片。叶片小,叶长4.7cm,宽2.3cm,椭圆形,浓绿色,先端急尖,叶基圆形,叶缘锯齿浅钝。花量中多,枣吊平均着花52.5朵,花序平均3.8朵。花小,花径6.1mm,昼开型。

生物学特性 树势较弱,萌芽率高,成枝力强。萌蘖力强,根蘖苗定植第二年开始结果,15年后进入盛果期,盛果期长,丰产,产量较稳定。坐果率高,枣头吊果率76.9%,2~3年生枝为76.1%,4年生枝为40.5%。在山西太谷地区,9月上旬果实着色,9月20日前后进入完熟期,果实生育期100d左右,为中熟品种类型。成熟期落果较严重。

果实性状 果个较小,扁柱形,纵径2.98cm,横径2.69cm,侧径2.30cm,单果重9.7g,大小整齐,果面光滑。果梗细,中长,梗洼中广较深。果顶微凹,柱头遗存。果皮中厚,紫红色。果肉厚,绿白色,肉质致密,较脆,甜味浓,汁液较少,鲜食、制干和加工蜜枣兼用,多以制干为主,且品质优异。鲜枣可食率96.3%,含可溶性固形物41.70%,总糖33.67%,酸0.36%,100g果肉维生素C含量 499.70mg;果皮含黄酮23.27mg/g,cAMP含量82.96μg/g。制干率57.0%,干枣含总糖74.5%,酸2.41%。干枣果实美观,肉厚且饱满,有弹性。果核小,纺锤形,核重0.36g,含仁率3.3%。

评价 该品种树体矮化,结果早,产量高且稳定。果个较小,但外形美观,品质优良,用途广泛,为制干和鲜食兼用的优良品种。该品种对气候适应性强,在山西、山东、河南、河北等地均表现良好。但对土壤肥水条件要求高,适宜北方土质肥沃的地区集约栽培。

Banzao

Source and Distribution The cultivar originated from Jishan County of Shanxi Province, and mainly spreads in Yaocun, Taoliang, Nanyang and Xiadi Villages of Jifeng Town. It is the dominant variety there.

Botanical Characters The small tree is half-spreading with a weak central leader trunk, dense branches and a natural-round crown. The reddish-brown 1-year-old shoots are 40.0 cm long. There are 4~5 permanent secondary branches with developed thorns. The medium-sized mother fruiting spurs can germinate 4~5 deciduous fruiting shoots. The small leaves are oval-shaped and dark green. The number of flowers is medium large, averaging 52.5 ones per deciduous fruiting shoot. The small blossoms have a diameter of 6.1 mm.

Biological Characters The tree has weak vigor, high germination rate, strong branching and suckering ability. The suckers have developed roots, blooming and bearing early. It enters the long full-bearing stage in the 15th year with a stable yield and high fruit set. It is a mid-ripening variety with serious fruit-dropping in maturing stage.

Fruit Characteristics The small fruit is flat column-shaped, averaging 9.7 g with a regular size. It has a smooth surface, medium-thick and purplish-red skin. The greenish-white thick flesh is tight-textured, crisp and strongly sweet, with less juice, suitable for fresh eating, processing dried fruits and candied fruits with an excellent quality. The percentage of edible part of fresh fruit is 96.3%, and SSC, TTS, TA and Vc is 41.70%, 33.67%, 0.36% and 499.70 mg per 100 g fresh fruit. The content of flavones and cAMP in mature fruit skin is 23.27 mg/g and 82.96 μg/g. The percentage of fresh fruits which can be made into dried ones is 57.0%, and the content of TTS and TA in dried fruit is 74.5% and 2.41%. Dried fruits have a beautiful appearance, thick, plump and whippy flesh. The small stone is spindle-shaped, averaging 0.36 g.

Evaluation The dwarfing plant of Banzao cultivar bears early with a high and stable yield. The small fruit has a beautiful appearance and an excellent quality with wide applications. It is a good drying and table variety.

兼 用
Multipurpose Varieties
品 种

晋 枣

品种来源及分布 又名吊枣、长枣、酒枣。分布于陕西和甘肃交界的泾河两岸坡地和塬边地带，主要分布于陕西的彬县、长武和甘肃的宁县、泾川、正宁、灵台等地，占当地枣树栽培面积的2/3左右，是陕甘两省著名的干鲜兼用优良品种。

植物学性状 树体高大，树姿直立，干性强，枝条较密，树冠圆锥形，主干皮裂深，块状，易剥落。枣头红褐色，平均长66.9cm，粗度1.07cm，节间长6.2cm，二次枝长26.7cm，平均生长7节。针刺发达。枣股大，圆柱形，抽吊力强，一般抽生枣吊3～6个。枣吊平均长23.1cm，着叶17片。叶片较大，椭圆形，绿色，先端钝尖，叶基圆楔形，叶缘锯齿浅钝。花量多，每花序着花5～9朵。花较大，花径7.5mm。

生物学特性 树势强旺，萌芽率高，成枝力强。嫁接苗结果较早，3年生结果株率可达100%，早丰性较强，10年左右进入盛果期，产量较高，成龄树一般株产鲜枣25～40kg，最高株产150kg，百年以上老树仍能正常结果，在管理条件较差、肥水条件不足时，表现有大小年结果现象。枣头吊果率37.2%，2～3年生枝为49.3%，4年生枝为2.9%。在山西太谷地区，9月20日左右进入脆熟期，10月初果实完熟，果实生育期120d左右，为晚熟品种类型。花期忌干热风和阴雨低温天气，果实成熟期遇雨易裂果浆烂。

果实性状 果实大，长卵形或圆柱形，纵径4.98cm，横径3.34cm，单果重25.7g，大小不整齐。果梗细，中长，梗洼广而浅。果顶微凹，柱头遗存，不明显。果皮薄，黄红色，果面不平滑。果点小而圆，分布较密，浅黄色。果肉厚，乳白色，肉质致密、酥脆，甜味浓，汁液较多，品质中上，适宜鲜食、制干和蜜枣加工。鲜枣可食率96.2%，含可溶性固形物30.00%，总糖25.80%，酸0.54%，100g果肉维生素C含量294.30mg。制干率45.6%，干枣含总糖64.67%，酸0.96%。果核较大，长纺锤形，纵径2.98cm，横径0.88cm，核重0.98g，核尖长，核纹浅，含仁率18.3%。

评价 该品种树体高大健壮，产量较高，果实大、皮薄、肉厚、核较大、味甜、适宜鲜食、制干和制作蜜枣、酒枣，主要用于鲜食和制干，品质较好。对土壤条件要求较高，花期和果实成熟期不耐阴雨，适宜土壤肥沃、干旱少雨的地区发展。

Jinzao

Source and Distribution The cultivar originated from the two banks of Jinghe River which is at the juncture of Shaanxi and Gansu Province. It mainly spreads in Binxian, Changwu in Shaanxi Province and Ningxian, Jingchuan, Zhengning, Lingtai in Gansu Province, which occupies 2/3 of the total jujube area there.

Botanical Characters The tree large is vertical with a strong central leader trunk, dense branches and a conical-shaped crown. The reddish-brown 1-year-old shoots are 66.9 cm long. The secondary branches are 26.7 cm long with developed thorns. The large column-shaped mother fruiting spurs can germinate 3～6 deciduous fruiting shoots. The large green leaves are oval-shaped. There are many flowers with a diameter of 7.5 mm.

Biological Characters The tree has strong vigor, high germination rate and strong branching ability. The tree enters the full-bearing stage in the 10th year with a high yield. It is a late-ripening variety. Dry-hot wind and rainy days with low temperature in blooming stage can cause serious problem. Fruit-cracking easily occurs.

Fruit Characteristics The large fruit is long oval-shaped or column-shaped, averaging 25.7 g with irregular sizes. It has yellowish-red thin skin and an unsmooth surface. The ivory-white thick flesh is tight-textured, crisp, juicy and strongly sweet, with an excellent quality for fresh-eating, dried fruits and candied fruits. The percentage of edible part of fresh fruit is 96.2%, and SSC, TTS, TA and Vc is 30.00%, 25.80%, 0.54% and 294.30 mg per 100 g fresh fruit. The rate of fresh fruits which can be made into dried ones is 45.6%, and the content of TTS and TA in dried fruit is 64.67% and 0.96%. The large stone is long spindle-shaped, averaging 0.98 g. The percentage of containing kernels is 18.3%.

Evaluation The large plant of Jinzao cultivar has strong vigor with a high yield. The large fruit has thin skin, thick flesh, a small stone and a sweet taste, suitable for fresh-eating, processing dried fruits, candied fruits and alcoholic fruits with a good quality. It can be developed in areas with fertile soils and light rainfall.

兼 用
Multipurpose Varieties
品 种

· 295 ·

中阳团枣

品种来源及分布 分布于山西中阳、柳林等县,栽培数量不多,与吕梁木枣混栽。

植物学性状 树体中大,树姿开张,干性弱,枝系较密,树冠自然半圆形。枣头生长势较强,平均长67.3cm,粗0.86cm,节间长6.5cm,蜡质少。着生永久性二次枝5~9个。二次枝平均长29.6cm,7节,弯曲度中等。针刺不发达。枣股中大,抽吊力中等,多抽生枣吊3个。枣吊长27.2cm,着叶16片。叶片中大,叶长6.8cm,叶宽3.3cm,卵圆形,绿色,先端渐尖,叶基圆形或心形,叶缘具钝锯齿。花量中等,花序平均着花4朵,花较大,花径6.4~7.3mm。

生物学特性 适应性和抗逆性较强,树势和发枝力中等。萌蘖力较强,根蘖生长健旺。多数第三年开始结果,坐果率较低,枣头、2~3年、4~6年和6年以上生枝吊果率分别为4.8%、49.0%、39.0%和5.4%。产量中等且不稳定。在山西太谷地区,9月中旬开始着色,9月下旬进入脆熟期,10月上旬成熟采收,果实生育期110d,为晚熟品种类型。采前落果和裂果严重,需适时采收。

果实性状 果实大,近圆形,纵径3.51cm,横径3.32cm。单果重18.3g,最大24.5g,大小较整齐。果肩斜圆,略耸起。梗洼窄,较深。果顶平圆,顶点微凹,柱头遗存,但不明显。果柄短,较粗。果皮较薄,紫红色。果肉厚,浅绿色,质地疏松,汁液中多,味甜,适宜鲜食和制干,品质中等。鲜枣可食率96.9%,含可溶性固形物28.80%,总糖24.60%,酸0.14%,100g果肉维生素C含量382.16mg。干枣含总糖67.23%,酸0.72%。果核较大,纺锤形,核重0.57g,核纹浅,核面较光滑,核尖短,含仁率66.7%。

评价 该品种适应性较强。果大,肉厚,果核较大,品质中等,可鲜食和制干兼用。但产量不高,且有采前易落果和裂果现象,不宜大面积栽培。

Zhongyangtuanzao

Source and Distribution The cultivar originated from and spreads in Zhongyang and Liulin County of Shanxi Province with a small quantity. It is mainly planted in mixture with Lvliangmuzao.

Botanical Characters The medium-sized tree is spreading with a weak central leader trunk, dense branches and a semi-round crown. The 1-year-old shoots have strong growth potential, 67.3 cm long. The internodes are 6.5 cm long with less wax and less-developed thorns. There are 5~9 permanent secondary branches, which are 29.6 cm long with 7 nodes of medium curvature. The mother fruiting spurs can germinate 3 deciduous fruiting shoots which are 27.2 cm long with 16 leaves. The medium-sized leaves are oval-shaped and green. The number of flowers is medium large, averaging 4 ones per inflorescence. The large blossoms have a diameter of 6.4~7.3 mm.

Biological Characters The tree has strong adaptability and strong resistance to adverse conditions with moderate vigor, medium branching ability, strong suckering ability, a medium and unstable yield. The suckers grow well and bear fruit in the 3rd year with low fruit set. In Taigu County of Shanxi Province, it begins coloring in mid-September, enters the crisp-maturing stage in late September and matures in early October with a fruit growth period of 110 d. It is a late-ripening variety with serious premature fruit-dropping and serious fruit-cracking. So harvest should be in time.

Fruit Characteristics The large oblong fruit has a vertical and cross diameter of 3.51 cm and 3.32 cm, averaging 18.3 g (maximum 24.5 g) with a regular size. It has a deflective-round shoulder (a little protuberant), a narrow and deep stalk cavity, a flat-round fruit apex, a remnant yet indistinct stigma, a short and thick stalk and light-red thin skin. The light-green flesh is thick, loose-textured and sweet, with medium juice. It has medium quality for fresh-eating and dried fruits. The percentage of edible part of fresh fruit is 96.9%, and SSC, TTS, TA and Vc is 28.80%, 24.60%, 0.14% and 382.16 mg per 100 g fresh fruit. The content of TTS and TA in dried fruit is 67.23% and 0.72%. The large spindle-shaped stone weighs 0.57 g.

Evaluation The cultivar has strong adaptability. The large fruit has thick and sweet flesh with medium quality for fresh-eating and dried fruits. Yet it has a low yield, serious premature fruit-dropping and fruit-cracking, which makes it not suitable for large-scale cultivation.

兼 用
Multipurpose Varieties
品 种

临汾团枣

品种来源及分布 分布于山西省临汾市尧都区的南永安、北永安、东张村、西张村、东孔郭、西孔郭等地，为当地主栽品种。栽培历史不详。

植物学性状 树体高大，树姿半开张，干性较强，枝条较密，树冠圆锥形。树干皮裂较浅、条状，老翘皮翘起的特征明显，易脱落。枣头红褐色，平均长60.0cm，粗0.88cm，节间长8.0cm。二次枝长31.0cm，平均6节。针刺发达。皮目中大，分布较密，圆形，凸起，开裂，灰白色。枣股中大，抽吊力较强，抽生枣吊2～5个，多为3～4个。枣吊平均长22.8cm，着叶13片。叶片中大，卵圆形，浓绿色，叶长6.9cm，宽3.4cm，先端钝尖，叶基偏斜，叶缘锯齿浅钝。花量较多，枣吊平均着花61朵，花序平均5.7朵。花较小，花径6.8mm。

生物学特性 树势强健，萌芽率较高，成枝力较强。结果较早，一般第二、三年开始结果。丰产，盛果期长，但产量不稳定。坐果率中等，枣头吊果率31.5%，2～3年生枝为38.8%，4年生枝为37.6%，主要坐果部位在枣吊的4～10节，占坐果总数的80%。在山西太谷地区，9月上旬果实着色，9月底脆熟，果实生育期115d左右，为晚熟品种类型。成熟期遇雨易裂果。

果实性状 果实大，椭圆形，纵径3.77cm，横径3.24cm，单果重18.3g，大小较整齐。果梗中长，较细，梗洼广而浅。果顶平，柱头遗存。果皮薄，浅红色，果面光滑。果肉厚，白色，肉质细脆，味甜，汁液较多，品质上等，鲜食、制干、加工蜜枣、酒枣兼用，以鲜食品质量优良，鲜枣耐贮藏。鲜枣可食率96.7%，含可溶性固形物28.80%，总糖25.31%，酸0.24%，100g果肉维生素C含量505.50mg；果皮含黄酮13.01mg/g，cAMP含量134.19μg/g。制干率46.5%，干枣含总糖65.20%。核较小，纺锤形，纵径2.11cm，横径0.88cm，核重0.61g，核尖短，核面粗糙。含仁率66.7%，种仁较饱满，多为单仁，偶有双仁。

评价 该品种耐旱，果实大而整齐，外形美观，肉质细脆多汁，味甜，维生素C含量较高，品质优良。适宜鲜食，也可制干。惟采前落果重，需适时采收。

Linfentuanzao

Source and Distribution The cultivar originated from Nanyongan, Beiyongan, Dongzhangcun, Xizhangcun, Dongkongguo and Xikongguo in Yaodu District of Linfen City in Shanxi Province. It is the dominant variety there.

Botanical Characters The large tree is half-spreading with a strong central leader trunk, dense branches and a conical-shaped crown. The reddish-brown 1-year-old shoots are 60.0 cm long and 0.88 cm thick with the internodes of 8.0 cm. The secondary branches are 31.0 cm long with 6 nodes and developed thorns. The medium-sized mother fruiting spurs can germinate 2～5 deciduous fruiting shoots which are 22.8 cm long with 13 leaves. The medium-sized leaves are oval-shaped and dark green. There are many small flowers with a diameter of 6.8 mm produced, averaging 61 ones per deciduous fruiting shoot and 5.7 ones per inflorescence.

Biological Characters The tree has strong vigor, high germination rate and strong branching ability. It bears early, generally in the 2nd or 3rd year after planting. It has a long full-bearing stage with a high yet unstable yield and medium fruit set. It is a late-ripening variety with serious fruit-cracking if it rains in the maturing stage.

Fruit Characteristics The large oval-shaped fruit weighs 18.3 g with a regular size. It has light-red thin skin and a smooth surface. The white thick flesh is delicate, crisp, sweet and juicy, with an excellent quality for fresh-eating, dried fruits, processing candied fruits and alcoholic fruits. The best quality is for fresh eating, and fresh fruits have a good storage character. The percentage of edible part of fresh fruit is 96.7%, and SSC, TTS, TA and Vc is 28.80%, 25.31%, 0.24% and 505.50 mg per 100 g fresh fruit. The rate of fresh fruits which can be made into dried ones is 46.5%. The small stone is spindle-shaped, averaging 0.61 g.

Evaluation The cultivar has strong tolerance to drought and weak tolerance to frost. The fruit has a large and regular size, an attractive appearance, crisp, juicy and sweet flesh with high content of Vc and an excellent quality. Yet it has serious premature fruit-dropping, so harvest should be in time.

兼 用
Multipurpose Varieties
品 种

榆 次 团 枣

品种来源及分布 分布于山西省晋中市榆次区东赵乡东赵、训峪村一带，为当地主栽品种之一。是地方古老品种，现仍有许多300多年生老树。

植物学性状 树体中大，树姿半开张，干性较弱，枝条稀疏，粗壮，树冠呈自然圆头形。树干皮裂块状。枣头红褐色，平均长89.3cm，粗1.01cm，节间长9.1cm，着生永久性二次枝5～6个。二次枝长31.9cm，平均6节。针刺不发达。枣股肥大，一般长4cm左右，粗1.2cm。抽吊力中等，抽生枣吊2～5个，平均3.3个。枣吊平均长26.9cm，着叶18片。叶片中大，叶长7.0cm，宽3.2cm，椭圆形，浓绿色，先端急尖，叶基圆形，叶缘锯齿钝。花量少，枣吊平均着花44.3朵，花序平均2.8朵。花较大，花径7.3mm，昼开型。

生物学特性 树势较强，萌发力弱，结果较早，第二年开始结果，一般15年左右进入盛果期。坐果率中等，枣头吊果率48.2%，2～3年生枝为73.9%，4年生枝为32.2%，主要坐果部位在枣吊的5～10节，占坐果总数的78.2%。在山西太谷地区，9月中旬果实着色，10月上旬果实成熟，果实生育期115d左右，为晚熟品种类型。

果实性状 果实大，圆柱形，纵径3.76cm，横径3.42cm，单果重18.2g，大小不整齐。果梗短而粗，梗洼广、中深。果顶微凹，柱头遗存。果皮中厚，赭红色，果面光滑。果肉厚，绿白色，肉质致密，味甜，汁液多，品质中上，可制干和鲜食。鲜枣可食率96.5%，含可溶性固形物27.00%，总糖21.86%，酸0.54%，100g果肉维生素C含量435.56mg，含水量70.1%；果皮含黄酮16.81mg/g，cAMP含量108.28μg/g。制干率45.6%，干枣含总糖62.02%，酸0.84%。核小，纺锤形，纵径2.22cm，横径0.85cm，核重0.63g，核尖中等长，核纹浅，核面粗糙，含仁率85.0%，种仁不饱满。

评价 该品种风土适应性强，较抗枣疯病。结果较早，产量较高，采前落果少，裂果较轻。果实大而均匀，果肉厚，可食率高，但制干率较低，品质中上，可鲜食和制干。

Yucituanzao

Source and Distribution The cultivar originated from Xunyucun in Dongzhaoxiang Village of Yuci District in the middle part of Shanxi Province. It is an old dominant variety there.

Botanical Characters The medium-sized tree is half-spreading with a weak central leader trunk, sparse and strong branches and a natural-round crown. The reddish-brown 1-year-old shoots are 89.3 cm long. There are 5～6 secondary branches which are 31.9 cm long with 6 nodes and less-developed thorns. The large mother fruiting spurs are 4 cm long and 1.2 cm thick which can germinate 2～5 deciduous fruiting shoots, averaging 26.9 cm long with 18 leaves. The medium-sized leaves are oval-shaped and dark green. The number of flowers is small, averaging 44.3 ones per deciduous fruiting shoot. The daytime-bloomed large blossoms have a diameter of 7.3 mm.

Biological Characters The tree has strong vigor and weak germination ability. It bears early, generally enters the full-bearing stage in the 15th year after planting with medium fruit set. In Taigu County of Shanxi Province, it begins coloring in mid-September and matures in early October. It is a late-ripening variety.

Fruit Characteristics The large column-shaped fruit weighs 18.2 g with irregular sizes. It has medium-thick and reddish-brown skin and a smooth surface. The greenish-white thick flesh is tight-textured, sweet and juicy, with a better than normal quality for dried fruits and fresh eating. The percentage of edible part of fresh fruit is 96.5%, and SSC, TTS, TA and Vc is 27.00%, 21.86%, 0.54% and 435.56 mg per 100 g fresh fruit. The water content is 70.1%. The rate of fresh fruits which can be made into dried ones is 45.55%, and the content of TTS and TA is 62.02% and 0.84%. The small stone is spindle-shaped, averaging 0.63 g.

Evaluation The cultivar has strong adaptability and strong resistance to jujube witches' broom, bearing early with a high yield, light premature fruit-dropping and fruit-cracking. The large fruit has a regular size and thick flesh with high edibility. Yet it has a low rate of fresh fruits which can be made into dried ones with a better than normal quality. It is suitable for fresh eating and dried fruits.

兼 用
Multipurpose Varieties
品 种

安阳团枣

品种来源及分布 原产和分布于河南省安阳市等地。

植物学性状 树体较大，树姿直立，树冠呈扇形。主干皮裂条状。枣头红褐色，年生长量76.1cm，节间长7.5cm，二次枝长43.7cm，6～8节。无针刺。枣股平均抽生枣吊4.0个。枣吊长24.1cm，着叶12片。叶片卵圆形，叶长6.5cm，叶宽3.0cm，深绿色，先端钝尖，叶基圆楔形，叶缘具钝齿。花量大，花序平均9朵。花中大，花径6.2mm。

生物学特性 树势强。坐果率较低，丰产性一般。在山西太谷地区，9月下旬果实着色，10月上旬成熟采收，果实生育期115d左右，为极晚熟品种类型。

果实性状 果实较小，圆形或卵圆形，纵径2.90cm，横径2.50cm，果重9.1g左右，大小较整齐。梗洼窄而较深。果顶凹，柱头遗存。果皮薄，浅红色，果面有隆起。肉质酥脆，味甜，汁液中多，品质上等，适宜鲜食或制干。鲜枣可食率94.4%，含可溶性固形物31.20%，总糖27.97%，酸0.48%，糖酸比58.27∶1，100g果肉维生素C含量387.95mg；果皮含黄酮7.07mg/g，cAMP含量170.48μg/g。干枣含总糖72.74%，酸1.44%。果核椭圆形，种仁较饱满，含仁率90.0%。

评价 该品种树体较大，树势强，丰产性一般，品质上等，可鲜食或制干。

Anyangtuanzao

Source and Distribution The cultivar originated from and spreads in Anyang of Henan Province.

Botanical Characters The large tree is vertical with a fan-shaped crown. The trunk bark has striped fissures. The reddish-brown 1-year-old shoots are 76.1 cm long, with the internodes of 7.5 cm. The secondary branches are 43.7 cm long with 6～8 nodes and no thorns. The mother fruiting spurs can germinate 4 deciduous fruiting shoots, which are 24.1 cm long with 12 leaves. The oval-shaped leaves are dark green, 6.5 cm long and 3.0 cm wide, with a bluntly-cuspate apex, a round-cuneiform base and a blunt saw-tooth pattern on the margin. There are many medium-sized flowers with a diameter of 6.2 mm produced, averaging 9 ones per inflorescence.

Biological Characters The tree has strong vigor with low fruit set and a medium yield. In Taigu County of Shanxi Province, it begins coloring in late September and matures in early October with a fruit growth period of 115 d. It is a late-ripening variety.

Fruit Characteristics The small oblong fruit has a vertical and cross diameter of 2.90 cm and 2.50 cm averaging 9.1 g with a regular size. It has a narrow and deep stalk cavity, a sunken fruit apex, a remnant stigma and light-red thin skin with some protuberances. The flesh is crisp, sweet and juicy. It has a superior quality for fresh-eating and dried fruits. The percentage of edible part of fresh fruit is 94.4%, and SSC, TTS, TA and Vc is 31.20%, 27.97%, 0.48% and 387.95 mg per 100 g fresh fruit. The SAR is 58.27∶1. The content of flavones and cAMP in mature fruit skin is 7.07 mg/g and 170.48 μg/g. The oval-shaped stone contains a well-developed kernel. The percentage of containing kernels is 90.0%.

Evaluation The large plant of Anyangtuanzao cultivar has strong tree vigor with a medium yield. It has a superior quality and is suitable for fresh-eating and making dried fruits.

兼 用
Multipurpose Varieties
品 种

延川狗头枣

品种来源及分布 分布于黄河中游的陕西延川县张家河乡（现延水关镇）张家河村一带，近年发展面积较大。

植物学性状 树体高大，树姿开张，树冠乱头形。树干皮裂条状，浅，易剥落。枣头紫褐色，平均长71.6cm，粗1.01cm，节间长7.8cm。二次枝长26.5cm，平均5节。针刺不发达。皮目小而圆，凸起，不开裂。枣股较粗大，圆锥形或圆柱形，抽吊力较强，一般抽生3～5个。枣吊平均长23.9cm，着叶15片。叶片中大，叶长5.4cm，叶宽2.1cm，卵圆形或卵状披针形，绿色，先端急尖，叶基偏斜，叶缘锯齿较粗。花量少，花小，花径6mm左右。

生物学特性 树势强，结果较早，栽植后一般第二年开始结果，成龄树一般株产鲜枣50kg左右，产量高而稳定，坐果率中等，枣头吊果率6.7%，2～3年生枝为113.5%，4年生枝为30.6%。在山西太谷地区，9月下旬果实成熟，果实生育期110d左右，为晚熟品种。果实成熟期遇雨易裂果。

果实性状 果实较大，卵圆形或长平顶锥形，似狗头状而得名，单果重18.1g，大小不整齐。果肩宽，果顶较窄，梗洼窄深，果面平滑。果皮中度厚，深红色。果点小而圆，分布密，较显著。果肉较厚，绿白色，肉质致密，细而脆，味甜，汁液较多，品质上等，制干和鲜食品质均佳。鲜枣可食率94.5%，含可溶性固形物25.00%，酸0.42%，100g果肉维生素C含量419.35mg；果皮含黄酮12.65mg/g，cAMP含量76.33μg/g。干枣含总糖67.90%，酸1.64%。核较大，纺锤形，纵径3.20cm，横径0.84cm，核重1.00g，核纹粗深，核尖细长，核面粗糙，含仁率6.7%。

评价 该品种适应性较差，要求土壤肥沃的条件。树体高大，树势强，结果早，丰产稳产。果实大，品质上等，鲜食和制干均可，但易裂果。

Yanchuangoutouzao

Source and Distribution The cultivar originated from and spreads in Zhuangtoucun of Zhangjiahe Village in Yanchuan County of Shaanxi Province, which is along the middle reaches of Yellow River.

Botanical Characters The large tree is spreading with an irregular crown. The purplish-brown 1-year-old shoots are 71.6 cm long and 1.01 cm thick with the internodes of 7.8 cm. The secondary branches are 26.5 cm long with 5 nodes and less-developed thorns. The small lenticels are round, protuberant and not cracked. The thick and large mother bearing spurs are conical or column-shaped, which can germinate 3～5 deciduous beraring shoots, averaging 23.9 cm long with 15 leaves. The medium-sized green leaves are oval-shaped or ovate-lanceolate. There are only a few flowers with a diameter of 6 mm produced.

Biological Characters The tree has strong vigor, bearing early (generally in the 2nd year after planting) with medium fruit set. The percentage of deciduous fruiting fruits to shoots of 1-year-old shoots, 2～3-year-old branches and 4-year-old ones is 6.7%, 113.5% and 30.6%. It has a high and stable yield. A mature tree has a yield of 50 kg. In Taigu County of Shanxi Province it matures in late September with a fruit growth period of 110 d. It is a late-ripening variety with serious fruit-cracking if it rains in the maturing stage.

Fruit Characteristics The large fruit is oval-shaped or long-conical-shaped with a flat apex, which looks like the head of a dog. That is why it is called 'Goutouzao' (Goutou means the head of a dog in Chinese). It weighs 18.1 g with irregular sizes. It has a wide shoulder, a narrow fruit apex, a narrow and deep stalk cavity, a smooth surface and medium-thick and dark-red skin with small, round, dense and distinct dots. The greenish-white thick flesh is tight-textured, delicate, crisp, sweet and juicy, with an excellent quality for dried fruits and fresh eating. The percentage of edible part of fresh fruit is 94.5%, and SSC, TA and Vc is 25.00%, 0.42% and 419.35 mg per 100 g fresh fruit. The content of flavones and cAMP in mature fruit skin is 12.65 mg/g and 76.33 μg/g. The small stone is spindle-shaped averaging 1.00 g.

Evaluation The cultivar has poor adaptability, so it needs fertile soil. The tree has a large size and strong vigor, bearing early with a high and stable yield. The large fruit has an excellent quality for fresh eating and dried fruits. Yet fruit-cracking easily occurs.

兼 用
Multipurpose Varieties
品种

敦 煌 大 枣

品种来源及分布 主要分布于甘肃敦煌，为当地的原产品种。敦煌以东的金塔、安西、酒泉等地也有少量栽种。100多年前引入新疆哈密市五堡乡，当地叫哈密大枣、五堡大枣，现已大面积发展，为主要产区。

植物学性状 树体较大，树姿半开张，枝叶稠密，树冠呈自然圆头形。主干浅灰褐色，皮裂纹深，呈宽条状，不易剥落。枣头黄褐色，平均长68.1cm，节间长6.2cm，蜡质多。二次枝发育健壮，平均长38.1cm，6节，弯曲度中等。针刺发达，2、3年后脱落。枣股圆锥形，抽生枣吊3～4个，多者5个。枣吊长18.8cm，着叶14片。叶片较小，椭圆形，绿色，先端渐尖，叶基截形，叶缘具整齐钝锯齿。花量少，每花序着花3朵，花朵较大，花径7.0mm，昼开型。

生物学特性 适应性强，抗寒、耐旱，抗病虫。树势和发枝力较强，树体寿命长。结果早，定植后2～3年开始结果，10年后进入盛果期，产量高而稳定，最高株产65kg以上。在山西太谷地区，4月中旬开始萌芽，5月下旬始花，9月中旬果实着色成熟，果实生育期100d左右，为中熟品种类型。成熟期落果较重。

果实性状 果实较大，圆形，纵径3.04cm，横径2.70cm，单果重15.6g，大小较整齐。果肩宽平，有数条宽深的沟棱。梗洼中广、深。果顶凹陷。果面不平整，有小块隆起。果皮较厚，紫红色。果肉浅绿色，肉质疏松，汁液中多，味甜酸，鲜食和制干品质中上。鲜枣含总糖20.00%，酸0.64%，100g果肉维生素C含量317.63mg。制干率47.0%以上；果皮含黄酮5.38mg/g，cAMP含量307.07μg/g。干枣含总糖67.90%，酸1.64%。果核较大，短纺锤形，核重0.52g。核内无种子。

评价 该品种适应性强，抗寒、耐旱，抗病虫。树体较高大，发枝力强，树体寿命长，结果早，丰产性好，惟成熟期易落果。果皮浓红美观，果肉厚，味甜酸，可鲜食、制干和加工蜜枣、酒枣等。

Dunhuangdazao

Source and Distribution The cultivar originated from and spreads in Dunhuang of Gansu Province. There are also some distributions in east Dunhuang. About 100 years ago, the cultivar was introduced into Wubao Village of Hami City in Xinjiang Province, which becomes a main producing area from then on.

Botanical Characters The large tree is half-spreading with dense branches and a natural-round crown. The yellowish-brown 1-year-old shoots are 68.1 cm long. The internodes are 6.2 cm long with much wax and developed thorns. The secondary branches grow well, averaging 38.1 cm long with 6 nodes of medium curvature. The conical mother fruiting spurs can germinate 3～4 deciduous fruiting shoots, which are 18.8 cm long with 14 leaves. The small green leaves are oval-shaped. The number of flowers is small, averaging 3 ones per inflorescence. The daytime-bloomed large blossoms have a diameter of 7.0 mm.

Biological Characters The tree has strong adaptability, strong tolerance to cold and drought, strong resistance to diseases and pests, strong vigor, strong branching ability and a long life. It bears early, generally in the 2nd or 3rd year after planting, and enters the full-bearing stage in the 10th year with a high and stale yield. The largest yield per tree is over 65 kg. It is a mid-ripening variety with serious fruit-dropping in maturing stage.

Fruit Characteristics The large round fruit weighs 15.6 g with a regular size. It has an unsmooth surface with small massive protuberances, and purplish-red thick skin. The light-green flesh is loose-textured, sour and sweet with medium juice. It has a better than normal quality for fresh-eating and dried fruits. The content of TTS, TA and Vc in fresh fruit is 20.00%, 0.64% and 317.63 mg per 100 g fresh fruit. The percentage of fresh fruits which can be made into dried ones is over 47.0%. The large stone is short spindle-shaped, averaging 0.52 g.

Evaluation The large plant of Dunhuangdazao cultivar has strong adaptability, strong tolerance to cold and drought, strong resistance to diseases and pests, and strong branching ability. It bears early with high productivity and has serious fruit-dropping in the maturing stage. It can be used for fresh-eating, making dried fruits, candied fruits and alcoholic fruits.

定襄星星枣

品种来源及分布 别名星星枣。分布于山西定襄的北社西一带，栽培数量不多，起源历史不详。

植物学性状 树体较大，树姿半开张，干性较强，枝系粗壮，树冠呈圆锥形。主干皮裂块状。枣头生长势强，平均长75.4cm，节间长10cm左右。针刺不发达。枣股大，抽吊力弱，平均抽生枣吊3.0个。枣吊长17.1cm，着叶12片。叶片大，叶长5.5cm，叶宽3.2cm，卵圆形，绿色，先端渐尖，先端尖圆，叶基心形或圆形，叶缘具浅而稀疏的钝齿。花量中等，花序平均着花4朵，花较大，花径7.5～8.0mm。

生物学特性 适应性强，要求栽培条件不严。树势强健，发枝力弱，枝系稀疏。结果较迟，一般第三年开始结果，15年进入盛果期。坐果率低，枣头、2～3年、4～6年和6年以上生枝吊果率分别为16.8%、45.5%、49.8%和17.3%。产量较低，且落果严重。在山西太谷地区，9月中旬果实成熟采收，果实生育期100d，为早中熟品种类型。成熟期遇雨裂果严重。

果实性状 果实中大，卵圆形，纵径3.52cm，横径2.86cm，单果重13.5g。果肩平圆，梗洼广、中深，果顶微凹。果面光滑，果皮紫红色，较薄。果肉中厚，浅绿色，质地疏松，汁液中多，味酸甜，适宜制干，也可鲜食，品质中等。鲜枣可食率94.1%，含可溶性固形物34.50%，含总糖27.20%，酸0.59%，100g果肉维生素C含量336.09mg，果皮含黄酮4.22mg/g，cAMP含量195.44μg/g。制干率54.0%，干枣含可溶性固形物78.10%，含总糖68.67%，酸1.40%。核大，纺锤形，核重0.80g，核尖长，核面粗糙，含仁率53.3%。

评价 该品种适土性强。结果较晚，产量较低。果实中大，果皮薄，肉质疏松，品质中等，可鲜食和制干。采前裂果和落果严重，不宜大面积生产栽培。

Dingxiangxingxingzao

Source and Distribution The cultivar originated from and spreads in Beishexi of Dingxiang County in Shanxi Province, with an unknown cultivation history.

Botanical Characters The large tree is half-spreading with a strong central leader trunk, thick and strong branches, and a conical crown. The 1-year-old shoots have strong growth potential, averaging 75.4 cm long. The internodes are 10 cm long with less-developed thorns. The large mother fruiting spurs can germinate 3 deciduous fruiting shoots which are 17.1 cm long with 12 leaves. The large leaves oval-shaped and green. The number of flowers is medium large, averaging 4 ones per inflorescence. The large blossoms have a diameter of 7.5～8.0 mm.

Biological Characters The tree has strong adaptability, strong vigor, weak branching ability and sparse branches, with a low demand on cultural conditions. It bears late, generally in the 3rd year after planting, and enters the full-bearing stage in the 15th year, with low fruit set. The percentage of fruits deciduous fruiting shoots to of 1-year-old shoots, 2～3-year-old branches, 4～6-year-old ones and over-6-year-old ones is 16.8%, 45.5%, 49.8% and 17.3%. It has a low yield and serious fruit-dropping. In Taigu County of Shanxi Province, it ripens in mid-September with a fruit growth period of 100 d. It is an early-mid-ripening variety with light fruit-cracking even if it rains in the maturing stage.

Fruit Characteristics The medium-sized fruit is oval-shaped with a vertical and cross diameter of 3.52 cm and 2.86 cm, averaging 13.5 g. It has a flat-round shoulder, a wide and medium-deep stalk cavity, a slightly-sunken fruit apex, a smooth surface and purplish-red thin skin. The medium-thick flesh is light-green, loose-textured, sour and sweet, with medium juice. It has medium quality for fresh-eating and dried fruits. The percentage of edible part of fresh fruit is 94.1%, and SSC, TTS, TA and Vc is 34.50%, 27.20%, 0.59% and 336.09 mg per 100 g fresh fruit. The content of flavones and cAMP in mature fruit skin is 4.22 mg/g and 195.44 μg/g. The percentage of fresh fruits which can be made into dried ones is 54.0%, and SSC, TTS and TA in dried fruit is 78.10%, 68.67% and 1.40%. The large spindle-shaped stone weighs 0.8 g, with a long apex and a rough surface.

Evaluation The cultivar has strong adaptability, bearing late with a low yield. The medium-sized fruit has thin skin, loose-textured and normal juicy flesh with medium quality, suitable for fresh-eating and dried fruits. Premature fruit-cracking and fruit-dropping are serious, so it should not be developed on a large scale.

兼 用
Multipurpose Varieties
品 种

洪赵小枣

品种来源及分布 分布于山西洪洞县赵城、稽村、许村一带，为当地原有的主栽品种。

植物学性状 树体中大，树姿半开张，干性强，枝条细而较稀，树冠乱头形。树干皮裂条状、中深，不易脱落。枣头红褐色，平均生长量67.0cm，粗1.02cm，节间长8.0cm，着生永久性二次枝4～5个。二次枝长23.5cm，平均7节。针刺不发达。皮目大而稀，椭圆形，凸起，开裂，灰白色。枣股中大，抽吊力强，平均抽生枣吊4.1个。枣吊平均长20.1cm，着叶13片。叶片较大，叶长7.6cm，叶宽3.1cm，椭圆形，绿色，先端钝尖，叶基圆楔形或偏斜形。花量较少，枣吊平均着花47.5朵，花序平均3.2朵。花较小，花径6.7mm，昼开型。

生物学特性 树势中等，萌发力弱。幼树结果早，一般第二年开始结果，15年后进入盛果期，盛果期长，丰产，产量稳定。坐果率中等，枣头吊果率23.2%，2～3年生枝为52.0%，4年生枝为30.0%，主要坐果部位在枣吊的7～12节，占坐果总数的65.6%。在山西太谷地区，9月底果实成熟，果实生育期110d左右，为晚熟品种类型。采前落果较严重。

果实性状 果实较小，圆柱形，纵径2.79cm，横径2.53cm，单果重9.4g，大小整齐。果梗中长，较细，梗洼中度广，较深。果顶平，柱头遗存。果皮较薄，深红色，果面光滑。果点较大，绿白色，肉质细而酥脆，味甜，汁液较多，品质上等，可鲜食和制干。鲜枣可食率96.0%，含可溶性固形物29.40%，总糖21.90%，酸0.48%，100g果肉维生素C含量501.20mg，含水量66.5%；果皮含黄酮19.09mg/g，cAMP含量125.46μg/g。制干率47.2%，干枣含总糖70.10%，酸1.84%。核较小，纺锤形，纵径1.65cm，横径0.56cm，核重0.38g，核尖短，核纹浅，含仁率10.0%，种仁饱满，多为单仁，偶有双仁。

评价 该品种适应性强，耐旱涝。结果早，丰产性好，产量稳定，品质上等，适宜鲜食和制干。但采前易落果。

Hongzhaoxiaozao

Source and Distribution The cultivar originated from Jicun and Xucun Village in Hongtong County of Shanxi Province. It is the native dominant veriety there with a small quantity.

Botanical Characters The medium-sized tree is half-spreading with a strong central leader trunk, thin and sparse branches and an irregular crown. The reddish-brown 1-year-old shoots are 67.0 cm long and 1.02 cm thick with the internodes of 8.0 cm. There are 4～5 permanent secondary branches which are 23.5 cm long with 7 nodes and less-developed thorns. The large grayish-white lenticels are sparsely distributed, oval-shaped, protuberant and cracked. The medium-sized mother fruiting spurs can germinate 4.1 deciduous fruiting shoots which are 20.1 cm long with 13 leaves. The large green leaves are oval-shaped. The number of flowers is small, averaging 47.5 ones per deciduous fruiting shoot. The daytime-bloomed small blossoms have a diameter of 6.7 mm.

Biological Characters The tree has moderate vigor and weak germination ability. Young trees bear early, generally in the 2nd year after planting and enter the long full-bearing stage in the 15th year with a high and stable yield and medium fruit set. It is a late-ripening variety with serious premature fruit-dropping.

Fruit Characteristics The small oblong fruit has a vertical and cross diameter of 2.79 cm and 2.53 cm, averaging 9.4 g with a regular size. It has dark-red thin skin and a smooth surface. The greenish-white flesh is delicate, crisp, sweet and juicy, with an excellent quality for dried fruits and fresh-eating. The percentage of edible part of fresh fruit is 96.0%, and SSC, TTS, TA and Vc is 29.40%, 21.90%, 0.48% and 501.20 mg per 100 g fresh fruit. The water content is 66.5%. The rate of fresh fruits which can be made into dried ones is 47.2%, and the content of TTS and TA in dried fruit is 70.10% and 1.84%. The small stone is spindle-shaped, averaging 0.38 g. The percentage of containing kernels is 10.0%. Most stones contain single kernel, sometimes double kernels.

Evaluation The cultivar has strong adaptability and strong tolerance to drought and water-logging, bearing early with a high and stable yield. The fruit has a good quality and is suitable for fresh-eating and dried fruits. Premature fruit-dropping easily occurs.

彬县圆枣

品种来源及分布 别名圆疙瘩枣、疙瘩枣、冬枣、冬疙瘩。分布于陕西、甘肃交界地带，以彬县、长武最多，镇原、灵台、庆阳、宁县、径川、崇信、平凉等地也有零星栽培。为当地原产的主栽品种之一。栽培历史悠久。

植物学性状 树体高大，树姿半开张，树冠多呈自然圆头形。主干灰褐色，皮裂纹深，纵宽条形，易剥落。枣头红褐色，平均长94.3cm，节间长6.8cm。二次枝生长健壮，平均长34.2cm，7节左右，弯曲度中等。针刺发达。枣股圆柱形，平均抽生枣吊4.0个。枣吊长20.4cm，着叶15片。叶片较小，叶长4.2cm，叶宽2.2cm，椭圆形，绿色，先端钝尖，叶基圆楔形，叶缘有粗钝整齐的锯齿。花序平均着花4朵。花较大，花径7.0mm，昼开型。

生物学特性 适应性较强，对肥水要求较高。树势强健，发枝力较强。产量高而稳定，枣头、2~3年和4~6年生枝吊果率分别为30.6%、26.1%和3.6%。一般定植后第三年开始结果，10年以后进入盛果期。在山西太谷地区，9月下旬果实进入完熟期采收，果实生育期113d左右，为晚熟品种类型。

果实性状 果实大，扁柱形或倒卵圆形，纵径3.83cm，横径3.48cm，单果重21.1g，大小整齐。果肩圆或平圆，梗洼深窄。果柄中粗，长5.1mm。果顶平或微凹，柱头遗存。果面粗糙，果皮薄，浅红色。果肉浅绿色，质地较致密，汁液多，甜味较浓，适宜鲜食和制干，品质上等。鲜枣可食率95.6%，含可溶性固形物27.60%，总糖22.98%，酸0.48%，100g果肉维生素C含量414.03mg，果皮含黄酮9.10mg/g，cAMP含量248.39μg/g。干枣含总糖66.62%，酸1.29%。果核较大，纺锤形，核重0.93g。含仁率11.7%。

评价 该品种适应性较强，对肥水要求高。树体强健，发枝力较强，结果早，产量高而稳定。果实大，成熟期晚，肉质脆，风味良好，品质上等，适宜鲜食和制干，可适量发展。

Binxianyuanzao

Source and Distribution The cultivar originated from and spreads in junctures of Shaanxi and Gansu Province with a long cultivation history. The main producing areas are Binxian County and Changwu County. There are also some distributions in Zhenyuan, Lingtai, Qingyang, Ningxian, Jingchuan, Chongxin and Pingliang.

Botanical Characters The large tree is half-spreading with a natural-round crown. The reddish-brown 1-year-old shoots are 94.3 cm long. The internodes are 6.8 cm long, with developed thorns. The secondary branches are 34.2 cm long with 7 nodes of medium curvature. The column-shaped mother fruiting spurs can germinate 4 deciduous fruiting shoots which are 20.4 cm long with 15 leaves. The small green leaves are oval-shaped. Each inflorescence has 4 flowers on average. The large blossoms have a diameter of 7.0 mm.

Biological Characters The tree has strong adaptability, a high demand on irrigation and fertilization, strong vigor and strong branching ability, with a high and stable yield. It generally bears in the 3rd year after planting and enters the full-bearing stage in the 10th year. In Taigu County of Shanxi Province, it matures in late September with a fruit growth period of 103 d. It is a late-ripening variety.

Fruit Characteristics The large fruit is flat column-shaped with a vertical and cross diameter of 3.83 cm and 3.48 cm, averaging 21.1 g with a regular size. It has a round or flat-round shoulder, a deep and narrow stalk cavity, a medium-thick stalk of 5.1 mm long, a flat or slightly-sunken fruit apex, a remnant stigma, a rough surface and light-red thin skin. The light-green flesh is tight-textured, strongly sweet and juicy. It has a good quality for fresh-eating and dried fruits. The percentage of edible part of fresh fruit is 95.6%, and SSC, TTS, TA and Vc is 27.60%, 22.98%, 0.48% and 414.03 mg per 100 g fresh fruit. The content of TTS and TA in dried fruit is 66.62% and 1.29%. The large spindle-shaped stone weighs 0.93 g, and the percentage of containing kernels is 11.7%.

Evaluation The cultivar has strong adaptability with strong vigor and strong branching ability. It bears early with a high and stable yield. The large fruit has crisp and delicious flesh with a good quality for fresh-eating and dried fruits. It can be developed on a proper scale.

兼 用
Multipurpose Varieties
品 种

鸣山大枣

品种来源及分布 分布于甘肃敦煌,为敦煌大枣优良的变异株系。1979年发现,1983年正式命名。

植物学性状 树体较大,树姿开张,枝条中密,树冠呈自然半圆形。树干皮裂较深、条状,易剥落。枣头红褐色,平均生长量65.9cm,粗0.90cm,节间长6.4cm,二次枝长26.7cm,平均6节,针刺较发达。皮目中大,分布稀,椭圆形,凸起,开裂。枣股小,圆锥形,抽吊力较强,一般抽生枣吊3~4个,多者达5个。枣吊长12~15cm,节间长1.6cm。叶片中大,叶长5.3cm,叶宽3.4cm,卵圆形,绿色,先端渐尖,叶基圆形或宽楔形,叶缘锯齿细。花较大,花径7mm左右,夜开型。

生物学特性 树势较强,生长势中等,萌芽率和成枝力较强。结果较早,一般第二、三年开始结果,盛果期树产量高而稳定。当年生枣头吊果率6.7%,2~3年生枝为48.9%,4年生枝为69.0%。在山西太谷地区,9月上旬果实进入脆熟期,9月中旬完熟,果实生育期90d左右,为早中熟品种类型。果实采前落果严重。

果实性状 果实大,圆柱形,纵径4.70cm,横径3.50cm,单果重23.9g,最大42.5g,大小不整齐。果梗长,较细,梗洼窄而深。果顶凹,柱头遗存。果皮厚,深红色,果面不平滑。果点大,近圆形,分布较密。果肉厚,肉质致密细脆,味甜,汁液多,适宜鲜食和制干,品质上等。鲜枣可食率96.2%,含可溶性固形物37.50%,总糖31.40%,酸0.54%,100g果肉维生素C含量396.20mg。制干率51.8%。核较大,纺锤形,核重0.9g,核纹深,核尖长,不含种仁。

评价 该品种适应性强,抗寒、耐旱。结果早,丰产。果实大,色泽鲜艳,果肉厚,汁液多,肉细脆,味甜,品质上等,皮较厚,耐贮运,可鲜食、制干、加工蜜枣和酒枣。是适宜无霜期短、干旱地区发展的大果型兼用良种。但采前落果严重。

Mingshandazao

Source and Distribution The cultivar originated from Dunhuang in Gansu Province and was discovered in 1979 and named in 1983. It is a fine variant strain from Dunhuangdazao.

Botanical Characters The large tree is spreading with medium-dense branches and a semi-round crown. The reddish-brown 1-year-old shoots are 65.9 cm long and 0.90 cm thick with the internodes of 6.4 cm long. The secondary branches are 26.7 cm long with 6 nodes and developed thorns. The medium-sized lenticels are sparsely-distributed, oval-shaped, protuberant and cracked. The small conical-shaped mother fruiting spurs can germinate 3~4 deciduous fruiting shoots which are 12~15 cm with the internodes of 1.6 cm long. The medium-sized green leaves are oval-shaped. The night-bloomed large blossoms have a diameter of 7.4 mm.

Biological Characters The tree has strong growth vigor, high germination rate and strong branching ability. It bears early, generally in the 2nd or 3rd year after planting with a high and stable yield in the full-beating stage. In Taigu County of Shanxi Province, it enters the crisp-maturing stage in early September and completely matures in mid-September. It is an early-mid-ripening variety with serious premature fruit-dropping.

Fruit Characteristics The large column-shaped fruit has a vertical and cross diameter of 4.70 cm and 3.50 cm, averaging 23.9 g (maximum 42.5 g) with irregular sizes. It has a thin and long stalk, a narrow and deep stalk cavity, a sunken fruit apex, a remnant stigma, dark-red thick skin and an unsmooth surface with large, nearly round and dense dots. The thick flesh is tight-textured, crisp, sweet and juicy, with a good quality for fresh-eating and dried fruits. The percentage of edible part of fresh fruit is 96.2%, and SSC, TTS, TA and Vc is 37.50%, 31.40%, 0.54% and 396.20 mg per 100 g fresh fruit. The rate of fresh fruits which can be made into dried ones is 51.8%. The large spindle-shaped stone weighs 0.9 g.

Evaluation The cultivar has strong adaptability, strong tolerance to cold and drought, bearing and maturing early with a high yield. The large fruit has a bright color, thick, juicy crisp and sweet flesh, with a good quality. It has thick skin, tolerant to storage and transport. It can be used for fresh-eating, processing dried fruits, candied fruits and alcoholic fruits. It is a table and processing variety which can be developed in areas with short frost-free stage and dry climate. Yet it has serious premature fruit-dropping.

兼用
Multipurpose Varieties
品种

· 315 ·

赞 新 大 枣

品种来源及分布 新疆阿克苏地区农科所在20世纪70年代从赞皇大枣品种中筛选的丰产、果个大的优良株系，1985年命名，在当地有一定推广面积。

植物学性状 树体中大，树姿半开张，干性较强，枝条较稀，粗壮，树冠呈自然半圆形。树皮条状开裂，易脱落。枣头红褐色，平均生长量76.7cm，粗1.21cm，节间长8.8cm。二次枝长35.0cm，平均生长7节，无针刺。皮目小，卵圆形。枣股抽吊力较弱，一般抽生枣吊2~4个。枣吊粗壮，一般长15cm左右。叶片大而厚，叶长5.3cm，叶宽3.2cm，卵圆形，浓绿色，先端钝尖，叶基心形，叶缘锯齿粗大。花量多，枣吊着花60~70朵，花序着花6朵左右。花大，花径8mm左右。

生物学特性 树势强旺，枣头萌发力中等，生长势强，成枝力中等。幼树结果早，一般嫁接苗第二年结果株率可达95%以上，1~2年生枝结果能力极强，吊果率高达100%，极丰产，5年生树株产鲜枣10kg左右。在山西太谷地区，9月中旬果实进入脆熟期，10月上旬果实完熟，果实生育期110d左右，比赞皇大枣晚熟5d左右，为晚熟品种类型。

果实性状 果实大，倒卵形，纵径4.20cm，横径3.70cm，单果重24.4g，最大35.2g，大小不整齐。果梗粗，较长，梗洼浅而广。果顶圆，柱头遗存。果皮较薄，深红色，果面平滑。果点小而密，圆形，浅黄色。果肉厚，绿白色，肉质致密，细脆，味甜微酸，汁液中多，可制干和鲜食，品质中上。鲜枣可食率96.9%，含总糖27.40%，酸0.43%，100g果肉维生素C含量425.68mg；果皮含黄酮28.47mg/g，cAMP含量80.49μg/g。制干率48.8%，干枣含总糖72.5%。核较大，纺锤形，核重0.76g，核尖较短，核面粗糙，不含种仁。

评价 该品种适应性强，管理简便，结果早，产量高而稳定，果实大，品质中上，为优良的制干和鲜食兼用品种。

Zanxindazao

Source and Distribution It is a fine strain with high productivity and large fruit size selected from Zanhuangdazao in 1970s by Agricultural Science Research Institute of Aksu District in Xinjiang Uygur Autonomous Pregion.

Botanical Characters The medium-sized tree is half-spreading with a strong central leader trunk, strong and sparse branches and a semi-round crown. The reddish-brown 1-year-old shoots are 76.7 cm long and 1.21 cm thick, with the internodes of 8.8 cm. The secondary branches are 35.0 cm long with 7 nodes and without thorns. The small lenticels are oval-shaped. The mother fruiting spurs can germinate 2~4 strong deciduous fruiting shoots which are 15 cm long. The large thick leaves are oval-shaped and dark green. There are many large flowers with a diameter of 8 mm produced, averaging 60~70 ones per deciduous fruiting shoot and 6 ones per inflorescence.

Biological Characters The tree has strong vigor, medium germination rate and medium branching ability. Young trees begin fruiting early. The rate of fruiting grafted seedlings in the 2nd year reaches 95% or more. It is highly productive with a yield of 10.2 kg for a 5-year-old tree. In Taigu County of Shanxi Province, it enters the crisp-maturing stage in mid-September and completely matures in early October, with a fruit growth period of 110 d, which is 5 days later than Zanhaungdazao. It is a late-ripening variety.

Fruit Characteristics The large obovate fruit has a vertical and cross diameter of 4.20 cm and 3.70 cm, averaging 24.4 g (maximum 35.2 g) with irregular sizes. It has dark-red thin skin and a smooth surface with small, round, dense and light-yellow dots. The greenish-white thick flesh is tight-textured, crisp, sweet and a little sour, with medium juice and a better than normal quality for fresh-eating and dried fruits. The percentage of edible part of fresh fruit is 96.9%, and the content of TTS, TA and Vc is 27.40%, 0.43% and 425.68 mg per 100 g fresh fruit. The rate of fresh fruits which can be made into dried ones is 48.8%, and the content of TTS in dried fruit is 72.5%. The small spindle-shaped stone weighs 0.76 g, with a short apex and a rough surface. There are no kernels in the stones.

Evaluation The cultivar has strong adaptability, bearing early with a high and stable yield and easy management. The large fruit has a better than normal quality. It is a good table and drying variety.

兼 用
Multipurpose Varieties
品 种

· 317 ·

核 桃 纹

品种来源及分布 别名团枣、圆红、大纹枣。栽培历史悠久，广泛分布在山东省西南部和河南省北部地区，为山东省巨野、菏泽、成武、济宁一带的主栽品种，数量占当地枣树的50%以上，河南的濮阳、范县、汉阳、内黄等地也有栽培。

植物学性状 树体较大，树姿开张，枝叶密度大，树冠呈圆柱形或自然圆头形。主干深灰褐色，树皮粗糙，小块裂，不易剥落。枣头黄褐色，较细软，平均长81.4cm，粗0.89cm，节间长8.3cm。二次枝较细，弯曲度大，呈弓背形，平均长24.4cm，5节。针刺发达。枣股圆柱形，平均抽生枣吊4.0个。枣吊长27.8cm，着叶13片。叶片椭圆形，深绿色，略有光泽，先端较短，圆或尖圆，叶基圆形。叶缘具粗浅的钝锯齿。花量多，花序平均着花8朵，昼开型。

生物学特性 对气候、土壤适应性强，在较贫瘠的粉沙质土或砾质沙壤都能生长良好，发枝力强，枣股寿命较短。花期能适应较低的空气湿度。易坐果，成龄结果树一般株产35～40kg。在山西太谷地区，4月中旬萌芽，5月下旬始花，9月下旬果实成熟采收，果实生育期115d，为晚熟品种类型。成熟期遇雨不易裂果，但生理落果较重。

果实性状 果实大，圆形或近球形，纵径3.06cm，横径2.92cm，单果重20.8g，大小不匀。果肩平。梗洼深狭。果顶广圆，顶洼中深、广，柱头遗存。果柄短，较细，长4.9mm。果面光滑平整。果皮呈紫红色。果点较大，肩部较稀，胴部密度中等，较明显。果肉浅绿色，肉质致密，较细稍脆，汁液中多，略具酸味，可鲜食、制干和加工蜜枣，品质中上。鲜枣可食率95.6%，含可溶性固形物32.40%，总糖30.12%，酸0.43%，100g果肉维生素C含量439.59mg；果皮含黄酮6.38mg/g，cAMP含量88.56μg/g。干制红枣肉质较松软，果皮皱褶，外观不艳丽。制干率55.0%左右，干枣含总糖72.91%，酸0.83%。果核大，纺锤形，核重0.92g，大果有发育不饱满的种仁，中小果一般不含种仁。

评价 该品种适应性强，较耐瘠薄，树体强健，较丰产。果个大，抗裂果能力强，鲜食、制干、加工均可，外观品质中等。

Hetaowen

Source and Distribution The cultivar spreads in southwest Shandong Province and north Henan Province, with a long cultivation history. It is the dominant variety in Juye, Heze, Chengwu and Jining in Shandong Province.

Botanical Characters The large tree is spreading with dense branches and a column-shaped or round crown. The soft yellowish-brown 1-year-old shoots are 81.4 cm long with developed thorns. The thin secondary branches are 24.4 cm long. The column-shaped mother fruiting spurs can germinate 4 deciduous fruiting shoots. The oval-shaped leaves are dark green and glossy. There are many flowers produced, averaging 8 ones per inflorescence.

Biological Characters The tree has strong adaptability, strong branching ability and a short life for mother fruiting spurs with strong tolerance to low humidity in blooming stage. It grows well in poor silty loam soil or gravelly sandy loam soil. It is a late-ripening variety with light fruit-cracking. Yet physiological fruit-dropping is serious.

Fruit Characteristics The large ball-shaped fruit weighs 20.8 g with irregular sizes. It has a smooth surface and purplish-red skin. The fruit dots are large and distinct, sparse on the shoulder and medium-dense on the bottom. The light-green flesh is tight-textured, delicate and crisp, a little sour with medium juice. It has a better than normal quality for fresh-eating, dried fruits and candied fruits. The percentage of edible part of fresh fruit is 95.6%, and SSC, TTS, TA and Vc is 32.40%, 30.12%, 0.43% and 439.59 mg per 100 g fresh fruit. The content of flavones and cAMP in mature fruit skin is 6.38 mg/g and 88.56 μg/g. The percentage of fresh fruits which can be made into dried ones is 55.0%. Dried fruits have soft flesh, crinkled skin and an unattractive appearance. The large spindle-shaped stone weighs 0.92 g. Large fruit contains a shriveled kernel, while small fruit contains no kernel.

Evaluation The cultivar has strong adaptability and strong tolerance to poor soils with strong vigor and a high yield. The large fruit has strong resistance to fruit-cracking and is suitable for fresh-eating, making dried fruits and processing. It has a medium quality for both exterior and internal parts.

兼 用
Multipurpose Varieties
品 种

临汾针葫芦

品种来源及分布 别名针葫芦。分布于山西临汾市尧都区西孔郭一带，栽培数量不多。

植物学性状 树体较大，树姿较直立，干性强，枝条中密，树冠自然伞形。主干皮裂条状。枣头红褐色，生长势较弱，平均长21.0cm，粗0.89cm，节间长9.5cm，蜡质少。二次枝平均长28.9cm，5节，弯曲度大。针刺细小，不发达。枣股中大，抽吊力中等，平均抽生枣吊4.0个。枣吊长20.5cm，着叶11片。叶片较大，叶长6.7cm，叶宽3.5cm，卵圆形，绿色，先端渐尖。叶基圆形。花量多，花序平均着花7朵，花较小，花径5.3～6.3mm，为昼开型。

生物学特性 风土适应性中等，耐旱力一般，抗霜力较弱。树势中等，发枝力较弱。结果较早，2～3年开始结果，15年后进入盛果期。枣头、2～3年、4～6年和6年以上生枝吊果率分别为15.5%、45.5%、7.50%和5.7%，产量中等。在山西太谷地区，9月上旬开始着色，9月下旬成熟采收，果实生育期107d左右，为晚熟品种类型。着色后落果较重，较易裂果。

果实性状 果实中大，长倒卵圆形，纵径3.85cm，横径2.63cm。单果重11.8g，最大16.0g，大小较整齐。果肩小，略耸起，梗洼窄、中深。果顶平圆，柱头遗存。果面光滑，果皮红色，中厚。果肉厚，浅绿色，质地致密，汁液中多，味甜，适宜鲜食和制干，品质中等。鲜枣可食率94.3%，含可溶性固形物34.20%，含总糖27.81%，酸0.46%，100g维生素C含量423.79mg，果皮含黄酮8.15mg/g，cAMP含量225.17μg/g。制干率51.0%，干枣含可溶性固形物75.90%，总糖61.80%，酸0.55%。果核中大，纺锤形，核重0.67g，核纹浅，核面较平滑，含仁率68.3%。

评价 该品种适应性中等。树体中大，树势中等。结果较早，产量中等。果实中大，适宜鲜食和制干，品质中上。裂果和采前落果严重。

Linfenzhenhulu

Source and Distribution The cultivar originated from and spreads in Xikongguo of Yaodu District in Linfen City of Shanxi Province, with a small quantity.

Botanical Characters The large tree is vertical with a strong central leader trunk, medium-dense branches and an umbrella-shaped crown. The reddish-brown 1-year-old shoots have weak growth potential, 21.0 cm long and 0.89 cm thick. The internodes are 9.5 cm long with less wax. The secondary branches are 28.9 cm long with 5 nodes of large curvature and with small less-developed thorns. The medium-sized mother fruiting spurs can germinate 4 deciduous fruiting shoots which are 20.5 cm long with 11 leaves. The large leaves are oval-shaped and green. There are many small flowers with a diameter of 5.3～6.3 mm produced, averaging 7 ones per inflorescence.

Biological Characters The tree has medium adaptability, medium tolerance to drought and weak tolerance to frost, with moderate vigor and weak branching ability. It bears early, generally in the 2nd or 3rd year after planting, and enters the full-bearing stage in the 15th year with a medium yield. In Taigu County of Shanxi Province, it begins coloring in early September and matures in late September, with a fruit growth period of 107 d. It is a late-ripening variety with serious fruit-dropping after coloring and serious fruit-cracking.

Fruit Characteristics The medium-sized fruit is long obovate, with a vertical and cross diameter of 3.85 cm and 2.63 cm, averaging 11.8 g with a regular size. It has a smooth surface and medium-thick red skin. The light-green flesh is thick, tight-textured and sweet, with medium juice. It has medium quality for fresh-eating and dried fruits. The percentage of edible part of fresh fruit is 94.3%, and SSC, TTS, TA and Vc is 34.20%, 27.81%, 0.46% and 423.79 mg per 100 g fresh fruit. The content of flavones and cAMP in mature fruit skin is 8.15 mg/g and 225.17 μg/g. The percentage of fresh fruits which can be made into dried ones is 51.0%, and SSC, TTS and TA in dried fruit is 75.90%, 61.80% and 0.55%. The medium-sized stone is spindle-shaped, averaging 0.67 g with shallow veins and a smooth surface. The percentage of containing kernels is 68.3%.

Evaluation The large plant of Linfenzhenhulu cultivar has medium adaptability and moderate vigor, bearing early with a medium yield. The large fruit has a better than normal quality for fresh-eating and dried fruits, with serious fruit-cracking and premature fruit-dropping.

兼 用
Multipurpose Varieties
品 种

太谷墩墩枣

品种来源及分布 分布于山西太谷的北汪乡和平遥县小胡、梁赵等地,栽培数量不多。

植物学性状 树体高大,树姿直立,干性中强,枝系密度中等,树冠呈圆锥形。主干皮裂条状。枣头红褐色,平均长66.6cm,粗0.89cm,节间长9.8cm,蜡质少。二次枝多数3节。针刺不发达。枣股中大,平均抽生枣吊4.0个,枣吊长25.6cm,着叶14片。叶片大,叶长7.5cm,叶宽3.2cm,卵圆形,绿色,先端渐尖,叶基圆形,叶缘具钝锯齿。花量较少,花序平均着花4朵,花大,花径6.8~8.0mm,昼开型。

生物学特性 适应性较强,对栽培条件要求不严,耐旱。树势强,发枝力中等,二次枝数量较少。结果较迟,一般第三年开始结果,10年后进入盛果期。坐果率较高,枣头、2~3年、4~6年和6年以上生枝吊果率分别为79.0%、62.8%、25.7%和21.8%。丰产,但不稳定。在山西太谷地区,8月下旬开始着色,9月下旬成熟采收,果实生育期110d左右,为晚熟品种类型。

果实性状 果实大,扁柱形,纵径3.75cm,横径3.57cm。单果重20.5g。果肩平圆,梗洼浅而广。果顶圆,顶点微凹。果皮较薄,红色。果肉厚,浅绿色,质地疏松,汁液中多,味甜略苦,可鲜食和制干,品质中等。鲜枣可食率97.4%,含可溶性固形物34.80%,总糖28.78%,酸0.63%,100g果肉维生素C含量438.51mg;果皮含黄酮5.66mg/g,cAMP含量130.94μg/g。干枣含总糖68.02%,酸1.26%。醉枣含可溶性固形物28.50%,总糖25.50%,酸0.84%。果核较小,纺锤形,核重0.54g,含仁率8.3%。

评价 该品种树体高大,树势强健,风土适性较强。结果较晚,进入结果期后产量较高。果实大,果肉厚,可食率高,味甜,略具苦味,品质中等,可鲜食和制干兼用。干枣果肉较松,不耐挤压,贮运性较差。

Taigudundunzao

Source and Distribution The cultivar originated from and spreads in Beiwang Village of Taigu County, and Xiaohu, Liangzhao Villages of Pingyao County in Shanxi Province, with a small quantity.

Botanical Characters The large tree is vertical with a medium-strong central leader trunk, medium-dense branches and a conical crown. The reddish-brown 1-year-old shoots are 66.6 cm long and 0.89 cm thick. The internodes are 9.8 cm long with less wax and less-developed thorns. There are 3 nodes on each secondary branch. The medium-large mother fruiting spurs can germinate 4 deciduous fruiting shoots which are 25.6 cm long with 14 leaves. The large leaves are oval-shaped and green. The number of flowers is small, averaging 4 ones per inflorescence. The daytime-bloomed blossoms have a diameter of 6.8~8.0 mm.

Biological Characters The tree has strong adaptability, strong vigor, medium branching ability and fewer secondary branches, with a low demand on cultural conditions and strong tolerance to drought. It bears late, generally in the 3rd year after planting, and enters the full-bearing stage in the 10th year with high fruit set and a high yet unstable yield. In Taigu County of Shanxi Province, its fruit begins coloring in late August and matures in late September with a fruit growth period of 110 d. It is a late-ripening variety.

Fruit Characteristics The large fruit is flat column-shaped, averaging 20.5 g. It has thin red skin. The light-green flesh is thick, loose-textured, sweet and a little bitter with medium juice. It has medium quality for fresh-eating and dried fruits. The percentage of edible part of fresh fruit is 97.4%, and SSC, TTS, TA and Vc is 34.80%, 28.78%, 0.63% and 438.51 mg per 100 g fresh fruit. The content of flavones and cAMP in mature fruit skin is 5.66 mg/g and 130.94 μg/g, and the content of TTS and TA in dried fruit is 68.02% and 1.26%. The content of SSC, TTS and TA in alcoholic jujubes is 28.50%, 25.50% and 0.84%. The small spindle-shaped stone weighs 0.54 g with the percentage of containing kernels of 8.3%.

Evaluation The large plant of Taigudundunzao cultivar has strong vigor and strong adaptability bearing late with a high yield in the bearing stage. The large fruit has thick, sweet and a little bitter flesh, with high edibility. It has medium quality and is suitable for fresh-eating and making dried fruits. Yet dried fruits are soft and is susceptible to extrusion and intolerant to storage and transport.

兼用
Multipurpose Varieties
品种

· 323 ·

彬县耙齿枣

品种来源及分布 别名耙齿枣、马牙枣。分布于陕西彬县、长武、大荔等地,为当地原产品种,栽培历史悠久,产地现有百年以上的结果大树,但数量较少。

植物学性状 树体高大,树姿开张,树冠呈偏斜形。主干灰褐色,皮裂深,呈条块状,不易剥落。枣头红褐色,被覆蜡质明显,平均长84.9cm,粗1.14cm,节间长7.7cm。二次枝长33.0cm,7节左右,弯曲度小。针刺发达。枣股圆柱形,平均抽生枣吊4.0个。枣吊长21.4cm,着叶15片。叶片中大,叶长5.4cm,叶宽2.5cm,椭圆形,较厚,绿色,较光亮,先端较钝,叶基圆形,叶缘具浅锐或浅钝锯齿。花量较多,每花序着花6朵,花较大,花径7.3mm。

生物学特性 适应性强,耐旱、耐瘠薄。树势强健,发枝力较强,产量高而稳定,枣头、2~3年和4~6年生枝吊果率分别为60.2%、43.1%和3.3%,一般盛果期树株产50kg左右。在山西太谷地区,9月下旬果实成熟采收,果实生育期110d左右,为晚熟品种类型。落果和裂果较轻。

果实性状 果实中大,卵圆形,纵径3.43cm,横径2.74cm,单果重11.4g,大小较整齐。果肩平圆,梗洼浅广。果顶凹陷,顶洼浅广或深广不等,常有较深的棱角而得名。果面平整,果皮厚,紫红色。果肉白色,质细较致密,脆甜,略带酸味,汁液中多,可鲜食和制干,品质中等。鲜枣可食率95.4%,含可溶性固形物28.20%,总糖21.60%,酸0.57%,100g果肉维生素C含量457.60mg,果皮含黄酮3.68mg/g,cAMP含量225.37μg/g。干枣含总糖72.50%,酸1.20%。果核大,纺锤形,核重0.53g,含仁率38.3%。

评价 该品种耐旱、耐瘠、抗裂果。树体高大强健,寿命长,产量高而稳定。果实中大,宜鲜食和制干,可在土壤贫瘠的河滩和山地栽种,并可作为培育抗裂果良种的种质材料。

Binxianpachizao

Source and Distribution The cultivar, also called Mayazao, originated from and spreads in Binxian, Changwu and Dali in Shaanxi Province, with a long cultivation history. There are still some old fruiting trees of over 100 years old in the original place. It is planted in scattered regions with a small quantity.

Botanical Characters The large tree is spreading with a deflective crown. The reddish-brown 1-year-old shoots are 84.9 cm long and 1.14 cm thick with distinct wax. The internodes are 7.7 cm long with developed thorns. The secondary branches are 33.0 cm long with 7 nodes of small curvature. The column-shaped mother fruiting spurs can germinate 4 deciduous fruiting shoots which are 21.4 cm long with 15 leaves. The medium-sized leaves are oval-shaped, thick, dark green and glossy. There are many large flowers with a diameter of 7.3 mm produced, averaging 6 ones per inflorescence.

Biological Characters The tree has strong adaptability, tolerant to drought and poor soils, with strong vigor, strong branching ability, a high and stable yield. A mature tree has a yield of about 50 kg on average. In Taigu County of Shanxi Province, it matures in late September with a fruit growth period of 110 d. It is a late-ripening variety with light fruit-dropping and fruit-cracking.

Fruit Characteristics The medium-sized fruit is oval-shaped, with a vertical and cross diameter of 3.43 cm and 2.74 cm, averaging 11.4 g with a regular size. It has a flat-round shoulder, a shallow and wide stalk cavity, a sunken fruit apex, a smooth surface and purplish-red thick skin. The white flesh is tight-textured, crisp and sweet with a little sour taste and medium juice. It has medium quality for fresh-eating and dried fruits. The percentage of edible part of fresh fruit is 95.4%, and SSC, TTS, TA and Vc is 28.20%, 21.60%, 0.57% and 457.60 mg per 100 g fresh fruit. The content of flavones and cAMP in mature fruit skin is 3.68 mg/g and 225.37 μg/g, and the content of TTS and TA in dried fruit is 72.50% and 1.20%. The large spindle-shaped stone weighs 0.53 g, and the percentage of containing kernels is 38.3%.

Evaluation The large plant of Binxianpachizao cultivar has strong tolerance to drought and poor soils, strong resistance to fruit-cracking, strong vigor, a long life, a high and stable yield. The medium-sized fruit is suitable for fresh-eating and making dried fruits. It can be planted in flood bed and mountainous areas with poor soils. It can also be considered as a breeding germplasm for varieties with strong resistance to fruit-cracking.

兼 用
Multipurpose Varieties
品种

·325·

玉田小枣

品种来源及分布 别名玉田金丝小枣，系地方名贵品种。起源历史较早，已栽培500年以上。分布于河北玉田和天津蓟县的丘陵地区。玉田孤村一带是集中产区，数量占当地枣树的90%以上。

植物学性状 树体中大，树姿半开张，枝叶较密，树冠圆头形。主干灰褐色，树皮粗糙，皮裂块状，易剥落。枣头黄褐色，平均长75.7cm，节间长7.6cm。二次枝中度弯曲，枝长32.7cm，6节左右。针刺易脱落。枣股圆柱形，平均抽生枣吊3.0个。枣吊长22.5cm，着叶12片。叶片卵圆形，叶长7.3cm，叶宽3.7cm，绿色，较光亮，先端渐尖，端部钝尖。叶基心形，叶缘钝锯齿。花量多，花序平均着花11朵，花较大，花径7.0mm，昼开型。

生物学特性 树势中等，发枝力强，成枝力达93.3%，结果较晚，坐果率中等，枣头、2~3年和4~6年生枝的吊果率分别为57.8%、79.2%和23.7%，花期环剥后产量较高。在山西太谷地区，9月下旬果实开始着色，10月上旬完熟采收，果实生育期115d左右，为晚熟品种类型。较抗裂果。

果实性状 果实较小，圆柱形，纵径2.89cm，横径2.36cm，单果重8.3g，最大11.0g，较整齐。果肩平斜，梗洼深、狭窄，果顶微凹，柱头遗存。果柄较长，平均8.4mm。果皮中厚，浅红色，富光泽。果肉浅绿色，质地疏松，汁液较多，味甘甜，品质极好，适宜鲜食和制干。鲜枣可食率95.9%，含可溶性固形物29.00%，总糖23.98%，酸0.78%，100g果肉维生素C含量459.27mg，果皮含黄酮4.32mg/g，cAMP含量131.19 μg/g。干枣含总糖62.28%，酸0.84%。果核较小，椭圆形，纵径1.50cm，横径0.71cm，核重0.34g，核纹浅，细短，含仁率10.0%。

评价 该品种适应性强，在丘陵瘠薄地亦能良好生长，丰产且产量稳定。果肉厚，干制红枣富有弹性，耐贮运，品质极上。

Yutianxiaozao

Source and Distribution The cultivar is a precious variety with a cultivation history of over 500 years. It spreads in Yutian of Hebei Province and in hilly areas of Jixian in Tianjin. Gucun Village in Yutian is the main producing area of this cultivar, the area of which occupies over 90% of the total jujube area there.

Botanical Characters The medium-sized tree is half-spreading with dense branches and a round crown. The yellowish-brown 1-year-old shoots are 75.7 cm long. The internodes are 7.6 cm long with easily falling-off thorns. The secondary branches are 32.7 cm long with 6 nodes. The column-shaped mother fruiting spurs can germinate 3 deciduous fruiting shoots which are 22.5 cm long with 12 leaves. The oval-shaped leaves are green and glossy. There are many large flowers with a diameter of 7.0 mm produced, averaging 11 ones per inflorescence. It blooms in the daytime.

Biological Characters The tree has moderate vigor and strong branching ability with a branching rate of 93.3%. It bears late with medium fruit set. The yield can be higher after girdling in blooming stage. In Taigu County of Shanxi Province, it begins coloring in late September and matures in early October with a fruit growth period of 115 d. It is a late-ripening variety with strong resistance to fruit-cracking.

Fruit Characteristics The small column-shaped fruit has a vertical and cross diameter of 2.89 cm and 2.36 cm, averaging 8.3 g (maximum 11.0 g) with a regular size. It has a flat-deflective shoulder, a deep and narrow stalk cavity, a slightly-sunken fruit apex, a remnant stigma, a stalk of 8.4 mm long and medium-thick and light-red glossy skin. The light-green flesh is loose-textured, sweet and juicy. It has a very good quality for fresh-eating and dried fruits. The percentage of edible part of fresh fruit is 95.9%, and SSC, TTS, TA and Vc is 29.00%, 23.98%, 0.78% and 459.27 mg per 100 g fresh fruit. The content of flavones and cAMP in mature fruit skin is 4.32 mg/g and 131.19 μg/g, and the content of TTS and TA in dried fruit is 62.28% and 0.84%. The small oval-shaped stone has a vertical and cross diameter of 1.50 cm and 0.71 cm, averaging 0.34 g with shallow, thin and short veins.

Evaluation The cultivar has strong adaptability. It grows well even in poor soils of hilly areas with a high and stable yield. The fruit has thick flesh. Dried fruits have whippy flesh, tolerant to storage and transport with a very good quality.

交 城 端 枣

品种来源及分布　别名端枣。分布于山西交城的边山一带，以奈林、覃村、夏家营等地栽培集中，数量较多，为当地主栽品种之一，数量占当地枣树总数的40%左右。

植物学性状　树体较小，树姿开张，干性弱，枝系较密，树冠呈自然乱头形。主干皮裂条状。枣头红褐色，生长势强，平均长86.0cm，粗1.03cm，节间长8.5cm，蜡质少。二次枝平均长39.4cm，8节，弯曲度中等。枣股中大，抽吊力较差，平均抽生枣吊3.0个。枣吊长20.2cm，着叶14片。叶片中大，叶长6.3cm，叶宽3.3cm，卵圆形，浅绿色，先端渐尖，叶基圆楔形。花量较多，花较大，每序平均着花6朵。

生物学特性　风土适应性较强，树势中等，发枝力较弱。坐果率高，丰产，结果稳定，枣头、2~3年、4~6年和6年以上生枝吊果率分别为48.6%、145.2%、74.6%和22.2%。在山西太谷地区，4月下旬萌芽，5月下旬始花，8月15日开始着色，9月中旬进入脆熟期，果实生育期100d，为中熟品种类型。采前落果严重和果实易软化。

果实性状　果实中大，圆形，纵径2.87cm，横径2.92cm，单果重12.5g，大小不均匀。果肩平圆，梗洼浅、中广。果顶圆，顶点微凹。果梗长0.6cm。果面光滑，果皮较薄，红色。果肉厚，浅绿色，质地疏松，汁液中多，味酸甜，适宜制干，也可鲜食，品质中等。鲜枣可食率93.4%，含可溶性固形物33.90%，总糖22.60%，酸0.65%，100g果肉维生素C含量429.50mg；果皮含黄酮19.96mg/g，cAMP含量120.35μg/g。干枣肉质松软，不耐挤压，贮运性能较差，含总糖70.10%，酸2.88%。果核较大，椭圆形，核重0.82g，含仁率20.0%。

评价　该品种适应性强，树体较矮小，树势中等，产量高而稳定，便于集约栽培。果实中大，品质中等，可鲜食和制干。

Jiaochengduanzao

Source and Distribution　The cultivar originated from and spreads in Bianshan areas of Jiaocheng County in Shanxi Province. It is concentrated in Nailin, Tancun and Xiaying with a large quantity. It is a dominant variety there which occupies 40% of the total jujube area.

Botanical Characters　The small tree is spreading with a weak central leader trunk, dense branches and an irregular crown. The reddish-brown 1-year-old shoots have strong growth potential, 86 cm long and 1.03 cm thick. The internodes are 8.5 cm long with less wax. The secondary branches are 39.4 cm long with 8 nodes of medium curvature. The medium-sized mother fruiting spurs can germinate 3 deciduous fruiting shoots, which are 20.2 cm long with 14 leaves. The medium-sized leaves are oval-shaped and light green. There are many large flowers produced, averaging 6 ones per inflorescence.

Biological Characters　The tree has strong adaptability, moderate vigor and weak branching ability, with high fruit set, a high and stale yield. In Taigu County of Shanxi Province, it germinates in late April, begins blooming in late May, begins coloring on August 15 and enters the crisp-maturing stage in mid-September with a fruit growth period of 100 d. It is a mid-ripening variety with serious premature fruit-dropping and easily softening flesh.

Fruit Characteristics　The medium-sized round fruit has a vertical and cross diameter of 2.87 cm and 2.92 cm, averaging 12.5 g with irregular sizes. It has a smooth surface and thin red skin. The light-green flesh is thick, loose-textured, sour and sweet, with medium juice. It can be used for fresh-eating and making dried fruits, with medium quality. The percentage of edible part of fresh fruit is 93.4%, and SSC, TTS, TA and Vc is 33.90%, 22.60%, 0.65% and 429.50 mg per 100 g fresh fruit. The content of flavones and cAMP in mature fruit skin is 19.96 mg/g and 120.35 μg/g. Dried fruits have soft flesh, intolerant to extrusion, storage and transport. The content of TTS and TA in dried fruit is 70.10% and 2.88%. The large oval-shaped stone weighs 0.82 g.

Evaluation　The small plant of Jiaochengduanzao cultivar has strong adaptability and strong vigor with a high and stable yield. The medium-sized fruit has medium quality for fresh-eating and dried fruits.

兼用
Multipurpose Varieties
品种

文 水 沙 枣

品种来源及分布　别名沙枣。分布于山西文水县，栽培数量不多。

植物学性状　树体较高大，树姿半开张，干性较强，枝系密度中等，树冠呈自然半圆形。主干条状皮裂。枣头红褐色，生长势中等，平均长65.8cm，粗0.89cm，节间长8.4cm，蜡质少。二次枝平均长25.4cm，5节。无针刺。枣股较小，抽吊力较强，平均抽生枣吊4.0个。枣吊长19.2cm。叶片大，叶长7.8cm，叶宽3.7cm，卵状披针形，浅绿色，先端渐尖，叶基圆形，叶缘具细密的钝锯齿。花量中等，花序平均着花4朵。花中大，蜜盘较大，昼开型。

生物学特性　适土性和生长势均较强，结果较晚，定植3年后开始结果，15年后进入盛果期。产量中等，坐果率较低，枣头、2~3年、4~6年和6年以上生枝吊果率分别为27.0%、78.3%、71.6%和9.6%。在山西太谷地区，4月中旬萌芽，5月下旬始花，9月上旬开始着色，9月下旬进入脆熟期，果实生育期约115d，为晚熟品种类型。采前落果和裂果严重。

果实性状　果实中大，长圆形或倒卵圆形，纵径3.20cm，横径2.89cm，单果重11.9g，最大20.0g，大小不整齐。果肩平，梗洼窄、中深。果面光滑。果皮薄，红色。果点大而明显。果肉厚，白色，质地较致密，汁液少，味酸，略有苦味，可鲜食和制干，品质中上。鲜枣可食率93.7%，含可溶性固形物27.00%，总糖18.70%，酸0.39%，100g果肉维生素C含量399.80mg；果皮含黄酮17.75mg/g，cAMP含量56.66μg/g。制干率44.0%，干枣含可溶性固形物75.20%，总糖71.90%，酸0.29%。果核较大，纺锤形，核重0.75g，核面不光滑，核内多无种仁。

评价　该品种适土性较强。结果较晚，产量中等。采前落果和裂果严重。果实较大，但不整齐，品质中上，可鲜食和制干。

Wenshuishazao

Source and Distribution　The cultivar originated from and spreads in Wenshui County of Shanxi Province with a small quantity.

Botanical Characters　The large tree is half-spreading with a strong central leader trunk, medium-dense branches and a semi-round crown. The reddish-brown 1-year-old shoots have moderate growth potential, 65.8 cm long and 0.89 cm thick. The internodes are 8.4 cm long with less wax and no thorns. The secondary branches are 25.4 cm long with 5 nodes. The small mother fruiting spurs can germinate 4 deciduous bearing shoots, which are 19.2 cm long. The large leaves are ovate-lanceolate and light green. The number of flowers is medium large, averaging 4 ones per inflorescence.

Biological Characters　The tree has strong adaptability and strong vigor with low fruit set and a medium yield. It bears late, generally in the 3rd year after planting, and enters the full-bearing stage in the 15th year. In Taigu County of Shanxi Province, it germinates in mid-April, begins blooming in late May, begins coloring in early September and enters the crisp-maturing stage in late September with a fruit growth period of 115 d. It is a late-ripening variety with serious premature fruit-dropping and fruit-cracking.

Fruit Characteristics　The medium-sized oblong fruit has a vertical and cross diameter of 3.20 and 2.89 cm, averaging 11.9 g (maximum 20 g) with irregular sizes. It has a flat shoulder, a medium-deep and narrow stalk cavity, a smooth surface and thin red skin with large and distinct dots. The white flesh is thick, tight-textured and sour, a little bitter with less juice. It has a better than normal quality for fresh-eating and dried fruits. The content of SSC, TTS, TA and Vc in fresh fruit is 27.00%, 18.70%, 0.39% and 399.80 mg per 100 g fresh fruit. The content of flavones and cAMP in mature fruit skin is 17.75 mg/g and 56.66 μg/g. The percentage of fresh fruits which can be made into dried ones is 44.0%, and SSC, TTS and TA in dried fruit is 75.20%, 71.90% and 0.29%. The large spindle-shaped stone weighs 0.75 g, with a rough surface. Most stones contain no kernels.

Evaluation　The cultivar has strong adaptability, bearing late with a medium yield, poor tolerance to frost, serious premature fruit-dropping and fruit-cracking. The large fruit has irregular sizes. It has a better than normal quality for fresh-eating and dried fruits.

兼用
Multipurpose Varieties
品种

黎 城 小 枣

品种来源及分布 分布于山西东南部的黎城县，栽培数量不多。

植物学性状 树体较大，树姿较直立，干性中等，树冠多呈伞形。主干皮裂条状。枣头黄褐色，生长势较强，平均长51.4cm，粗1.01cm，节间长8.2cm，蜡质少。二次枝平均长29.6cm，5节，弯曲度大。针刺发达。枣股较小，圆锥形，平均抽生枣吊4.0个。枣吊长22.4cm左右，着叶13片。叶片中大，叶长6.3cm，叶宽3.1cm，卵圆形，合抱，绿色，先端渐尖，叶基圆形，叶缘具钝锯齿。花量大，花序平均着花11朵，花较大，花径6.9mm。

生物学特性 适应性强，耐旱，对栽培条件要求不严，平原、山地和水浇地、旱地均可栽培。树势强健，发枝力强，冠形紧凑。定植第二年开始结果，坐果率高，枣头、2～3年、4～6年和6年以上生枝吊果率分别为73.7%、129.5%、104.7%和27.4%，进入盛果期产量稳定，株产22.5kg左右。在山西太谷地区，9月上旬果实开始着色，9月下旬进入完熟期采收，果实生育期107d左右，为晚熟品种类型。成熟期遇雨裂果严重。

果实性状 果实较小，卵圆形，纵径2.54cm，横径2.22cm，单果重7.3g，最大8.5g，大小整齐。果肩平圆，梗洼深、窄。果顶微凹。果面光滑。果肉较厚，浅绿色，肉质致密，汁多，味甜，适宜鲜食和制干，品质中上。鲜枣可食率95.6%，含可溶性固形物33.60%，总糖26.71%，100g果肉维生素C含量365.31mg；果皮含黄酮10.65mg/g，cAMP含量41.72μg/g。制干率52.2%，干枣总糖69.73%，酸1.52%。果核中大，纺锤形，核重0.32g，含仁率35.0%。

评价 该品种耐旱、风土适应性强。树体强健，结果较早，坐果稳定，产量较高。果实较小，品质中上，干鲜兼用。成熟期遇雨裂果严重。

Lichengxiaozao

Source and Distribution The cultivar originated from and spreads in Licheng County of southeast Shanxi Province with a small quantity.

Botanical Characters The large tree is vertical with a medium-strong central leader trunk and an umbrella-shaped crown. The yellowish-brown 1-year-old shoots have strong growth potential, 51.4 cm long and 1.01 cm thick. The internodes are 8.2 cm long with less wax and developed thorns. The secondary branches are 29.6 cm long with 5 nodes of large curvature. The small conical mother fruiting spurs can germinate 4 deciduous bearing shoots which are 22.4 cm long with 13 leaves. The medium-sized leaves are oval-shaped and green, bending toward the center. There are many large flowers with a diameter of 6.9 mm produced, averaging 11 ones per inflorescence.

Biological Characters The tree has strong adaptability, strong vigor, strong branching ability and a compact canopy, tolerant to drought with a low demand on cultural conditions. It grows well in plain fields, hilly areas, irrigated land and dry land. It generally bears in the 2nd year after planting with high fruit set. A tree in its full-bearing stage has a stable yield of 22.5 kg. In Taigu County of Shanxi Province, it begins coloring in early September and matures in late September. It is a late-ripening variety with serious fruit-cracking.

Fruit Characteristics The small oval-shaped fruit has a vertical and cross diameter of 2.54 cm and 2.22 cm, averaging 7.3 g (maximum 8.5 g) with a regular size. It has a flat-round shoulder, a deep and narrow stalk cavity, a slightly-sunken fruit apex and a smooth surface. The light-green flesh is thick, tight-textured, sweet and juicy. It has a better than normal quality for fresh-eating and dried fruits. The percentage of edible part of fresh fruit is 95.6% and SSC, TTS and Vc is 33.60%, 26.71% and 365.31 mg per 100 g fresh fruit. The percentage of fresh fruits which can be made into dried ones is 52.2%, and the content of TTS and TA in dried fruit is 69.73% and 1.52%. The medium-sized stone is spindle-shaped, averaging 0.32 g.

Evaluation The cultivar has strong vigor, strong adaptability and strong tolerance to drought, bearing early with a high and stable yield. The small fruit has a better than normal quality for fresh-eating and dried fruits. Fruit-cracking is serious if it rains in the maturing stage.

兼 用
Multipurpose Varieties
品 种

· 333 ·

黎城大马枣

品种来源及分布 别名大马枣。分布于山西黎城县，栽培数量不多。

植物学性状 树体高大，树姿直立，干性强，枝系较密，树冠呈圆锥形。主干呈条状皮裂。枣头红褐色，生长势较强，平均长43.5cm，节间长6.5cm，无蜡质。针刺发达。二次枝平均长25.6cm，6节左右，弯曲度较小。枣股较大，抽吊力较强，平均抽生枣吊4.0个。枣吊长19.5cm，着叶14片。叶片较小，叶长6.1cm，叶宽2.4cm，椭圆形，深绿色，先端渐尖，叶基偏斜形，叶缘具钝锯齿。花量多，花序平均着花11朵，花小，花径5.5～6.2mm。

生物学特性 风土适应性和抗逆性均强，要求栽培条件不严。树势强健，发枝力较强，易形成枝叶较稠密的树冠。定植后第二、三年开始结果，坐果能力较强，产量中等，枣头、2～3年、4～6年和6年以上生枝吊果率分别为50.5%、91.1%、108.1%和21.8%。在山西太谷地区，4月中旬萌芽，5月下旬始花，9月下旬开始着色，10月中旬进入脆熟期，果实生育期120d以上，为极晚熟品种类型。采前落果和裂果较轻。

果实性状 果实较小，圆形，纵径2.45cm，横径2.49cm，单果重6.8g，大小整齐。果肩平圆，梗洼浅、广。果顶平圆。果面光滑。果皮厚，紫红色。果肉浅绿色，质地致密，汁液少，味较甜，可鲜食和制干，但品质差。鲜枣可食率91.2%，含可溶性固形物31.50%，总糖24.50%，酸0.61%，100g果肉维生素C含量470.40mg；果皮含黄酮39.94mg/g，cAMP含量51.03μg/g。制干率53.0%，干枣含总糖65.10%，酸1.74%。果核中大，椭圆形，核重0.60g，含仁率100%。

评价 该品种适应性和抗逆性均强，树体高大，树势强健。结果较早，产量中等，采前落果、裂果轻。果个小，果肉薄，可鲜食和制干。果核较大，可食率低，品质差，不宜商品栽培。但因果实成熟期极晚、含仁率高，在培育晚熟品种和砧木选择上有一定的应用价值。

Lichengdamazao

Source and Distribution The cultivar originated from and spreads in Licheng County of Shanxi Province, with a small quantity.

Botanical Characters The large tree is vertical with a strong central leader trunk, dense branches and a conical crown. The reddish-brown 1-year-old shoots have strong growth potential, averaging 43.5 cm long. The internodes are 6.5 cm long, without wax and with developed thorns. The secondary branches are 25.6 cm long with 6 nodes of small curvature. The large mother fruiting spurs can germinate 4 deciduous fruiting shoots which are 19.5 cm long with 14 leaves. The small leaves are 6.1 cm long and 2.4 cm wide, oval-shaped and dark green. There are many small flowers with a diameter of 5.5～6.2 mm produced, averaging 11 ones per inflorescence.

Biological Characters The tree has strong adaptability, strong resistance to adverse conditions, a low demand on cultural conditions, strong vigor and strong branching ability. It generally bears in the 2nd or 3rd year after planting with high fruit set and a medium yield. In Taigu County of Shanxi Province, it enters the crisp-maturing stage in mid-October. It is an extremely late-ripening variety with light premature fruit-dropping and fruit-cracking.

Fruit Characteristics The small round fruit weighs 6.8 g with a regular size. It has a smooth surface and purplish-red thick skin. The light-green flesh is tight-textured and sweet with less juice. It can be used for fresh-eating and making dried fruits, yet with a low quality. The percentage of edible part of fresh fruit is 91.2%, and SSC, TTS, TA and Vc is 31.50%, 24.50%, 0.61% and 470.40 mg per 100 g fresh fruit. The percentage of fresh fruits which can be made into dried ones is 53.0%, and the content of TTS and TA in dried fruit is 65.10% and 1.74%. The large oval-shaped stone weighs 0.60 g, and the percentage of containing kernels is 100%.

Evaluation The large plant of Lichengdamazao cultivar has strong adaptability, strong resistance to adverse conditions. It has small and a regular size, thin flesh, low edibility and a poor quality and is unsuitable for commercial production. It has a high percentage of containing kernels, which makes it valuable in breeding of late-ripening varieties and rootstocks.

兼 用
Multipurpose Varieties
品 种

夏县圆脆枣

品种来源及分布 别名圆脆枣。分布于山西夏县，栽培数量不多。

植物学性状 树体较大，树姿开张，干性中强，层性明显，枝系较密，树冠自然圆头形。主干皮裂条状。枣头黄褐色，生长势中等，平均长57.5cm，粗0.95cm，节间长6.5cm。针刺不发达。枣股中大，抽吊力较差，平均抽生枣吊3.0个。枣吊细，平均长15.4cm，着叶12片。叶片较小，叶长5.0cm，叶宽2.2cm，叶基圆形。花量较多，花序平均着花7朵，花较小，花径5.6～6.7mm，昼开型。

生物学特性 适应性强，抗枣疯病，要求栽培条件不严。树势和发枝力较强。定植第三年开始结果，10年后进入盛果期。坐果率高，枣头、2～3年、4～6年和6年以上生枝吊果率分别为78.0%、106.1%、82.0%和54.3%。产量高而稳定，成龄树株产33.3kg，最高40kg。在山西太谷地区，9月中旬开始着色，9月下旬进入脆熟期，果实生育期115d左右，为晚熟品种类型。裂果轻，采前落果少。

果实性状 果实较小，圆柱形，纵径3.01cm，横径2.46cm。单果重9.6g，最大果重13.0g，大小较整齐。果肩小，圆斜。梗洼窄、中深。果顶平圆。果面光滑。果皮中厚，紫红色。果肉厚，浅绿色，质地致密，汁液少，味甜，可鲜食和制干，品质中上。鲜枣可食率94.2%，含可溶性固形物32.40%，含总糖27.67%，酸0.65%，100g果肉维生素C含量343.61mg，果皮含黄酮3.85mg/g，cAMP含量63.95μg/g。制干率56.0%，干枣含可溶性固形物71.90%，含总糖60.17%，酸1.10%。果核较大，纺锤形，纵径1.99cm，横径0.81cm，核重0.55g，核纹浅，核面较光滑，含仁率11.7%，不饱满，偶有双仁。

评价 该品种适应性强，落果和裂果轻，产量高而稳定。果实肉质致密，可鲜食和制干，品质中上。

Xiaxianyuancuizao

Source and Distribution The cultivar originated from and spreads in Xiaxian County of Shanxi Province with a small quantity.

Botanical Characters The large tree is spreading with a medium-strong central leader trunk, dense branches and a natural-round crown. The yellowish-brown 1-year-old shoots have moderate growth potential, 57.5 cm long and 0.95 cm thick. The internodes are 6.5 cm long with less-developed thorns. The medium-sized mother bearing spurs can germinate 3 deciduous fruiting shoots which are 15.4 cm long with 12 leaves. The small leaves are 5.0 cm long and 2.2 cm wide. There are many small flowers with a diameter of 5.6～6.7 mm produced, averaging 7 ones per inflorescence.

Biological Characters The tree has strong adaptability, strong resistance to jujube witches' broom and a low demand on cultural conditions, with strong vigor and strong branching ability. It generally bears in the 3rd year after planting, and enters the full-bearing stage in the 10th year with high fruit set, a high and stable yield. In Taigu County of Shanxi Province, it begins coloring in mid-September and enters the crisp-maturing stage in late September. It is a late-ripening variety with light fruit-cracking and premature fruit-dropping.

Fruit Characteristics The small column-shaped fruit has a vertical and cross diameter of 3.01 cm and 2.46 cm, averaging 9.6 g (maximum 13.0 g) with a regular size. It has a small shoulder, a narrow and medium-deep stalk cavity, a flat-round fruit apex, a smooth surface, medium-thick and purplish-red skin. The light-green flesh is thick, tight-textured and sweet with less juice. It has a better than normal quality for fresh-eating and dried fruits. The percentage of edible part of fresh fruit is 94.2%, and SSC, TTS, TA and Vc is 32.40%, 27.67%, 0.65% and 343.61 mg per 100 g fresh fruit. The content of flavones and cAMP in mature fruit skin is 3.85 mg/g and 63.95 μg/g. The percentage of fresh fruits which can be made into dried ones is 56.0%, and SSC, TTS and TA in dried fruit is 71.90%, 60.17% and 1.10%. The large spindle-shaped stone has a vertical and cross diameter of 1.99 cm and 0.81 cm, averaging 0.55 g, with shallow veins, a smooth surface and a shriveled kernel. The percentage of containing kernels is 11.7%, sometimes double kernels.

Evaluation The cultivar has strong adaptability, light fruit-dropping and light fruit-cracking with a high and stable yield. The fruit has a better than normal quality for fresh-eating and dried fruits.

马连小枣

品种来源及分布　别名车头小枣、铃枣。分布于河北南部的枣强、故城、冀县和山东北部的武城、夏津等地。枣强孟屯乡王枣林、李枣林、孙枣林、邢枣林一带为集中产区，有1 000年以上的栽培历史。

植物学性状　树体较大，树姿半开张，枝叶较密，树冠多呈伞形。树干灰黑色，树皮粗糙，裂缝浅，呈不规则的宽条状，容易剥落。枣头红褐色或紫褐色，稍有光泽，平均枝长63.0cm。二次枝平均长31.4cm，6～8节，弯曲度中等。无针刺。枣股圆柱形，平均抽生枣吊3个。枣吊长21.3cm，着叶12片。叶片卵圆形，叶长6.8cm，叶宽3.4cm，浓绿色，反卷，先端渐尖、尖圆。叶基圆形。叶缘平展，具钝锯齿。花量大，花序平均着花9朵。

生物学特性　适应性强，在黏壤土和壤土上表现最好。耐旱涝、盐碱能力强，较耐瘠薄。树势强，发枝力中等，结果早，枣头、2～3年和4～6年生枝吊果率分别为44.7%、35.9%和9.3%，定植后2年开始结果，丰产性能中等，平均株产10kg，最高株产50kg。在山西太谷地区，9月中旬果实成熟采收，果实生育期100d左右，为中熟品种类型。较抗裂果。

果实性状　果实较小，圆柱形，纵径3.06cm，横径2.43cm，单果重8.9g，最大10.3g，大小较整齐。果肩圆整，梗洼中深、广。果顶平圆，顶点微凹，残柱小，遗存。果面平滑。果皮浅红色、中厚。果肉浅绿色，质地致密，汁液多，甜味浓，略具酸味，适宜鲜食和制干，品质上等。鲜枣可食率96.1%，含可溶性固形物33%左右，完熟期可达38.90%，总糖29.89%，酸0.74%。制干率53.4%，干枣含总糖67.45%，酸1.22%，果形饱满有弹性，耐贮运。果核中大，椭圆形，核重0.35g，含仁率91.7%。

评价　该品种适应性强。树体强健，经济寿命长，较丰产。果实品质优良，成熟期裂果轻，适宜鲜食和制干，为干鲜兼用的优良品种。

Malianxiaozao

Source and Distribution　The cultivar spreads in Zaoqiang, Gucheng, Jixian in south Hebei Province, and Wucheng, Xiajin in north Shandong Province. Mengtun Village of Zaoqiang County are the central producing areas of this cultivar. It has a cultivation history of over 1 000 years.

Botanical Characters　The large tree is half-spreading with dense branches and an umbrella-shaped crown. The reddish or purplish-brown 1-year-old shoots are a little glossy, 63.0 cm long without thorns. The secondary branches are 31.4 cm long with 6～8 nodes of medium curvature. The column-shaped mother fruiting spurs can germinate 3 deciduous fruiting shoots which are 21.3 cm long with 12 leaves. The oval-shaped dark-green leaves bend toward the back. The number of flowers is large, averaging 9 ones per inflorescence.

Biological Characters　The tree has strong adaptability, strong vigor and medium branching ability, tolerant to water-logging, saline-alkaline and poor soils. It grows best in clay loam soil and loam soil. It generally bears in the 2nd year after planting with a medium yield. A mature tree has a yield of 10 kg. It is a mid-ripening variety with strong resistance to fruit-cracking.

Fruit Characteristics　The small column-shaped fruit weighs 8.9 g with a regular size. It has a smooth surface, medium-thick and light-red skin. The light-green flesh is tight-textured, juicy, strongly sweet and a little sour. It has an excellent quality for fresh-eating and dried fruits. The percentage of edible part of fresh fruit is 96.1%, and SSC, TTS and TA is 33.00%, 29.89% and 0.74%. SSC in fresh fruit of the complete-maturing stage reaches as high as 38.90%. The percentage of fresh fruits which can be made into dried ones is 53.4%, and the content of TTS and TA in dried fruit is 67.45% and 1.22%. Dried fruits have plump and whippy flesh, tolerant to storage and transport. The medium-sized stone is oval-shaped, averaging 0.35 g.

Evaluation　The cultivar has strong adaptability, strong vigor and a long economic life with a high yield. The fruit has an excellent quality for fresh-eating and dried fruits. Fruit-cracking seldom occurs in the maturing stage.

兼 用
Multipurpose Varieties
品 种

· 339 ·

定襄油荷枣

品种来源及分布 别名油荷枣。分布在山西省定襄县白村乡、受录乡，原平市子干乡，忻州市忻口乡、高成乡等地，占当地枣树总数的80%左右。栽培历史300年以上。

植物学性状 树体较大，树姿直立，干性较强，树冠呈圆柱形。主干灰褐色，皮裂呈不规则块状。枣头黄褐色，长54.8cm，粗0.85cm，节间长6.6cm，蜡质少。针刺不发达。枣股圆锥形，抽吊力强，平均抽生枣吊5.0个。枣吊长17.0cm，着叶10片。叶片椭圆形，绿色，反卷，叶面被蜡质，富光泽，先端渐尖，顶端凹尖，叶基部圆形或偏斜形。叶缘具钝锯齿。花量少，花序平均着花3朵。

生物学特性 适应性强，耐旱、抗风、抗霜冻。树体强健，发枝力中等。结果早，栽后第二年开始结果，15年左右进入盛果期，一般株产35kg以上。在山西太谷地区，9月下旬果实成熟采收，果实生育期110d左右，为晚熟品种类型。成熟期遇雨易裂果。

果实性状 果实中大，长圆形，纵径3.37cm，横径2.52cm。单果重10.4g，最大23.0g，大小较整齐。果肩平圆，略耸起。梗洼浅而窄。果顶渐细，顶点尖圆或平圆，柱头遗存。果柄细而长，长2.5mm。果皮中厚，紫红色，有光泽。果肉厚，脆熟期呈淡绿色，完熟后呈乳黄色，质地疏松，汁液少，味酸，适宜鲜食和制干，品质中上。鲜枣含可溶性固形物29.10%，100g果肉维生素C含量392.48mg。制干率59.1%。果核中大，倒纺锤形，核重0.68g，含仁率36.4%。

评价 该品种耐旱、抗霜冻，适应性较强。树体较大，树势强健，结果早，产量中等。果实中大，外形美观，品质中等，但易裂果，是干鲜兼用的优良品种，适于北方秋雨少的地区栽培发展。

Dingxiangyouhezao

Source and Distribution The cultivar originated from and spreads in Baicun and Shoulu Village of Dingxiang County, Zigan Village of Yuanping City, Xinkou and Gaocheng Village of Xinzhou City in Shanxi Province. The area of this variety occupies 80% of the total jujube area there, with a cultivation history of over 300 years.

Botanical Characters The large tree is vertical with a strong central leader trunk and a column-shaped crown. The grayish-brown trunk bark has irregular massive fissures. The yellowish-brown 1-year-old shoots are 54.8 cm long and 0.85 cm thick. The internodes are 6.6 cm long with less wax and less-developed thorns. The conical mother fruiting spurs can germinate 5 deciduous fruiting shoots which are 17.0 cm long with 10 leaves. The oval-shaped leaves are green and glossy, bending toward the back, with a gradually-cuspate apex, a round or deflective base and a blunt saw-tooth pattern on the margin. The number of flowers is small, averaging 3 per inflorescence.

Biological Characters The tree has strong adaptability, tolerant to drought, wind and frost, with strong vigor and medium branching ability. It bears early, generally in the 2nd year after planting, and enters the full-bearing stage in the 15th year with a yield of 35 kg per tree. In Taigu County of Shanxi Province, it matures in late September, with a fruit growth period of 110 d. It is a late-ripening variety with serious fruit-cracking if it rains in the maturing stage.

Fruit Characteristics The medium-sized oblong fruit has a vertical and cross diameter of 3.37 cm and 2.52 cm, averaging 10.4 g (maximum 23.0 g) with a regular size. It has a flat-round shoulder (a little protuberant), a shallow and narrow stalk cavity, a gradually-thinner fruit apex, a remnant stigma, a thin stalk of 2.5 mm long, medium-thick and purplish-red, glossy skin. The flesh is loose-textured, thick and sour with less juice. Its flesh is light green in the crisp-maturing stage and turns creamy-yellow after complete maturing. It has a better than normal quality for fresh-eating and dried fruits. The content of SSC and Vc in fresh fruit is 29.10% and 392.48 mg per 100 g fresh fruit. The percentage of fresh fruits which can be made into dried ones is 59.1%. The medium-sized stone is inverted spindle-shaped, averaging 0.68 g. The percentage of containing kernels is 36.4%.

Evaluation The large plant of Dingxiangyouhezao cultivar has strong tolerance to drought and frost with strong adaptability and strong vigor, bearing early with a medium yield. The medium-sized fruit has an attractive appearance and medium quality and is suitable for fresh-eating and making dried fruits. Yet fruit-cracking easily occurs. It can be developed in northern areas with short growth period and light rainfall in autumn.

兼 用
Multipurpose Varieties
品 种

八 升 胡

品种来源及分布 分布于陕西大荔县的八渔、苏村、官池等乡。

植物学性状 树体较高大，树姿开张，树冠自然半圆形。主干黑褐色，皮裂纹中深，裂片中大，呈条状或不规则形，不易剥落。枣头红褐色，平均长31.0cm，粗0.38cm，节间长4.8cm，最长7.0cm，蜡质多。二次枝健壮，3～8节。针刺较发达。枣股圆柱形，一般抽生枣吊2～4个，枣吊长17.5cm，着叶15片。叶片小，叶长3.0cm，叶宽1.5cm，椭圆形，绿色，有光泽，先端渐尖，先端尖圆，叶基圆形，叶缘锯齿尖钝不一。花量少，花较大，昼开型。

生物学特性 适应性强，耐旱、耐涝、耐瘠薄，在地下水位较高的沙壤土生长结果良好。树势强健，成枝力差。结果早，一般成龄树株产45kg左右，丰产稳产，枣吊平均结果0.55个。在山西太谷地区，4月中旬萌芽，5月下旬始花，9月上旬成熟，果实生育期约90d，为早熟品种类型。易裂果。

果实性状 果实较小，扁圆形，纵径2.91cm，横径2.38cm，单果重9.4g，最大14.3g，大小不整齐。果肩凸，梗洼中深、广。果顶凹陷。果面平整，果皮中厚，红色。果点中大，分布较密。果肉浅绿色，质地较致密，汁液多，味酸甜，可鲜食和制干，品质中等。鲜枣可食率93.8%，含可溶性固形物25.40%，总糖22.39%，酸0.40%，100g果肉维生素C含量312.44mg；果皮含黄酮9.60mg/g，cAMP含量182.89μg/g。果核大，椭圆形，侧面略扁，核重0.58g，含仁率10.0%。

评价 该品种适应性强，结果早，丰产稳产。果实肉质脆甜，品质中等，既可鲜食，又能制干，是早熟的鲜食、制干兼用良种。但果个小，不抗裂果，可在成熟期少雨的地区适当发展。

Bashenghu

Source and Distribution The cultivar originated from and spreads in Bayu, Sucun and Guanchi Village of Dali County in Shaanxi Province.

Botanical Characters The large tree is spreading with a semi-round crown. The blackish-brown trunk bark has irregular, medium-deep and medium-sized striped fissures, uneasily shelled off. The reddish-brown 1-year-old shoots are 31.0 cm long and 0.38 cm thick. The internodes are 4.8 cm (maximum 7.0 cm) long with much wax and developed thorns. There are 3～8 nodes on the strong secondary branches. The column-shaped mother fruiting spurs can germinate 2～4 deciduous fruiting shoots, which are 17.5 cm long with 15 leaves. The small oval-shaped leaves are 3.0 cm long and 1.5 cm wide, green and glossy with a gradually-cuspate apex, a round base and a sharp or blunt saw-tooth pattern on the margin. The number of flowers is small. The daytime-bloomed blossoms have a large size.

Biological Characters The tree has strong adaptability, tolerant to drought, water-logging and poor soils, with strong vigor, poor branching ability, a high and stable yield. It bears early and grows well in sandy loam soil with higher underground water level. The deciduous fruiting shoot bears 0.55 fruits on average. The average yield per tree is about 45 kg. In Taigu County of Shanxi Province, it germinates in mid-April, begins blooming in late May and matures in early September with a fruit growth period of 90 d. It is an early-ripening variety with serious fruit-cracking.

Fruit Characteristics The small oblate fruit has a vertical and cross diameter of 2.91 cm and 2.38 cm, averaging 9.4 g (maximum 14.3 g) with irregular sizes. It has a protuberant shoulder, a medium-deep and wide stalk cavity, a sunken fruit apex, a smooth surface and medium-thick red skin with dense medium-sized dots. The light-green flesh is tight-textured, sour, sweet and juicy. It has medium quality for fresh-eating and dried fruits. The percentage of edible part of fresh fruit is 93.8%, and SSC, TTS, TA and Vc is 25.40%, 22.39%, 0.40% and 312.44 mg per 100 g fresh fruit. The content of flavones and cAMP in mature fruit skin is 9.60 mg/g and 182.89 μg/g. The large oval-shaped stone is a little flattened on the lateral sides, averaging 0.58 g. The percentage of containing kernels is 10.0%.

Evaluation The cultivar has strong adaptability bearing early with a high and stable yield. The fruit has crisp and sweet flesh, with medium quality for fresh-eating and dried fruits. It is a fine early-ripening variety with a small fruit size and poor resistance to fruit-cracking. It can be developed in areas with light rainfall in the maturing stage.

兼 用
Multipurpose Varieties
品 种

· 343 ·

嵩 县 大 枣

品种来源及分布 分布于河南省嵩县的各乡镇，为当地主栽品种，数量占当地枣树的80%左右。

植物学性状 树体高大，树姿半开张，枝系较密，树冠呈圆柱形。主干灰褐色，皮较粗糙，条裂，裂纹浅，易剥落。枣头紫褐色，平均长72.2cm，粗0.92cm，节间长7.4cm，无蜡质。针刺短小。枣股较细，圆柱形，一般抽生枣吊3个。枣吊长22.3cm，平均着叶14片。叶片中大，椭圆形，浅绿色，先端渐尖，叶基圆楔形，叶缘钝锯齿。花量多，花序平均着花7朵，萼片反卷，花较大，花径6.8mm，夜开型。

生物学特性 适应性和抗逆性较强，耐旱、耐瘠薄。树势和发枝力较强。结果较晚，定植后3年开始结果，10年生开始进入盛果期。坐果率较高，枣吊平均结果0.48个，最多达4个。成龄树株产50kg左右。在山西太谷地区，10月上旬果实成熟采收，成熟期较一致，果实生育期117d左右，为极晚熟品种类型。采前落果极轻，裂果少。

果实性状 果实较大，卵圆形，纵径4.66cm，横径3.12cm，单果重16.3g，最大25.4g，大小较整齐。果肩平圆。梗洼广、中深。果顶平圆，顶点略凹陷。果皮中厚，浅红色。果肉浅绿色，质地酥脆，汁液多，味甜略酸，可鲜食和制干，品质上等。鲜枣可食率96.4%，含可溶性固形物24.60%，100g果肉维生素C含量306.45mg；果皮含黄酮4.10mg/g，cAMP含量168.95μg/g。制干率43.0%，干枣含总糖71.30%，酸2.10%。果核中大，倒纺锤形，核重0.59g。含仁率6.7%。

评价 该品种抗逆性强，耐旱、耐瘠。果大，味甜，品质上等，产量高而稳定，适宜鲜食、制干，也宜作蜜枣等产品，抗裂果能力强，可发展栽培。

Songxiandazao

Source and Distribution The cultivar originated from and spreads in villages of Songxian County in Henan Province. It is the dominant variety there, which occupies 80% of the total jujube area.

Botanical Characters The large tree is half-spreading with dense branches and a column-shaped crown. The grayish-brown trunk bark has rough, shallow, striped fissures, easily shelled off. The purplish-brown 1-year-old shoots are 72.2 cm long and 0.92 cm thick. The internodes are 7.4 cm long without wax and with short, small thorns. The column-shaped mother fruiting spurs can germinate 3 deciduous fruiting shoots which are 22.3 cm long with 14 leaves. The medium-sized leaves are oval-shaped and light green, with a gradually-cuspate apex, a round-cuneiform base and a blunt saw-tooth pattern on the margin. There are many large flowers with a diameter of 6.8 mm produced, averaging 7 ones per inflorescence. It blooms at night.

Biological Characters The tree has strong adaptability, strong resistance to adverse conditions, strong vigor and strong branching ability, tolerant to drought and poor soils. It bears late, generally in the 3rd year after planting, and enters the full-bearing stage in the 10th year with high fruit set. The deciduous fruiting shoot bears 0.48 fruits on average (4 at most). A mature tree has a yield of 50 kg. In Taigu County of Shanxi Province, it matures in early October, with a fruit growth period of 117 d. It is an extremely late-ripening variety with very light premature fruit-dropping and light fruit-cracking.

Fruit Characteristics The large oval-shaped fruit has a vertical and cross diameter of 4.66 cm and 3.12 cm, averaging 16.3 g (maximum 25.4 g) with a regular size. It has a flat-round shoulder, a medium-deep and wide stalk cavity, a flat-round fruit apex, medium-thick and light-red skin. The light-green flesh is crisp, juicy, sweet and a little sour. It has a good quality for fresh-eating and dried fruits. The percentage of edible part of fresh fruit is 96.4%, and the content of SSC and Vc is 24.60% and 306.45 mg per 100 g fresh fruit. The content of flavones and cAMP in mature fruit skin is 4.10 mg/g and 168.95 μg/g. The percentage of fresh fruits which can be made into dried ones is 43.0%, and the content of TTS and TA in dried fruit is 71.30% and 2.10%. The medium-sized stone is inverted spindle-shaped, averaging 0.59 g. The percentage of containing kernels is 6.7%.

Evaluation The cultivar has strong resistance to adverse conditions and strong vigor, tolerant to drought and poor soils. The large fruit has sweet flesh with a good quality for fresh-eating, dried fruits and candied fruits. It has a high and stable yield with strong resistance to fruit-cracking.

兼 用
Multipurpose Varieties
品 种

· 345 ·

洪赵葫芦枣

品种来源及分布 原产和分布于山西省洪洞县稽村乡许村一带，栽培数量不多。

植物学性状 树体较大，树姿开张，干性中强，树冠呈自然乱头形。主干皮裂条状。枣头红褐色，平均长39.0cm，粗0.9cm，节间长9.4cm，无蜡质。二次枝平均长27.4cm，5节，弯曲度小。针刺发达。枣股较大，抽生枣吊3个。枣吊长19.3cm，着叶10片。叶片大，叶长7.9cm，叶宽3.9cm，卵圆形，绿色，较光亮，先端渐尖，叶基心形，叶缘具钝锯齿。花量较多，花较大，花序平均着花6朵，夜开型。

生物学特性 适应性强，要求栽培条件不严。树势强健，发枝力强。结果较早，一般第二、三年开始结果，第五年后进入盛果期，丰产性强，坐果率高，枣头、2~3年、4~6年和6年以上生枝吊果率分别为62.8%、88.1%、110.2%和27.0%。在山西太谷地区，4月中旬萌芽，5月下旬始花，9月20日前后进入脆熟期，果实生育期100 d左右，为中晚熟品种类型。采前落果少，抗裂果。

果实性状 果实较大，倒卵形，侧面略扁，纵径3.64cm，横径2.67cm，单果重12.3g，果个大小整齐。果肩平圆，梗洼浅、中广，果顶平圆。果面光滑，果皮紫红色，较薄。果肉厚，浅绿色，质地致密，汁少，味甜，可鲜食和制干，品质中等。鲜枣可食率95.1%，含可溶性固形物31.20%，总糖24.82%，酸0.54%，100g果肉维生素C含量700.43mg；果皮含黄酮5.33mg/g，cAMP含量95.37μg/g。干枣含总糖59.33%，酸1.32%。果核较大，纺锤形，核重0.60g，含仁率46.7%。

评价 该品种适应性强，树体较大，树势强健，早实丰产，产量较稳定。果实较大，适宜鲜食和制干，品质中等。成熟期落果少，抗裂果。

Hongzhaohuluzao

Source and Distribution The cultivar originated from and spreads in Xucun of Jicun Village in Hongtong County of Shanxi Province, with a small quantity.

Botanical Characters The large tree is spreading with a medium-strong central leader trunk and an irregular crown. The trunk bark has striped fissures. The reddish-brown 1-year-old shoots are 39 cm long and 0.9 cm thick. The internodes are 9.4 cm long without wax and with developed thorns. The secondary branches are 27.4 cm long with 5 nodes of small curvature. The large mother fruiting spurs can germinate 3 deciduous bearing shoots which are 19.3 cm long with 10 leaves. The large oval-shaped leaves are 7.9 cm long and 3.9 cm wide, green and glossy, with a gradually-cuspate apex, a heart-shaped base and a blunt sawtooth pattern on the margin. There are many large flowers produced, averaging 6 ones per inflorescence. It blooms at night.

Biological Characters The tree has strong adaptability, a low demand on cultural conditions, strong vigor and strong branching ability. It bears early, generally in the 2nd or 3rd year after planting, and enters the full-bearing stage in the 5th year with high fruit set and high productivity. The percentage of fruits to deciduous fruiting shoots of 1-year-old shoots, 2~3-year-old branches, 4~6-year-old ones and over-6-year-old ones is 62.8%, 88.1%, 110.2% and 27.0%. In Taigu County of Shanxi Province, it germinates in mid-April, begins blooming in late May and enters the crisp-maturing stage around September 20 with a fruit growth period of 100 d. It is a mid-late-ripening variety with light fruit-dropping and strong resistance to premature fruit-cracking.

Fruit Characteristics The large obovate fruit is a little flattened on the lateral sides, with a vertical and cross diameter of 3.64 cm and 2.67 cm, averaging 12.3 g with a regular size. It has a flat-round shoulder, a shallow and medium-wide stalk cavity, a flat-round fruit apex, a smooth surface and purplish-red thin skin. The light-green flesh is thick, tight-textured and sweet with less juice. It has medium quality for fresh-eating and dried fruits. The percentage of edible part of fresh fruit is 95.1%, and SSC, TTS, TA and Vc is 31.20%, 24.82%, 0.54% and 700.43 mg per 100 g fresh fruit. The content of flavones and cAMP in mature fruit skin is 5.33 mg/g and 95.37 μg/g, and the content of TTS and TA in dried fruit is 59.33% and 1.32%. The large spindle-shaped stone weighs 0.60 g with the percentage of containing kernels of 46.7%.

Evaluation The large plant of Hongzhaohuluzao cultivar has strong adaptability and strong vigor, bearing early with a high and stable yield, light fruit-dropping. The large fruit has medium quality for fresh-eating and dried fruits. Fruit-cracking seldom occurs in the maturing stage.

兼用
Multipurpose Varieties
品种

民 勤 小 枣

品种来源及分布　分布于甘肃省民勤县薛百乡，武威、永昌也有零星栽培，为当地原有的主栽品种。

植物学性状　树体中大，树姿开张，树冠偏斜形。主干灰褐色，皮裂纹中深，呈窄条状纵裂，不易剥落。枣头红褐色，平均长56.0cm，粗0.79cm，节间长6cm左右。二次枝发育健壮，平均长23.1cm，6节。针刺发达。枣股圆柱形，一般抽生枣吊3～4个，最多5个。枣吊长16.3cm，着生叶片13片。叶片小，卵圆形，绿色，先端尖圆，叶基圆形，叶缘具细匀的锯齿，齿角钝圆，较整齐。花量少，每花序着花2朵，花较大，昼开型。

生物学特性　适应性强，抗寒、抗风、耐旱、耐瘠薄、耐盐碱，喜沙质壤土。树势中等，发枝力较强，树体寿命长。定植后2、3年开始结果，10年后进入盛果期，产量较高，而且稳定，一般株产25kg。在山西太谷地区，4月中旬萌芽，5月底始花，9月下旬成熟，果实生育期110d左右，为晚熟品种类型。

果实性状　果实较小，圆柱形，纵径2.96cm，横径2.69cm，单果重9.5g，大小不整齐。果肩平圆。梗洼窄、中深。果顶凹陷，柱头遗存。果面光滑，果皮红色，较中厚。果肉浅绿色，质地酥脆，汁液较多，味甜酸，可鲜食和制干，品质上等。鲜枣可食率98.1%，含可溶性固形物35.24%，总糖35.24%，酸0.49%，100g果肉维生素C含量418.58mg；果皮含黄酮2.55mg/g，cAMP含量207.45μg/g。制干率50%以上，干枣含总糖69.15%，酸0.83%。果核小，倒卵形，核重0.18g。核内多无种仁。

评价　该品种适应性强，树势中庸，结果早，产量高而稳定。果实小，肉质细脆，糖分高，宜鲜食和制干，制干率较高，干枣耐贮运，适于我国西北干旱地区发展。

Minqinxiaozao

Source and Distribution　The cultivar spreads in Minqin of Gansu Province, with some distributions in Wuwei and Yongchang. It is the native dominant variety there.

Botanical Characters　The medium-sized tree is spreading with a deflective crown. The grayish-brown trunk bark has vertical, medium-deep, narrow-striped fissures, uneasily shelled off. The reddish-brown 1-year-old shoots are 56.0 cm long and 0.79 cm thick. The internodes are 6.0 cm long with developed thorns. The strong secondary branches are 23.1 cm long with 6 nodes. The column-shaped mother fruiting spurs can germinate 3～4 (5 at most) deciduous bearing shoots which are 16.3 cm long with 13 leaves. The small green leaves are oval-shaped, with a cuspate-round apex, a round base and a thin, blunt-round saw-tooth pattern on the margin. The number of flowers is small, averaging 2 ones per inflorescence. The large blossoms bloom in the daytime.

Biological Characters　The tree has strong adaptability, tolerant to cold, wind, drought, poor soils and saline-alkaline, with moderate vigor, strong branching ability and a long life. It prefers sandy loam soil. The tree generally bears in the 2nd or 3rd year after planting, and enters the full-bearing stage in the 10th year with a high and stable yield. The average yield per tree is about 25 kg. In Taigu County of Shanxi Province, it germinates in mid-April, begins blooming in late May and matures in late September with a fruit growth period of 110 d. It is a late-ripening variety.

Fruit Characteristics　The small round fruit has a vertical and cross diameter of 2.96 cm and 2.69 cm, averaging 9.5 g with irregular sizes. It has a flat-round shoulder, a medium-deep and narrow stalk cavity, a sunken fruit apex, a remnant stigma, a smooth surface and medium-thick red skin. The light-green flesh is crisp, juicy, sour and sweet. It has a good quality for fresh-eating and dried fruits. The percentage of edible part of fresh fruit is 98.1%, and SSC, TTS, TA and Vc is 35.24%, 35.24%, 0.49% and 418.58 mg per 100 g fresh fruit. The content of flavones and cAMP in mature fruit skin is 2.55 mg/g and 207.45 μg/g. The percentage of fresh fruits which can be made into dried ones is over 50.0%. The small obovate stone weighs 0.18 g. Most stones contain no kernels.

Evaluation　The cultivar has strong adaptability and moderate vigor, bearing early with a high and stable yield. The small fruit has delicate and crisp flesh with high sugar content. It is suitable for fresh-eating and making dried fruits. It has a high rate of making dried fruits from fresh ones. Dried fruits are tolerant to storage and transport. It can be developed in northwestern China with dry climate.

兼用
Multipurpose Varieties
品种

佳县牙枣

品种来源及分布　别名倒卵形牙枣、牙枣。原产和分布于陕西佳县城关小会坪、峪口等地，数量不多，多零星栽种。

植物学性状　树体中大，树姿半开张，树冠呈伞形。主干黄褐色，皮裂纹深，呈不规则宽条状，不易剥落。枣头红褐色，长势强，平均长67.8cm，粗1.08cm。二次枝长23.2cm，5节左右，弯曲度中等。针刺易脱落。枣股平均抽生枣吊3.0个。枣吊长22.1cm，着叶13片。叶片中大，椭圆形，绿色，较光亮，先端渐尖，先端尖锐。叶基圆楔形。叶缘具钝锯齿。

生物学特性　适应性强，耐瘠薄，对土质要求不严。树势和发枝力较强，产量中等而稳定。枣吊平均结果0.56个。在山西太谷地区，4月中旬萌芽，5月下旬始花，9月下旬果实成熟，果实生育期110d左右，为晚熟品种类型。

果实性状　果实中大，倒卵圆形，略偏斜，纵径3.25cm，横径2.43cm，单果重11.5g，大小较整齐。果肩圆，梗洼、环洼窄浅。果顶广圆，顶点凹陷。果柄短，中粗，平均长3.3mm。果面光滑，果皮厚，果点小，密度大，较明显。果肉浅绿色，质地致密，汁液少，味酸甜，可鲜食和制干，品质中上。鲜枣可食率95.1%，含可溶性固形物30.40%，可溶性糖20.07%，酸0.44%，100g果肉维生素C含量408.27mg；果皮含黄酮3.36mg/g，cAMP含量199.73 μg/g。果核较大，纺锤形，核重0.56g，含仁率低，仅8.3%。

评价　该品种适应性强，耐瘠薄，对土质要求不严。产量中等、稳定。果实中大，肉质致密，可鲜食和制干，品质中上。

Jiaxianyazao

Source and Distribution　The cultivar, also called Daoluanxingyazao or Yazao, originated from and spreads in Xiaohuiping and Yukou of Chengguan in Jiaxian County of Shaanxi Province, with a small quantity.

Botanical Characters　The medium-sized tree is half-spreading with an umbrella-shaped crown. The yellowish-brown trunk bark has irregular, deep, wide-striped fissures, uneasily shelled off. The reddish-brown 1-year-old shoots have strong growth potential, 67.8 cm long and 1.08 cm thick, with easily falling-off thorns. The secondary branches are 23.2 cm long with 5 nodes of medium curvature. The mother fruiting spurs can germinate 3 deciduous bearing shoots which are 22.1 cm long with 13 leaves. The medium-sized leaves are oval-shaped, green and glossy, with a gradually-cuspate apex, a round-cuneiform base and a blunt saw-tooth pattern on the margin.

Biological Characters　The tree has strong adaptability, tolerant to poor soils, with a low demand on cultural conditions, strong vigor, strong branching ability, a medium and stable yield. It grows well even in flood plains. The deciduous fruiting shoot bears 0.56 fruits on average. In Taigu County of Shanxi Province, it germinates in mid-April, begins blooming in late May and matures in late September with a fruit growth period of 110 d. It is a late-ripening variety with strong resistance to fruit-cracking during rainy days in the maturing stage.

Fruit Characteristics　The medium-sized obovate fruit is a little deflective, with a vertical and cross diameter of 3.25 cm and 2.43 cm, averaging 11.5 g with a regular size. It has a round shoulder, a narrow and shallow stalk cavity, a wide-round fruit apex, a medium-thick short stalk of 3.3 mm long, a smooth surface and thick skin with small, dense and distinct dots. The light-green flesh is tight-textured, sour and sweet with less juice. It has a better than normal quality for fresh-eating and dried fruits. The percentage of edible part of fresh fruit is 95.1%, and SSC, TTS, TA and Vc is 30.40%, 20.07%, 0.44% and 408.27 mg per 100 g fresh fruit. The content of flavones and cAMP in mature fruit skin is 3.36 mg/g and 199.73 μg/g. The large spindle-shaped stone weighs 0.56 g. It has a low percentage of containing kernels which is only 8.3%.

Evaluation　The cultivar has strong adaptability and strong tolerance to poor soils with a medium and stable yield. The medium-sized fruit has tight-textured flesh with a better than normal quality for fresh-eating and dried fruits.

兼 用
Multipurpose Varieties
品 种

广 洋 枣

品种来源及分布 别名圆铃枣、小圆铃。原产和分布于河南省西南部的镇平县、方城县等枣区，为当地主栽品种，约占栽培面积的90%。栽培历史2 000年以上。

植物学性状 树体较大，树姿半开张，树冠呈自然圆头形。主干灰褐色，表面粗糙，皮裂块状，不易剥落。枣头红褐色，平均长61.7cm，粗1.07cm，节间长5.6cm。二次枝长27.7cm，着生10节左右，弯曲度中等。针刺发达。枣股圆柱形，平均抽生枣吊4.0个。枣吊长20.7cm，着叶17片。叶片较小，叶长4.5cm，叶宽2.4cm，卵圆形，浓绿色，先端渐尖，叶基圆形，叶缘钝锯齿。花序平均着花6朵。

生物学特性 适应性和抗逆性均强，耐瘠薄。树势强，产量中等，枣头和2～3年生枝吊果率分别为13.6%和20.2%。在山西太谷地区，10月上旬果实成熟采收，果实生育期117d左右，为极晚熟品种类型。果实成熟一致，抗裂果。

果实性状 果实中大，圆柱形，纵径3.07cm，横径2.70cm，单果重10.8g，大小整齐。果肩宽圆，梗洼中深、窄。果顶微凹。果柄细，长约5.5mm。果皮薄，浅红色。果肉白色，质地较致密，汁液中多，较细脆，味酸甜，适宜鲜食、制干和加工，品质上等。鲜枣可食率94.4%，含总糖35.03%，酸0.78%，100g果肉维生素C含量477.35mg，果皮含黄酮6.00mg/g，cAMP含量49.60μg/g。干枣肉质松软，有弹性，含总糖76.21%，酸1.43%。果核中大，纺锤形，核重0.6g左右，含仁率23.3%。

评价 该品种适应性强，耐瘠薄。树体强健，经济寿命长。产量较稳定。果实中大，裂果轻，品质上等，可用于鲜食、制干和加工。适宜气候温暖、生长期雨量较多的地区栽培。

Guangyangzao

Source and Distribution The cultivar, also called Yuanlingzao or Xiaoyuanling, originated from and spreads in Zhenping and Fangcheng of southwest Henan Province. It is the dominant variety there, which occupies 90% of total jujube area. It has a cultivation history of over 2 000 years.

Botanical Characters The large tree is half-spreading with a natural-round crown. The grayish-brown trunk bark has rough massive fissures, uneasily shelled off. The reddish-brown 1-year-old shoots are 61.7 cm long and 1.07 cm thick. The internodes are 5.6 cm long with developed thorns. The secondary branches are 27.7 cm long with 10 nodes of medium curvature. The column-shaped mother fruiting spurs can germinate 4 deciduous bearing shoots which are 20.7 cm long with 17 leaves. The small leaves are 4.5 cm long and 2.4 cm wide, oval-shaped and dark green, with a gradually-cuspate apex, a round base and a blunt saw-tooth pattern on the margin. Each inflorescence has 6 flowers.

Biological Characters The tree has strong adaptability and strong resistance to adverse conditions, tolerant to poor soils, with strong vigor and a medium yield. The percentage of fruits deciduous fruiting shoots to of 1-year-old shoots and 2 or 3-year-old branches is 13.6% and 20.2%. In Taigu County of Shanxi Province, it matures in early October, with a fruit growth period of 117 d. It is an extremely late-ripening variety with strong resistance to fruit-racking.

Fruit Characteristics The medium-sized fruit is column-shaped with a vertical and cross diameter of 3.07 cm and 2.70 cm, averaging 10.8 g with a regular size. It has a wide-round shoulder, a medium-deep and narrow stalk cavity, a slightly-sunken fruit apex, a thin stalk of 5.5 mm long and light-red thin skin. The white flesh is tight-textured, crisp, juicy, sour and sweet. It has a good quality for fresh-eating, dried fruits and processing. The percentage of edible part of fresh fruit is 94.4%, and the content of TTS, TA and Vc is 35.03%, 0.78% and 477.35 mg per 100 g fresh fruit. The content of flavones and cAMP in mature fruit skin is 6.00 mg/g and 49.60 μg/g. Dried fruits have soft and whippy flesh, and the content of TTS and TA is 76.21% and 1.43%. The medium-sized stone is spindle-shaped, averaging 0.6 g. The percentage of containing kernels is 23.3%.

Evaluation The cultivar has strong adaptability, tolerant to poor soils with strong vigor, a long economic life, a high and stable yield. The large fruit has light fruit-cracking and an excellent quality for fresh-eating, dried fruits and processing. It can be developed in warm areas with much rainfall in the growing season.

兼 用
Multipurpose Varieties
品 种

溆浦观音枣

品种来源及分布 别名观音枣。主要分布于湖南西部的溆浦县，该县双井乡圹湾村、祖市天乡四门村为集中产地，低庄镇连山村也有栽培。

植物学性状 树体较大，树姿开张，树冠偏斜形。主干深褐色，皮裂呈纵条状，横裂不明显，容易小片剥落。枣头紫褐色，枝长77.5cm，粗0.98cm，节间长8.0cm，蜡质多。二次枝长31.5cm，着生6节左右，弯曲度小。针刺不发达。枣股圆柱形，平均抽生枣吊4.0个。枣吊长23.0cm，着叶15片。叶片小，椭圆形，合抱，绿色，色泽较浅，先端钝尖，基部圆楔形，叶缘具钝锯齿。花量少，花序平均着花4朵，花中大，昼开型。

生物学特性 适应性强。树势中等，发枝力强。结果较晚，一般定植后3年开始结果。盛果期产量高，株产可达75kg。坐果较稳定，自然落果少，2～3年和3年以上枝的吊果率分别为115.4%和16.3%。在山西太谷地区，9月下旬果实成熟，果实生育期110d左右，为晚熟品种类型。

果实性状 果个较小，卵圆形。纵径2.74cm，横径1.97cm，单果重9.3g。果肩平圆，梗洼浅广，顶点平，胴部中腰细瘦凹陷，形如盘坐的菩萨塑像，具观赏价值。果皮中厚，紫红色。果肉浅绿色，质地酥脆，汁液多，甜味浓，可鲜食和制干，品质上等。鲜枣可食率90.9%，含可溶性固形物33.60%，总糖31.98%，酸0.67%，100g果肉维生素C含量373.49mg，果皮含黄酮4.71mg/g，cAMP含量88.69μg/g。制干率49.1%，干枣含总糖70.24%，酸1.40%。果核纺锤形，略弯曲，核重0.85g，含仁率16.7%。

评价 该品种适应性强，树体较大，产量高而稳定，自然落果少。果实较小，形状奇特，鲜食、制干和观赏栽培均可。

Xupuguanyinzao

Source and Distribution The cultivar mainly spreads in Xupu County of western Hunan Province. Kuangwancun of Shuangjing Village and Simencun of Zushitian Village in the county are the main producing areas of this cultivar.

Botanical Characters The large tree is spreading with a deflective crown. The dark-brown trunk bark has vertical striped fissures, easily peeled off in small pieces. The purplish-brown 1-year-old shoots are 77.5 cm long and 0.98 cm thick. The internodes are 8.0 cm long with much wax and no thorns. The secondary branches are 31.5 cm long with 6 nodes of small curvature. The column-shaped mother fruiting spurs can germinate 4 deciduous fruiting shoots which are 23.0 cm long with 15 leaves. The small leaves are oval-shaped and light green, with a bluntly-cuspate apex, a round-cuneiform base and a blunt saw-tooth pattern on the margin. The number of flowers is small, averaging 4 ones per inflorescence. The daytime-bloomed blossoms are of medium size.

Biological Characters The tree has strong adaptability, moderate vigor and strong branching ability. It bears late with a high yield in the full-bearing stage, stable fruit set and light natural fruit-dropping. The percentage of fruits deciduous fruiting shoots to of 2～3-year-old branches and over-3-year-old ones is 115.4% and 16.3%. It generally bears in the 2nd year after planting. A mature tree has a yield of about 75 kg. In Taigu County of Shanxi Province, it matures in late September with a fruit growth period of 110 d. It is a late-ripening variety.

Fruit Characteristics The small oval-shaped fruit has a vertical and cross diameter of 2.74 cm and 1.97 cm, averaging 9.3 g. It has a flat-round shoulder, a shallow and wide stalk cavity, a flat fruit apex, medium-thick and purplish-red skin. The middle part of the fruit is thin and sunken, which makes it like the figure of Buddha, high in ornamental value. The light-green flesh is crisp and juicy, strongly sweet. It has a good quality for fresh-eating and dried fruits. The percentage of edible part of fresh fruit is 90.9%, and SSC, TTS, TA and Vc is 33.60%, 31.98%, 0.67% and 373.49 mg per 100 g fresh fruit. The content of flavones and cAMP in mature fruit skin is 4.71 mg/g and 88.69 μg/g. The percentage of fresh fruits which can be made into dried ones is 49.1%. The spindle-shaped stone weighs 0.85 g with the percentage of containing kernels of 16.7%.

Evaluation The large plant of Xupuguanyinzao cultivar has strong adaptability, a high and stable yield with light natural fruit-dropping. The small fruit has a special shape and is suitable for fresh-eating, making dried fruits and also for ornament.

兼 用
Multipurpose Varieties
品种

· 355 ·

大 荔 面 枣

品种来源及分布　原产和分布于陕西大荔的石槽乡马坊、王马一带，蒲城、澄城等地也有零星栽培。

植物学性状　树体较大，树姿开张，树冠呈偏斜形。主干灰褐色或黑褐色，皮裂纹深，呈不规则条状，不易剥落。枣头红褐色，平均长45.8cm，粗0.74cm，节间长6.4cm，蜡质少。二次枝发育健壮，平均长20.2cm，5节左右，弯曲度中等。针刺不发达，细短，质软易落。枣股圆柱形，粗壮，平均抽生枣吊3.0个。枣吊长23.7cm，着叶18片。叶片中大，卵圆形，叶厚，深绿色，较光亮，先端渐尖，先端尖圆，叶基圆形，叶缘锯齿浅钝。花量少，每花序着花2朵，花大，萼片短。

生物学特性　适应性强，在沙壤土条件下生长良好。树势强健，发枝力强，容易更新复壮。产量高而稳定，枣吊平均结果0.35个。在山西太谷地区，4月中旬萌芽，5月下旬始花，9月中旬果实成熟，果实生育期104d，为中熟品种类型。果实抗裂果。

果实性状　果实较大，卵圆形或扁圆形，纵径3.29cm，横径2.78cm，单果重13.5g，大小较整齐。果肩宽大，略耸起。梗洼窄、中深。果顶凹陷。果面光滑。果皮中厚，紫红色。果点小，不明显。果肉浅绿色，质地较致密，汁液中多，味甜，宜鲜食，也可制干，品质中上。鲜枣可食率95.2%，含可溶性固形物33.00%，总糖35.74%，酸0.30%；果皮含黄酮3.73mg/g，cAMP含量186.45μg/g。干枣含总糖64.80%，酸0.84%。果核较大，椭圆形，核重0.65g，含仁率6.7%。

评价　该品种适应性强，产量高而稳定，抗裂果，果实较大，肉质较致密，可鲜食和制干，品质中上。

Dalimianzao

Source and Distribution　The cultivar originated from and spreads in Mafang and Wangma of Shicao Village in Dali County of Shaanxi Province. There are also some distributions in Pucheng and Chengcheng.

Botanical Characters　The large tree is spreading with a deflective crown. The trunk bark is grayish brown or blackish brown with irregular deep striped fissures, uneasily shelled off. The reddish-brown 1-year-old shoots are 45.8 cm long and 0.74 cm thick. The internodes are 6.4 cm long with less wax. The soft thorns are less-developed, thin and short, easily falling off. The strong secondary branches are 20.2 cm long with 5 nodes of medium curvature. The column-shaped mother fruiting spurs can germinate 3 deciduous fruiting shoots which are 23.7 cm long with 18 leaves. The medium-sized leaves are thick, oval-shaped, dark green and glossy, with a gradually-cuspate apex, a round base and a shallow, blunt saw-tooth pattern on the margin. The number of flowers is small, averaging 2 ones per inflorescence. The large blossoms have short sepals.

Biological Characters　The tree has strong adaptability, strong vigor and strong branching ability, easily regenerating and invigorating with a high and stable yield. It grows well in sandy soil. The deciduous fruiting shoot bears 0.35 fruits on average. In Taigu County of Shanxi Province, it germinates in mid-April, begins blooming in late May and matures in mid-September, with a fruit growth period of 104 d. It is a mid-late-ripening variety with strong resistance to fruit-cracking.

Fruit Characteristics　The large fruit is oval-shaped or oblate with a vertical and cross diameter of 3.29 cm and 2.78 cm, averaging 13.5 g with a regular size. It has a wide and a little protuberant shoulder, a narrow and medium-deep stalk cavity, a slightly-sunken fruit apex, a smooth surface, medium-thick and purplish-red skin with small and indistinct dots. The light-green flesh is tight-textured and sweet with medium juice. It is suitable for fresh-eating, and can also be used for making dried fruits with a better quality. The percentage of edible part of fresh fruit is 95.2%, and SSC, TTS and TA is 33.00%, 35.74% and 0.30%. The content of flavones and cAMP in mature fruit skin is 3.73 mg/g and 186.45 μg/g, and the content of TTS and TA in dried fruit is 64.80% and 0.84%. The large oval-shaped stone weighs 0.65 g, and the percentage of containing kernels is 6.7%.

Evaluation　The cultivar has strong adaptability and a high and stable yield. The large fruit has tight-textured flesh and strong resistance to fruit-cracking and is suitable for fresh-eating and making dried fruits with a better than normal quality.

· 357 ·

保 德 小 枣

品种来源及分布 原产和分布于山西保德黄河沿岸的杨家湾、韩家川等村，约占当地枣树的20%。

植物学性状 树体中大，树姿开张，干性中强，枝系中密较粗壮，树冠呈圆形。枣头红褐色，较粗壮，萌发力中强，平均长61.0cm，粗1.02cm，节间长8.5cm，蜡质少。针刺不发达。枣股较大，圆锥形，抽吊力较弱，平均抽生枣吊3.0个。枣吊长20.8cm，平均着叶11片。叶片大，叶长7.2cm，叶宽3.1cm，卵圆形，绿色，较光亮，先端渐尖，叶基偏斜形，叶缘具钝锯齿。花量多，花序平均着花7朵。花朵蜜盘较大，为昼开型。

生物学特性 适应性强，耐旱，对栽培条件要求不严，水地、旱地均可栽植。树势较强，树冠较密。结果较早，定植后第二、三年开始结果。成龄树产量较高，且稳定，坐果率高，枣头、2～3年、4～6年和6年以上生枝吊果率分别为54.5%、67.4%、63.6%和11.0%。在山西太谷地区，4月下旬萌芽，6月上旬始花，9月上旬着色，9月底成熟采收，果实生育期115d，为晚熟品种类型。

果实性状 果实中大，圆柱形或平顶锥形，纵径3.48cm，横径2.57cm，单果重12.5g，大小整齐。果肩平圆，梗洼狭、浅。果顶凹陷。果面粗糙，果皮红色，较厚。果肉厚，浅绿色，质地致密，汁液中多，味酸，鲜食和制干均可，品质中等。鲜枣可食率97.4%，含可溶性固形物33.60%，酸1.02%，100g果肉维生素C含量415.10mg。干枣含总糖67.20%，酸1.02%。果核较小，纺锤形，核重0.33g。含仁率仅1.7%。

评价 该品种适应性强，耐旱。产量高而稳定。果实中大，大小整齐，可鲜食和制干，品质中等。

Baodexiaozao

Source and Distribution The cultivar originated from and spreads in Yangjiawan and Hanjiachuan Village of Baode County in Shanxi Province, which occupies 20% of the total jujube area there.

Botanical Characters The medium-sized tree is spreading with a medium-strong central leader trunk, medium-dense strong branches and a round crown. The reddish-brown 1-year-old shoots have medium germination ability, 61.0 cm long and 1.02 cm thick. The internodes are 8.5 cm long, with less wax and less-developed thorns. The large conical mother fruiting spurs can germinate 3 deciduous fruiting shoots, which are 20.8 cm long with 11 leaves. The large oval-shaped leaves are 7.2 cm long and 3.1 cm wide, green and glossy, with a gradually-cuspate apex, a deflective base and a blunt saw-tooth pattern on the margin. There are many large flowers produced, averaging 7 per inflorescence. It blooms in the daytime.

Biological Characters The tree has strong adaptability, strong tolerance to drought, strong vigor and a compact canopy, with a low demand on cultural conditions. It grows well in irrigated land and in dry land. The tree bears early, generally in the 2nd or 3rd year after planting, with high fruit set, a high and stable yield. The percentage of fruits to deciduous bearing shoots of 1-year-old shoots, 2～3-year-old branches, 4～6-year-old ones and over-6-year-old ones is 54.5%, 67.4%, 63.6% and 11.0%. In Taigu County of Shanxi Province, it germinates in late April, begins blooming in early June, begins coloring in early September and matures in late September, with a fruit growth period of 115 d. It is an extremely late-ripening variety.

Fruit Characteristics The medium-sized fruit is column-shaped, with a vertical and cross diameter of 3.48 cm and 2.57 cm, averaging 12.5 g with regular sizes. It has a flat-round shoulder, a narrow and shallow stalk cavity, a slightly-sunken fruit apex, a rough surface and thick red skin. The light-green flesh is thick, tight-textured and sour, with medium juice. It has medium quality for fresh-eating and dried fruits. The percentage of edible part of fresh fruit is 97.4%, and SSC, TA and Vc is 33.60%, 1.02% and 415.10 mg per 100 g fresh fruit. The content of TTS and TA in dried fruit is 67.20% and 1.02%. The small spindle-shaped stone weighs 0.33 g, and the percentage of containing kernels is only 1.7%.

Evaluation The cultivar has strong adaptability and strong tolerance to drought, with a high and stable yield. The medium-sized fruit has regular size and is suitable for fresh-eating and making dried fruits, with medium quality.

兼用 品种
Multipurpose Varieties

溆浦香枣

品种来源及分布 别名湘西香枣。原产于湖南溆浦的双井乡和低庄镇各村,以双井分水口和塘湾村为集中产地。

植物学性状 树体较大,树姿开张,干性较弱,枝叶较稀,树冠呈偏斜形。树干灰褐色,呈条状剥落。枣头红褐色,年平均生长量51.0cm,节间长5.4cm,阳面有少量灰色蜡质。二次枝平均长20.6cm,5～7节,弯曲度中等。针刺不发达,长0.3～0.7cm。枣股圆锥形,略歪斜,平均抽生枣吊3.9个。枣吊长25.8cm,着叶16片。叶片较小,绿色,卵状披针形。先端急尖,叶基圆楔形,叶缘锯齿粗。花量少,花序平均着花2朵。花小,花径5.7mm,昼开型。

生物学特性 树势中等,萌蘖力弱。结果较晚,坐果稳定,丰产性强。定植后4年开始结果,10年前后进入盛果期。成龄树一般株产50 kg。在山西太谷地区,9月下旬果实成熟,果实生育期110d左右,为中晚熟品种类型。生理落果较轻。

果实性状 果实小,椭圆形或卵圆形,纵径3.21cm,横径2.52cm,单果重6.9g,大果重17.5g,大小较整齐。果肩平圆。梗洼浅、中广,果顶平圆。果皮厚,紫红色。果肉质地较紧密,汁液少,味甜,有浓郁香味,品质中上,可鲜食或制干。鲜枣可食率88.4%,含总糖20.96%,含酸0.27%,100g果肉维生素C含量421.06mg;果皮含黄酮51.07mg/g,cAMP含量401.30μg/g。制干率37.4%,干枣含总糖70.96%,酸0.44%。果核较大,椭圆形,核重0.80 g,核内无种子。

评价 该品种适应性广,抗逆性强,产量高而稳定。果实较小,可食率低,肉细汁少,有芳香味,品质中上,可鲜食或制干。

Xupuxiangzao

Source and Distribution The cultivar originated from Shuangjing Village and Dizhuang Town in Xupu of Hunan Province. It mainly spreads in Fenshuikou and Tangwancun.

Botanical Characters The large tree is spreading with a weak central leader trunk, sparse branches and leaves, and a deflective crown. The grayish-brown trunk bark has striped fissures. The reddish-brown 1-year-old shoots are 51.0 cm long with the internodes of 5.4 cm long. The secondary branches are 20.6 cm long with 5～7 nodes of medium curvature and less-developed thorns of 0.3～0.7 cm long. The conical mother fruiting spurs can germinate 3.9 deciduous fruiting shoots which are 25.8 cm long with 16 leaves. The small green leaves are ovate-lanceolate, with a sharply-cuspate apex, a round-cuneiform base and a thick saw-tooth pattern on the margin. The number of flowers is small, averaging 2 ones per inflorescence. The daytime-bloomed small blossoms have a diameter of 5.7 mm.

Biological Characters The tree has moderate vigor and weak suckering ability. It bears late, generally in the 4th year after planting and enters the full-bearing stage in the 10th year with stable fruit set and a high yield. A mature tree has a yield of 50 kg. In Taigu County of Shanxi Province, it matures in late September with a fruit growth period of 110 d. It is a mid-late-ripening variety with light physiological fruit-dropping.

Fruit Characteristics The small oblong fruit has a vertical and cross diameter of 3.21 cm and 2.52 cm, averaging 6.9 g (maximum 17.5 g) with a regular size. It has a flat-round shoulder, a shallow and medium-wide stalk cavity, a flat-round fruit apex and purplish-red thick skin. The tight-textured flesh has medium juice, a sweet taste and strong aroma. It has a better than normal quality for fresh-eating or dried fruits. The percentage of edible part of fresh fruit is 88.4%, and the content of TA and Vc is 0.27% and 421.06 mg per 100 g fresh fruit. The content of flavones and cAMP in mature fruit skin is 51.07 mg/g and 401.30 μg/g. The percentage of fresh fruits which can be made into dried ones is 37.4%. The content of TTS and TA in dried fruit is 70.96% and 0.44%. The large oval-shaped stone weighs 0.80 g without kernels.

Evaluation The cultivar has strong adaptability, strong resistance to adverse conditions and a high and stable yield. The small fruit has delicate and less juicy flesh with strong aroma. It has a better than normal quality for fresh-eating and dried fruits.

兼 用
Multipurpose Varieties
品 种

大荔林檎枣

品种来源及分布 原产和分布于陕西大荔的石槽、苏村等地，少量栽培。

植物学性状 树体中大，树姿半开张，树冠呈伞形。主干灰褐色，皮裂纹较浅，呈不规则块状，不易剥落。枣头红褐色，阳面灰褐色，较粗壮，平均长84.4cm，粗1.05cm，节间长8.6cm，蜡质多。二次枝平均长29.4cm，6节。针刺细短，质软易落。枣股粗大，圆柱形或圆锥形，平均抽生枣吊4.0个。枣吊长29.4cm，着叶16片。叶片较小，椭圆形，绿色，有光泽，先端渐尖，先端尖圆，叶基圆楔形，叶缘锯齿短、钝、锐相间，不整齐。花量少，每花序着花3朵，花较大，萼片短。

生物学特性 适应性较强，耐瘠薄，在沙质土壤生长良好。树势较强，发枝力中等，枣吊平均坐果0.59个，产量中等。在产地，4月中旬萌芽，5月下旬始花，9月下旬着色成熟，果实生育期110d左右，为晚熟品种类型。

果实性状 果实较大，长圆形，后期果近圆形，纵径3.69cm，横径3.22cm，单果重15.25g，大小整齐。果肩平圆，有数条凸起的辐射沟棱，延伸到胴部的中部或下部。梗洼中深、广，果顶凹陷。果面光滑，果皮薄，紫红色。果肉疏松，汁液少，味甜，可鲜食、制干和加工蜜枣，品质中等。鲜枣可食率91.7%，100g果肉维生素C含量410.28mg；果皮含黄酮10.33mg/g，cAMP含量492.62μg/g。果核大，短梭形或近圆形，核重1.27g，含仁率20%左右。

评价 该品种适应性较强，耐瘠薄，树体较大，树势较强，产量中等。果实大，肉质粗松，可食率较低，品质中等，可鲜食、制干和加工蜜饯。

Dalilinqinzao

Source and Distribution The cultivar originated from and spreads in Shicao and Sucun Village of Dali County in Shaanxi Province with a small quantity.

Botanical Characters The medium-sized tree is half-spreading with an umbrella-shaped crown. The grayish-brown trunk bark has irregular, shallow, massive fissures, uneasily shelled off. The reddish-brown 1-year-old shoots are 84.4 cm long and 1.05 cm thick, with grayish-brown sun-side. The internodes are 8.6 cm long, with much wax. The soft thorns are thin and short, easily falling off. The secondary branches are 29.4 cm long with 6 nodes. The column-shaped or conical mother fruiting spurs can germinate 4 deciduous fruiting shoots, which are 29.4 cm long with 16 leaves. The small oval-shaped leaves are green and glossy, with a gradually-cuspate apex, a round-cuneiform base and a short, bluntly-cuspate saw-tooth pattern on the margin. The number of flowers is small, averaging 3 per inflorescence. The large blossoms have short sepals.

Biological Characters The tree has strong adaptability, strong tolerance to poor soils, strong vigor and medium branching ability. It grows well in sandy soil. The deciduous bearing shoot bears 0.59 fruits on average, with a medium yield. In the original places, it germinates in mid-April, begins blooming in late May and matures in late September, with a fruit growth period of 110 d. It is a late-ripening variety.

Fruit Characteristics The large oblong fruit has a vertical and cross diameter of 3.69 cm and 3.22 cm, averaging 15.25 g with regular sizes. It has a flat-round shoulder with several protuberant ridges stretching to the middle part or to the bottom. It has a medium-deep and wide stalk cavity, a slightly-sunken fruit apex, a smooth surface and purplish-red thin skin. The flesh is loose-textured and sweet, with less juice. It can be used for fresh-eating, making dried fruits and candied fruits, with medium quality. The percentage of edible part of fresh fruit is 91.7%, and the content of Vc is 410.28 mg per 100 g fresh fruit. The content of flavones and cAMP in fruit skin is 10.33 mg/g and 492.62 μg/g. The large stone is short shuttle-shaped or nearly round, averaging 1.27 g. The percentage of containing kernels is 20%.

Evaluation The large plant of Dalilinqinzao cultivar has strong adaptability, strong tolerance to poor soils, strong vigor and a medium yield. The large fruit has loose-textured flesh with low edibility. It has medium quality for fresh-eating, dried fruits and candied fruits.

兼 用
Multipurpose Varieties
品 种

彬县酸疙瘩

品种来源及分布 别名酸疙瘩。原产陕西彬县的城关、水帘、南玉子等地。栽培历史悠久。

植物学性状 树体高大，树姿半开张，枝叶密度中等，树冠呈自然圆头形。主干皮裂深，裂片大，条状，容易剥落。枣头红褐色，长87.3cm，节间长7.7cm，蜡质少。二次枝长34.6cm，平均5节，弯曲度小。针刺发达。每枣股抽生枣吊4.0个。枣吊长21.7cm，着生叶片15片。叶片中大，叶长5.4cm，叶宽2.6cm，卵圆形，浓绿色，有光泽，先端锐尖，叶基圆形，叶缘具浅钝锯齿。花量较多，每花序着花6朵，花中大，花径6.0mm，昼开型。

生物学特性 适应性较广，树势强，发枝力中等。坐果率中等，枣头、2～3年和4～6年生枝吊果率分别为25.4%、38.7%和3.5%，产量中等，较稳定，成龄树一般株产50kg左右。在山西太谷地区，9月下旬果实成熟采收，果实生育期105d左右，为中晚熟品种类型。果实成熟期落果轻。

果实性状 果实较大，圆柱形或长圆形，纵径4.44cm，横径2.57cm，单果重15.5g，大小不整齐。果肩、果顶平圆。梗洼浅平，环洼大而浅。果顶略大，顶点凹陷。果面平整，皮薄。果肉浅绿色，质地酥脆，汁液较多，味酸甜，可鲜食和制干，品质中等。鲜枣可食率93.6%，含可溶性固形物33.00%，总糖26.26%，酸0.56%，100g果肉维生素C含量252.15mg；果皮含黄酮9.19mg/g，cAMP含量207.96μg/g。干枣含总糖73.02%，酸1.26%。果核大，纺锤形，核重1.0g，含仁率58.3%。

评价 该品种适应性较强，树体强健，产量中等。果实大，适宜鲜食和制干，品质中等。

Binxiansuangeda

Source and Distribution The cultivar originated from Chengguan, Shuilian and Nanyuzi in Binxian County of Shaanxi Province with a long cultivation history.

Botanical Characters The large tree is half-spreading with medium-dense branches and a natural-round crown. The trunk bark has deep striped fissures, easily shelled off. The reddish-brown 1-year-old shoots are 87.3 cm long. The internodes are 7.7 cm long with less wax and developed thorns. The secondary branches are 34.6 cm long with 5 nodes of small curvature. The mother fruiting spurs can germinate 4 deciduous fruiting shoots which are 21.7 cm long with 15 leaves. The medium-sized leaves are oval-shaped, 5.4 cm long and 2.6 cm wide, dark green and glossy, with a sharply-cuspate apex, a round base and a shallow, blunt saw-tooth pattern on the margin. There are many medium-sized flowers with a diameter of 6.0 mm produced, averaging 6 ones per inflorescence. It blooms in the daytime.

Biological Characters The tree has strong adaptability, strong vigor and medium branching ability, with medium fruit set, a medium and stable yield. The percentage of fruits to deciduous fruiting shoots of 1-year-old shoots, 2～3-year-old branches and 4～6-year-old ones is 25.4%, 38.7% and 3.5%. A mature tree has a yield of about 50 kg on average. In Taigu County of Shanxi Province, it ripens in late September with a fruit growth period of 105 d. It is a late-ripening variety with light fruit-dropping and fruit-cracking.

Fruit Characteristics The large column-shaped fruit has a vertical and cross diameter of 4.44 cm and 2.57 cm, averaging 15.5 g with irregular sizes. It has a flat-round shoulder and fruit apex, a shallow stalk cavity, a sunken top point, a smooth surface and thin skin. The light-green flesh is crisp, juicy, sour and sweet. It has medium quality for fresh-eating and dried fruits. The percentage of edible part of fresh fruit is 93.6%, and SSC, TTS, TA and Vc is 33.00%, 26.26%, 0.56% and 252.15 mg per 100 g fresh fruit. The content of flavones and cAMP in mature fruit skin is 9.19 mg/g and 207.96 μg/g, and the content of TTS and TA in dried fruit is 73.02% and 1.26%. The large spindle-shaped fruit stone weighs 1.0 g and the percentage of containing kernels is 58.3%.

Evaluation The cultivar has strong adaptability and strong vigor with medium productivity. The large fruit has medium quality for fresh-eating and dried fruits.

兼用
Multipurpose Varieties
品种

·365·

亚腰长红

品种来源及分布 别名笨枣、滑皮枣、青瓢躺枣、晚熟躺枣、枕头枣、马尾枣、长枣。是长红枣品种群的重要品种，栽培历史悠久，分布很广。鲁中南山区的宁阳、曲阜、泗水、邹县有大片集中栽培，相邻的兖州、微山、济宁、滕州、泰安、长清以及黄河以北的德州、惠民，河北沧州、衡水都有零星分布，至今山东庆云的周尹村，还有一株留传隋代的古树，栽培历史1 300年以上。

植物学性状 树体高大，树姿半开张，干性强，树冠呈自然圆头形或长圆形，树干浅灰褐色，裂纹较浅，呈不规则的宽条状。枣头红褐色，平均长81.5cm，阳面被覆灰白蜡质浮皮。二次枝平均长32.9cm，4～8节，弯曲度中等。针刺发达，长2.0～2.5cm，不易脱落。枣股圆柱形，平均抽生枣吊3.7个。枣吊长20.2cm，着叶14片。叶片椭圆形，叶长5.0cm，叶宽2.3cm，深绿色，富光泽，先端钝尖，叶基圆或广楔形，叶缘具粗锯齿，齿尖圆。花量较多，花序平均着花6朵。花中大，花径6.0mm，为昼开型。

生物学特性 树势强旺，发枝力强。结果年龄较晚，定植后3、4年开始结果，进入盛果期后产量高而稳定，株产50kg左右。在山西太谷地区，果实9月中旬开始着色，9月底成熟，果实生育期110d左右，为晚熟品种类型。成熟期遇雨易裂果。

果实性状 果实较小，长柱形，侧面略扁，中腰稍细瘦，纵径3.91cm，横径2.08cm，侧径1.89cm，单果重9.2g，大小整齐。果肩平圆，梗洼窄、深，果顶平圆，微凹呈一字沟纹。果面平滑，果皮红色，富光泽。果点小，分布稀疏，不明显。果肉绿白色，质地酥脆，汁液多，味甜，品质上等，适宜鲜食或制干。鲜枣可食率95.8%，果皮含黄酮16.19mg/g，cAMP含量146.70μg/g。制干率48.0%，干枣含总糖66.47%，酸0.91%。果核小，纺锤形，平均重0.39g。种仁不饱满，含仁率16.7%。

评价 该品种丰产性好，果实小，品质上等，适宜鲜食或制干。果实抗裂果能力较差。

Yayaochanghong

Source and Distribution The cultivar originated from and spreads in Dezhou City of Shandong Province.

Botanical Characters The large tree is half-spreading with a strong central leader trunk and a round or oblong crown. The grayish-brown trunk bark has shallow wide-striped fissures. The reddish-brown 1-year-old shoots are 81.5 cm long with the internodes of 32.9 cm long. There are generally 4～8 nodes of medium curvature. The thorns are well-developed, 2.0～2.5 cm long, uneasily falling off. The column-shaped mother fruiting spurs can germinate 3.7 deciduous fruiting shoots which are 20.2 cm long with 14 leaves. The oval-shaped leaves are dark green and glossy, 5.0 cm long and 2.3 cm wide, with a bluntly cuspate apex, a round or wide-cuneiform base and a thick saw-tooth pattern on the margin. There are many medium-sized flowers with a diameter of 6.0 mm produced, averaging 6 ones per inflorescence. It blooms in the daytime.

Biological Characters The tree has strong vigor and strong branching ability. It bears late, generally in the 3rd or 4th year after planting, with a high and stable yield in the full-bearing stage. A tree has a yield of 50 kg on average. In Taigu County of Shanxi Province, it begins coloring in mid-September and matures in late September with a fruit growth period of 110 d. It is a late-ripening variety with serious fruit-cracking if it rains in the maturing stage.

Fruit Characteristics The small column-shaped fruit is a little bit flat on the lateral sides and thin in the middle part, with a vertical and cross diameter of 3.91 cm and 2.08 cm, averaging 9.2 g with a regular size. It has a flat-round shoulder, a narrow and deep stalk cavity, a pointed fruit apex, a smooth surface and red glossy skin with small and sparse dots. The greenish-white flesh is crisp, juicy and sweet. It has an excellent quality for fresh-eating or dried fruits. The percentage of edible part of fresh fruit is 95.8%. The content of flavones and cAMP in mature fruit skin is 16.19 mg/g and 146.70 μg/g. The percentage of fresh fruits which can be made into dried ones is 48.0%. The content of TTS and TA in dried fruit is 66.47% and 0.91%. The small spindle-shaped stone weighs 0.39 g with a shriveled kernel. The percentage of containing kernels is 16.7%.

Evaluation The cultivar has high yield and a small fruit size. It has an excellent quality and is suitable for fresh-eating and making dried fruits. Yet it has poor resistance to fruit-cracking.

兼 用
Multipurpose Varieties
品 种

· 367 ·

泰安酥圆铃

品种来源及分布 别名酥圆铃、团铃、圆铃。原产和分布山东泰安一带，数量不多，零星栽种。

植物学性状 树体较大，树姿开张，树冠多呈自然乱头形。树干灰黑色，皮裂小块状，不易剥落。枣头黄褐色，平均枝长89.4cm，粗1.00cm，节间长7.8cm，蜡质少。针刺不发达。枣股圆锥形，平均抽生枣吊4.0个。枣吊短，较细，长28.0cm，着叶17片。叶片中大，卵圆形，浓绿色，少光泽，先端渐尖较长，叶基圆楔形，叶缘钝锯齿，先端尖圆。花量大，花序平均着花8朵。花中大，花径5.5～6.0mm，昼开型。

生物学特性 适应性强，耐干旱、瘠薄。树势和发枝力较强。结果较早，定植后3年开始结果，7～8年进入盛果期，产量高而稳定，枣吊平均结果0.42个。在山西太谷地区，9月下旬果实着色成熟，果实生育期113d，为晚熟品种类型。抗裂果，采前落果轻。

果实性状 果实较小，近圆形，纵径3.09cm，横径3.00cm，单果重8.3g，大小不整齐。果肩平圆，微凹。梗洼中深、广。果顶微凹。果面不平，有块状起伏。果皮紫红色，果肉厚，质地致密，汁液中多，甜味浓，略具酸味，制干、鲜食均可，品质中上。鲜枣含可溶性固形物34.00%，总糖24.19%，100g果肉维生素C含量390.39mg；果皮含黄酮8.54mg/g，cAMP含量159.46μg/g。果核较小，纺锤形，核重0.32g。大果核内偶有种仁。

评价 该品种适应性较强，耐旱瘠薄，产量高而稳定。果实较小，肉质硬脆略酥，可鲜食和制干，品质中上。

Taiansuyuanling

Source and Distribution The cultivar, also called Tuanling or Yuanling, originated from and spreads in Taian of Shandong Province with a small quantity. It is mainly planted in scattered regions.

Botanical Characters The large tree is spreading with an irregular crown. The grayish-black trunk bark has small massive fissures, uneasily shelled off. The yellowish-brown 1-year-old shoots are 89.4 cm long and 1.00 cm thick. The internodes are 7.8 cm long with less wax and less-developed thorns. The conical mother fruiting spurs can germinate 4 short and thin deciduous fruiting shoots which are 28.0 cm long with 17 leaves. The medium-sized leaves are oval-shaped and dark green, less glossy, with a long gradually-cuspate apex, a round-cuneiform base and a blunt saw-tooth pattern on the margin. There are many medium-sized flowers with a diameter of 5.5～6.0 mm produced, averaging 8 ones per inflorescence. It blooms in the daytime.

Biological Characters The tree has strong adaptability, strong vigor and strong branching ability, tolerant to drought and poor soils. It bears early, generally in the 3rd year after planting, and enters the full-bearing stage in the 7th or 8th year, with a high and stable yield. The deciduous fruiting shoot bears 0.42 fruits on average. In Taigu County of Shanxi Province, it matures in late September with a fruit growth period of 113 d. It is a late-ripening variety with strong resistance to fruit-cracking and light premature fruit-dropping.

Fruit Characteristics The small oval-shaped fruit has a vertical and cross diameter of 3.09 cm and 3.00 cm, averaging 8.3 g with irregular sizes. It has a flat-round shoulder (slightly-sunken), a medium-deep and wide stalk cavity, a slightly-sunken fruit apex, an unsmooth surface with some protuberances and purplish-red skin. The thick flesh is tight-textured, a little sour and strongly sweet, with medium juice. It has a better than normal quality for fresh-eating and dried fruits. The content of SSC, TTS and Vc in fresh fruit is 34.00%, 24.19% and 390.39 mg per 100 g fresh fruit. The content of flavones and cAMP in mature fruit skin is 8.54 mg/g and 159.46 μg/g. The small spindle-shaped stone weighs 0.32 g. Stones in large fruits occasionally contain a kernel.

Evaluation The cultivar has strong adaptability, tolerant to drought and poor soils with a high and stable yield. The srnall fruit has hard and crisp flesh. It has a better than normal quality suitable for fresh-eating and making dried fruits.

兼 用
Multipurpose Varieties
品 种

太谷端子枣

品种来源及分布　别名端子枣。原产和分布于山西太谷枣区，栽培数量不多。

植物学性状　树体高大，树姿较直立，干性强，树冠呈圆锥形。主干皮裂块状。枣头红褐色，平均长40.0cm，粗0.79cm，节间长10.3cm，蜡质少。针刺发达。二次枝平均长30.9cm，6节左右，弯曲度中等。枣股中大，平均抽生枣吊4.0个。枣吊长21.0cm，着叶13片。叶片中大，叶长6.5cm，叶宽3.2cm，卵圆形，浓绿色，较光亮，先端渐尖，叶基圆形，叶缘具钝锯齿。花量较多，花较小，花序平均着花6朵。

生物学特性　适土性强，树势强旺，发枝力较强，根蘖萌生力强，生长健旺。结果较早，第二、三年开始结果，产量高较稳定，枣吊平均坐果0.5个。在山西太谷地区，4月中旬萌芽，5月下旬始花，9月中旬进入脆熟期，果实生育期100d左右，为中熟品种类型。成熟期遇雨裂果严重。

果实性状　果实中大，倒卵圆形，纵径3.19cm，横径2.95cm，单果重12.1g，大小整齐。果肩平圆，梗洼中深、广，果顶平圆。果面平滑，果皮中厚，红色。果肉较厚，浅绿色，质地疏松，汁液少，味酸甜，可鲜食和制干，但品质较差。鲜枣可食率95.0%，含可溶性固形物32.40%，总糖27.56%，酸0.76%，100g果肉维生素C含量488.54mg；果皮含黄酮11.22mg/g，cAMP含量57.62μg/g。干枣含可溶性糖67.53%，酸1.35%。果核较大，纺锤形，核重0.61g，含仁率61.7%。

评价　该品种适土性强，树体高大强健，产量高，较稳定。可用于鲜食和制干，但品质较差，不抗裂果。

Taiguduanzizao

Source and Distribution　The cultivar originated from and spreads in the jujube producing areas of Taigu County of Shanxi Province with a small quantity.

Botanical Characters　The large tree is vertical with a strong central leader trunk and a conical crown. The trunk bark has massive fissures. The reddish-brown 1-year-old shoots are 40.0 cm long and 0.79 cm thick. The internodes are 10.3 cm long with less wax and developed thorns. The secondary branches are 30.9 cm long with 6 nodes of medium curvature. The medium-sized mother fruiting spurs can germinate 4 deciduous fruiting shoots which are 21.0 cm long with 13 leaves. The medium-sized leaves are oval-shaped, 6.5 cm long and 3.2 cm wide, dark green and glossy, with a gradually-cuspate apex, a round base and a blunt saw-tooth pattern on the margin. There are many small flowers produced, averaging 6 ones per inflorescence.

Biological Characters　The tree has strong adaptability, strong vigor, strong branching ability and strong suckering ability. It bears early, generally in the 2nd or 3rd year after planting with a high and stale yield. The deciduous fruiting shoot bears 0.5 fruits on average. In Taigu County of Shanxi Province, it germinates in mid-April, begins blooming in late May and enters the crisp-maturing stage in mid-September with a fruit growth period of 100 d. It is a mid-ripening variety with serious fruit-cracking if it rains in the maturing stage.

Fruit Characteristics　The medium-sized obovate fruit has a vertical and cross diameter of 3.19 cm and 2.95 cm, averaging 12.1 g with a regular size. It has a flat-round shoulder, a medium-deep and wide stalk cavity, a flat-round fruit apex, a smooth surface and medium-thick red skin. The light-green flesh is thick, loose-textured, sour and sweet with less juice. It can be used for fresh-eating and making dried fruits, yet with a low quality. The percentage of edible part of fresh fruit is 95.0%, and SSC, TTS, TA and Vc is 32.40%, 27.56%, 0.76% and 488.54 mg per 100 g fresh fruit. The content of flavones and cAMP in mature fruit skin is 11.22 mg/g and 57.62 μg/g, and the content of TTS and TA in dried fruit is 67.53% and 1.35%. The large spindle-shaped stone weighs 0.61 g and the percentage of containing kernels is 61.7%.

Evaluation　The large plant of Taiguduanzizao cultivar has strong adaptability and strong vigor with a high and stable yield, and serious fruit-cracking. It can be used for fresh-eating and making dried fruits with a low quality.

新郑鸡蛋枣

品种来源及分布 原产和分布于河南省新郑市等地。

植物学性状 树体高大，树姿较开张，树冠乱头形。主干皮裂条状。枣头黄褐色，平均长81.5cm，节间长6.4cm，蜡质多。二次枝长31.7cm，5～8节，弯曲度小。针刺发达。枣股平均抽生枣吊3.4个。枣吊长20.8cm，着叶14片。叶片较小，叶长4.2cm，叶宽2.8cm，卵圆形，平展，先端锐尖，叶基心形，叶缘锯齿细。花量少，花序平均着花3朵，花较小，花径5.9mm。

生物学特性 树势强，萌芽率和成枝力强，结果较早，一般定植第二年结果，10年左右进入盛果期，产量较低，枣头、2～3年和4～6年生枝的吊果率分别为55.4%、24.7%和2.0%。在山西太谷地区，9月下旬果实进入脆熟期，10月上旬成熟采收，果实生育期115d以上，属极晚熟品种。成熟期遇雨易裂果。

果实性状 果个大，卵圆形，纵径4.25cm，横径3.50cm，单果重21.8g，大小整齐。果皮薄，紫红色，果面平滑。果点大，分布稀疏。梗洼窄而深。果顶微凹，柱头遗存，不明显。肉质细嫩酥脆，味甜，汁液中多，品质上，可鲜食或制干。鲜枣可食率96.8%，含可溶性固形物27.00%，总糖25.02%，酸0.45%，糖酸比55.6∶1，100g果肉维生素C含量354.89mg；果皮含黄酮7.48mg/g，cAMP含量141.50μg/g。干枣含总糖65.34%，酸0.90%。核较小，椭圆形，纵径1.70cm，横径0.70cm，核重0.70g，种仁不饱满，含仁率31.7%。

评价 该品种树体高大，树势强，产量较低。果个大，品质优异，适宜鲜食或制干。果实抗裂果能力较差。

Xinzhengjidanzao

Source and Distribution The cultivar originated from and spreads in Xinzheng of Henan Province.

Botanical Characters The large tree is spreading with an irregular crown. The trunk bark has striped fissures. The yellowish-brown 1-year-old shoots are 81.5 cm long with the internodes of 6.4 cm long, and much wax. The secondary branches are 31.7 cm long with 5～8 nodes of small curvature and developed thorns. The mother fruiting spurs can germinate 3.4 deciduous fruiting shoots, which are 20.8 cm long with 14 leaves. The small leaves are 4.2 cm long and 2.8 cm wide, oval-shaped and flat, with a sharply-cuspate apex, a heart-shaped base and a thin saw-tooth pattern on the margin. The number of flowers is small, averaging 3 ones per inflorescence. The small blossoms have a diameter of 5.9 mm.

Biological Characters The tree has strong vigor, strong germination and branching ability. It bears early, generally in the 2nd year after planting and enters the full-bearing stage in the 10th year, with a low yield. The percentage of fruits to deciduous fruiting shoots of 1-year-old shoots, 2～3-year-old branches and 4～6-year-old ones is 55.4%, 24.7% and 2.0%. In Taigu County of Shanxi Province, it enters the crisp-maturing stage in late September and is harvested in early October with a fruit growth period of 115 d. It is an extremely late-ripening variety with serious fruit-cracking if it rains in the maturing stage.

Fruit Characteristics The large oval-shaped fruit has a vertical and cross diameter of 4.25 cm and 3.50 cm, averaging 21.8 g with a regular size. It has purplish-red thin skin, a smooth surface with large and sparse dots, a narrow and deep stalk cavity, a slightly sunken fruit apex and a remnant stigma. The flesh is crisp, sweet and juicy. It has a good quality for fresh-eating and dried fruits. The percentage of edible part of fresh fruit is 96.8%, and SSC, TTS, TA and Vc is 27.00%, 25.02%, 0.45% and 354.89 mg per 100 g fresh fruit. The SAR is 55.6∶1. The content of flavones and cAMP in mature fruit skin is 7.48 mg/g and 141.50 μg/g. The content of TTS and TA in dried fruit is 65.34% and 0.9%. The small oval-shaped stone has a vertical and cross diameter of 1.70 cm and 0.70 cm averaging 0.70 g with a shriveled kernel. The percentage of containing kernels is 31.7%.

Evaluation The large plant of Xinzhengjidanzao cultivar has strong tree vigor. The large fruit has a good quality for fresh-eating and dried fruits. It has poor resistance to fruit-cracking.

兼 用
Multipurpose Varieties
品 种

清 徐 圆 枣

品种来源及分布　原产和分布于山西清徐县。栽培数量不多。

植物学性状　树体较大，树姿开张，干性中强，枝系粗壮较密，树冠呈自然圆头形。主干皮裂条状。枣头粗壮，红褐色，平均长79.0cm，粗1.09cm，节间长10.0cm，蜡质多。针刺不发达。枣股中大，抽吊力较强，平均抽生枣吊4.0个。枣吊长14.5cm，着叶9片。叶片较大，叶长6.8cm，叶宽2.8cm，椭圆形，绿色，先端渐尖，叶基圆形。花量大，花序平均着花9朵，花朵较大，花径6.6～7.5mm，为昼开型。

生物学特性　树势旺，发枝力强。结果较迟，产量中等，2～3年生枝坐果能力较强，枣头、2～3年、4～6年和6年以上生枝吊果率分别为0、103.4%、58.1%和6.8%。在山西太谷地区，9月下旬果实成熟采收，果实生育期110d左右，为晚熟品种类型。成熟期落果严重，遇雨易裂果。

果实性状　果实较大，卵圆形，纵径3.12cm，横径3.10cm，单果重14.6g，大小较整齐。果肩圆斜，梗洼浅而广，果顶平圆。果面平滑。果皮薄，红色。果肉中厚，白色，质地疏松，汁液中多，味酸甜，适宜鲜食和制干，品质中等。鲜枣可食率95.3%，含可溶性固形物33.00%，总糖24.76%，酸0.56%，100g果肉维生素C含量264.95mg。完熟期果皮含黄酮6.71mg/g，cAMP含量216.25μg/g。干枣含总糖69.57%，酸1.58%。果核中大，倒纺锤形，核重0.68g，含仁率68.3%。

评价　该品种耐旱，抗枣疯病。结果晚，产量中等。果实较大，品质中等，可鲜食和制干。但采前落果、裂果严重。

Qingxuyuanzao

Source and Distribution　The cultivar originated from and spreads in Bianshan Village of Qingxu County in Shanxi Province, with a small quantity.

Botanical Characters　The large tree is spreading with a medium-strong central leader trunk, dense branches and a natural-round crown. The trunk bark has striped fissures. The thick reddish-brown 1-year-old shoots are 79.0 cm long and 1.09 cm thick. The internodes are 10.0 cm long with much wax and less-developed thorns. The medium-sized mother fruiting spurs can germinate 4 deciduous fruiting shoots, which are 14.5 cm long with 9 leaves. The large leaves are 6.8 cm long and 2.8 cm wide, oval-shaped and green, with a gradually-cuspate apex and a round base. There are many large flowers with a diameter of 6.6～7.5 mm produced, averaging 9 per inflorescence. It blooms in the daytime.

Biological Characters　The tree has strong vigor and strong branching ability, bearing late with a medium yield. 2～3-year-old branches have strong fruiting ability. The percentage of fruits to deciduous fruiting shoots of 1-year-old shoots, 2～3-year-old branches, 4～6-year-old ones and over-6-year-old ones is 0, 103.4%, 58.1% and 6.8%. In Taigu County of Shanxi Province, it matures in late September with a fruit growth period of 110 d. It is a late-ripening variety with serious fruit-dropping in the maturing stage and serious fruit-cracking if it rains.

Fruit Characteristics　The large nearly-round fruit has a vertical and cross diameter of 3.12 cm and 3.10 cm, averaging 14.6 g with a regular size. It has a deflective-round shoulder, a shallow and wide stalk cavity, a flat-round fruit apex, a smooth surface and thin red skin. The medium-thick flesh is white, loose-textured, sour and sweet with medium juice. It has medium quality for fresh-eating and dried fruits. The percentage of edible part of fresh fruit is 95.3%, and SSC, TTS, TA and Vc is 33.00%, 24.76%, 0.56% and 264.95 mg per 100 g fresh fruit. The content of flavones and cAMP in completely-mature fruit skin is 6.71 mg/g and 216.25 μg/g. The content of TTS and TA in dried fruit is 69.57% and 1.58%. The medium-sized stone is inverted spindle-shaped, averaging 0.68 g. The percentage of containing kernels is 68.3%.

Evaluation　The cultivar has strong tolerance to drought, and strong resistance to jujube witches' broom. It bears late with a medium yield. The large fruit has medium quality for fresh-eating and dried fruits. Yet both premature fruit-dropping and fruit-cracking are serious.

兼 用
Multipurpose Varieties
品 种

大荔墩墩枣

品种来源及分布 别名墩墩枣。原产和分布于陕西华阴的华山乡和大荔的马坊等地。

植物学性状 树体中大，树姿开张，树冠圆锥形。主干灰褐色，皮裂较深，呈宽条状，容易剥落。枣头红褐色，长62.2cm左右，粗1.10cm，节间长7.7cm。二次枝长33.1cm，7节左右。针刺不发达。枣股圆锥形，平均抽生枣吊4个。枣吊长22.7cm，着叶16片。叶片较小，叶长5.0cm，叶宽2.4cm，椭圆形，绿色，有光泽，先端尖圆或微凹，叶基圆形，叶缘上卷，具钝锯齿。花量中等，花序平均着花4朵，花小，花径5.0mm。

生物学特性 适应性中等，在壤土和沙壤土中栽植表现良好。树势中等，发枝力弱，枝系较稳定。枣头、2～3年和4～6年生枝吊果率分别为12.9%、25.0%和13.0%。产量中等，成龄树株产50kg左右。在山西太谷地区，9月中旬果实着色，9月下旬成熟，果实发育期110d，为晚熟品种类型。

果实性状 果实大，扁圆形，纵径3.84cm，横径3.28cm，单果重19.6g，大小较整齐。果肩平圆，向一侧偏斜，梗洼深窄。果顶微凹。果面粗糙，果皮厚，浅红色。果肉浅绿色，质地疏松，汁液中等多，味甜，鲜食和制干均可，品质中等。鲜枣可食率95.3%，含总糖24.93%，100g果肉维生素C含量372.29mg；果皮含黄酮19.49mg/g，cAMP含量219.26μg/g。干枣含总糖68.34%，酸1.03%。果核较大，倒纺锤形，核重0.92g，含仁率13.3%。

评价 该品种适应性中等，要求土壤条件较好。树势较强，产量中等，果实大，品质中等，可鲜食和制干。

Dalidundunzao

Source and Distribution The cultivar originated from and spreads in Huashan Village of Huayin County and Mafang Village of Dali County in Shaanxi Province.

Botanical Characters The medium-sized tree is spreading with a conical crown. The grayish-brown trunk bark has deep wide-striped fissures, easily shelled off. The reddish-brown 1-year-old shoots are 62.2 cm long and 1.10 cm thick. The internodes are 7.7 cm long with less-developed thorns. The secondary branches are 33.1 cm long with 7 nodes. The conical mother fruiting spurs can germinate 4 deciduous fruiting shoots which are 22.7 cm long with 16 leaves. The small oval-shaped leaves are 5.0 cm long and 2.4 cm wide, green and glossy, with a cuspate-round apex, a round base and a blunt saw-tooth pattern on the rolling-up margin. The number of flowers is medium large, averaging 4 ones per inflorescence. The small blossoms have a diameter of 5.0 mm.

Biological Characters The tree has medium adaptability, moderate vigor, weak branching ability, stable branches and medium yield. It grows well in loam soil and sandy loam soil. The percentage of fruits to deciduous fruiting shoots of 1-year-old shoots, 2～3-year-old branches and 4～6-year-old ones is 12.9%, 25.0% and 13.0%. A mature tree has a yield of 50 kg on average. In Taigu County of Shanxi Province, it begins coloring in mid-September and matures in late September with a fruit growth period of 110 d. It is a late-ripening variety.

Fruit Characteristics The large oblate fruit has a vertical and cross diameter of 3.84 cm and 3.28 cm, averaging 19.6 g with a regular size. It has a flat-round shoulder, a deep and narrow stalk cavity, a slightly-sunken fruit apex, a rough surface and light-red thick skin. The light-green flesh is loose-textured and sweet with medium juice. It has medium quality for fresh-eating and dried fruits. The percentage of edible part of fresh fruit is 95.3%, and the content of TTS and Vc is 24.93% and 372.29 mg per 100 g fresh fruit. The content of flavones and cAMP in mature fruit skin is 19.49 mg/g and 219.26 μg/g, and the content of TTS and TA in dried fruit is 68.34% and 1.03%. The large stone is inverted spindle-shaped, averaging 0.92 g. The percentage of containing kernels is 13.3%.

Evaluation The cultivar has medium adaptability with a high demand on soils, moderate vigor and medium productivity. The large fruit has medium quality for fresh-eating and dried fruits.

兼 用
Multipurpose Varieties
品 种

献县圆小枣

品种来源及分布 原产和分布于河北献县。

植物学性状 树体中大，树姿较开张，树冠呈圆头形。主干皮裂条状。枣头黄褐色，平均长74.1cm，节间长7.1cm，蜡质少。二次枝长33.9cm，5～8节，弯曲度中等。针刺发达。枣股平均抽生枣吊3.3个。枣吊长21.3cm，着叶15片。叶片中等大，叶长5.9cm，叶宽2.7cm，卵圆形，部分叶片反卷，先端锐尖，叶基心形，叶缘具钝锯齿。花量多，花序平均着花11朵，花较大，花径6.6mm。

生物学特性 树势强，萌芽率和成枝力强。幼龄枝坐果率较高，枣头、2～3年、4～6年和6年生以上枝的吊果率分别为82.5%、108.5%、17.0%和26.5%。一般定植第三年结果，10年左右进入盛果期，产量较低。在山西太谷地区，9月下旬果实进入脆熟期，10月上旬成熟采收，果实生育期115d以上，属极晚熟品种。成熟期遇雨裂果较轻。

果实性状 果个小，卵圆形，纵径2.67cm，横径2.07cm，单果重6.3g，大小较整齐。果皮中等厚，浅红色，果面平滑。果点小而稀疏。梗洼窄深。果顶平圆，柱头遗存，不明显。肉质较致密，味甜，汁液多，品质上，适宜鲜食或制干。鲜枣可食率95.4%，含可溶性固形物33.00%，总糖26.27%，100g果肉维生素C含量473.35mg；果皮含黄酮10.36mg/g，cAMP含量66.14μg/g。制干率55.9%，干枣含总糖68.97%，酸1.09%。果核小，椭圆形，平均重0.29g。种仁较饱满，含仁率86.7%。

评价 该品种树体中大，树势强，产量较低。果个小，品质上等，适宜鲜食或制干。果实抗裂果能力较强。

Xianxianyuanxiaozao

Source and Distribution The cultivar originated from and spreads in Xianxian County of Hebei Province.

Botanical Characters The medium-sized tree is spreading with a round crown. The trunk bark has striped fissures. The yellowish-brown 1-year-old shoots are 74.1 cm long with the internodes of 7.1 cm long less wax and developed thorns. The secondary branches are 33.9 cm long with 5～8 nodes of medium curvature. The mother fruiting spurs can germinate 3.3 deciduous fruiting shoots which are 21.3 cm long with 15 leaves. The medium-sized leaves are oval-shaped, 5.9 cm long and 2.7 cm wide, with a sharply-cuspate apex, a heart-shaped base and a blunt saw-tooth pattern on the margin. There are many large flowers with a diameter of 6.6 mm produced, averaging 11 ones per inflorescence.

Biological Characters The tree has strong vigor, strong germination and branching ability. Young branches have high fruit set. The percentage of fruits to deciduous fruiting shoots of 1-year-old shoots, 2～3-year-old branches, 4～6-year-old ones and over-6-year-old ones is 82.5%, 108.5%, 17.0% and 26.5%. It generally bears in the 3rd year after planting and enters the full-bearing stage in the 10th year with a low yield. In Taigu County of Shanxi Province, it enters the crisp-maturing stage in late September and matures in early October with a fruit growth period of 115 d. It is an extremely late-ripening variety with light fruit-cracking even if it rains in the maturing stage.

Fruit Characteristics The small oval-shaped fruit has a vertical and cross diameter of 2.67 cm and 2.07 cm, averaging 6.3 g with a regular size. It has medium-thick and light-red skin, a smooth surface with small and sparse dots, a narrow and deep stalk cavity, a flat-round fruit apex and a remnant stigma. The tight-textured flesh is sweet and juicy. It has a good quality for fresh-eating and dried fruits. The percentage of edible part of fresh fruit is 95.4%, and SSC, TTS and Vc is 33.00%, 26.27% and 473.35 mg per 100 g fresh fruit. The content of flavones and cAMP in mature fruit skin is 10.36 mg/g and 66.14 μg/g. The small oval-shaped stone weighs 0.29 g with a well-developed kernel. The percentage of containing kernels is 86.7%.

Evaluation The medium-sized plant of Xianxianyuanxiaozao cultivar has strong tree vigor with a low yield. The small fruit has a superior quality and is suitable for fresh-eating and making dried fruits. It has strong resistance to fruit-cracking.

兼 用
Multipurpose Varieties
品 种

· 379 ·

沧 县 傻 枣

品种来源及分布　原产和分布于河北沧县黄递铺、辛庄和枣强县等地。数量较少，多与其他品种混栽。

植物学性状　树体中大，树姿半开张，枝叶较密，树冠呈自然半圆形。主干灰黑色，树皮粗糙，裂纹条状，易剥落。枣头紫褐色，枝长80.6cm，粗1.11cm，节间长6.5cm，蜡质多。二次枝长37.8cm，节间长6cm，7节，弯曲度大。针刺发达。枣股圆锥形，平均抽生枣吊3个。枣吊细短，长20.1cm，着叶11片。叶片卵圆形，叶长6.2cm、宽3.4cm，浓绿色，先端渐尖，叶基圆形或心形，叶缘具钝锯齿。属多花型，每花序着花13朵，花朵大。

生物学特性　适应性强，耐旱、耐涝、耐盐碱。树势和发枝力均强。坐果率高，产量稳定，自然落果轻，枣头和2~3年生枝吊果率分别为24.8%和48.3%。在山西太谷地区，10月上旬采收，果实生育期120d，为极晚熟品种类型。抗裂果。

果实性状　果实较小，圆柱形，纵径3.19cm，横径2.23cm，单果重7.9g，大小整齐。果肩平，梗洼中广、深。果顶平圆，先端凹下，柱头遗存。果面平滑光亮，果皮厚，红色。果肉浅绿色，质地致密，汁液少，味酸甜，品质中等，可鲜食和制干。鲜枣可食率92.7%，含可溶性固形物28.00%，100g果肉维生素C含量545.28mg，果皮含黄酮4.34mg/g，cAMP含量257.77μg/g。制干率57.7%，干枣含总糖60.44%，酸1.23%。果核大，纺锤形，核重0.58，含仁率90.0%，但不饱满。

评价　该品种适应性强，耐旱涝和盐碱，抗裂果，产量较高，品质中等，鲜食和制干兼用。

Cangxianshazao

Source and Distribution　The cultivar originated from and spreads in Huangdipu and Xinzhuang of Cangxian County and in Zaoqiang County of Hebei Province. It has a small quantity, mostly planted in mixture with other varieties.

Botanical Characters　The medium-sized tree is half-spreading with dense branches and a semi-round crown. The grayish-black trunk bark has rough striped fissures, easily shelled off. The purplish-brown 1-year-old shoots are 80.6 cm long and 1.11 cm thick. The internodes are 6.5 cm long with much wax and developed thorns. The secondary branches are 37.8 cm long with 7 nodes of large curvature. The internodes on the secondary branches are 6 cm long. The conical mother fruiting spurs can germinate 3 deciduous fruiting shoots which are 20.1 cm long with 11 leaves. The leaves are oval-shaped and dark green, 6.2 cm long and 3.4 cm wide, with a gradually-cuspate apex, a round or heart-shaped base and a blunt saw-tooth pattern on the margin. There are many large flowers produced, averaging 13 ones per inflorescence.

Biological Characters　The tree has strong adaptability, tolerant to drought, water-logging and saline-alkaline, with strong vigor, strong branching ability, high fruit set and a stable yield. The percentage of fruits to deciduous fruiting shoots of 1-year-old shoots and 2~3-year-old branches is 24.84% and 48.3%. In Taigu County of Shanxi Province, it matures in mid-October, with a fruit growth period of 125 d. It is an extremely late-ripening variety with light natural fruit-dropping and strong resistance to fruit-cracking.

Fruit Characteristics　The small column-shaped fruit has a vertical and cross diameter of 3.19 cm and 2.23 cm, averaging 7.9 g with a regular size. It has a flat shoulder, a medium-wide and deep stalk cavity, a flat-round fruit apex, a remnant stigma, a smooth and glossy surface and thick red skin. The light-green flesh is tight-textured, sour and sweet with less juice. It has medium quality for fresh-eating and dried fruits. The percentage of edible part of fresh fruit is 92.7%, and SSC and Vc is 28.00% and 545.28 mg per 100 g fresh fruit. The content of flavones and cAMP in mature fruit skin is 4.34 mg/g and 257.77 μg/g. The large spindle-shaped stone weighs 0.58 g, and the percentage of containing kernels is 90.0%. Most kernels are shriveling.

Evaluation　The cultivar has strong adaptability, tolerant to drought, water-logging and saline-alkaline with a high yield. The fruit has medium quality and is suitable for fresh-eating and making dried fruits.

兼用
Multipurpose Varieties
品种

· 381 ·

斑　枣

品种来源及分布　分布于河北保定及石家庄地区，以望都县数量较多。

植物学性状　树体高大，树姿半开张，干性较强，树冠圆头形。主干灰黑色，树皮呈长条状深裂，不易剥落。枣头红褐色，平均长92.6cm，粗1.06cm，节间长9.3cm，蜡质少。针刺发达。枣股圆锥形，平均抽生枣吊3.0个。枣吊平均长24.2cm，着叶15片。叶片卵圆形，深绿色，有光泽，先端锐尖，叶基心形。叶缘钝锯齿。花序平均着花7朵。

生物学特性　抗风和耐涝力强，对土壤适应性强，以黏壤土生长最好。树势强盛，发枝力中等，坐果率较高，枣头、2～3年和3年以上枝的吊果率分别为8.0%、111.7%和87.3%。栽植后2年开始结果，15年左右进入盛果期，产量较稳定，盛果期树株产15kg，最高株产50kg。在山西太谷地区，10月中旬成熟采收，果实生育期120d以上，属极晚熟品种类型。

果实性状　果实中大，扁圆形，单果重13.2g，大小较整齐。果肩平圆，梗洼深广。果柄长4.1mm。果顶微凹。果皮较厚，紫红色，有光泽。果肉浅绿色，质地较致密，汁液中多，味甜，品质中等，可鲜食和制干。鲜枣可食率93.6%，含可溶性固形物31.80%，总糖29.51%，酸0.59%，100g果肉维生素C含量376.00mg；果皮含黄酮5.47mg/g，cAMP含量388.36μg/g。果核大，椭圆形，核重0.85g，含仁率15.5%。

评价　该品种风土适应性强，坐果稳定，产量中等，品质中等，鲜食和制干兼用，但果核大，可食率低。

Banzao

Source and Distribution　The cultivar spreads in Baoding and Shijiazhuang of Hebei Province.

Botanical Characters　The large tree is half-spreading with a strong central leader trunk and a round crown. The grayish-black trunk bark has deep striped fissures, uneasily shelled off. The reddish-brown 1-year-old shoots are 92.6 cm long and 1.06 cm thick. The internodes are 9.3 cm long with less wax and developed thorns. The conical mother fruiting spurs can germinate 3 deciduous fruiting shoots which are 24.2 cm long with 15 leaves. The oval-shaped leaves are dark green and glossy, with a sharply-cuspate apex, a heart-shaped base and a blunt saw-tooth pattern on the margin. Each inflorescence has 7 flowers.

Biological Characters　The tree has strong vigor, medium branching ability and strong adaptability, with strong tolerance to wind and water-logging. It grows best in clay loam soil with high fruit set. The percentage of fruits to deciduous fruiting shoots of 1-year-old shoots, 2～3-year-old branches and over-3-year-old ones is 8.0%, 111.7% and 87.3%. It generally bears in the 2nd year after planting and enters the full-bearing stage in the 15th year, with a stable yield. A tree in its full bearing stage can produce 15 kg (maximum 50 kg) of fresh jujubes. In Taigu County of Shanxi Province, it matures in mid-October with a fruit growth period of over 120 d. It is an extremely late-ripening variety.

Fruit Characteristics　The medium-sized oblate fruit weighs 13.2 g with a regular size . It has a flat-round shoulder, a deep and wide stalk cavity, a stalk of 4.1 mm long, a slightly-sunken fruit apex and thick, purplish-red and glossy skin. The light-green flesh is tight-textured, juicy and sweet. It has a good quality for fresh-eating and dried fruits. The percentage of edible part of fresh fruit is 93.6%, and SSC, TTS, TA and Vc is 31.80%, 29.51%, 0.59% and 376.00 mg per 100 g fresh fruit. The content of flavones and cAMP in mature fruit skin is 5.47 mg/g and 388.36 μg/g. The large oval-shaped stone weighs 0.85 g with the percentage of containing kernels of 15.5%.

Evaluation　The cultivar has strong adaptability, bearing early with stable fruit set and medium productivity. The fruit has a good quality for fresh-eating and dried fruits. Yet it has a large stone, which greatly reduces the edibility.

北 京 笨 枣

品种来源及分布 原产和分布在北京海淀区北安河一带。

植物学性状 树体较大，树姿开张，枝系较密，树冠呈乱头形。主干黑褐色，树皮裂片大，块状剥落，枣头阳面褐色，枣头平均长42.3cm，粗0.85cm，节间长6.4cm，蜡质少。二次枝平均长21.2cm，5节，弯曲度中等，无针刺。枣股圆柱形，抽生枣吊3～5个。枣吊长17.5cm，着叶9片。叶片中大，椭圆形，绿色，先端钝尖，叶基截形，叶缘锯齿粗。花量多，花序平均着花10朵。

生物学特性 风土适应性强，树势和发枝力较强，坐果率高而稳定，枣头、2～3年、4～6年和6年以上生枝吊果率分别为70.7%、165.9%、168.3%和77.3%，丰产性好。在山西太谷地区，4月下旬萌芽，5月底始花，9月中旬果实成熟，果实生育期95～100d，为早中熟品种类型。

果实性状 果实较小，卵圆形，纵径2.81cm，横径2.43cm，单果重9.0g，大小整齐。果肩平，梗洼中深、广，果顶平圆。果面平整有光泽，果点大而稀疏，较明显。果皮红色，中厚。果肉浅绿色，质地较致密，较细，汁液多，味极甜，宜鲜食和制干，品质中上等。鲜枣可食率97.4%，含可溶性固形物30.40%，总糖28.25%，酸0.39%，100g果肉维生素C含量467.70mg；果皮含黄酮9.53mg/g，cAMP含量60.48μg/g。果核较小，椭圆形，核重0.23g，核内多有种子。

评价 该品种树冠较大，树势较强，丰产性好。果实肉质致密较硬，汁多，味甜，鲜食制干均可。

Beijingbenzao

Source and Distribution The cultivar originated from and spreads in Beianhe of Haidian District in Beijing.

Botanical Characters The large tree is spreading with dense branches and an irregular crown. The blackish-brown trunk bark has large massive fissures. The 1-year-old shoots have a brown sun-side, 42.3 cm long and 0.85 cm thick. The internodes are 6.4 cm long with less wax and no thorns. The secondary branches are 21.2 cm long with 5 nodes of medium curvature. The column-shaped mother fruiting spurs can germinate 3～5 deciduous fruiting shoots which are 17.5 cm long with 9 leaves. The medium-sized leaves are oval-shaped and green, with a blunt-cuspate apex, a truncate base and a thick saw-tooth pattern on the margin. There are many flowers produced, averaging 10 ones per inflorescence.

Biological Characters The tree has strong adaptability, strong vigor and strong branching ability, with high and stable fruit set and a high yield. The percentage of fruits to deciduous fruiting shoots of 1-year-old shoots, 2～3-year-old branches, 4～6-year-old ones and over-6-year-old ones is 70.7%, 165.9%, 168.3% and 77.3%. In Taigu County of Shanxi Province, it germinates in late April, begins blooming in late May and matures in mid-September with a fruit growth period of 95～100 d. It is an early-mid-ripening variety.

Fruit Characteristics The small oval-shaped fruit has a vertical and cross diameter of 2.81 cm and 2.43 cm, averaging 9.0 g with a regular size. It has a flat shoulder, a medium-deep and wide stalk cavity, a flat-round fruit apex, a smooth and glossy surface, and medium-thick red skin. The light-green flesh is tight-textured, delicate and juicy, strongly sweet. It has a better than normal quality for fresh-eating and dried fruits. The percentage of edible part of fresh fruit is 97.4%, and SSC, TTS, TA and Vc is 30.40%, 28.25%, 0.39% and 467.70 mg per 100 g fresh fruit. The content of flavones and cAMP in mature fruit skin is 9.53 mg/g and 60.48 μg/g. The small oval-shaped stone weighs 0.23 g. Most stones contain a kernel.

Evaluation The cultivar has a large crown, strong vigor and high productivity. The fruit has tight-textured, hard and sweet flesh with medium juice and is suitable for fresh-eating and making dried fruits.

兼 用 品 种
Multipurpose Varieties

衡阳珍珠枣

品种来源及分布 别名珍珠枣。原产湖南衡山、衡东两地。

植物学性状 树体较大，树姿开张，树冠偏斜形。主干灰褐色，粗糙，皮裂条状，易剥落。枣头红褐色，平均长57.9cm，粗0.86cm，节间平均长6.6cm，蜡质多。二次枝长26.4cm，节数6节，弯曲度中等。针刺不发达。枣股圆柱形或圆锥形，平均抽生枣吊4.0个。枣吊长25.2cm，着叶17片。叶片小，椭圆形，绿色，光亮，先端急尖，叶基圆楔形，叶缘锯齿小，齿角钝。花量较多，枣吊平均着花6朵，花朵小，花径4.9mm。

生物学特性 适应性强，树势强旺。结果早，坐果稳定，产量高，定植后2年开始结果，20年左右达到盛果期，一般株产鲜枣40kg左右，枣吊平均结果0.91个。在山西太谷地区，9月中旬果实成熟，果实生育期105d左右，为早中熟品种类型。自然落果少，大小年不明显，抗裂果。

果实性状 果实小，卵圆形，纵径2.29cm，横径1.98cm，单果重5.3g，最大10.9g。果肩圆整，梗洼浅、窄。果顶略凹下。果皮薄，红色。果肉浅绿色，质地较致密，细脆，汁液少，味甜，品质中等，鲜食制干兼用。鲜枣可食率90.2%，含可溶性固形物23.92%，酸0.45%，100g果肉维生素C含量396.00mg，果皮含黄酮6.19mg/g，cAMP含量222.35μg/g。干枣含总糖55.00%。果核较小，椭圆形，核重0.52g，含仁率90.0%。

评价 该品种树势健壮，适应性强，产量高而稳定，抗裂果。果个小，果实肉质细脆，甜酸适口，品质中等，可鲜食和制干。

Hengyangzhenzhuzao

Source and Distribution The cultivar originated from Hengshan and Hengdong in Hunan Province.

Botanical Characters The large tree is spreading with a deflective crown. The grayish-brown trunk bark has rough striped fissures, easily shelled off. The reddish-brown 1-year-old shoots are 57.9 cm long and 0.86 cm thick. The internodes are 6.6 cm long with much wax and less-developed thorns. The secondary branches are 26.4 cm long with 6 nodes of medium curvature. The conical or column-shaped mother fruiting spurs can germinate 4 deciduous fruiting shoots which are 25.2 cm long with 17 leaves. The small oval-shaped leaves are green and glossy, with a sharply-cuspate apex, a round-cuneiform base and a small, blunt saw-tooth pattern on the margin. There are many small flowers with a diameter of 4.9 mm produced, averaging 6 ones per inflorescence.

Biological Characters The tree has strong adaptability and strong vigor, bearing early with stable fruit set and a high yield. It generally bears in the 2nd year after planting, and enters the full-bearing stage in the 20th year. The deciduous fruiting shoot bears 0.91 fruits on average. A tree has a yield of 40 kg on average. In Taigu County of Shanxi Province, it matures in mid-September with a fruit growth period of 105 d. It is an early-mid-ripening variety with light natural fruit-dropping, light alternate bearing and strong resistance to fruit-cracking.

Fruit Characteristics The small ovoid fruit has a vertical and cross diameter of 2.29 cm and 1.98 cm, averaging 5.3 g (maximum 10.9 g). It has a round shoulder, a shallow and narrow stalk cavity, a slightly-sunken fruit apex and thin red skin. The light-green flesh is tight-textured, crisp and sweet, with less juice. It has medium quality for fresh-eating and dried fruits. The content of SSC, TA and Vc in fresh fruit is 23.92%, 0.45% and 396 mg per 100 g fresh fruit. The content of flavones and cAMP in mature fruit skin is 6.19 mg/g and 222.35 μg/g. The content of TTS in dried fruit is 55.00%. The small oval-shaped stone weighs 0.52 g, and the percentage of containing kernels is 90.0%.

Evaluation The cultivar has strong adaptability, strong vigor, a high and stable yield and strong resistance to fruit-cracking. The fruit has crisp and tasteful flesh and is medium quality for fresh-eating and dried fruits.

兼 用
Multipurpose Varieties
品 种

镇平九月寒

品种来源及分布 原产于河南省镇平县王岗乡厚碾盘村，数量极少。

植物学性状 树体较大，树姿半开张，树冠呈伞形。主干灰褐色，皮粗糙，条裂，不易脱落。枣头灰褐色，粗壮，长79.6cm，粗1.09cm，节间长7.3cm，蜡质少。二次枝较细弱，略弯曲，6节左右。针刺脱落。枣股圆柱形，平均抽生枣吊4.0个。枣吊平均长23.6cm，着生15片叶。叶片卵状披针形，平展，绿色，先端渐尖、尖圆，叶基圆楔形，叶缘粗锯齿。花序平均着花7朵。

生物学特性 适应性强，较耐瘠薄。树势中庸。产量较高，采前落果少，以2～3年生枝坐果为主，枣头、2～3年和3年以上枝的吊果率分别为2.0%、103.7%和31.2%。成熟期裂果轻。在山西太谷地区，9月中旬果实着色，10月上旬完熟采收，果实生育期125d左右，为极晚熟品种类型。

果实性状 果实较大，卵圆形，纵径3.03cm，横径2.41cm，单果重14.5g，较整齐。果肩平圆，梗洼浅窄。果顶平，顶点微凹。果皮薄，浅红色。果点中大，显著，密度中等。果柄长3.1mm。果肉浅绿色，质地较致密，汁液中多，味甜，可鲜食和制干，品质中等。鲜枣可食率96.0%，100g果肉维生素C含量291.33mg；果皮含黄酮2.74mg/g，cAMP含量93.97μg/g。果核中大，纺锤形，核重0.58g，沟纹浅，含仁率8.4%。

评价 该品种适应性较广，耐瘠薄土壤条件，产量较高，果实品质中等，可鲜食和制干。

Zhenpingjiuyuehan

Source and Distribution The cultivar originated from Hounianpancun of Wanggang Village in Zhenping of Henan Province with a very small quantity.

Botanical Characters The large tree is half-spreading with an umbrella-shaped crown. The grayish-brown trunk bark has rough striped fissures, uneasily shelled off. The grayish-brown 1-year-old shoots are 79.6 cm long and 1.09 cm thick. The internodes are 7.3 cm long with less wax and falling-off thorns. The thin and weak secondary branches have 6 nodes. The column-shaped mother fruiting spurs can germinate 4 deciduous fruiting shoots which are 23.6 cm long with 15 leaves. The medium-thick green leaves are ovate-lanceolate and flat, with a gradually-cuspate apex, a round-cuneiform base and a thick saw-tooth pattern on the margin. Each inflorescence has 7 flowers.

Biological Characters The tree is strongly adaptable and tolerant to poor soils, with moderate vigor, a high yield and light premature fruit-dropping. The 2 or 3-year-old branches bear the most fruit. The percentage of fruits to deciduous fruiting shoots of 1-year-old shoots, 2 or 3-year-old branches and over-3-year-old ones is 2.0%, 103.7% and 31.2%. In Taigu County of Shanxi Province, it begins coloring in mid-September and matures in early October with a fruit growth period of 125 d. It is an extremely late-ripening variety.

Fruit Characteristics The large oval-shaped fruit has a vertical and cross diameter of 3.03 cm and 2.41 cm, averaging 14.5 g with a regular size. It has a flat-round shoulder, a shallow and narrow stalk cavity, a stalk of 3.1 mm long, a flat fruit apex and light-red thin skin with medium-large, distinct and medium-dense dots. The light-green flesh is tight-textured, sweet and juicy. It has medium quality for fresh-eating and dried fruits. The percentage of edible part of fresh fruit is 96.0%, and the content of Vc is 291.33 mg per 100 g fresh fruit. The content of flavones and cAMP in mature fruit skin is 2.74 mg/g and 93.97 μg/g. The medium-sized stone is spindle-shaped, averaging 0.58 g. The percentage of containing kernels is 8.4%.

Evaluation The cultivar has strong adaptability, tolerant to poor soils with high productivity. The fruit has medium quality for fresh-eating and dried fruits.

兼 用
Multipurpose Varieties
品 种

溆浦沙糖枣

品种来源及分布　别名沙糖枣。原产于湖南溆浦的祖市天乡，以四门、松溪村数量较多。

植物学性状　树体高大，树姿开张下垂，枝系较密，树冠呈乱头形。主干浅褐色，粗糙，树皮呈小块状剥落。枣头黄褐色，平均长78.9cm，粗0.95cm，节间长6.8cm，表面被有灰色蜡质。二次枝粗短，平均长28.5cm，7节，弯曲度中等，无针刺。枣股圆柱形，平均抽生枣吊4个。枣吊长23.3cm，着叶16片。叶片小，椭圆形，先端渐尖，叶基圆契形，叶缘锯齿钝、细小。花量较多，花序平均着花5朵，昼开型。

生物学特性　抗逆性强，树势强健，发枝力强，枝叶密集，丰产性能好，稳定。定植后2年开始结果，15年左右进入盛果期。在山西太谷地区，4月中旬萌芽，5月下旬始花，9月下旬果实成熟。果实生育期115d，为晚熟品种类型。成熟期抗裂果。

果实性状　果实小，卵圆形，侧面较扁，纵径3.23cm，横径2.49cm，单果重7.2g，大小不整齐。果肩平圆。梗洼中深、广，果顶平。果皮薄，光滑，赭红色。果肉浅绿色，质地酥脆，汁液多，味甜酸，可鲜食和制干，品质中等。鲜枣可食率90.1%，含总糖19.50%，酸0.58%；果皮含黄酮39.30mg/g，cAMP含量150.47μg/g。制干率44.3%，干枣含总糖62.11%，酸1.38%。果核大，纺锤形，核重0.71g，含仁率3.4%。

评价　该品种抗逆性强，产量高而稳定。果实小，不整齐，肉质酥脆，核大，品质中等，可鲜食和制干。

Xupushatangzao

Source and Distribution　The cultivar originated from Zushitian Village of Xupu in Hunan Province. The largest quantity is in Simen and Songxicun.

Botanical Characters　The large tree is spreading with dense branches and an irregular crown. The light-brown trunk bark has small, rough, massive fissures. The yellowish-brown 1-year-old shoots are 78.9 cm long and 0.95 cm thick. The internodes are 6.8 cm long with some gray wax and no thorns. The thick and short secondary branches are 28.5 cm long with 7 nodes of medium curvature. The column-shaped mother fruiting spurs can germinate 4 deciduous fruiting shoots which are 23.3 cm long with 16 leaves. The small oval-shaped leaves have a gradually-cuspate apex, a round-cuneiform base and a thin, blunt saw-tooth pattern on the margin. There are many flowers produced, averaging 5 ones per inflorescence. It blooms in the daytime.

Biological Characters　The tree has strong resistance to adverse conditions, strong vigor and strong branching ability, with a high and stable yield. It generally bears in the 2nd year after planting, and enters the full-bearing stage in the 15th year. In Taigu County of Shanxi Province, it germinates in mid-April, begins blooming in late May and matures in late September with a fruit growth period of 115 d. It is a late-ripening variety with strong resistance to fruit-cracking in the maturing stage.

Fruit Characteristics　The small oval-shaped fruit is a little flattened on the lateral sides, with a vertical and cross diameter of 3.23 cm and 2.49 cm, averaging 7.2 g with irregular sizes. It has a flat-round shoulder, a medium-deep and wide stalk cavity, a flat fruit apex and thin, smooth and brownish-red skin. The light-green flesh is crisp, juicy, sour and sweet. It has medium quality for fresh-eating and dried fruits. The percentage of edible part of fresh fruit is 90.1%, and the content of TTS and TA is 19.50% and 0.58%. The content of flavones and cAMP in mature fruit skin is 39.30 mg/g and 150.47 μg/g. The percentage of fresh fruits which can be made into dried ones is 44.3%. The large spindle-shaped stone weighs 0.71 g, and the percentage of containing kernels is 3.4%.

Evaluation　The cultivar has strong resistance to adverse conditions with a high and stable yield. The small fruit has irregular sizes, soft and crisp flesh and a large stone. It has medium quality for fresh-eating and dried fruits.

兼 用
Multipurpose Varieties
品 种

洪 赵 脆 枣

品种来源及分布　原产和分布于山西洪洞县的赵城等地。

植物学性状　树体较大，树姿较直立，树冠呈伞形。主干皮裂条状。枣头红褐色，平均长63.5cm，节间长7.2cm。二次枝长32.2cm，5～9节。针刺不发达。枣股平均抽生枣吊3.9个。枣吊长18.4cm，着叶13片。叶片中大，卵圆形，两侧略向叶面合抱，先端钝尖，叶基圆楔形，叶缘具钝锯齿。花量中多，花序平均着花5朵，花较大，花径6.4mm。

生物学特性　树势中庸，萌芽率和成枝力强。结果较早，一般定植第二年结果，第三年有一定产量，盛果期产量和坐果率一般，枣头、2～3年和4～6年生枝的吊果率分别为39.3%、88.0%和66.5%。在山西太谷地区，9月中旬果实进入脆熟期，下旬成熟采收，果实生育期110d以上，为中晚熟品种类型。成熟期遇雨裂果较少。

果实性状　果个中大，卵圆形，纵径3.02cm，横径2.70cm，单果重10.7g，大小整齐。果皮中厚，紫红色，果面平滑。梗洼中广、较浅。果顶平，柱头遗存，不明显。肉质疏松，味酸甜，汁液多，品质中等，适宜鲜食或制干。鲜枣可食率97.2%，含可溶性固形物35.40%，总糖27.97%，酸0.71%，糖酸比39.39∶1，100g果肉维生素C含量361.11mg；果皮含黄酮2.96mg/g，cAMP含量160.72μg/g。干枣含总糖70.51%，酸0.88%。核较小，纺锤形，核重0.29g，种仁不饱满，含仁率13.3%。

评价　该品种果个中大，品质中等，适宜鲜食或制干。果实抗裂果能力较强。

Hongzhaocuizao

Source and Distribution　The cultivar originated from and spreads in Zhaocheng of Hongdong County in Shanxi Province.

Botanical Characters　The large tree is vertical with an umbrella-shaped crown. The trunk bark has striped fissures. The reddish-brown 1-year-old shoots are 63.5 cm long with the internodes of 7.2 cm. The secondary branches are 32.2 cm long, with 5～9 nodes and less-developed thorns. The mother fruiting spurs can germinate 3.9 deciduous fruiting shoots which are 18.4 cm long with 13 leaves. The medium-sized leaves are oval-shaped, with a bluntly-cuspate apex, a round-cuneiform base and a blunt saw-tooth pattern on the margin. The number of flowers is medium large, averaging 5 ones per inflorescence. The large blossoms have a diameter of 6.4 mm.

Biological Characters　The tree has moderate vigor, strong germination and branching ability, yet with low fruit set. The percentage of fruits to deciduous fruiting shoots of 1-year-old shoots, 2～3-year-old branches and 4～6-year-old ones is 39.3%, 88.0% and 66.5%. It bears early, generally in the 2nd year after planting and there is certain yield in the 3rd year. It has moderate productivity. In Taigu County of Shanxi Province, it enters the crisp-maturing stage in mid-September and ripens in late September with a fruit growth period of 103 d. It is a mid-late-ripening variety with light fruit-cracking even if it rains in the maturing stage.

Fruit Characteristics　The medium-sized fruit is oval-shaped, with a vertical and cross diameter of 3.02 cm and 2.70 cm, averaging 10.7 g with a regular size. It has medium-thick and light-red skin, a smooth surface, a medium-wide and shallow stalk cavity, a flat fruit apex and a remnant stigma. The loose-textured flesh is sour, sweet and juicy. It has medium quality for fresh-eating and dried fruits. The percentage of edible part of fresh fruit is 97.2%, and SSC, TTS, TA and Vc is 35.40%, 27.97%, 0.71% and 361.11 mg per 100 g fresh fruit. The SAR is 39.39∶1. The content of flavones and cAMP in mature fruit skin is 2.96 mg/g and 160.72 μg/g. The content of TTS and TA in dried fruit is 70.51% and 0.88%. The small spindle-shaped stone weighs 0.29 g with a shriveled kernel. The percentage of containing kernels is 13.3%.

Evaluation　The cultivar has a medium-large fruit size and medium quality and is suitable for fresh-eating and making dried fruits. The fruit has strong resistance to fruit-cracking.

兼 用
Multipurpose Varieties
品种

·393·

河津水枣

品种来源及分布　原产和分布于山西河津赵家乡北里村一带，栽培数量不多。

植物学性状　树体较小，树姿开张，枝系稀疏，树冠呈自然乱头形。主干皮裂条状。枣头红褐色，生长势弱，平均长70.1cm，节间长7.1cm。针刺不发达。枣股圆锥形，平均抽生枣吊3.7个。枣吊长25.3cm，着叶15片。叶片较小，中厚，椭圆形，浅绿色，先端渐尖，叶基圆形，叶缘具钝锯齿。花量较少，每花序着花3.8朵。

生物学特性　适应性较差，要求深厚肥沃的土壤。树势和发枝力弱，坐果较稳定，产量中等。在山西太谷地区，4月中旬萌芽，5月下旬开花，9月中旬果实成熟采收，果实生育期约100d左右，为中熟品种类型。

果实性状　果实较小，扁柱形，纵径3.15cm，横径2.45cm，侧径2.08cm，单果重8.08g，大小整齐。果肩平。梗洼狭窄、中深。果顶平圆。柱头遗存。果柄细短。果面光滑，果皮红色，中厚，果肉厚，浅绿色，质地疏松，汁液多，味酸甜，可鲜食和制干，品质中上。鲜枣可食率95.4%，含可溶性固形物31.80%，总糖23.10%，酸0.45%，100g果肉维生素C含量369.01mg；果皮含黄酮20.05mg/g，cAMP含量34.37μg/g。干枣含可溶性固形物66%，总糖59.10%，酸1.01%。果核小，纺锤形，核重0.37g。

评价　该品种树体小，树势弱，适土性差。果实较小，肉质细，较松脆，制干鲜食均可，品质中上，适宜土壤条件好的地区集约栽培。

Hejinshuizao

Source and Distribution　The cultivar originated from and spreads in Beilicun of Zhaojia Village in Hejin City of Shanxi Province, with a small quantity.

Botanical Characters　The small tree is spreading with sparse branches and an irregular crown. The trunk bark has striped fissures. The reddish-brown 1-year-old shoots are 70.1 cm long with weak growth potential. The internodes are 7.1 cm long with less-developed thorns. The conical mother fruiting spurs can germinate 3.7 deciduous fruiting shoots which are 25.3 cm long with 15 leaves. The small light-green leaves are medium-thick and oval-shaped, with a gradually-cuspate apex, a round base and a blunt saw-tooth pattern on the margin. The number of flowers is small, averaging 3.8 ones per inflorescence.

Biological Characters　The tree has poor adaptability, weak vigor and weak branching ability with a medium and stable yield. It requires deep and fertile soils. In Taigu County of Shanxi Province, it germinates in mid-April, begins blooming in late May and matures in mid-September with a fruit growth period of 100 d. It is a mid-ripening variety.

Fruit Characteristics　The small fruit is flat column-shaped, with a vertical and cross diameter of 3.15 cm and 2.45 cm, averaging 8.08 g with a regular size. It has a flat shoulder, a narrow and medium-deep stalk cavity, a flat-round fruit apex, a remnant stigma, a short and thin stalk, a smooth surface and medium-thick red skin. The light-green flesh is thick, loose-textured, sour, sweet and juicy. It has a better than normal quality for fresh-eating and dried fruits. The percentage of edible part of fresh fruit is 95.4%, and SSC, TTS, TA and Vc is 31.80%, 23.10%, 0.45% and 369.01 mg per 100 g fresh fruit. The content of flavones and cAMP in mature fruit skin is 20.05 mg/g and 34.37 μg/g, and the content of SSC, TTS and TA in dried fruit is 66%, 59.10% and 1.01%. The small spindle-shaped stone weighs 0.37 g.

Evaluation　The small plant of Hejinshuizao cultivar has weak vigor and poor adaptability. The small fruit has delicate and crisp flesh with a better quality for fresh-eating and dried fruits. It is suitable for intensive-planting in areas with rich soils.

兼 用
Multipurpose Varieties
品 种

中宁大红枣

品种来源及分布 原产于宁夏中宁的曹桥、鸣沙、枣园堡一带，多零星栽培，当地有百年左右的大树。

植物学性状 树体较大，树姿半开张，干性中强，树冠圆锥形。主干淡灰褐色，皮裂中深，不规则条形。枣头紫褐色，平均长74.8cm，粗1.01cm，节间长7.3cm，蜡质少。针刺发达。枣股圆锥形，平均着生枣吊4.0个。枣吊长23.6cm，着叶19片。叶片中大，卵圆形或长卵圆形，绿色，先端渐尖，叶基圆形或楔形，叶缘具钝锯齿。花量较少，花序平均着花3朵，花大，花径8.0mm。

生物学特性 风土适应性强，耐旱、耐瘠薄，且耐低温。树势中等，枣头延续生长力强，发枝力弱，定植后2年开始结果，成龄树产量高而稳定，平均吊果率35.10%。在山西太谷地区，9月中旬成熟采收，果实生育期100d左右，为早中熟品种类型。

果实性状 果实大，近圆形，果实纵径2.90cm，横径2.50cm，单果重15.5g，大小整齐。果肩平，梗洼中深、窄。果顶微凹。果面平滑光亮。果皮浅红色。果肉浅绿色，质地酥脆，汁液多，味甜，可鲜食和制干，品质上等。鲜枣可食率95.7%，含可溶性固形物32.00%，总糖16.26%。果核中大，倒卵形，核重0.67g，含仁率13.3%。

评价 该品种适应性强，树体较大，丰产稳产。果实大，品质上等，为较好的早中熟鲜食、制干兼用品种，适宜气温较低的北方地区发展栽培。

Zhongningdahongzao

Source and Distribution The cultivar originated from Caoqiao, Mingsha and Zaoyuanbao in Zhongning County of Ningxia Province. It is planted in scattered regions. There are still some trees of about 100 years old in the local areas.

Botanical Characters The large tree is half-spreading, with a medium-strong central leader trunk and a conical crown. The dark grayish-brown trunk bark has irregular, medium-deep striped fissures. The purplish-brown 1-year-old shoots are 74.8 cm long and 1.01 cm thick. The internodes are 7.3 cm long with less wax and developed thorns. The conical mother fruiting spurs can germinate 4 deciduous fruiting shoots which are 23.6 cm long with 19 leaves. The medium-sized green leaves are oval-shaped or long oval-shaped, with a gradually-cuspate apex, a round or cuneiform-shaped base and a blunt saw-tooth pattern on the margin. The number of flowers is small, averaging 3 ones per inflorescence. The large blossoms have a diameter of 8.0 mm.

Biological Characters The tree has strong adaptability, tolerant to drought, cold and poor soils, with moderate vigor, weak branching ability, strong growth potential for 1-year-old shoots, a high and stable yield. It generally bears in the 2nd year after planting. The percentage of fruits to deciduous fruiting shoots is 35.10% on average. In Taigu County of Shanxi Province, it matures in mid-September with a fruit growth period of 100 d. It is an early-mid-ripening variety.

Fruit Characteristics The large oblong fruit has a vertical and cross diameter of 2.90 cm and 2.50 cm, averaging 15.5 g with a regular size. It has a flat shoulder, a medium-deep and narrow stalk cavity, a slightly-sunken fruit apex, a smooth and glossy surface, and light-red skin. The light-green flesh is crisp, juicy and sweet. It has a good quality for fresh-eating and dried fruits. The percentage of edible part of fresh fruit is 95.7%, and SSC and TTS is 32.00% and 16.26%. The medium-sized obovate stone weighs 0.67 g and the percentage of containing kernels is 13.3%.

Evaluation The cultivar has strong adaptability, strong vigor, a high and stable yield. The large fruit has a good quality for fresh-eating and dried fruits. It is a good early-mid-ripening variety and can be developed in areas of north China with lower temperature.

兼 用
Multipurpose Varieties
品 种

蒲城绵枣

品种来源及分布　原产和分布在陕西蒲城、合阳等地，仅有零星栽培。

植物学性状　树体中大，树姿较开张，树冠呈圆头形。主干皮裂块状。枣头紫褐色，平均长89.8cm，节间长7.1cm，蜡质少。二次枝平均长31.6cm，5～9节，弯曲度中等。针刺不发达。枣股平均抽生枣吊3.3个。枣吊长22.4cm，着叶16片。叶片卵圆形，平均长4.8cm，叶宽2.5cm，浅绿色，先端锐尖，叶基圆形，叶缘锯齿浅钝。花量少，花中大，花序平均着花3朵。

生物学特性　树势中等，发枝力强。结果少，产量偏低，枣头、2～3年和4～6年生枝的吊果率分别为28.0%、41.8%和8.2%。在山西太谷地区，10月上旬成熟，果实生长期120d左右，为极晚熟品种类型。成熟期遇雨极易裂果。

果实性状　果实大，近圆形，纵径3.45cm，横径3.15cm，单果重17.3g，大小整齐。果肩平圆。梗洼中深、广。果顶平圆，顶点凹陷。果面光滑，果点中等密度，较明显。果皮中厚，赭红色。果肉浅绿色，质地致密，汁液多，味酸甜，适宜鲜食或制干，品质较好。鲜枣可食率94.3%，含可溶性固形物37.20%，总糖28.80%，酸0.57%，100g果肉维生素C含量446.33mg；果皮含黄酮7.09mg/g，cAMP含量222.28μg/g。干枣含总糖64.13%，含酸0.73%。果核大，椭圆形，核重0.99g，种仁不饱满，含仁率56.7%。

评价　该品种树体中大，产量较低。果实大，质地致密，味酸甜，品质好，适宜鲜食或制干。

Puchengmianzao

Source and Distribution　The cultivar originated from and spreads in Pucheng and Heyang of Shaanxi Province in scattered regions.

Botanical Characters　The medium-sized tree is spreading with a round crown. The trunk bark has massive fissures. The purplish-brown 1-year-old shoots are 89.8 cm long with the internodes of 7.1 cm and less wax. The secondary branches are 31.6 cm long with 5～9 nodes of medium curvature and less-developed thorns. The mother fruiting spurs can germinate 3.3 deciduous fruiting shoots which are 22.4 cm long with 16 leaves. The leaves are oval-shaped and light green, 4.8 cm long and 2.5 cm wide, with a sharply-cuspate apex, a round base and a blunt saw-tooth pattern on the margin. The number of flowers is small, averaging 3 ones per inflorescence. The blossoms are medium-sized.

Biological Characters　The tree has moderate vigor and strong branching ability, with a low yield. The percentage of frutis to deciduous frutitng shoots of 1-year-old shoots, 2～3-year-old branches and 4～6-year-old ones is 28.0%, 41.8% and 8.2%. In Taigu County of Shanxi Province, it matures in early October with a fruit growth period of 120 d. It is an extremely late-ripening variety with serious fruit-cracking if it rains in the maturing stage.

Fruit Characteristics　The large oval-shaped fruit has a vertical and cross diameter of 3.45 cm and 3.15 cm, averaging 17.3 g with a regular size. It has a flat-round shoulder, a medium-deep and wide stalk cavity, a flat-round fruit apex, a smooth surface and medium-thick red skin. The light-green flesh is tight-textured, sour, sweet and juicy. It has a good quality for fresh-eating and dried fruits. The percentage of edible part of fresh fruit is 94.3%, and SSC, TTS, TA and Vc is 37.20%, 28.80%, 0.57% and 446.33 mg per 100 g fresh fruit. The content of flavones and cAMP in mature fruit skin is 7.09 mg/g and 222.28 μg/g. The content of TTS and TA in dried fruit is 64.13% and 0.73%. The large oval-shaped stone weighs 0.99 g with a shriveled kernel. The percentage of containing kernels is 56.7%.

Evaluation　The medium-sized plant of Puchengmianzao cultivar has a low yield. The large fruit has tight-textured, sour and sweet flesh and is suitable for fresh-eating and making dried fruits.

兼 用
Multipurpose Varieties
品 种

沧县小枣

品种来源及分布 原产和分布于河北沧州地区沧县等地。

植物学性状 树体中大，树姿较直立，树冠自然圆头形。主干皮裂块状。枣头黄绿色，年平均生长量74.6cm，节间长8.0cm，蜡质多。二次枝长35.0cm，5～7节。无针刺。枣吊长20.9cm，着叶14片。叶片椭圆形，叶长5.8cm，叶宽2.6cm，浅绿色，部分叶面反卷，先端锐尖，叶基圆楔形，叶缘具钝锯齿。花量中多，花序平均着花4朵，花较大，花径6.5mm。

生物学特性 树势较强。幼龄枝坐果率较高，成龄树产量较低，枣头、2～3年、4～6年和6年以上生枝的吊果率分别为98.8%、109.7%和18.7%。在山西太谷地区，9月下旬果实进入白熟期，10月上旬开始成熟采收，果实生育期125d左右，属极晚熟品种。果实抗裂果能力较差。

果实性状 果个较小，圆柱形，纵径2.90cm，横径2.30cm，单果重7.8g，大小整齐。果皮薄，浅红色，果面平滑。梗洼窄而深。果顶平，柱头遗存。肉质致密，味甜，汁液中多，品质上，适宜鲜食或制干。鲜枣可食率95.5%，含可溶性固形物31.20%，总糖29.52%，酸0.79%，100g果肉维生素C含量381.40mg；果皮含黄酮8.51mg/g，cAMP含量94.30μg/g。干枣含总糖65.62%，酸0.82%。核较小，纺锤形，核重0.35g。种仁较饱满，含仁率86.7%。

评价 该品种果个较小，品质优良，适宜鲜食或制干。抗裂果能力较差，成熟期应注意防雨。

Cangxianxiaozao

Source and Distribution The cultivar originated from and spreads in Cangxian County of Hebei Province.

Botanical Characters The medium-sized tree is vertical with a natural-round crown. The trunk bark has massive fissures. The yellowish-green 1-year-old shoots are 74.6 cm long with the internodes of 8.0 cm and much wax. The secondary branches are 35.0 cm long with 5～7 nodes and no thorns. The deciduous fruiting shoots are 20.9 cm long with 14 leaves. The leaves are oval-shaped and light green, 5.8 cm long and 2.6 cm wide, with a sharply-cuspate apex, a round-cuneiform base and a blunt saw-tooth pattern on the margin. The number of flowers is medium large, averaging 4 ones per inflorescence. The large blossoms have a diameter of 6.5 mm.

Biological Characters The tree has strong vigor with high fruit set for young branches and a low yield. The percentage of fruits to deciduous fruiting shoots of 1-year-old shoots, 2～3-year-old branches and 4～6-year-old ones is 98.8%, 109.7% and 18.7%. In Taigu County of Shanxi Province, it enters the white-maturing stage in late September and matures in early October with a fruit growth period of 125 d. It is an extremely late-ripening variety with poor resistance to fruit-cracking.

Fruit Characteristics The small column-shaped fruit has a vertical and cross diameter of 2.90 cm and 2.30 cm, averaging 7.8 g a with regular size. The fruit has light-red thin skin, a smooth surface, a shallow and narrow stalk cavity, a flat fruit apex and a remnant stigma. The tight-textured flesh is sweet with medium juice. It has a superior quality for fresh-eating and dried fruits. The percentage of edible part of fresh fruit is 95.5%, and SSC, TTS, TA and Vc is 31.20%, 29.52%, 0.79% and 381.40 mg per 100 g fresh fruit. The content of flavones and cAMP in mature fruit skin is 8.51 mg/g and 94.30 μg/g. The content of TTS and TA in dried fruit is 65.62% and 0.82%. The small spindle-shaped stone weighs 0.35 g, with a well-developed kernel. The percentage of containing kernels is 86.7%.

Evaluation The cultivar has a small fruit size and a superior quality and is suitable for fresh-eating and dried fruits. Yet it has poor resistance to fruit-cracking, so rain protection should be paid much attention to in the maturing stage.

兼 用
Multipurpose Varieties
品 种

蒲城圆梨枣

品种来源及分布　原产和分布于陕西蒲城县。

植物学性状　树体高大，树姿半开张，树冠呈圆头形。主干皮裂条状。枣头红褐色，平均长71.4cm，节间长7.0cm，蜡质少。二次枝平均长43.2cm，7~9节，弯曲度中等。针刺不发达。枣股平均抽生枣吊3.3个。枣吊长25.2cm，着叶14片。叶片中大，卵圆形，绿色，部分叶片呈合抱状，先端锐尖，叶基偏斜形，叶缘具锐锯齿。花量多，花序平均着花6朵，花中大，花径6.3mm。

生物学特性　树势中庸，萌芽率和成枝力较高，结果较晚，定植第三年结果，10年左右进入盛果期，较丰产，枣吊平均结果0.26个。在山西太谷地区，10月上旬果实成熟采收，果实生育期113d左右，为极晚熟品种类型。成熟期遇雨裂果轻。

果实性状　果个大，短柱形或近圆形，纵径3.69cm，横径3.24cm，单果重19.0g，最大25g以上，大小不整齐。果皮薄，红色，果面平滑。梗洼浅而广。果顶凹陷，柱头宿存。肉质疏松，汁液多，味极甜，品质极好，鲜食、制干兼用。鲜枣可食率96.5%，含可溶性固形物36.0%，总糖29.92%，酸0.78%，100g果肉维生素C含量426.92mg。干枣含总糖64.68%，酸1.17%。核中大，纺锤形，平均重0.66g。种仁较饱满，含仁率65.0%。

评价　该品种树体高大，树势中庸，结果晚，较丰产，抗裂果能力极强。果个大，味极甜，品质极好，鲜食和制干兼用。

Puchengyuanlizao

Source and Distribution　The cultivar originated from and spreads in Pucheng County of Shaanxi Province.

Botanical Characters　The large tree is half-spreading with a round crown. The trunk bark has striped fissures. The reddish-brown 1-year-old shoots are 71.4 cm long, with the internodes of 7.0 cm long less wax and less-developed thorns. The secondary branches are 43.2 cm long with 7~9 nodes of medium curvature. The mother fruiting spurs can germinate 3.3 deciduous fruiting shoots which are 25.2 cm long with 14 leaves. The medium-sized green leaves are oval-shaped, with a sharply-cuspate apex, a deflective base and a sharp saw-tooth pattern on the margin. There are many medium-sized flowers with a diameter of 6.3 mm produced, averaging 6 ones per inflorescence.

Biological Characters　The tree has moderate vigor, strong germination and branching ability. It bears late, generally in the 3rd year after planting and enters the full-bearing stage in the 10th year with a high yield. The deciduous fruiting shoot bears 0.26 fruits on average. In Taigu County of Shanxi Province, it matures in early October with a fruit growth period of 113 d. It is an extremely late-ripening variety with light fruit-cracking even if it rains in the maturing stage.

Fruit Characteristics　The large obovate fruit has a vertical and cross diameter of 3.69 cm and 3.24 cm, averaging 19.0 g (maximum over 25 g) with irregular sizes. It has thin red skin, a smooth surface, a shallow and wide stalk cavity, a slightly-sunken fruit apex and a remnant stigma. The loose-textured flesh is strongly sweet and juicy. It has an excellent quality for fresh-eating and dried fruits. The percentage of edible part of fresh fruit is 96.5%, and SSC, TTS, TA and Vc is 36.0%, 29.92%, 0.78% and 426.92 mg per 100 g fresh fruit. The content of TTS and TA in dried fruit is 64.68% and 1.17%. The medium-sized stone is spindle-shaped, averaging 0.66 g with a well-developed kernel. The percentage of containing kernels is 65.0%.

Evaluation　The large plant of Puchengyuanlizao cultivar has moderate tree vigor, bearing late with a high yield. The fruit has very strong resistance to fruit-cracking. It has a large fruit size and sweet flesh with very good quality and is suitable for both fresh-eating and making dried fruits.

兼 用
Multipurpose Varieties
品 种

· 403 ·

献县小小枣

品种来源及分布 原产和分布于河北献县等地。

植物学性状 树体中大，树姿半开张，树冠自然圆头形。主干条状皮裂。枣头黄褐色，平均长73.5cm，节间长7.4cm。二次枝长36.0cm，5～7节。针刺基本退化。枣股平均抽生枣吊3.2个。枣吊长21.1cm，着叶13片。叶片中等大，叶长6.2cm，叶宽2.7cm，卵圆形，部分叶片反卷，先端锐尖，叶基圆楔形，叶缘具钝锯齿。花量多，花序平均着花7朵，花大，花径6.4mm。

生物学特性 树势强，萌芽率和成枝力强。结果较早，定植第二年结果，10年左右进入盛果期，产量中等，坐果率一般，枣头、2～3年和4～6年生枝的吊果率分别为58.4%、86.6%和5.7%。在山西太谷地区，9月下旬果实进入脆熟期，10月上旬成熟采收，果实生育期119d左右，为极晚熟品种类型。成熟期遇雨裂果严重。

果实性状 果个小，卵圆形或近圆形，纵径2.63cm，横径2.28cm，单果重7.0g，大小较整齐。果皮薄，红色，果面平滑。梗洼窄而深。果顶微凹，柱头遗存，不明显。肉质疏松，味甜，汁液多，品质优良，适宜鲜食和制干。鲜枣可食率94.6%，含可溶性固形物34.20%，总糖23.58%，酸0.70%，糖酸比33.69∶1，100g果肉维生素C含量443.62mg。制干率53.9%，干枣含总糖61.72%，酸1.14%。核较小，椭圆形，核重0.38g。种仁饱满，含仁率16.7%。

评价 该品种果个小，但品质好，适宜鲜食或制干。成熟期注意防雨。

Xianxianxiaoxiaozao

Source and Distribution The cultivar originated from and spreads in Xianxian County of Hebei Province.

Botanical Characters The medium-sized tree is half-spreading with a natural-round crown. The trunk bark has striped fissures. The yellowish-brown 1-year-old shoots are 73.5 cm long, with the internodes of 7.4 cm. The secondary branches are 36.0 cm long with 5～7 nodes and very few thorns. The mother fruiting spurs can germinate 3.2 deciduous fruiting shoots which are 21.1 cm long with 13 leaves. The medium-sized leaves are oval-shaped, 6.2 cm long and 2.7 cm wide, with a sharply-cuspate apex, a round-cuneiform base and a blunt saw-tooth pattern on the margin. There are many large flowers with a diameter of 6.4 mm produced, averaging 7 ones per inflorescence.

Biological Characters The tree has strong vigor, strong germination and branching ability, with medium fruit set. The percentage of fruits to deciduous fruiting shoots of 1-year-old shoots, 2～3-year-old branches and 4～6-year-old ones is 58.4%, 86.6% and 5.7%. It bears early, generally in the 2nd year after planting and enters the full-bearing stage in the 10th year with a medium yield. In Taigu County of Shanxi Province, it enters the crisp-maturing stage in late September and matures in early October with a fruit growth period of 119 d. It is an extremely late-ripening variety with serious fruit-cracking if it rains in the maturing stage.

Fruit Characteristics The small oval-shaped fruit has a vertical and cross diameter of 2.63 cm and 2.28 cm, averaging 7.0 g with a regular size. It has thin red skin, a smooth surface, a narrow and deep stalk cavity, a slightly-sunken fruit apex and a remnant stigma. The loose-textured flesh is sweet and juicy. It has a good quality for fresh-eating and dried fruits. The percentage of edible part of fresh fruit is 94.6%, and SSC, TTS, TA and Vc is 34.20%, 23.58%, 0.70% and 443.62 mg per 100 g fresh fruit. The SAR is 33.69∶1. The medium-sized stone is oval-shaped, averaging 0.38 g with a well-developed kernel. The percentage of containing kernels is 16.7%.

Evaluation The cultivar has a small fruit size and a good quality and is suitable for fresh-eating and making dried fruits. Rain protection should be paid much attention to in the maturing stage.

兼 用
Multipurpose Varieties
品种

· 405 ·

沧县屯子枣

品种来源及分布 原产和分布于河北沧州地区沧县等地。

植物学性状 树体中大，树姿半开张，树冠自然圆头形。枣头黄褐色，平均长68.5cm，节间长7.8cm。二次枝长33.7cm，5～7节，弯曲度中等。针刺不发达。枣股平均抽生枣吊3.0个。枣吊长22.6cm，着叶13片。叶片长5.7cm，叶宽3.1cm，卵圆形，部分叶片反卷，先端锐尖，叶基截形，叶缘具钝锯齿。花量多，花序平均着花8朵。花中大，花径6.3mm。

生物学特性 树势中等。产量一般，坐果率较高，枣头、2～3年、4～6年和6年以上生枝的吊果率分别为98.8%、109.7%和26.0%。在山西太谷地区，9月下旬果实进入脆熟期，10月上旬成熟采收，果实生育期119d左右，属极晚熟品种。成熟期遇雨易裂果。

果实性状 果个较小，卵圆形，纵径2.46cm，横径2.21cm，单果重8.4g，大小整齐。果皮薄，浅红色，果面光滑。梗洼浅而广。果顶平，柱头遗存，不明显。肉质致密，味酸甜，汁液中多，品质上等，适宜鲜食或制干。鲜枣可食率97.4%，含可溶性固形物33.00%，总糖24.36%，酸0.64%，100g果肉维生素C含量499.88mg；果皮含黄酮2.67mg/g，cAMP含量146.18μg/g。干枣含可溶性糖60.35%，酸0.84%。果核较小，椭圆形，核重0.22g。含仁率1.7%。

评价 该品种树体中大，生长势中庸，果个较小，品质上等，适宜鲜食或制干，成熟期遇雨易裂果。

Cangxiantunzizao

Source and Distribution The cultivar originated from and spreads in Cangxian County of Hebei Province.

Botanical Characters The medium-sized tree is half-spreading with a natural-round crown. The yellowish-brown 1-year-old shoots are 68.5 cm long, with the internodes of 7.8 cm. The secondary branches are 33.7 cm long with 5～7 nodes of medium curvature and less-developed thorns. The mother fruiting spurs can germinate 3 deciduous fruiting shoots which are 22.6 cm long with 13 leaves. The oval-shaped leaves are 5.7 cm long and 3.1 cm wide, with a sharply-cuspate apex, a truncate base and a blunt saw-tooth pattern on the margin. There are many medium-sized flowers with a diameter of 6.3 mm produced, averaging 8 ones per inflorescence.

Biological Characters The tree has moderate vigor with high fruit set. The percentage of fruits to deciduous fruiting shoots of 1-year-old shoots, 2～3-year-old branches and 4～6-year-old ones is 98.8%, 109.7% and 26.0%. In Taigu County of Shanxi Province, it enters the crisp-maturing stage in late September and matures in early October with a fruit growth period of 119 d. It is a late-ripening variety with serious fruit-cracking if it rains in the maturing stage.

Fruit Characteristics The small oval-shaped fruit has a vertical and cross diameter of 2.46 cm and 2.21 cm, averaging 8.4 g with a regular size. It has light-red thin skin, a smooth surface, a shallow and wide stalk cavity, a flat fruit apex and a remnant stigma. The tight-textured flesh is sweet and sour with medium juice. It has a superior quality for fresh-eating and dried fruits. The percentage of edible part of fresh fruit is 97.4%, and SSC, TTS, TA and Vc is 33.00%, 24.36%, 0.64% and 499.88 mg per 100 g fresh fruit. The content of flavones and cAMP in mature fruit skin is 2.67 mg/g and 146.18 μg/g. The content of soluble sugar and TA in dried fruit is 60.35% and 0.84%. The small spindle-shaped stone weighs 0.22 g. The percentage of containing kernels is 1.7%.

Evaluation The medium-sized plant of Cangxiantunzizao cultivar has moderate tree vigor. The small fruit has a superior quality and is suitable for fresh-eating and making dried fruits. Fruit-cracking easily occurs if it rains in the maturing stage.

兼 用
Multipurpose Varieties
品 种

· 407 ·

灌 阳 短 枣

品种来源及分布 别名米枣、糖枣、珠枣、短枣。原产于广西的全县、灌阳、临桂等地。

植物学性状 树体较大，树姿开张，枝叶较稀，树冠呈乱头形。树干灰褐色，皮裂呈块状。枣头紫褐色，年生长量62.5cm，节间长5.9cm，蜡质少。二次枝平均长23.3cm，5～7节，弯曲度中等，无针刺。枣股平均抽生枣吊3.8个。枣吊长21.6cm，着生叶片13片。叶片中大，椭圆形，先端急尖，叶基圆楔形，叶缘具细锯齿。花量多，花序平均着生6朵。花较小，花径5.9mm。

生物学特性 树势强，发枝力较弱。产量高而稳定，定植后第二年开始结果，盛果期树一般株产75kg，枣吊平均结果0.82个。在山西太谷地区，10月初果实成熟，果实生长期120d左右，为极晚熟品种类型。

果实性状 果实较小，圆形，纵径2.34cm，横径1.98cm，单果重7.5g，大小整齐。果肩平圆，梗洼中深、广，果顶平。果皮厚，红色，果面光滑。果肉浅绿色，质地较致密，汁液较多，味甜，品质上，鲜食制干兼用。鲜枣可食率90.9%，含总糖26.50%，还原糖含量20.80%，100g果肉维生素C含量538.3mg；果皮含黄酮6.53mg/g，cAMP含量186.06μg/g。制干率48.0%，干枣含总糖60.60%，酸0.76%。果核大，椭圆形，核重0.68g。种仁较饱满，含仁率70%左右。

评价 该品种树体较大，树姿开张，树势强，产量高而稳定。果实较小，味甜，品质上等，鲜食、制干兼用。

Guanyangduanzao

Source and Distribution The cultivar, also called Mizao, Tangzao or Zhuzao, originated from and spreads in Guanyang and Lingui of Guangxi Province

Botanical Characters The large tree is spreading with sparse branches and an irregular crown. The grayish-brown trunk bark has massive fissures. The purplish-brown 1-year-old shoots are 62.5 cm long, with the internodes of 5.9 cm and less wax. The secondary branches are 23.3 cm long with 5～7 nodes of medium curvature and no thorns. The mother fruiting spurs can germinate 3.8 deciduous fruiting shoots which are 21.6 cm long with 13 leaves. The medium-sized leaves are oval-shaped, with a sharply-cuspate apex, a round-cuneiform base and a thin saw-tooth pattern on the margin. There are many small flowers with a diameter of 5.9 mm produced, averaging 6 ones per inflorescence.

Biological Characters The tree has strong vigor and weak branching ability. It bears early, generally in the 2nd year after planting, with a high and stable yield. The deciduous fruiting shoot bears 0.82 fruits on average. Trees in full-bearing stage have an output of 75 kg per plant. In Taigu County of Shanxi Province, it matures in early October, with a fruit growth period of 120 d. It is an extremely late-ripening variety.

Fruit Characteristics The small round fruit has a vertical and cross diameter of 2.34 cm and 1.98 cm, averaging 7.5 g with a regular size. It has a flat-round shoulder, a medium-deep and wide stalk cavity, a flat fruit apex, thick red skin and a smooth surface. The light-green flesh is tight-textured, sweet and juicy. It has a good quality for fresh-eating and dried fruits. The percentage of edible part of fresh fruit is 90.9%, and the content of TTS, reducing sugar and Vc is 26.50%, 20.80% and 538.3 mg per 100 g fresh fruit. The content of flavones and cAMP in mature fruit skin is 6.53 mg/g and 186.06 μg/g. The percentage of fresh fruits which can be made into dried ones is 48.0%. The large oval-shaped stone weighs 0.68 g, with a well-developed kernel. The percentage of containing kernels is 70%.

Evaluation The large plant of Guanyangduanzao cultivar is spreading with strong tree vigor. It has a high and stable yield. The small fruit has a sweet taste and is suitable for fresh-eating and dried fruits.

兼 用
Multipurpose Varieties
品 种

大荔知枣

品种来源及分布 别名迟枣、质枣、稚枣。原产和分布于陕西大荔蒲城小园等村。

植物学性状 树体较大，树姿半开张，树冠乱头形。主干皮裂块状。枣头红褐色，平均长95.0cm，节间长7.5cm。二次枝长28.2cm，4~6节。针刺发达。枣股平均抽生枣吊3个。枣吊长21.8cm，着叶16片。叶片椭圆形，叶长4.9cm，叶宽2.4cm，先端钝尖，叶基圆形，叶缘锯齿尖锐。花量少，花序平均着花3朵，花中大，花径6.1mm。

生物学特性 树势较强，一般定植第三年结果，10年左右进入盛果期。丰产性一般，枣头、2~3年和4~6年生枝的吊果率分别为80.0%、56.8%和10.0%。在山西太谷地区，9月中旬果实进入脆熟期，10月初成熟采收，果实生育期117d左右，为极晚熟品种类型。成熟期遇雨易裂果。

果实性状 果个较大，扁圆形，纵径3.66cm，横径3.03cm，单果重14.0g，大小整齐。果皮中厚，红色，果面有隆起。果点小。梗洼窄而浅。果顶微凹，柱头残存。肉质致密，味甜，汁液中多，品质中等，适宜鲜食或制干。鲜枣可食率94.9%，含可溶性固形物30.90%，总糖25.66%，酸0.67%，糖酸比38.30∶1，100g果肉维生素C含量237.62mg；果皮含黄酮5.16mg/g，cAMP含量269.38μg/g。干枣总糖含量67.57%，酸1.58%。果核较大，纺锤形，核重0.72g，含仁率28.3%。

评价 该品种丰产性一般，果个中大，品质中等，适宜鲜食或制干。

Dalizhizao

Source and Distribution The cultivar originated from and spreads in Dali County of Shaanxi Province.

Botanical Characters The large tree is half-spreading with an irregular crown. The trunk bark has massive fissures. The reddish-brown 1-year-old shoots are 95.0 cm long with the internodes of 7.5 cm. The secondary branches are 28.2 cm long with 4~6 nodes and developed thorns. The mother fruiting spurs can germinate 3 deciduous fruiting shoots which are 21.8 cm long with 16 leaves. The oval-shaped leaves are 4.9 cm long and 2.4 cm wide, with a bluntly-cuspate apex, a round base and a sharp saw-tooth pattern on the margin. The number of flowers is small, averaging 3 ones per inflorescence. The medium-sized blossoms have a diameter of 6.1 mm.

Biological Characters The tree has strong vigor. It generally bears in the 3rd year after planting and enters the full-bearing stage in the 10th year, with a medium yield. The percentage of fruits to deciduous fruiting shoots of 1-year-old shoots, 2~3-year-old branches and 4~6-year-old ones is 80.0%, 56.8% and 10.0%. In Taigu County of Shanxi Province, it enters the crisp-maturing stage in mid-September and matures in early October with a fruit growth period of 117 d. It is an extremely late-ripening variety with poor resistance to fruit-cracking.

Fruit Characteristics The large oblate fruit has a vertical and cross diameter of 3.66 cm and 3.03 cm, averaging 14.0 g with a regular size. It has medium-thick red skin with small dots and some protuberances, a shallow and narrow stalk cavity, a slightly-sunken fruit apex and a remnant stigma. The tight-textured flesh is sweet with medium juice. It has medium quality for fresh-eating and dried fruits. The percentage of edible part of fresh fruit is 94.9%, and SSC, TTS, TA and Vc is 30.90%, 25.66%, 0.67% and 237.62 mg per 100 g fresh fruit. The SAR is 38.30∶1. The content of flavones and cAMP in mature fruit skin is 5.16 mg/g and 269.38 μg/g. The content of TTS and TA in dried fruit is 67.57% and 1.58%. The large spindle-shaped stone weighs 0.72 g. The percentage of containing kernels is 28.3%.

Evaluation The cultivar has a medium yield, medium-large fruit size and medium quality and is suitable for fresh-eating and making dried fruits.

兼 用
Multipurpose Varieties
品 种

· 411 ·

香山小白枣

品种来源及分布 别名香山白枣、小白枣。原产和分布于北京海淀区香山附近，数量不多。

植物学性状 树体较大，树姿半开张，树冠呈自然圆头形。树干灰褐色，树皮裂片小，块状，易剥落。枣头阳面褐红色，阴面红褐色，平均长68.0cm，蜡质少。二次枝平均长27.2cm，6~8节，弯曲度中等，无针刺。枣股平均抽生枣吊5.0个，枣吊长25.7cm，着生叶片16片左右。叶片椭圆形，绿色，先端急尖，叶基圆形。叶缘锯齿细小。花量较多，花序平均着花7朵，花中大，花径6.0mm。

生物学特性 适应性较强，树势强健。定植后第二年开始结果，10年左右进入盛果期，产量一般。在山西太谷地区，果实9月下旬成熟，生长期110d左右，为晚熟品种类型。

果实性状 果实小，长圆形或卵圆形，纵径2.93cm，横径2.29cm，单果重7.1g，最大10.0 g以上，大小不整齐。果肩平圆，梗洼浅、中广。果顶平圆。果面光滑平整，果皮薄，浅红色。果肉白色，质地酥脆，汁液多，味甜，品质好，可鲜食或制干。鲜枣可食率96.9%，含可溶性固形物24.00%，总糖23.42%，酸0.34%，100g果肉维生素C含量129.10mg。核小，纺锤形，核重0.22g，种仁饱满，含仁率33.3%。

评价 该品种树体较大，风土适应性较强。果实小，质脆味甜，品质优异，可鲜食和制干。

Xiangshanxiaobaizao

Source and Distribution The cultivar, also called Xiangshanbaizao or Xiaobaizao, originated from and spreads in Xiangshan of Beijing with a small quantity.

Botanical Characters The large tree is half-spreading with a natural-round crown. The grayish-brown trunk bark has small massive fissures, easily shelled off. The reddish-brown 1-year-old shoots are 68 cm long with less wax. The secondary branches are 27.2 cm long with 6~8 nodes of medium curvature and no thorns. The mother fruiting spurs can germinate 5 deciduous fruiting shoots which are 25.7 cm long with 16 leaves. The green leaves are oval-shaped, with a sharply-cuspate apex, a round base and a thin, small saw-tooth pattern on the margin. There are many medium-sized flowers with a diameter of 6.0 mm produced averaging 7 ones per inflorescence.

Biological Characters The tree has strong vigor and strong adaptability. It generally bears in the 2nd year after planting and enters the full-bearing stage in the 10th year with a medium yield. In Taigu County of Shanxi Province, it matures in late September with a fruit growth period of 110 d. It is a late-ripening variety with light fruit-cracking even if it rains in the maturing stage.

Fruit Characteristics The small fruit is oblong or oval-shaped, with a vertical and cross diameter of 2.93 cm and 2.29 cm, averaging 7.1 g (maximum 10.0 g) with irregular sizes. It has a flat-round shoulder, a shallow and medium-wide stalk cavity, a flat-round fruit apex, a smooth surface and light-red thin skin. The white flesh is crisp, sweet and juicy. It has a good quality for fresh-eating and dried fruits. The percentage of edible part of fresh fruit is 96.9%, and SSC, TTS, TA and Vc is 24.00%, 23.42%, 0.34% and 129.10 mg per 100 g fresh fruit. The small spindle-shaped stone weighs 0.22 g, with a well-developed kernel. The percentage of containing kernels is 33.3%.

Evaluation The large plant of Xiangshanxiaobaizao cultivar has strong adaptability. The small fruit has crisp and sweet flesh with a good quality for fresh-eating and dried fruits.

中 草 笨 枣

品种来源及分布　原产和分布于陕西大荔县官池镇中草、小元等村。

植物学性状　树体中大，树姿开张，树冠呈乱头形。主干皮裂条状。枣头红褐色，平均长82.7cm，节间长6.7cm，无蜡质。二次枝平均长26.1cm，4~6节，弯曲度中等。针刺发达。枣股平均抽生枣吊3.6个。枣吊长19.3cm，着叶15片。叶片小，卵圆形，深绿色，平展，先端锐尖，叶基圆形，叶缘具钝锯齿。花量少，花序平均着花3朵。

生物学特性　树势强，萌芽率和成枝力较高，结果较晚，定植第四年结果，10年左右进入盛果期，较丰产，枣吊平均结果0.71个。在山西太谷地区，10月上旬果实成熟采收，果实生育期120d左右，为极晚熟品种类型。成熟期遇雨裂果轻。

果实性状　果个大，短柱形，纵径3.60cm，横径3.15cm，单果重17.9g，最大25g以上，大小较整齐。果皮薄，红色，果面粗糙。梗洼浅，中等广。果顶凹陷，柱头宿存。肉质较致密，汁液中多，味甜，品质中等，鲜食、制干兼用。鲜枣可食率95.8%，含可溶性固形物27.60%，总糖22.19%，酸0.51%，100g果肉维生素C含量315.97mg；果皮含黄酮7.51mg/g，cAMP含量261.95μg/g。干枣含总糖62.08%，酸1.05%。核较大，纺锤形，平均重0.75g。种仁饱满，含仁率61.7%。

评价　该品种树体中大，树势强，结果较晚，较丰产，抗裂果。果个大，品质中等，鲜食、制干兼用。

Zhongcaobenzao

Source and Distribution　The cultivar originated from and spreads in Zhongcao and Xiaoyuan Village of Dali County in Shaanxi Province.

Botanical Characters　The medium-sized tree is spreading with an irregular crown. The trunk bark has striped fissures. The reddish-brown 1-year-old shoots are 82.7 cm long, with the internodes of 6.7 cm, no wax and developed thorns. The secondary branches are 26.1 cm long with 4~6 nodes of medium curvature. The mother fruiting spurs can germinate 3.6 deciduous fruiting shoots which are 19.3 cm long with 15 leaves. The small oval-shaped leaves are dark green and flat, with a sharply-cuspate apex, a round base and a blunt saw-tooth pattern on the margin. The number of flowers is small, averaging 3 ones per inflorescence.

Biological Characters　The tree has strong vigor, strong germination and branching ability. It bears very late, generally in the 4th year after planting and enters the full-bearing stage in the 10th year with a high yield. The deciduous fruiting shoot bears 0.71 fruits on average. In Taigu County of Shanxi Province, it matures in early October with a fruit growth period of 120 d. It is an extremely late-ripening variety with light fruit-cracking even if it rains in the maturing stage.

Fruit Characteristics　The large column-shaped fruit has a vertical and cross diameter of 3.60 cm and 3.15 cm, averaging 17.9 g (maximum over 25 g) with a regular size. It has thin red skin, a rough surface, a shallow and medium-wide stalk cavity, a slightly-sunken fruit apex and a remnant stigma. The tight-textured flesh is sweet and juicy. It has medium quality for fresh-eating and dried fruits. The percentage of edible part of fresh fruit is 95.8%, and SSC, TTS, TA and Vc is 27.60%, 22.19%, 0.51% and 315.97 mg per 100 g fresh fruit. The content of flavones and cAMP in mature fruit skin is 7.51 mg/g and 261.95 μg/g. The large spindle-shaped stone weighs 0.75 g with a well-developed kernel. The percentage of containing kernels is 61.7%.

Evaluation　The medium-sized plant of Zhongcaobenzao cultivar has strong tree vigor bearing late with a high yield. The fruit has strong resistance to fruit-cracking. It has a large fruit size and medium quality and is suitable for fresh-eating and making dried fruits.

兼 用
Multipurpose Varieties
品 种

沧县长小枣

品种来源及分布 原产和分布于河北沧州地区沧县等地。

植物学性状 树体中大，树姿半开张，树冠自然圆头形。主干皮裂条状。枣头黄褐色，萌发力强，年生长量72.9cm，节间长7.1cm，二次枝长34.9cm，5～7节。针刺不发达。枣股平均抽生枣吊4个。枣吊长19.8cm，着叶14片。叶片卵圆形，叶长6.0cm，叶宽2.7cm，浅绿色，先端钝尖，叶基圆楔形，叶缘具钝锯齿。花量多，花序平均着花9朵。花较大，花径6.5mm。

生物学特性 树势强。坐果率较高，但丰产性一般，枣头、2～3年和4～6年生枝的吊果率分别为90.8%、110.4%和10.5%。在山西太谷地区，9月下旬果实着色，10月上旬成熟采收，果实生育期119d左右，为极晚熟品种类型，抗裂果能力极强。

果实性状 果实小，卵圆形，纵径2.82cm，横径2.29cm，果重7.6g左右，大小整齐。梗洼窄而较深。果顶凹，柱头遗存。果皮薄，浅红色，果面光滑。肉质致密，味甜，汁液中多，品质上等，适宜鲜食或制干。鲜枣可食率96.7%，含可溶性固形物35.40%，总糖24.04%，酸0.63%，糖酸比38.16∶1，100g果肉维生素C含量490.59mg；果皮含黄酮2.62mg/g，cAMP含量138.65μg/g。果核纺锤形，核重0.25g，大多核内无种仁，含仁率16.7%。

评价 该品种果实小，品质上等，可鲜食或制干，果实抗裂果能力极强。

Cangxianchangxiaozao

Source and Distribution The cultivar originated from and spreads in Cangxian County of Hebei Province.

Botanical Characters The medium-sized tree is half-spreading with a natural-round crown. The trunk bark has striped fissures. The yellowish-brown 1-year-old shoots are 72.9 cm long with strong germination ability. The internodes are 7.1 cm long. The secondary branches are 34.9 cm long with 5～7 nodes and less-developed thorns. The mother fruiting spurs can germinate 4 deciduous fruiting shoots which are 19.8 cm long with 14 leaves. The leaves are oval-shaped and light green, 6.0 cm long and 2.7 cm wide, with a bluntly-cuspate apex, a round-cuneiform base and á blunt saw-tooth pattern on the margin. There are many large flowers with a diameter of 6.5 mm produced, averaging 9 ones per inflorescence.

Biological Characters The tree has strong vigor with high fruit set and a medium yield. The percentage of fruits to deciduous fruiting shoots of 1-year-old shoots, 2～3-year-old branches and 4～6-year-old ones is 90.8%, 110.4% and 10.5%. In Taigu County of Shanxi Province, it begins coloring in late September and matures in early October with a fruit growth period of 119 d. It is a late-ripening variety with strong resistance to fruit-cracking.

Fruit Characteristics The small oval-shaped fruit has a vertical and cross diameter of 2.82 cm and 2.29 cm, averaging 7.6 g with a regular size. It has a narrow and deep stalk cavity, a sunken fruit apex, a remnant stigma, light-red thin skin and a smooth surface. The tight-textured flesh is sweet and juicy. It has a superior quality for fresh-eating and dried fruits. The percentage of edible part of fresh fruit is 96.7%, and SSC, TTS, TA and Vc is 35.40%, 24.04%, 0.63% and 490.59 mg per 100 g fresh fruit. The SAR is 38.16∶1. The content of flavones and cAMP in mature fruit skin is 2.62 mg/g and 138.65 μg/g. The spindle-shaped stone weighs 0.25 g. Most stones contain no kernels, and the percentage of containing kernels is 16.7%.

Evaluation The cultivar has a small fruit size and a superior quality, suitable for fresh-eating and making dried fruits. The fruit has strong resistance to fruit-cracking.

兼 用
Multipurpose Varieties
品 种

· 417 ·

新郑齐头白

品种来源及分布　别名齐头白、白枣。原产于河南新郑的孟庄，数量不多。

植物学性状　树体中大，树姿直立，树冠呈自然圆头形。主干灰褐色，皮裂块状，裂纹深，不易剥落。枣头黄褐色，较粗壮，长72.2cm，粗0.92cm。二次枝较短，生长较弱。针刺不发达。枣股圆柱形，平均着生枣吊4.0个。枣吊长22.1cm，着叶12片。叶片椭圆形，浅绿色，先端渐尖，叶基圆楔形，叶缘具细锯齿。花量多，花序平均着花6朵，花较大，花径7.0mm。

生物学特性　风土适应性中等。树势中等，幼树结果早，丰产性强，产量高而稳定。在山西太谷地区，9月下旬果实成熟采收，果实生育期113d左右，为晚熟品种类型。成熟期裂果较轻。

果实性状　果实中大，近圆形，单果重13.0g，大小较整齐。果肩平圆，果顶平。果面平滑。白熟期果皮色泽洁白，着色后转呈浅红色。果肉白色，质地酥脆，汁液多，味甜，适于鲜食和制干，品质上等。鲜枣可食率96.6%，含可溶性固形物30.10%，100g果肉维生素C含量450.69mg；果皮含黄酮8.10mg/g，cAMP含量171.95μg/g。制干率43.3%。果核较小，卵形，核重0.44g，含仁率93.3%，种子较饱满。

评价　该品种风土适应性中等，结果早，产量高。果实色泽艳丽，品质优良，制干鲜食兼用良种，可适当发展。

Xinzhengqitoubai

Source and Distribution　The cultivar, also called Baizao, originated from Mengzhuang Village of Xinzheng City in Henan Province, with a small quantity.

Botanical Characters　The medium-sized tree is vertical with a natural-round crown. The grayish-brown trunk bark has deep massive fissures, uneasily shelled off. The yellowish-brown 1-year-old shoots are 72.2 cm long and 0.92 cm thick with less-developed thorns. The secondary branches are short and weak. The column-shaped mother fruiting spurs can germinate 4 deciduous fruiting shoots, which are 22.1 cm long with 12 leaves. The oval-shaped and light-green leaves have a gradually-cuspate apex, a round-cuneiform base and a thin saw-tooth pattern on the margin. There are many large flowers with a diameter of 7.0 mm produced, averaging 6 ones per inflorescence.

Biological Characters　The tree has mdoerate adaptability and moderate vigor. Young trees bear early with a high and stable yield. In Taigu County of Shanxi Province, it matures in late September with a fruit growth period of 113 d. It is a late-ripening variety with light fruit-cracking in the maturing stage.

Fruit Characteristics　The medium-sized fruit is oval-shaped, averaging 13.0 g with a regular size. It has a flat-round shoulder, a flat fruit apex and a smooth surface. It is pure-white in the white-maturing stage and turns light-red after coloring. The white flesh is crisp, juicy and sweet. It has a good quality for fresh-eating and dried fruits. The percentage of edible part of fresh fruit is 96.6%, and the content of SSC and Vc is 30.10% and 450.69 mg per 100 g fresh fruit. The content of flavones and cAMP in mature fruit skin is 8.10 mg/g and 171.95 μg/g. The percentage of fresh fruits which can be made into dried ones is 43.3%. The small oval-shaped stone weighs 0.44 g, with the percentage of containing kernels of 93.3%. Most kernels are well-developed.

Evaluation　The has moderate adaptability, bearing early with a high yield. The fruit has an attractive appearance and an excellent quality and is suitable for fresh-eating and dried fruits. It can be developed on a proper scale.

兼 用
Multipurpose Varieties
品 种

· 419 ·

定襄小枣

品种来源及分布 原产和分布于山西北部的原平和定襄，为当地主栽品种，以定襄的北社、南社、里域等地栽培集中。

植物学性状 树体高大，树姿直立，枝系粗壮，树冠呈伞形。主干呈条状皮裂。枣头生长势中等，平均长69.5cm，粗1.09cm，节间长7.0cm，无蜡质。二次枝平均长26.8cm，节数6节，弯曲度小，无针刺。枣股中大，抽吊力较强，平均抽生枣吊4.0个。枣吊长20.4cm，着叶11片。叶片卵圆形，绿色，较光亮，先端渐尖，叶基圆楔形，叶缘具钝锯齿。花量较多，花序平均着花8朵。

生物学特性 适应性强，抗晚霜。树势强健，寿命长。产量高，但不够稳定。在山西太谷地区，9月下旬成熟，果实生育期110d，为晚熟品种类型。

果实性状 果实中大，卵圆形，纵径2.86cm，横径2.39cm，单果重10.8g，大小较整齐。果皮中厚，红色。果肉细脆，汁多，味酸甜，适宜制干，也可鲜食，品质中等。鲜枣可食率96.1%，含总糖26.33%，酸0.60%，100g果肉维生素C含量463.91mg；果皮含黄酮15.75mg/g，cAMP含量193.73μg/g。干枣含总糖66.80%。果核较小，纺锤形，核重0.42g，含仁率31.7%。

评价 该品种适应性强，树体高大，健壮，寿命长。产量高。果实中大，大小整齐。品质中等，可鲜食和制干兼用。

Dingxiangxiaozao

Source and Distribution The cultivar originated from and spreads in Yuanping and Dingxiang of north Shanxi Province. It is the dominant variety there, which is mainly concentrated in Beishe, Nanshe and Liyu in Dingxiang County.

Botanical Characters The large tree is vertical with strong branches and an umbrella-shaped crown. The trunk bark has striped fissures. The 1-year-old shoots have mdoerate growth potential, 69.5 cm long and 1.09 cm thick. The internodes are 7.0 cm long without wax and thorns. The secondary branches are 26.8 cm long with 6 nodes of small curvature. The medium-sized mother fruiting spurs can germinate 4 deciduous fruiting shoots which are 20.4 cm long with 11 leaves. The oval-shaped leaves are green and glossy, with a gradually-cuspate apex, a round-cuneiform base and a blunt saw-tooth pattern on the margin. There are many flowers produced, averaging 8 ones per inflorescence.

Biological Characters The tree has strong adaptability, strong tolerance to late-frost, strong vigor and a long life, with a high yet unstable yield. In Taigu County of Shanxi Province, it matures in late September with a fruit growth period of 110 d. It is a late-ripening variety.

Fruit Characteristics The medium-sized fruit is oval-shaped, with a vertical and cross diameter of 2.86 cm and 2.39 cm, averaging 10.8 g with a regular size. It has medium-thick red skin. The flesh is crisp, juicy, sour and sweet. It has medium quality for dried fruits and fresh-eating. The percentage of edible part of fresh fruit is 96.1%, and the content of TTS, TA and Vc is 26.33%, 0.60% and 463.91 mg per 100 g fresh fruit. The content of flavones and cAMP in mature fruit skin is 15.75 mg/g and 193.73 μg/g, and the content of TTS in dried fruit is 66.80%. The small spindle-shaped stone weighs 0.42 g, and the percentage of containing kernels is 31.7%.

Evaluation The large plant of Dingxiangxiaozao cultivar has strong adaptability and strong vigor, with a long life and high productivity. The medium-sized fruit has a regular size. It is suitable for fresh-eating and making dried fruits.

兼 用
Multipurpose Varieties
品种

南 京 鸭 枣

品种来源及分布　别名鸭枣。原产于江苏南京市郊，多零星栽培。

植物学性状　树体较小，树姿较开张，树冠呈伞形。主干灰色，树皮裂纹细浅、块状。枣头红褐色，平均长72.8cm，粗0.97cm，节间长7.6cm，蜡质厚。二次枝平均长25.5cm，5节，弯曲度中等。针刺短，不发达，但数量多。枣股短小，抽吊力强，平均抽生枣吊4.0个。枣吊长19.3cm，着叶11片。叶片较小，卵状披针形，绿色，较光亮。花量多，花中大，花序平均着花4朵，夜开型。

生物学特性　树势较强，产量高，结果稳定，坐果率极高，枣头、2～3年和3年以上枝的吊果率分别为111.4%、220.4%和71.0%。在山西太谷地区，9月下旬果实成熟，果实生育期115d左右，为晚熟品种类型。

果实性状　果实中大，圆形，纵径2.76cm，横径2.62cm，单果重12.8g。果面光滑，红色，果点小，中密，不明显。果肩平圆。梗洼深，中广。果顶略凹，柱头残存。果肉致密，汁液少，味酸甜，品质中等，可鲜食和制干。鲜枣可食率93.2%，含总糖22.45%，酸0.50%，100g果肉维生素C含量410.59mg；果皮含黄酮19.61mg/g，cAMP含量438.05μg/g。制干率44.1%，干枣含糖66.75%，酸1.31%。果核较大，椭圆形，核重0.87g。含仁率20.0%。

评价　该品种树体较小，适应性较强，产量高而稳定。品质中等，果核较大。

Nanjingyazao

Source and Distribution　The cultivar originated from the suburb areas of Nanjing City in Jiangsu Province. It is planted in scattered regions.

Botanical Characters　The small tree is spreading with an umbrella-shaped crown. The gray trunk bark has shallow massive fissures. The reddish-brown 1-year-old shoots are 72.8 cm long and 0.97 cm thick. The internodes are 7.6 cm long with much wax. There are many short and less-developed thorns. The secondary branches are 25.5 cm long with 5 nodes of medium curvature. The short mother fruiting spurs can germinate 4 ones deciduous fruiting shoots which are 19.3 cm long with 11 leaves. The small green leaves are ovate-lanceolate and glossy. There are many medium-sized flowers produced, averaging 4 ones per inflorescence. The blossom blooms at night.

Biological Characters　The tree has strong vigor, with very high fruit set, a high and stable yield. The percentage of fruits to deciduous fruiting shoots to fruits of 1-year-old shoots, 2～3-year-old branches and over-3-year-old ones is 111.4%, 220.4% and 71.0%. In Taigu County of Shanxi Province, it matures in mid-September with a fruit growth period of 115 d. It is a late-ripening variety.

Fruit Characteristics　The medium-sized round fruit has a vertical and cross diameter of 2.76 cm and 2.62 cm, averaging 12.8 g. It has a smooth surface, red skin with small and medium-dense dots, a flat-round shoulder, a deep and medium-wide stalk cavity, a slightly-sunken fruit apex and a remnant stigma. The tight-textured flesh is sour and sweet with less juice. It has medium quality for fresh-eating and dried fruits. The percentage of edible part of fresh fruit is 93.2%, and the content of TTS, TA and Vc is 22.45%, 0.50% and 410.59 mg per 100 g fresh fruit. The content of flavones and cAMP in mature fruit skin is 19.61 mg/g and 438.05 μg/g. The large oval-shaped stone weighs 0.87 g, with the percentage of containing kernels of 20.0%.

Evaluation　The small plant of Nanjingyazao cultivar has strong adaptability, a high and stable yield. The fruit has a large stone and low edibility with medium quality.

兼 用
Multipurpose Varieties
品 种

献县酸枣

品种来源及分布 原产和分布于河北献县。

植物学性状 树体较小，树姿较开张，树冠呈圆锥形。主干皮裂条状。枣头红褐色，平均长68.7cm，节间长5.3cm。二次枝长21.2cm，5～7节。针刺发达。枣股平均抽生枣吊3.5个。枣吊长15.5cm，着叶13片。叶片小，叶长3.7cm，叶宽1.9cm，卵圆形，先端锐尖，叶基圆形，叶缘具钝锯齿。花量中多，花序平均着花6朵，花中大，花径6.1mm。

生物学特性 树势强，萌芽率和成枝力中等。定植第三年结果，成龄树坐果率高，枣头、2～3年和4～6年生枝的吊果率分别为115.8%、169.4%和51.9%，但果实生理落果极为严重，产量一般。在山西太谷地区，9月下旬果实成熟，果实生育期109d以上，为晚熟品种类型。

果实性状 果实极小，圆形，纵径1.74cm，横径1.65cm，单果重2.1g，大小整齐。果皮薄，紫红色，果面平滑。果点小，分布较密。梗洼中广、较浅。果顶平，柱头遗存，不明显。肉质疏松，味酸甜，汁液少，品质中等。鲜枣可食率86.7%，含可溶性固形物32.78%，总糖22.78%，100g果肉维生素C含量485.87mg；果皮含黄酮40.96mg/g，cAMP含量46.39μg/g。干枣含总糖65.15%，酸1.89%。果核椭圆形，核重0.28g。种仁较饱满，含仁率15.0%。

评价 该品种果个极小，可食率极低，品质中等，适宜鲜食或制干。

Xianxiansuanzao

Source and Distribution The cultivar originated from and spreads in Xianxian County of Hebei Province.

Botanical Characters The small tree is spreading with a conical crown. The trunk bark has striped fissures. The reddish-brown 1-year-old shoots are 68.7 cm long with the internodes of 5.3 cm long. The secondary branches are 21.2 cm long with 5～7 nodes and developed thorns. The mother fruiting spurs can germinate 3.5 deciduous fruiting shoots which are 15.5 cm long with 13 leaves. The small oval-shaped leaves are 3.7 cm long and 1.9 cm wide, with a sharply-cuspate apex, a round base and a blunt saw-tooth pattern on the margin. The number of flowers is medium large, averaging 6 ones per inflorescence. The medium-sized blossoms have a diameter of 6.1 mm.

Biological Characters The tree has strong vigor, medium germination and branching ability. It generally bears in the 3rd year after planting, with high fruit set. The percentage of fruits to deciduous fruiting shoots of 1-year-old shoots, 2～3-year-old branches and 4～6-year-old ones is 115.8%, 169.4% and 51.9%. It has serious physiological fruit-drop, so the yield is just medium. In Taigu County of Shanxi Province, it matures in late September with a fruit growth period of 109 d. It is a late-ripening variety.

Fruit Characteristics The small round fruit has a vertical and cross diameter of 1.74 cm and 1.65 cm, averaging 2.1 g with a regular size. It has purplish-red thin skin, a smooth surface with small and dense dots, a medium-wide and shallow stalk cavity, a flat fruit apex and a remnant stigma. The loose-textured flesh is sour and sweet with less juice. It has medium quality. The percentage of edible part of fresh fruit is 86.7%, and SSC, TTS and Vc is 32.78%, 22.78% and 485.87 mg per 100 g fresh fruit. The content of flavones and cAMP in mature fruit skin is 40.96 mg/g and 46.39 μg/g. The oval-shaped stone weighs 0.28 g, with a well-developed kernel. The percentage of containing kernels is 15.0%.

Evaluation The cultivar has a very small fruit size and medium quality and is suitable for fresh-eating and making dried fruits.

兼 用
Multipurpose Varieties
品种

· 425 ·

赞皇长枣

品种来源及分布 原产和分布于河北赞皇县。

植物学性状 树体高大，树姿半开张，树冠呈乱头形。主干条状皮裂。针刺不发达。枣股平均抽生枣吊3.5个。枣吊长28.4cm，着叶13片。叶片中等大，卵圆形，深绿色，平展，先端急尖，叶基圆楔形，叶缘具钝锯齿。花量中多，花序平均着花4朵，花较大，花径6.4mm。

生物学特性 树势强，萌芽率和成枝力较高，结果较早，定植第二年结果，10年左右进入盛果期，丰产性较强，而且产量较稳定，枣吊平均结果0.70个。在山西太谷地区，9月下旬果实成熟采收，果实生育期109d左右，为晚熟品种类型。成熟期遇雨裂果较轻。

果实性状 果个大，圆柱形，纵径3.85cm，横径3.21cm，单果重17.9g，最大20g以上，大小不整齐。果皮中厚，红色，果面平滑。梗洼浅而广。果顶平圆，柱头残存。肉质疏松，汁液中多，味酸甜，品质中等，适宜鲜食或制干。鲜枣可食率94.6%，含可溶性固形物21.00%，总糖18.58%，酸0.37%，100g果肉维生素C含量408.27mg；果皮含黄酮32.16mg/g，cAMP含量44.56 μg/g。核较大，纺锤形，平均重0.97g。核内不含种子。

评价 该品种树体高大，树势强，结果较早，丰产稳产，抗裂果。果个大，品质中等，适宜鲜食或制干。

Zanhuangchangzao

Source and Distribution The cultivar originated from and spreads in Zanhuang County of Hebei Province.

Botanical Characters The large tree is half-spreading with an irregular crown. The trunk bark has striped fissures. The thorns are less-developed. The mother fruiting spurs can germinate 3.5 deciduous fruiting shoots which are 28.4 cm long with 13 leaves. The medium-sized leaves are oval-shaped, dark green and flat, with a sharply-cuspate apex, a round-cuneiform base and a blunt saw-tooth pattern on the margin. The number of flowers is medium large, averaging 4 ones per inflorescence. The large blossoms have a diameter of 6.4 mm.

Biological Characters The tree has strong vigor, strong germination and branching ability. It bears early, generally in the 2nd year after planting and enters the full-bearing stage in the 10th year with a high and stable yield. The deciduous fruiting shoot bears 0.70 fruits on average. In Taigu County of Shanxi Province, it matures in late September with a fruit growth period of 109 d. It is a late-ripening variety with light fruit-cracking even if it rains in the maturing stage.

Fruit Characteristics The large column-shaped fruit has a vertical and cross diameter of 3.85 cm and 3.21 cm, averaging 17.9 g (maximum over 20 g) with irregular sizes. It has medium-thick red skin, a smooth surface, a shallow and wide stalk cavity, a flat-round fruit apex and a remnant stigma. The loose-textured flesh is sour, sweet and juicy. It has medium quality for fresh-eating and dried fruits. The percentage of edible part of fresh fruit is 94.6%, and SSC, TTS, TA and Vc is 21.00%, 18.58%, 0.37% and 408.27 mg per 100 g fresh fruit. The content of flavones and cAMP in mature fruit skin is 32.16 mg/g and 44.56 μg /g. The large spindle-shaped stone weighs 0.97 g without kernel.

Evaluation The large plant of Zanhuangchangzao cultivar has strong tree vigor, bearing early with a high and stable yield. The fruit has strong resistance to fruit-cracking. It has a large fruit size and medium quality and is suitable for fresh-eating and making dried fruits.

兼用
Multipurpose Varieties
品种

· 427 ·

大荔小圆枣

品种来源及分布　别名小圆枣。原产和分布于陕西大荔的官池乡，数量不多。

植物学性状　树体较大，树姿开张，树冠自然乱头形。主干皮裂块状。枣头红褐色，平均长70.4cm，节间长6.0cm，蜡质多。二次枝平均长20.0cm，3～7节，弯曲度中等。针刺不发达。枣股平均抽生枣吊3.5个。枣吊长18.2cm，着叶16片。叶椭圆形，平展，较光亮。先端钝尖，叶基圆形，叶缘具钝锯齿。花量中多，花序平均着花4朵。花中大，花径6.1mm。

生物学特性　树势强，产量较高而稳定，成龄树枣吊平均结果0.52个。在山西太谷地区，9月上旬着色，9月下旬果实成熟，果实生育期约107d，为晚熟品种类型。

果实性状　果实中大，圆形，纵径2.20cm，横径2.17cm，单果重10.7g，最大12.8g，大小不整齐。果肩平圆，梗洼窄、中广。果顶圆，顶点微凹。果面光滑，果皮厚，紫红色。果肉浅绿色，质地酥脆，汁液较多，味甜，品质中等，宜制干和鲜食。鲜枣可食率94.8%，含可溶性固形物25%～30%，总糖22.92%。果核大，椭圆形，核重0.56g。大多核内无种子，含仁率16.7%。

评价　该品种树体较大，树势强，产量较高而稳定。果实中大，质地酥脆，品质中等，可鲜食和制干。

Dalixiaoyuanzao

Source and Distribution　The cultivar originated from and spreads in Guanchi Village of Dali County in Shaanxi Province.

Botanical Characters　The large tree is spreading with an irregular crown. The trunk bark has massive fissures. The reddish-brown 1-year-old shoots are 70.4 cm long, with the internodes of 6.0 cm long, much wax and less-developed thorns. The secondary branches are 20.0 cm long with 3～7 nodes of medium curvature. The mother fruiting spurs can germinate 3.5 deciduous fruiting shoots which are 18.2 cm long with 16 leaves. The oval-shaped leaves are flat and glossy, with a bluntly-cuspate apex, a round base and a blunt saw-tooth pattern on the margin. The number of flowers is medium large, averaging 4 ones per inflorescence. The medium-sized blossoms have a diameter of 6.1 mm.

Biological Characters　The tree has strong vigor with a high and stable yield. The deciduous fruiting shoot of a mature tree bears 0.52 fruits on average. In Taigu County of Shanxi Province, it begins coloring in late September with a fruit growth period of 107 d. It is a mid-ripening variety.

Fruit Characteristics　The medium-sized round fruit has a vertical and cross diameter of 2.20 cm and 2.17 cm, averaging 10.7 g (maximum 12.8 g) with irregular sizes. It has a flat-round shoulder, a medium-deep stalk cavity, a round fruit apex, a smooth surface and purplish-red thick skin. The light-green flesh is crisp, sweet and juicy. It has medium quality for fresh-eating and dried fruits. The percentage of edible part of fresh fruit is 94.8%, and SSC and TTS is 25%～30% and 22.92% respectively. The large oval-shaped stone weighs 0.56 g. Most stones contain no kernels, and the percentage of containing kernels is 16.7%.

Evaluation　The large plant of Dalixiaoyuanzao cultivar has strong tree vigor with a high and stable yield. The medium-sized fruit has tight-textured flesh with medium quality for fresh-eating and dried fruits.

兼 用
Multipurpose Varieties
品 种

新 郑 大 枣

品种来源及分布 原产和分布于河南省新郑市。

植物学性状 树体高大，树姿开张，树冠呈半圆形。主干皮裂条状。枣头黄褐色，平均长73.9cm，节间长7.5cm。二次枝长31.2cm，6～8节。针刺不发达。枣股平均抽生枣吊3.4个。枣吊长19.4cm，着叶11片。叶片中大，椭圆形，绿色，平展，先端急尖，叶基圆楔形，叶缘具钝锯齿。花量多，花序平均着花7朵，花中大，花径6.2mm。

生物学特性 树势中庸，萌芽率和成枝力较高，结果较早，定植第二年结果，10年左右进入盛果期，产量中等，枣吊平均结果0.94个。在山西太谷地区，9月下旬果实成熟采收，果实生育期110d左右，为晚熟品种类型。成熟期遇雨易裂果。

果实性状 果个中大，卵圆形，单果重14.1g，大小较整齐。果皮浅红色，果面平滑。梗洼中深、广。柱头残存。肉质酥脆，汁液多，味甜，品质好，适宜鲜食或制干。鲜枣可食率95.7%，含可溶性固形物25.00%，100g果肉维生素C含量519.42mg；果皮含黄酮6.09mg/g，cAMP含量79.72μg/g。干枣含总糖61.28%，酸0.73%。核中大，纺锤形，平均重0.61g。核内不含种子。

评价 该品种树体高大，树势中庸，结果较早，产量中等，抗裂果能力较差。果个中大，品质好，适宜鲜食或制干。

Xinzhengdazao

Source and Distribution The cultivar originated from and spreads in Xinzheng City of Henan Province.

Botanical Characters The large tree is spreading with a semi-round crown. The trunk bark has striped fissures. The yellowish-brown 1-year-old shoots are 73.9 cm long, with the internodes of 7.5 cm long and less-developed thorns. The secondary branches are 31.2 cm long with 6～8 nodes. The mother fruiting spurs can germinate 3.4 deciduous fruiting shoots, which are 19.4 cm long with 11 leaves. The medium-sized leaves are oval-shaped, green and flat, with a sharply-cuspate apex, a round-cuneiform base and a blunt saw-tooth pattern on the margin. There are many medium-sized flowers with a diameter of 6.2 mm produced, averaging 7 ones per inflorescence.

Biological Characters The tree has moderate vigor, strong germination and branching ability. It bears early, generally in the 2nd year after planting and enters the full-bearing stage in the 10th year with a medium yield. The deciduous fruiting shoot bears 0.94 fruits on average. In Taigu County of Shanxi Province, it matures in late September with a fruit growth period of 110 d. It is a late-ripening variety with serious fruit-cracking if it rains in the maturing stage.

Fruit Characteristics The large oblong fruit weighs 14.1 g on average with a regular size. It has light red skin, a smooth surface, a medium-deep and wide stalk cavity and a remnant stigma. The flesh is crisp, juicy and sweet. It has a good quality for fresh-eating and dried fruits. The percentage of edible part of fresh fruit is 95.7%, and SSC and Vc is 25.00% and 519.42 mg per 100 g fresh fruit. The content of flavones and cAMP in mature fruit skin is 6.09 mg/g and 79.72 μg/g. The medium-sized stone is spindle-shaped, averaging 0.61 g, without kernel.

Evaluation The large plant of Xinzhengdazao cultivar has moderate tree vigor, bearing early with a medium yield. The medium-sized fruit has a good quality and is suitable for fresh-eating and making dried fruits. It has poor resistance to fruit-cracking.

兼 用
Multipurpose Varieties
品 种

姑 苏 小 枣

品种来源及分布 原产和分布于江苏苏州市。

植物学性状 树体较小，树姿半开张，树冠自然圆头形。主干皮裂条状。枣头黄褐色，年平均生长量73.5cm，节间长8.7cm。二次枝平均长32.1cm，5～7节，弯曲度小。针刺基本退化。枣股平均抽生枣吊3.0个。枣吊长20.8cm，着叶11片。叶片大，椭圆形，两侧略向叶面合抱，先端急尖，叶基圆楔形，叶缘具钝锯齿。花量多，花序平均着花13朵，花大，花径6.6mm。

生物学特性 树势中庸，萌芽率和成枝力强，一般定植第二年结果，第三年有一定产量，较丰产，成龄树平均吊果率93.4%。在山西太谷地区，9月下旬果实成熟采收，果实生育期109d以上，为晚熟品种类型。成熟期遇雨易裂果。

果实性状 果个中大，卵圆形，纵径2.09cm，横径1.86cm，单果重12.0g，大小较整齐。果皮中厚，浅红色，果面平滑。梗洼中深、广。果顶平，柱头残存。肉质酥脆，味甜，汁液多，品质好，适宜鲜食或制干。鲜枣可食率95.8%，含总糖22.54%，酸0.34%；果皮含黄酮2.11mg/g，cAMP含量144.71μg/g。干枣含总糖59.15%，酸1.16%。核较小，纺锤形，核重0.5g，种仁不饱满，含仁率8.3%。

评价 该品种树体较小，树势中庸，较丰产。果个中大，品质优异，适宜鲜食或制干，唯果实抗裂果能力较差。

Gusuxiaozao

Source and Distribution The cultivar originated from and spreads in Suzhou City of Jiangsu Province.

Botanical Characters The small tree is half-spreading with a natural-round crown. The trunk bark has striped fissures. The yellowish-brown 1-year-old shoots are 73.5 cm long with the internodes of 8.7 cm long. The secondary branches are 32.1 cm long with 5～7 nodes of small curvature and almost degraded thorns. The mother fruiting spurs can germinate 3 deciduous fruiting shoots which are 20.8 cm long with 11 leaves. The large oval-shaped leaves have a sharply-cuspate apex, a round-cuneiform base and a blunt saw-tooth pattern on the margin. There are many large flowers with a diameter of 6.6 mm produced, averaging 13 ones per inflorescence.

Biological Characters The tree has moderate vigor, high germination rate, strong branching ability and high productivity. It generally bears in the 2nd year after planting and there is certain yield in the 3rd year. The percentage of fruits to deciduous fruiting shoots is 93.4% on average. In Taigu County of Shanxi Province, it matures in late September, with a fruit growth period of 109 d. It is a late-ripening variety with poor resistance to fruit-cracking.

Fruit Characteristics The medium-sized fruit is oval-shaped, with a vertical and cross diameter of 2.09 cm and 1.86 cm, averaging 12.0 g with a regular size. It has medium-thick and light-red skin, a smooth surface, a medium-deep and wide stalk cavity, a flat fruit apex and a remnant stigma. The flesh is delicate, crisp, sweet and juicy. It has a good quality for fresh-eating and dried fruits. The percentage of edible part of fresh fruit is 95.8%, and the content of TTS and TA is 22.54% and 0.34%. The content of flavones and cAMP in mature fruit skin is 2.11 mg/g and 144.71 μg/g. The small spindle-shaped stone weighs 0.5 g with a shriveled kernel. The percentage of containing kernels is 8.3%.

Evaluation The small plant of Gusuxiaozao cultivar has moderate tree vigor and a medium yield. The medium-sized fruit has an excellent quality and is suitable for fresh-eating and making dried fruits. Yet it has poor resistance to fruit-crack.

兼 用
Multipurpose Varieties
品 种

· 433 ·

延川跌牙枣

品种来源及分布 原产和分布于陕西延川县。

植物学性状 树体较小，树姿开张，树冠圆柱形。主干皮裂条状。枣头浅灰色或红褐色，平均长49.5cm，节间长5.9cm。二次枝长24.8cm，5～8节。针刺基本退化。枣股平均抽生枣吊3.7个。枣吊长27.6cm，着叶18片。叶片中等大，椭圆形，深绿色，两侧略向叶面合抱，先端急尖，叶基圆楔形，叶缘具钝锯齿。花量多，花序平均着花5朵，花中大，花径6.0mm。

生物学特性 树势中庸，萌芽率和成枝力中等，结果较早，一般定植第二年结果，10年左右进入盛果期，产量中等，枣吊平均结果0.17个。在山西太谷地区，9月下旬果实成熟采收，果实生育期110d以上，为晚熟品种类型。

果实性状 果个大，圆柱形，纵径3.31cm，横径2.81cm，单果重17.4g，大小较整齐。果皮中厚，红色，果面平滑。梗洼浅而广。果顶微凹，柱头脱落。肉质致密，味甜，汁液多，品质中等，适宜鲜食或制干。鲜枣可食率97.1%，含可溶性固形物22.00%，总糖21.94%。制干率41.8%，干枣含总糖67.52%，酸1.36%。果核中大，纺锤形，平均重0.50g。含仁率8.3%，种仁不饱满。

评价 该品种树体较小，树势中庸，结果较早，产量中等，果个大，品质中等，适宜鲜食或制干。

Yanchuandieyazao

Source and Distribution The cultivar originated from and spreads in Yanchuan County of Shaanxi Province.

Botanical Characters The small tree is spreading with a column-shaped crown. The trunk bark has striped fissures. The 1-year-old shoots are light gray or reddish brown, 49.5 cm long with the internodes of 5.9 cm and degraded thorns. The secondary branches are 24.8 cm long with 5～8 nodes. The mother fruiting spurs can germinate 3.7 deciduous fruiting shoots which are 27.6 cm long with 18 leaves. The medium-sized leaves are oval-shaped and dark green, with a sharply-cuspate apex, a round-cuneiform base and a blunt saw-tooth pattern on the margin. There are many medium-sized flowers with a diameter of 6.0 mm produced, averaging 5 ones per inflorescence.

Biological Characters The tree has moderate vigor, medium germination and branching ability. It bears early, generally in the 2nd year after planting and enters the full-bearing stage in the 10th year with a medium yield. The deciduous fruiting shoot bears 0.17 fruits on average. In Taigu County of Shanxi Province, it matures in late September with a fruit growth period of 110 d. It is a late-ripening variety.

Fruit Characteristics The large column-shaped fruit has a vertical and cross diameter of 3.31 cm and 2.81 cm, averaging 17.4 g with a regular size. It has medium-thick red skin, a smooth surface, a shallow and wide stalk cavity, a slightly-sunken fruit apex and a falling-off stigma. The tight-textured flesh is sweet and juicy. It has a good quality for fresh-eating and dried fruits. The percentage of edible part of fresh fruit is 97.1%, and SSC and TTS is 22.00% and 21.94%. The small spindle-shaped stone weighs 0.50 g, with a shriveled kernel. The percentage of containing kernels is 8.3%.

Evaluation The small plant of Yanchuandieyazao cultivar has moderate tree vigor, bearing early with a medium yield. The large fruit has a normal quality and is suitable for fresh-eating and making dried fruits.

兼 用
Multipurpose Varieties
品 种

喀什噶尔小枣

品种来源及分布　别名长枣、喀什小枣。集中分布于新疆喀什噶尔平原绿洲地带。

植物学性状　树体中大，树姿半开张，树冠多呈圆头形。主干灰褐色，粗糙，皮裂呈块状，不规则，容易剥落。枣头紫褐色，平均长63.7cm，粗0.82cm，节间长4.8cm，无蜡质。二次枝平均长17.4cm，6节左右，弯曲度中等。针刺发达。枣股圆柱形，平均抽生枣吊4.0个。枣吊长24.3cm，着叶18片。叶片小，椭圆形，浓绿色，先端渐尖，叶基偏斜形。叶缘波状，钝齿。花量较多，每序平均着花7朵，花小，花径6.5mm。

生物学特性　适应性强，耐旱，较耐盐碱。树势较强，枣吊平均结果0.6个，产量中等。在山西太谷地区，9月下旬果实成熟，果实生育期110d左右，为晚熟品种类型。

果实性状　果实极小，卵圆形，纵径2.13cm，横径1.68cm，单果重3.70g，大小较整齐。果肩平圆，梗洼中深、广，果顶微凹，果面平整，果皮浅红色。果肉浅绿色，质地疏松，汁多，味酸，品质中等，可鲜食和制干。鲜枣可食率88.9%。果核大，核重0.41g，含仁率60%左右。

评价　该品种适应性强，树体较高大，产量中等，果实小，味酸，可食率极低，品质中等。

Kashigeerxiaozao

Source and Distribution　The cultivar, also called Changzao, mainly spreads in oases of Kashigeer Plain in Xinjiang Province.

Botanical Characters　The medium-sized tree is half-spreading with a round crown. The grayish-brown trunk bark has rough, irregular, massive fissures, easily shelled off. The purplish-brown 1-year-old shoots are 63.7 cm long and 0.82 cm thick. The internodes are 4.8 cm long, without wax and with developed thorns. The secondary branches are 17.4 cm long with 6 nodes of medium curvature. The column-shaped mother fruiting spurs can germinate 4 deciduous fruiting shoots which are 24.3 cm long with 18 leaves. The small dark-green leaves are oval-shaped, with a gradually-cuspate apex, a deflective base and a blunt saw-tooth pattern on the wavy margin. There are many small flowers with a diameter of 6.5 mm produced, averaging 7 ones per inflorescence.

Biological Characters　The tree has strong adaptability, tolerant to drought and saline-alkaline, with strong vigor and a medium yield. The deciduous fruiting shoot bears 0.6 fruits on average. In Taigu County of Shanxi Province, it matures in late September with a fruit growth period of 110 d. It is a late-ripening variety.

Fruit Characteristics　The oval-shaped fruit has a very small size, with a vertical and cross diameter of 2.13 cm and 1.68 cm, averaging 3.70 g with a regular size. It has a flat-round shoulder, a medium-deep and wide stalk cavity, a slightly-sunken fruit apex, a smooth surface and light-red skin. The light-green flesh is loose-textured, sour and juicy. It has medium quality for fresh-eating and dried fruits. The percentage of edible part of fresh fruit is 88.9%. The large stone weighs 0.41 g, with the percentage of containing kernels of 60%.

Evaluation　The large plant of Kashigeerxiaozao cultivar has strong adaptability and a medium yield. The small fruit has a sour taste and low edibility with medium quality for fresh-eating and dried fruits.

兼用
Multipurpose Varieties
品种

新郑大马牙

品种来源及分布 原产和分布于河南省新郑枣区，其他地区也有少量分布。

植物学性状 树体较大，树姿半开张，树冠呈自然半圆头形。主干皮裂块状。枣头黄褐色，长79.6cm，粗度0.92cm，节间长8.5cm，蜡质多。二次枝长27.5cm，5节左右，弯曲度中等。针刺自然脱落。每枣股抽生枣吊4.0个。枣吊长20.9cm，着叶12片。叶片椭圆形，绿色，较光亮，先端渐尖，先端尖凹。叶基偏斜形，叶缘具粗锯齿。花量大，花序平均着花11朵。

生物学特性 适应性强，耐旱，耐瘠薄。产量较低，坐果率低，枣吊平均结果仅0.18个。在山西太谷地区，9月下旬成熟采收，成熟期不整齐，果实生育期约117d，为极晚熟品种类型。

果实性状 果实大，圆锥形，果顶向一侧偏斜。单果重18.7g，大小较整齐。果肉质地酥脆，汁液多，味酸甜，品质较好，鲜食、制干兼用。鲜枣可食率96.5%，含可溶性固形物27.60%，总糖25.98%，酸0.36%。干枣含总糖67.23%，酸0.96%。果核中大，长纺锤形，核重0.65g，核内含种仁。

评价 该品种适应性强，果实大，品质较好，制干和鲜食兼用。但产量较低，需采取环剥等栽培措施促进坐果。

Xinzhengdamaya

Source and Distribution The cultivar originated from and spreads in Xinzheng of Henan Province with some small quantity in other places.

Botanical Characters The large tree is half-spreading with a semi-round crown. The trunk bark has massive fissures. The yellowish-brown 1-year-old shoots are 79.6 cm long and 0.92 cm thick. The internodes are 8.5 cm long with much wax and naturally falling-off thorns. The secondary branches are 27.5 cm long with 5 nodes of medium curvature. The mother fruiting spurs can germinate 4 deciduous fruiting shoots which are 20.9 cm long with 12 leaves. The oval-shaped leaves are green and glossy, with a gradually-cuspate apex, a deflective base and a thick saw-tooth pattern on the margin. There are many flowers produced, averaging 11 ones per inflorescence.

Biological Characters The tree has strong adaptability, tolerant to drought and poor soils, with low fruit set and a low yield. The deciduous fruiting shoot bears only 0.18 fruits on average. In Taigu County of Shanxi Province, it matures in late September with a fruit growth period of 117 d. It is a late-ripening variety.

Fruit Characteristics The large conical fruit weighs 18.7 g with a regular size. The flesh is crisp, juicy, sour and sweet. It has a good quality for fresh-eating and dried fruits. The percentage of edible part of fresh fruit is 96.5%, and SSC, TTS and TA is 27.60%, 25.98% and 0.36%. The medium-sized stone is long spindle-shaped, averaging 0.65 g with a kernel.

Evaluation The cultivar has strong adaptability. The large fruit has a good quality for fresh-eating and dried fruits. Yet it has low productivity. Some cultural techniques such as girdling should be carried out to improve fruit set.

兼 用
Multipurpose Varieties
品 种

· 439 ·

太谷黑叶枣

品种来源及分布 别名没心红、黑叶枣。分布于山西汾河中游的清徐、交城、榆次、太谷、祁县等地。是地方古老品种之一，栽培数量较少。

植物学性状 树体高大，树姿半开张，干性较强，枝条中密，生长粗壮，树冠呈自然圆头形。树干皮裂呈条状。枣头黄褐色，平均长52.4cm，粗1.05cm，节间长9.6cm。二次枝长34.1cm，平均6节。针刺不发达。枣股中大，抽吊力较强，平均抽生枣吊3.9个。枣吊平均长21.0cm，着叶14片。叶片中大，叶长6.7cm，叶宽2.8cm，椭圆形，浓绿色，先端急尖，叶基圆楔形，叶缘锯齿钝。花量少，枣吊平均着花33.4朵。花朵大，花径8.0mm，夜开型。

生物学特性 树势较强，萌芽率低，成枝力较强。坐果率高，枣头吊果率52.6%，2~3年生枝为85.2%，4年生枝为48.7%，主要坐果部位在枣吊的3~9节，占坐果总数的91.3%。丰产，产量稳定。在山西太谷地区，9月下旬果实成熟，果实生育期110d左右，为晚熟品种类型。成熟期遇雨裂果较严重。

果实性状 果实中大，圆柱形，纵径3.87cm，横径3.15cm，单果重15.4g，大小不整齐。果梗中长、粗，梗洼中广、深。果顶平，柱头遗存。果皮薄，紫红色，果面光滑，果点大而稀疏。果肉厚，绿白色，肉质细，较松，味甜略酸，汁液较多，品质中上，可鲜食、制干和加工酒枣。鲜枣含可溶性固形物30.60%，总糖25.50%，酸0.66%，100g果肉维生素C含量462.05mg，含水量66.1%；果皮含黄酮8.64mg/g，cAMP含量84.88μg/g。制干率49.5%，干枣含总糖69.16%，酸1.00%。大果果核大，纺锤形，纵径2.39cm，横径0.83cm，核重0.71g，核尖中等长，核面粗糙，含仁率18.3%；中小果核小，软化可食，故把黑叶枣叫"没心红"。

评价 该品种适应性强。树势较强，丰产稳产。果实肉厚，品质较好，为干鲜兼用品种。采前落果轻，但不抗裂果。适于秋雨少的地区发展。

Taiguheiyezao

Source and Distribution The cultivar originated from and spreads in Qingxu, Jiaocheng, Yuci, Taigu and Qixian along the middle reaches of Fenhe River in Shanxi Province. It is an old local variety with a small quantity.

Botanical Characters The large tree is half-spreading with a strong central leader trunk, medium-dense and strong branches and a natural-round crown. The yellowish-brown 1-year-old shoots are 52.4 cm long and 1.05 cm thick with the internodes of 9.6 cm. The secondary branches are 34.1 cm long with 6 nodes and less-developed thorns. The medium-sized mother fruiting spurs can germinate 3.9 deciduous fruiting shoots which are 21.0 cm long with 14 leaves. The medium-sized leaves are oval-shaped and dark green. The number of flowers is small, averaging 33.4 ones per deciduous fruiting shoot. The night-bloomed large blossoms have a diameter of 8.0 mm.

Biological Characters The tree has strong vigor, low germination rate and strong branching ability with high fruit set and a high and stable yield. In Taigu County of Shanxi Province, it matures in late September with a fruit growth period of 100 d. It is a late-ripening variety with serious fruit-cracking if it rains in the maturing stage.

Fruit Characteristics The medium-sized fruit is column-shaped, averaging 15.4 g with irregular sizes. It has purplish-red thin skin and a smooth surface. The greenish-white thick flesh is loose-textured, juicy, sweet and a little sour, with a better than normal quality for fresh-eating, dried fruits and alcoholic fruits. The percentage of edible part of fresh fruit is 95.4%, and SSC, TTS, TA and Vc is 30.60%, 25.50%, 0.66% and 462.05 mg per 100 g fresh fruit. The water content is 66.1%. The rate of fresh fruits which can be made into dried ones is 49.5%, and the content of TTS and TA in dried fruit is 69.16% and 1.00%. The small stone is spindle-shaped, averaging 0.71 g.

Evaluation The cultivar has strong adaptability, poor tolerance to late frost and strong vigor, with a high and stable yield. The fruit has thick flesh, a small stone, high edibility and a good quality. It is a good table and drying variety with light premature fruit-dropping and poor resistance to fruit-cracking. It can be developed in areas with light rainfall and without late-frost in autumn.

兼 用
Multipurpose Varieties
品 种

曲阜猴头枣

品种来源及分布 原产和分布于山东曲阜等地。

植物学性状 树体中大，树姿直立，树冠呈伞形。主干皮裂条状。枣头黄褐色，年生长量79.9cm，节间长7.9cm，二次枝长28.2cm，5～7节。无针刺。枣股平均抽生枣吊4.0个。枣吊长23.1cm，着叶16片。叶片椭圆形，绿色，先端尖凹，叶基偏斜形，叶缘具钝锯齿。花量多，花序平均8朵。花中大，花径6.0mm。

生物学特性 树势中等。较丰产，抗裂果能力较强。在山西太谷地区，9月下旬果实着色成熟，果实生育期113d左右，为晚熟品种类型。

果实性状 果实中大，圆锥形或倒卵圆形，纵径3.70cm，横径2.50cm，果重11.1g左右，大小较整齐。梗洼浅、中广。果顶平，柱头脱落。果皮中厚，红色，果面光滑。肉质酥脆，味甜，汁液中多，品质中上，鲜食、制干兼用。鲜枣可食率94.8%，总糖含量26.25%，酸0.31%，糖酸比84.68∶1，100g果肉维生素C含量325.97mg；果皮含黄酮3.97mg/g，cAMP含量84.30μg/g。干枣含总糖67.08%，酸0.66%。果核较大，纺锤形，核重0.58g，核内含饱满种仁。

评价 该品种较丰产，抗裂果能力较强，果实中大，品质中上，为鲜食制干兼用品种。

Qufuhoutouzao

Source and Distribution The cultivar originated from and spreads in Qufu of Shandong Province.

Botanical Characters The medium-sized tree is vertical with an umbrella-shaped crown. The trunk bark has striped fissures. The yellowish-brown 1-year-old shoots are 79.9 cm long with the internodes of 7.9 cm. The secondary branches are 28.2 cm long with 5～7 nodes and no thorns. The mother fruiting spurs can germinate 4 deciduous fruiting shoots which are 23.1 cm long with 16 leaves. The green leaves are oval-shaped, with a sharply-sunken apex, a deflective base and a blunt saw-tooth pattern on the margin. There are many medium-sized flowers with a diameter of 6.0 mm produced, averaging 8 ones per inflorescence.

Biological Characters The tree has moderate vigor with a high yield. The fruit has strong resistance to fruit-cracking. In Taigu County of Shanxi Province, it matures in late September with a fruit growth period of 113 d. It is a late-ripening variety.

Fruit Characteristics The medium-sized conical fruit has a vertical and cross diameter of 3.70 cm and 2.50 cm, averaging 11.1 g with a regular size. It has a shallow and medium-wide stalk cavity, a flat fruit apex, a falling-off stigma, medium-thick red skin and a smooth surface. The flesh is crisp, sweet and juicy. It has a superior quality for fresh-eating and dried fruits. The percentage of edible part of fresh fruit is 94.8%, and the content of TTS, TA and Vc is 26.25%, 0.31% and 325.97 mg per 100 g fresh fruit. The SAR is 84.68∶1. The content of flavones and cAMP in mature fruit skin is 3.97 mg/g and 84.30 μg/g. The large spindle-shaped stone weighs 0.58 g, with a well-developed kernel.

Evaluation The cultivar has a high yield and strong resistance to fruit-cracking. The medium-sized fruit has a superior quality and is suitable for fresh-eating and making dried fruits.

兼用
Multipurpose Varieties
品种

蜜枣品种 Candied Varieties

蜜枣品种 Candied Varieties

沙漠老枣树（宁夏　中卫）
Old Jujube Trees in the Desert (Zhongwei, Ningxia)

灌阳长枣

品种来源及分布 别名牛奶枣。原产和分布于广西灌阳，为当地古老主栽品种，占当地栽培总面积的98%以上。

植物学性状 树体较高大，树姿开张，干性较强，枝条较密，树冠自然圆头形或半圆形，树干皮裂块状、较深，易剥落。枣头紫褐色，平均长69.4cm，粗0.92cm，节间长9.1cm，二次枝长25.4cm，平均5节。针刺不发达。皮目圆形，凸起，开裂。枣股中大，圆柱形，抽吊力强，抽生枣吊3～6个，平均4个。枣吊平均长19.1cm，着叶13片。叶片中大，叶长5.6cm，叶宽2.5cm，椭圆形，浓绿色，先端急尖，叶基圆形或楔形，叶缘锯齿粗钝，密度中等。花量多，枣吊平均着花77.5朵，最多达132朵。

生物学特性 树势强，成枝力中等，结果较早，一般栽后2～3年开始结果，15年左右进入盛果期。盛果期长。丰产，产量稳定，坐果率高，枣头枝吊果率48.2%，2～3年生枝为73.9%，4年生枝32.2%。在山西太谷地区，9月中旬果实开始着色，9月下旬脆熟，果实生育期110～120d，属晚熟品种类型。

果实性状 果实较大，圆锥形或圆柱形，果顶多向一侧歪斜，果实纵径4.20～7.00cm，横径2.20～2.70cm，单果重14.7g，大小较整齐。果梗长、中粗，梗洼窄而深。果顶尖，柱头遗存，不明显。果皮较薄，浅红色，果面不平滑。果点小而密，分布整齐。果肉厚，黄白色，肉质较细，稍松脆，味甜，汁液少，适宜加工蜜枣，也可鲜食和制干，蜜枣品质上等，鲜食和干枣品质中等。白熟期果实含可溶性固形物18.00%。脆熟期鲜枣可食率96.9%，含可溶性固形物28.00%，总糖22.09%，酸0.39%，100g果肉维生素C含量380.54mg；成熟期果皮含黄酮3.70mg/g，cAMP含量18.75μg/g。制干率40.0%。核小，长纺锤形，扁平稍弯曲，核重0.46g，核纹细浅，核尖细长，核面不粗糙，种仁发育不良。

评价 该品种风土适应性强，在原产地土层深厚的平川、河滩壤土和砂壤土，或土层浅薄的丘陵、山坡石灰质黏壤土和壤土上均能生长结果。着色期易落果，遇雨易裂果。结果早，高产稳产。果实较大，肉质细，略松脆，汁液少，味甜，加工蜜枣，品质优良。适宜南方蜜枣产区栽培。

Guanyangchangzao

Source and Distribution The cultivar originated from and spreads in Guanyang of Guangxi Province. It is an old dominant variety there which occupies 98% of the total jujube area.

Botanical Characters The large tree is spreading with a strong central leader trunk, dense branches and a natural-round or semi-round crown. The purplish-brown 1-year-old shoots are 69.4 cm long. The secondary branches are 25.4 cm long with less-developed thorns. The medium-sized column-shaped mother fruiting spurs can germinate 3～6 deciduous fruiting shoots which are 19.1 cm long with 13 leaves. The medium-sized leaves are oval-shaped and dark green. There are many flowers produced, averaging 77.5 ones per deciduous fruiting shoot.

Biological Characters The tree has strong vigor and medium branching ability. It bears early and enters the long full-bearing stage in the 15th year with high fruit set and a high and stable yield. It is a late-ripening variety.

Fruit Characteristics The large column-shaped fruit weighs 14.7 g on average with a regular size. The fruit apex is deflective toward one side. It has light-red thin skin and an unsmooth surface. The yellowish-white thick flesh is loose-texturedm, crisp and sweet, with less juice. It can be used for processing candied fruits, fresh eating and making dried fruits. Candied fruits have an excellent quality, while that for fresh eating and dried fruits is just medium. The SSC in fruits of white-maturing stage is 18%. The percentage of edible part of fresh fruit in crisp-maturing stage is 96.9%, and SSC, TTS, TA and Vc is 28.00%, 22.09%, 0.39% and 380.54 mg per 100 g fresh fruit. The small stone is long spindle-shaped averaging 0.46 g.

Evaluation The cultivar has strong adaptability. Fruit-dropping easily occurs in the coloring stage and fruit-cracking is serious. It bears early with a high and stable yield. The large fruit has delicate, crisp and sweet flesh with less juice. Candied fruits have a very good quality. It can be developed in candied-jujube production areas in sourthern China.

蜜 枣
Candied Varieties
品 种

连县木枣

品种来源及分布　分布于广东连州市东北部的星子、大路边乡等，为当地原有乡土和主栽品种。栽培历史400年以上。

植物学性状　树体中大，树姿较直立，树冠呈自然半圆形。主干灰黑色，皮裂条状，易剥落。枣头红褐色，平均长58.3cm，粗0.80cm，节间长6.7cm，二次枝长27.7cm，6节左右。针刺不发达。皮目大，圆形，凸起，开裂。枣股中大，圆柱形，抽吊力中等，一般抽生枣吊3～4个。枣吊平均长22.75cm，着叶15片。叶片中大，叶长5.2cm，叶宽2.8cm，卵圆形，绿色，先端急尖，叶基圆形，叶缘锯齿粗，较密。花量多，每花序着花7朵左右。花大，花径7.5mm左右，夜开型。

生物学特性　树势中庸，成枝力中等。结果较迟，定植后一般第三年开始结果。20年生树达盛果期。坐果率高，丰产。在山西太谷地区，8月底果实进入白熟期，10月上旬成熟，果实生育期115d左右，属晚熟品种。成熟期落果少，不裂果。

果实性状　果实中大，长椭圆形，纵径3.80～5.00cm，横径2.20～3.10cm，单果重12.1g，大小较整齐。果梗长而粗，梗洼广而浅。果顶平或微凹，柱头不明显。果皮中厚，浅红色。果点中大，不明显。果肉较厚，白绿色，肉质略松，味甜，汁液中多，鲜食品质中上，加工蜜枣品质优良。鲜枣可食率95.1%，含总糖24.14%，酸0.41%，100g果肉维生素C含量480.85mg。核较小，纺锤形，核重0.59g，核纹浅，核尖中长，核内不含种仁。

评价　该品种适应性较强，高产稳产，果实中大，肉质松软，干物质较多，为优良的蜜枣品种。

Lianxianmuzao

Source and Distribution　The cultivar spreads in Xingzi and Dalubianxiang of northeastern Lianzhou City in Guangdong Province. It is the native dominant variety there with a history of over 400 years.

Botanical Characters　The medium-sized tree is vertical with a semi-round crown. The grayish-black trunk bark has striped fissures, easily shelled off. The reddish-brown 1-year-old shoots are 58.3 cm long and 0.80 cm thick with the internodes of 6.7 cm. The secondary branches are 27.7 cm long with 6 nodes and less-developed thorns. The large lenticels are round, protuberant and cracked. The medium-sized column-shaped mother fruiting spurs can germinate 3～4 deciduous fruiting shoots which are 22.75 cm long with 15 leaves. The medium-sized green leaves are oval-shaped, 5.2 cm long and 2.8 cm wide, with a sharply-cuspate apex, a round base and a thick and dense saw-tooth pattern on the margin. There are many large flowers with a diameter of 7.5 mm produced, averaging 7 ones per inflorescence. It blooms at night.

Biological Characters　The tree has moderate vigor and medium branching ability. It bears late, generally in the 3rd year after planting and enters the full-bearing stage in the 20th year, with high fruit set and a high yield. In Taigu County of Shanxi Province, it enters the white-maturing stage at the end of August and matures in early October with a fruit growth period of 115 d. It is a late-ripening variety with light fruit-dropping and without fruit-cracking in maturing stage.

Fruit Characteristics　The medium-sized fruit is long oval-shaped, with a vertical and cross diameter of 3.80～5.00 cm and 2.20～3.10 cm, averaging 12.1 g with regular sizes. It has a long and thick stalk, a wide and shallow stalk cavity, a flat or slightly sunken fruit apex, an indistinct stigma and medium-thick and light-red skin with medium-large and indistinct dots. The greenish-white thick flesh is loose-textured and sweet, with medium juice and a better than normal quality for fresh eating. Yet it has an excellent quality for candied fruits. The percentage of edible part of fresh fruit is 95.1%, and the content of TTS, TA and Vc is 24.14%, 0.41% and 480.85 mgper 100 g fresh fruit. The small spindle-shaped stone has shallow veins and a medium-long apex. There are no kernels in the stones.

Evaluation　The cultivar has strong adaptability with a high and stable yield. The medium-sized fruit has soft flesh with many dry matters. It is a good variety for processing candied fruits.

蜜 枣
Candied Varieties
品 种

· 449 ·

义乌大枣

品种来源及分布 别名大枣。原产浙江东阳市，分布于义乌、东阳等地，为当地主栽品种。有700多年的栽培历史。

植物学性状 树体较大，树姿开张，干性较强，树冠偏斜形。树干皮裂浅、条状，较平滑，不易剥落。枣头紫褐色，平均长70.7cm，粗0.90cm，节间长9.5cm。二次枝长37.2cm，平均7节。针刺不发达。皮目小，椭圆形，凸起，开裂。枣股中大，圆柱形，最长可达2.5cm，直径1cm，抽吊力中等，一般抽生枣吊3～4个。枣吊平均长28.7cm，着叶16片。叶片大，叶长5.6cm，叶宽2.5cm，椭圆形，绿色，先端急尖，叶基圆楔形，叶缘锯齿粗浅。花量多，每花序着花5～9朵。花中大，花径7mm左右，夜开型。

生物学特性 树势强健，萌芽率和成枝力中等，结果较早，一般第二年开始结果，10年后进入盛果期，产量较高，生理落果轻，2～3年生枝吊果率为94.5%，4年生枝53.8%。在山西太谷地区，9月上旬果实进入白熟期，10月上旬果实成熟，果实生育期120d左右，属极晚熟品种类型。

果实性状 果实大，圆柱形或长圆形，纵径3.80cm，横径2.70cm，单果重15.9g，大小整齐。果梗中粗，较短，梗洼窄，中深。果顶宽平或微凹，柱头遗存。果皮较薄，赭红色，果面不平。果点小而密，不明显。果肉厚，乳白色，肉质稍松，汁液少，适宜加工蜜枣，品质上等。白熟期含可溶性固形物13.10%，100g维生素C含量503.20mg；成熟期鲜枣可食率96.2%，含可溶性固形物30.00%，总糖27.98%，酸0.50%，100g果肉维生素C含量527.16mg；果皮含黄酮21.50mg/g，cAMP含量622.61µg/g。核中大，纺锤形，稍弯曲，纵径2.10cm，横径0.80cm，核重0.60g，核尖较短，核纹中深，核面较粗糙，含仁率98.3%，种仁饱满。

评价 该品种适应性强，结果较早，产量高，但不稳定。果实大，肉质较松，汁液少，干物质多，加工蜜枣品质优异，为优良的加工品种，适宜蜜枣加工区栽培。

Yiwudazao

Source and Distribution The cultivar originated from Dongyang City of Zhejiang Province and mainly spreads in Yiwu and Dongyang. It is the dominant variety there with a history of over 700 years.

Botanical Characters The large tree is spreading with a strong central leader trunk and a deflective crown. The yellowish-brown 1-year-old shoots are 70.7 cm long and 0.90 cm thick with the internodes of 9.5 cm. The secondary branches are 37.2 cm long with 7 nodes and less-developed thorns. The medium-sized column-shaped mother fruiting spurs are 2.5 cm long at most with a diameter of 1 cm. They can germinate 3～4 deciduous fruiting shoots which are 28.7 cm long with 16 leaves. The large green leaves are oval-shaped. There are many medium-sized flowers with a diameter of 7 mm produced, averaging 5～9 ones per inflorescence.

Biological Characters The tree has strong vigor, medium germination and branching ability. It bears early, generally in the 2nd year after planting and enters the full-bearing stage in the 10th year with a high yield. The physiological fruit-dropping is light. It is an extremely late-ripening variety.

Fruit Characteristics The large fruit is oblong or column-shaped, with a vertical and cross diameter of 3.80 cm and 2.70 cm, averaging 15.9 g with a regular size. It has a short and medium-thick stalk, a narrow and medium-deep stalk cavity, a flat or slightly sunken fruit apex, a remnant stigma, reddish-brown thin skin and an unsmooth surface with small, dense and indistinct dots. The ivory-white thick flesh is loose-textured, with less juice. It is suitable for processing candied fruits with an excellent quality. The content of SSC and Vc in fruits of white-maturing stage is 13.10% and 503.20 mg per 100 g fresh fruit. The percentage of edible part of mature fresh fruit is 96.2%, and the content of SSC, TTS, TA and Vc is 30.00%, 27.98%, 0.50% and 527.16 mg per 100g fresh fruit. The medium-sized spindle-shaped stone is a little crooked, averaging 0.60 g.

Evaluation The cultivar has strong adaptability, bearing early with a high yet unstable yield. The large fruit has loose-textured flesh with less juice and many dry matters. It is a good variety for processing candied fruits with an excellent quality. It can be developed in candied-jujube production areas.

鹅子枣

品种来源及分布　别名鹅蛋枣。原产于浙江义乌的后宅、福田、平畴等地。系义乌大枣自然变异株系。

植物学性状　树体高大，树姿半开张，树冠自然圆头形。主干皮裂块状。枣头平均长73.6cm，粗0.90cm，节间长7.1cm，蜡质多。二次枝平均长33.8cm，8节，弯曲度中等。针刺发达。枣股平均抽生枣吊4.0个。叶片卵圆形，绿色，较光亮，先端急尖，叶基偏斜形。叶缘锯齿细锐。花量中多，花序平均着花4朵。

生物学特性　定植后2、3年结果，丰产性较差，枣吊平均结果0.11个。在山西太谷地区，4月中旬萌芽，6月上旬始花，9月下旬成熟，果实生育期111d，为晚熟品种类型。

果实性状　果实大，圆形，纵径3.32cm，横径2.90cm，单果重17.9g，最大35.0g，大小较整齐。果肩平，梗洼中深、广。果皮光滑，红色。果点中大，稀疏。果顶平，柱头脱落。果肉质地酥脆，汁液少，味酸甜，鲜枣品质中上，宜加工蜜枣。鲜枣可食率95.2%，含总糖19.32%，酸0.54%，100g果肉维生素C含量491.10mg。果核较大，平均重0.86g，含仁率43.3%。

评价　该品种系义乌大枣的自然变异株系，结果早，丰产性中等。果实大，较美观，为优良蜜枣品种。

Ezizao

Source and Distribution　The cultivar, also called Edanzao, originated from Houzhai, Futian and Pingchou of Yiwu in Zhejiang Province. It is a natural mutation from Yiwudazao.

Botanical Characters　The large tree is half-spreading with a natural-round crown. The trunk bark has massive fissures. The 1-year-old shoots are 73.6 cm long and 0.9 cm thick. The internodes are 7.1 cm long with much wax and developed thorns. The secondary branches are 33.8 cm long with 8 nodes of medium curvature. The mother fruiting spurs can germinate 4 deciduous fruiting shoots. The oval-shaped leaves are green and glossy, with a sharply-cuspate apex, a deflective base and a thin, sharp saw-tooth pattern on the margin. The number of flowers is medium large, averaging 4 ones per inflorescence.

Biological Characters　It generally bears in the 2nd or 3rd year after planting with poor productivity. The deciduous fruiting shoot bears 0.11 fruits on average. In Taigu County of Shanxi Province, it germinates in mid-April, begins blooming in early June and matures in late September with a fruit growth period of 111 d. It is a late-ripening variety.

Fruit Characteristics　The large oblong fruit has a vertical and cross diameter of 3.32 cm and 2.90 cm, averaging 17.9 g (maximum 35.0 g) with a regular size. It has a flat shoulder, a medium-deep and wide stalk cavity, a smooth and red surface with medium-large sparse dots, a flat fruit apex and a falling-off stigma. The flesh is crisp, sour and sweet, with less juice. It has a better than normal quality for candied fruits. The percentage of edible part of fresh fruit is 95.2%, and the content of TTS, TA and Vc is 19.32%, 0.54% and 491.10 mg per 100 g fresh fruit. The large stone weighs 0.86 g, and the percentage of containing kernels is 43.3%.

Evaluation　The cultivar is a natural mutation from Yiwudazao. It bears early with poor productivity. The large fruit has attractive appearance. It is a good variety for processing candied fruits.

蜜 枣
Candied Varieties
品 种

郎溪牛奶枣

品种来源及分布　主要分布在安徽郎溪的凌笪、花树、候树、独山和广德的下寺、施村、梅村、刘达等地。为当地原产的主栽品种，栽种面积占当地枣树的80%左右。栽培历史悠久，并以制作蜜枣品质优良著称。

植物学性状　树体较大，树姿开张，干性中等，树冠多自然圆头形。主干灰褐色，主干裂纹较细，条状，不易剥落。枣头紫红色，平均长53.6cm，粗0.50cm，节间长8.1cm，蜡质少。二次枝长24.4cm，5节，弯曲度中等。针刺较发达。枣股圆柱形，平均抽生枣吊4.0个。枣吊长27.9cm，着生叶片14片。叶片中大，较宽，卵圆形，中厚，绿色，先端渐尖，叶基圆楔形，叶缘具钝锯齿。花量少，为少花型，花序平均着花2朵。

生物学特性　适应性强，具有耐旱、抗风等特性。树势较强。嫁接苗当年即可开花结果。盛果期产量高，无大小年现象，枣吊平均结果0.56个。在山西太谷地区，4月中旬萌芽，5月下旬始花，9月下旬进入脆熟期，果实生育期120d左右，为极晚熟品种类型。遇雨有裂果现象。

果实性状　果实中大，长柱形，纵径3.93cm，横径2.28cm，单果重11.9g，大小整齐。果肩平，梗洼中等深广，果顶尖。果面光滑，果皮厚，脆熟期赭红色，果肉浅绿色，较紧密，汁液中多，适宜制作蜜枣。鲜枣可食率94.7%，含可溶性固形物29.40%，总糖20.37%，酸0.42%，100g果肉维生素C含量467.74mg；果皮含黄酮19.61mg/g，cAMP含量438.05μg/g。蜜枣成品透明度高，渗糖匀透，品质上等。果核中大，纺锤形，核重0.63g。大多核内无种子。

评价　该品种适应性较强，耐旱，抗风。树体强健，产量高而稳定。果实中大，制成蜜枣外形整齐美观，果面富糖霜，品质上等，为优良的蜜枣品种。

Langxiniunaizao

Source and Distribution　The cultivar mainly spreads in Lingda, Huashu, Houshu, Dushan in Langxi, and Xiasi, Shicun, Meicun, Liuda in Guangde, Anhui Province. It is the native dominant variety there, which occupies 80% of the total jujube area. The cultivar is famous for its candied fruits. It has a long cultivation history.

Botanical Characters　The large tree is spreading with a moderate central leader trunk and a natural-round crown. The purplish-red 1-year-old shoots are 53.6 cm long and 0.50 cm thick. The internodes are 8.1 cm long with less wax and developed thorns. The secondary branches are 24.4 cm long with 5 nodes of medium curvature. The column-shaped mother fruiting spurs can germinate 4 deciduous fruiting shoots which are 27.9 cm long with 14 leaves. The wide leaves are green and oval-shaped, medium-large and medium-thick. The number of flowers is small, averaging 2 ones per inflorescence.

Biological Characters　The tree has strong adaptability, strong tolerance to drought, wind and cold, with strong vigor and a high yield, without alternate bearing. Grafting seedlings bloom and bear fruit in the year of planting. The deciduous fruiting shoot bears 0.56 fruits on average. In Taigu County of Shanxi Province, it enters the crisp-maturing stage in late September. It is an extremely late-ripening variety with serious fruit-cracking if it rains.

Fruit Characteristics　The medium-sized fruit is column-shaped, with a vertical and cross diameter of 3.93 cm and 2.28 cm, averaging 11.9 g with a regular size. It has a flat shoulder, a medium-deep and medium-wide stalk cavity, a pointed fruit apex, a smooth surface, and brownish-red thick skin in the crisp-maturing stage. The light-green flesh is tight-textured with medium juice, suitable for processing candied fruits. The percentage of edible part of fresh fruit is 94.7%, and SSC, TTS, TA and Vc is 29.40%, 20.37%, 0.42% and 467.74 mg per 100 g fresh fruit. Candied fruits have high transparency and uniform sugar permeability with a good quality. The medium-sized spindle-shaped stone weighs 0.63 g. Most stones contain no kernels.

Evaluation　The cultivar has strong adaptability, strong tolerance to drought, cold and wind, with strong vigor, a long life, a high and stable yield. The fruit has a medium large size. Candied fruits of this cultivar have regular and beautiful appearance with rich sugar-frost on the surface and a good quality. It is a good variety for processing candied fruits.

蜜 枣
Candied Varieties
品 种

兰溪马枣

品种来源及分布 别名尖头马枣、马枣、牛奶枣。分布于浙江的兰溪、义乌、东阳、永康等地，为当地的主栽品种之一，栽培历史悠久。

植物学性状 树体较小，树姿半开张，树冠呈伞形。主干皮裂条状。枣头黄褐色，平均长72.9cm，节间长8.4cm。二次枝长34.8cm，5～7节，弯曲度中等。针刺发达。枣股平均抽生枣吊3.8个。枣吊长25.6cm，着叶14片。叶片中大，椭圆形，先端急尖，叶基圆楔形，叶缘具钝锯齿。花量少，花序平均着花3朵，花小，花径5.9mm。

生物学特性 风土适应性特强，水地和旱地都能较好地生长结果。树势中庸，成枝力较弱。定植第二年结果，10年左右进入盛果期，产量较高，平均吊果率102.0%。在山西太谷地区，9月下旬果实进入脆熟期，10月上旬成熟采收，果实生育期120d以上，属极晚熟品种。成熟期遇雨裂果少。

果实性状 果个较小，长柱形，纵径3.85cm，横径2.38cm，单果重8.5g，大小较整齐。果皮中厚，浅红色，果面平滑。梗洼深、中广。果顶尖，柱头残存。肉质较致密，味甜，汁液中多，品质中等，适宜制作蜜枣和南枣。鲜枣可食率92.0%，含总糖20.52%，酸0.36%，糖酸比57.00∶1，100g果肉维生素C含量502.02mg；果皮含黄酮46.56mg/g，cAMP含量190.60μg/g。核较大，纺锤形，核重0.68g。种仁不饱满，含仁率48.3%。

评价 该品种适应性强，树势中庸，结果早，坐果多，产量高而稳定，果实小，品质中等，适宜制作蜜枣和南枣。抗裂果能力较强。

Lanximazao

Source and Distribution The cultivar, also called Jiantoumazao, Mazao or Niunaizao, mainly spreads in Lanxi, Yiwu, Dongyang and Yongkang in Zhejiang Province with a long cultivation history.

Botanical Characters The small tree is half-spreading with an umbrella-shaped crown. The trunk bark has striped fissures. The yellowish-brown 1-year-old shoots are 72.9 cm long with the internodes of 8.4 cm. The secondary branches are 34.8 cm long with 5～7 nodes of medium curvature and developed thorns. The mother fruiting spurs can germinate 3.8 deciduous fruiting shoots which are 25.6 cm long with 14 leaves. The medium-sized leaves are oval-shaped, with a sharply-cuspate apex, a round-cuneiform base and a blunt saw-tooth pattern on the margin. The number of flowers is small, averaging 3 ones per inflorescence. The small blossoms have a diameter of 5.9 mm.

Biological Characters The tree has strong adaptability to various conditions. It grows well along paddy land in south China and in dry land of north China. It has moderate vigor and weak branching ability. The tree generally bears in the 2nd year after planting and enters the full-bearing stage in the 10th year, with a high yield. The percentage of fruits to deciduous fruiting shoots is 102.0% on average. In Taigu County of Shanxi Province, it enters the crisp-maturing stage in late September and matures in early October with a fruit growth period of over 120 d. It is an extremely late-ripening variety with light fruit-cracking even if it rains in the maturing stage.

Fruit Characteristics The small conical fruit has a vertical and cross diameter of 3.85 cm and 2.38 cm, averaging 8.5 g with a regular size. It has medium-thick and light-red skin, a smooth surface, a deep and medium-wide stalk cavity, a pointed fruit apex and a remnant stigma. The tight-textured flesh is sweet and juicy. It has medium quality for candied fruits and dried fruits. The percentage of edible part of fresh fruit is 92.0%, and the content of TTS, TA and Vc is 20.52%, 0.36% and 502.02 mg per 100g fresh fruit. The SAR is 57.00∶1. The content of flavones and cAMP in mature fruit skin is 46.56 mg/g and 190.60 μg/g. The large spindle-shaped stone weighs 0.68 g with a shriveled kernel. The percentage of containing kernels is 48.3%.

Evaluation The cultivar has strong adaptability and moderate tree vigor, bearing early with high fruit set, a high and stable yield. The small fruit has medium quality and is suitable for processing candied fruits and dried fruits. It has strong resistance to fruit-cracking.

蜜 枣
Candied Varieties
品 种

· 457 ·

南 京 枣

品种来源及分布 别名京枣、南京枕头枣。产于浙江兰溪,为当地的主栽品种,有300年以上的栽培历史。

植物学性状 树体较大,树姿开张,干性较强,树冠乱头形。主干灰褐色,树皮裂纹较深,不规则细纵条纹,不易剥落。枣头粗壮,紫褐色,平均长64.8cm,粗0.91cm,节间长8.7cm,蜡质少。二次枝长32.5cm,5节,弯曲度中等。针刺不发达。枣股中大,圆柱形或圆锥形,平均抽生枣吊4.0个。枣吊长32.1cm,平均着叶16片。叶片中大,卵状披针形,深绿色,中厚,先端渐尖,叶基偏斜形,叶缘具粗浅的钝锯齿。花量中多,花序平均着花4朵,夜开型。

生物学特性 适应性中等,耐旱、耐涝,不耐瘠薄。树势和发枝力中等,定植后2、3年开始结果,坐果不稳定,花期遇多雨降温天气,坐果不良。在山西太谷地区,4月中旬萌芽,5月下旬始花,9月上旬进入白熟期开始采收加工蜜枣,白熟果生育期90d左右,为晚熟品种类型。

果实性状 果实大,圆柱形,纵径3.40cm,横径2.90cm,单果重14.0g,最大39.5g,大小较整齐。果肩平。梗洼浅、中广。果顶凹陷,顶洼浅广,果面平整,稍有粗糙感,光泽较差,果皮白熟期呈乳白色,果点小,圆形,稀疏不明显,完熟后呈暗紫红色。果肉白色,质地致密,较脆,汁液中多,味甜,适宜制作蜜枣,品质上等。鲜枣可食率95.4%,白熟果含可溶性固形物14.80%,100g果肉维生素C含量496.60mg。果核中大,长纺锤形,核重0.65g。核内多含有1粒饱满的种子。

评价 该品种耐旱涝,不耐瘠薄。树势中等,产量较高但不稳定,花期忌阴雨低温。果实大,外观好,皮肉色浅,质细,为优良蜜枣品种。

Nanjingzao

Source and Distribution The cultivar originated from Lanxi of Zhejiang Province. It is the dominant variety there, with a cultivation history of over 300 years.

Botanical Characters The large tree is spreading with a strong central leader trunk and an irregular crown. The grayish-brown trunk bark has irregular, deep, striped fissures, uneasily shelled off. The strong purplish-brown 1-year-old shoots are 64.8 cm long and 0.91 cm thick. The internodes are 8.7 cm long with less wax and less-developed thorns. The secondary branches are 32.5 cm long with 5 nodes of medium curvature. The conical or column-shaped mother fruiting spurs can germinate 4 deciduous fruiting shoots which are 32.1 cm long with 16 leaves. The medium-sized leaves are ovate-lanceolate, dark green and medium thick. The number of flowers is medium large, averaging 4 ones per inflorescence. It blooms at night.

Biological Characters The tree has medium adaptability, strong tolerance to drought and water-logging, poor tolerance to infertile soils, with moderate vigor and medium branching ability. It generally bears in the 2nd or 3rd year after planting with unstable fruit set. In Taigu County of Shanxi Province, it germinates in mid-April and begins blooming in late May. The fruit is harvested in white-maturing stage in early September. It has a fruit growth period of 90 d for the white-maturing fruit. It is a late-ripening variety.

Fruit Characteristics The large oblong fruit has a vertical and cross diameter of 3.40 cm and 2.0 cm, averaging 14.0 g with a regular size. It has a flat shoulder, a shallow and medium-wide stalk cavity, a sunken fruit apex, a smooth and a little rough surface, with poor gloss. The skin is ivory-white in the white-maturing stage. It turns dark purplish-red after complete maturing. The white flesh is tight-textured and crisp with medium juice. It has a good quality for processing candied fruits. The fruit has a sweet taste for fresh-eating. The percentage of edible part of fresh fruit is 95.4%, and the content of SSC and Vc in the white-maturing fruit is 14.80% and 496.60 mg per 100 g fresh fruit. The medium-sized stone is long spindle-shaped averaging 0.65 g.

Evaluation The cultivar has strong tolerance to drought and water-logging, yet intolerant to infertile soils, with moderate vigor and a high yet unstable yield. Rain and low temperature in the blooming stage are unfavorable to the plants. The large fruit has an attractive appearance and light colored, delicate flesh. It is a good variety for processing candied fruits.

蜜 枣
Candied Varieties
品 种

· 459 ·

歙县马枣

品种来源及分布 别名马头枣。主要分布于安徽歙县的杞梓里区、岔口区、深渡区，以杞梓里区英坑乡金竹村最多，为当地原产品种。栽培历史近800年。

植物学性状 树体较大，树姿开张，树冠偏斜形。主干灰褐色，树皮裂片较小，成条状。枣头红褐色，生长势强，平均长72.0cm，粗0.78cm，节间长8.4cm，蜡质多。二次枝平均长32.2cm。针刺发达。枣股圆锥形，平均抽生枣吊5.0个。枣吊长28.5cm，平均着生叶片18片。叶片中大，卵圆形，绿色，先端渐尖，叶基偏斜形，叶缘具锐锯齿。花量少，花序平均着花3朵。

生物学特性 适应性较强，耐干旱、耐瘠薄。树势强健，发枝力较强。定植后3年开始结果，盛果期产量较稳定，枣吊平均坐果3个，最多5个。在山西太谷地区，4月下旬萌芽，5月下旬始花，8月下旬进入白熟期，9月上旬着色进入脆熟期，果实生育期100d左右，为中熟品种类型。

果实性状 果实较大，长柱形，纵径3.56cm，横径2.20cm，单果重13.0g，大小较整齐。果肩平，梗洼中深、广，果顶突尖，柱头残存。果面光滑，果皮较薄，红色。果点小，分布较稀。果肉浅绿色，质地酥脆，汁液中多，适宜制作蜜枣，品质上等。鲜枣可食率95.5%，白熟期含总糖11.80%，酸0.22%，100g果肉维生素C含量426.95mg；果皮含黄酮2.96mg/g，cAMP含量105.52μg/g。果核纺锤形，核重0.58g，核内一般不含种子。

评价 该品种适应性较强，产量高而稳定。果实较大，白熟期皮、肉浅绿色，质松，少汁，制作蜜枣果形整齐美观，透明度高，肉厚核小，为中熟的优良蜜枣品种。

Shexianmazao

Source and Distribution The cultivar, also called Matouzao, mainly spreads in Qiziliqu, Chakouqu and Shenduqu of Shexian County in Anhui Province. It is the native variety in Jinzhucun of Yingkeng Village in Qiziliqu, with the largest quantity there. It has a cultivation history of 800 years.

Botanical Characters The large tree is spreading with a deflective crown. The grayish-brown trunk bark has small striped fissures. The reddish-brown 1-year-old shoots have strong growth potential, 72.0 cm long and 0.78 cm thick. The internodes are 8.4 cm long with much wax and developed thorns. The secondary branches are 32.2 cm long. The conical mother fruiting spurs can germinate 5 deciduous fruiting shoots which are 28.5 cm long with 18 leaves. The medium-sized leaves are green and oval-shaped, with a gradually-cuspate apex, a deflective base and a sharp saw-tooth pattern on the margin. The number of flowers is small, averaging 3 ones per inflorescence.

Biological Characters The tree has strong adaptability, strong tolerance to drought and poor soils, strong vigor and strong branching ability. It generally bears in the 3rd year after planting, with a stable yield. The deciduous fruiting shoot bears 3 fruits (5 at most). In Taigu County of Shanxi Province, it germinates in late April, begins blooming in late May, enters the white-maturing stage in late August, begins coloring and enters the crisp-maturing stage in early September with a fruit growth period of 100 d. It is a mid-ripening variety.

Fruit Characteristics The large conical fruit has a vertical and cross diameter of 3.56 cm and 2.20 cm, averaging 13.0 g with a regular size. It has a flat shoulder, a medium-deep and wide stalk cavity, a sharply-pointed fruit apex, a remnant stigma, a smooth surface, and thin red skin with small and sparse dots. The light-green flesh is crisp, with medium juice. It has a good quality for candied fruits. The percentage of edible part of fresh fruit is 95.5%, and the content of TTS, TA and Vc is 11.80%, 0.22% and 426.95 mg per 100 g fresh fruit. The content of flavones and cAMP in mature fruit skin is 2.96 mg/g and 105.52 μg/g. The large spindle-shaped stone weighs 0.58 g. Most stones contain no kernels.

Evaluation The cultivar has strong adaptability and a high yet unstable yield. The large fruit has light-green skin and flesh in white-maturing stage. The thick flesh is loose-texture with medium juice. Candied fruits of this variety have irregular and beautiful appearance, with a high transparency and a small stone. It is a good mid-ripening variety for processing candied fruits.

蜜 枣
Candied Varieties
品 种

苏南白蒲枣

品种来源及分布 广泛分布于江苏南部和上海市郊，主产吴县、无锡、溧阳、宜兴一带，为当地重要的主栽品种。

植物学性状 树体较大，树姿半开张，干性强，层性明显，树冠呈圆锥形。主干皮裂条状。枣头棕褐色，平均长62.8cm，粗0.95cm，节间长8.2cm，蜡质少。二次枝平均长26.9cm，5节，弯曲度中等。针刺发达。枣股圆柱形，平均抽生枣吊3.0个。枣吊长24.4cm，着叶15片。叶片大，椭圆形，绿色或深绿色，较光亮，先端渐尖，叶基圆楔形，叶缘具钝锯齿。花量少，花序平均着花3朵，花大，花径8.0mm左右。

生物学特性 适应性强。树势强，发枝力中等。结果早，嫁接后第二年即有一定产量，成龄树枣吊平均结果0.5个。在山西太谷地区，4月中旬萌芽，6月初始花，10月上旬果实着色成熟，果实生育期120d左右，为极晚熟品种类型。裂果较重。在苏南、上海等产地，为避免裂果损失，都在白熟期采收加工蜜枣，故有"白蒲枣"之称。

果实性状 果实中大，长圆形，单果重10.0g，大小整齐。果肩较小，平圆。梗洼浅平。果顶尖。果柄细，长3mm左右。果面平滑光洁，富光泽。果皮较薄，浅红色。果点中大，中密，不明显。果肉浅绿色，质地酥脆，汁液多，宜制作蜜枣和鲜食。鲜枣可食率97.2%，白熟期可溶性固形物14.00%，总糖11.70%，酸0.42%，100g果肉维生素C含量520.80mg。果核小，纺锤形，平均重0.28g。核内常有种子。

评价 该品种适应性强，树体强健，结果早，丰产稳产。果实中大，肉质细脆，甜味较淡，维生素C含量较高，为晚熟蜜枣优良品种，适宜蜜枣产区引种栽培。

Sunanbaipuzao

Source and Distribution The cultivar spreads in south Jiangsu Province, and in suburbs of Shanghai. It is mainly produced in Wuxian, Wuxi, Liyang and Yixing areas. It is an important dominant variety there.

Botanical Characters The large tree is half-spreading with a strong central leader trunk and a conical crown. The trunk bark has striped fissures. The brown 1-year-old shoots are 62.8 cm long and 0.95 cm thick. The internodes are 8.2 cm long with less wax and developed thorns. The secondary branches are 26.9 cm long with 5 nodes of medium curvature. The column-shaped mother fruiting spurs can germinate 3 deciduous fruiting shoots which are 24.4 cm long with 15 leaves. The large oval-shaped leaves are green or dark green, glossy, with a gradually-cuspate apex, a round-cuneiform base and a blunt saw-tooth pattern on the margin. The number of flowers is small, averaging 3 ones per inflorescence. The large blossoms have a diameter of about 8.0 mm.

Biological Characters The tree has strong adaptability, strong vigor and medium branching ability. It bears early, generally in the 2nd year after planting. The deciduous fruiting shoot bears 0.5 fruits on average. In Taigu County of Shanxi Province, it germinates in mid-April, begins blooming in early June and begins coloring and matures in early October with a fruit growth period of 120 d. It is an extremely late-ripening variety with serious fruit-cracking. In south Jiangsu Province and in Shanghai, it is harvested in white-maturing stage in order to avoid losses caused by fruit-cracking. Then it is processed into candied fruits.

Fruit Characteristics The medium-sized oblong fruit weighs 10.0 g with a regular size. It has a small flat-round shoulder, a shallow stalk cavity, a thin stalk of 3 mm long, a smooth and bright surface, and light-red thin skin with indistinct medium-sized and medium-dense dots. The light-green flesh is crisp and juicy. It is suitable for processing candied fruits and for fresh-eating. The percentage of edible part of fresh fruit is 97.2%, and SSC, TTS, TA and Vc in white-maturing stage is 14.00%, 11.70%, 0.42% and 520.80 mg per 100 g fresh fruit. The small spindle-shaped stone weighs 0.28 g. Most stones contain a kernel.

Evaluation The cultivar has strong adaptability and strong vigor, bearing early with a high and stable yield. The medium-sized fruit has delicate and crisp flesh with a light-sweet taste and a high content of Vc. It is a good late-ripening variety for processing candied fruits. It can be developed in the production areas of candied jujubes.

蜜 枣
Candied Varieties
品 种

定 襄 山 枣

品种来源及分布 别名山枣。原产和分布在山西定襄的横山、龙弯、虎山，原平的同河沿岸及五台的阳白等地，栽培面积占当地枣树的40%以上，栽培历史约1 000年左右。

植物学性状 树体中大，树姿半开张，干性较强，树冠呈自然圆头形。主干深灰色，皮呈块状不规则纵裂，较易剥落。枣头红褐色，平均长65.7cm，节间长7.0cm，蜡质少。二次枝长24.7cm，6节左右，弯曲度中等。针刺发达。枣股圆柱形，抽生枣吊4.0个。枣吊长16.8cm，着叶13片。叶片卵圆形，绿色，叶面富有光泽，先端渐尖，先端钝圆。叶基圆形。叶缘具钝锯齿。花量少，花序平均着花3朵。

生物学特性 适应性强，耐干旱瘠薄，抗风。树势较强，根蘖萌生力强。栽后1、2年开始结果，15年左右进入盛果期，吊果率16.4%，一般株产50kg左右。在山西太谷地区，9月下旬果实成熟，果实生育期118d，为晚熟品种类型。

果实性状 果实较大，倒卵圆形，纵径3.00cm，横径2.45cm，单果重15.7g，大小整齐。果肩圆平，梗洼浅、中广。果顶渐细，先端平圆，柱头残存。果柄长1.2mm。果面平整，果皮紫红色，有光泽。果肉脆熟期淡绿色，完熟期转呈浅黄色，质地致密，味酸甜，品质中等，可加工用。鲜枣可食率94.1%，100g果肉维生素C含量373.13mg；果皮含黄酮4.22mg/g，cAMP含量447.41μg/g。果核大，核重0.93g，两端呈不对称的纺锤形，含仁率76.7%。

评价 该品种适应性强，耐旱，耐瘠，抗风。树体健壮，寿命长，结果早，产量高而稳定。果实较大，核大，品质中等，可加工用，不宜做经济栽培。

Dingxiangshanzao

Source and Distribution The cultivar originated from and spreads in Hengshan, Longwan, Hushan of Dingxiang County, in the costal areas of Tonghe River in Yuanping County, and in Yangbai Village of Wutai County in Shanxi Province. The cultivation area of this cultivar occupies over 40% of the total jujube area in those places with a cultivation history of about 1 000 years.

Botanical Characters The medium-sized tree is half-spreading with a strong central leader trunk and a natural-round crown. The dark-gray trunk bark has irregular, vertical, massive fissures, easily shelled off. The reddish-brown 1-year-old shoots are 65.7 cm long. The internodes are 7.0 cm long with less wax and developed thorns. The secondary branches are 24.7 cm long with 6 nodes of medium curvature. The column-shaped mother fruiting spurs can germinate 4 deciduous fruiting shoots which are 16.8 cm long with 13 leaves. The oval-shaped leaves are green and glossy, with a gradually-cuspate apex, a round base and a blunt saw-tooth pattern on the margin. The number of flowers is small, averaging 3 ones per inflorescence.

Biological Characters The tree has strong adaptability, strong tolerance to drought, poor soils and wind, with strong vigor and strong suckering ability. It generally bears in the 1st or 2nd year after planting, and enters the full-bearing stage in the 15th year. The percentage of deciduous fruiting shoots to fruit is 16.40% on average. The yield per tree is about 50 kg on average. In Taigu County of Shanxi Province, it matures in late September with a fruit growth period of 108 d. It is a late-ripening variety.

Fruit Characteristics The large obovate fruit has a vertical and cross diameter of 3.00 cm and 2.45 cm, averaging 15.7 g. It has a flat-round shoulder, a shallow and medium-wide stalk cavity, a gradually-thinner fruit apex with a flat-round top-point, a remnant stigma, a smooth surface and purplish-red glossy skin. The tight-textured, sour and sweet flesh is light-green in the crisp-maturing stage and light-yellow after complete maturing. It can be used for processing with medium quality. The percentage of edible part of fresh fruit is 94.1%, and the content of Vc is 373.13 mg per 100 g fresh fruit. The content of flavones and cAMP in mature fruit skin is 4.22 mg/g and 447.41 μg/g. The large spindle-shaped stone weighs 0.93 g. The percentage of containing kernels is 76.7%.

Evaluation The cultivar has strong adaptability, strong tolerance to drought, poor soils and wind, strong vigor and a long life. It bears early with a high and stable yield. The large fruit has a large stone with medium quality. It is unsuitable for commercial cultivation.

蜜 枣
Candied Varieties
品 种

· 465 ·

大荔水枣

品种来源及分布　主要原产和分布于陕西大荔县的小元、北丁、西营、三教等地。栽培历史不详。

植物学性状　树体中大，树姿开张，干性较弱，枝条中密，树冠乱头形。树干皮裂浅、块状，不易脱落。枣头红褐色，平均长60.4cm，粗1.03cm，节间长6.8cm，着生永久性二次枝5～7个，二次枝长32.4cm，平均6节。针刺发达。皮目小，椭圆形，凸起，开裂。枣股较小，抽吊力中等，抽生枣吊2～5个。枣吊平均长18.8cm，着叶16片。叶片小而较厚，叶长4.6cm，叶宽2.4cm，卵圆形，浓绿色，先端钝尖，叶基圆形或楔形，叶缘锯齿钝。花量较少。花中大，花径7.5mm左右，夜开型。

生物学特性　树势中庸，萌芽率高，成枝力中等。结果较早，一般栽后第二年开始结果，10年后进入盛果期。丰产，产量稳定，坐果率高，枣头吊果率74.1%，2～3年生枝为77.2%，4年生枝为4.5%。在山西太谷地区，9月初果实着色，9月中旬果实脆熟，果实生育期100d，为早中熟品种类型。成熟期遇雨易裂果，采前落果较重。

果实性状　果实大，卵圆形，纵径4.13cm，横径3.25cm，单果重18.4g，大小较整齐。果梗中长，梗洼中广、深。果顶凹，柱头遗存，不明显。果皮中厚，红色，果面有隆起。果点小而圆，分布较稀。果肉厚，绿白色，肉质细，较松，味甜，汁液较少，品质中上，可鲜食、制干和加工蜜枣，多以蜜枣加工为主。鲜枣可食率95.8%，含可溶性固形物28.80%，总糖21.47%，酸0.55%，100g果肉维生素C含量368.98mg；果皮含黄酮16.22mg/g，cAMP含量175.82μg/g。制干率50.0%，干枣含总糖60.71%，酸0.84%。核小，纺锤形，核重0.77g，核尖短，核面较粗糙，含仁率18.3%。

评价　该品种适应性较强，沙性土壤表现更好。树势中庸，结果早，高产稳产。果实大，肉质略疏松，宜制干和加工蜜枣，品质上等，但易裂果，可在气候干旱、成熟期降雨较少的地区适度发展。

Dalishuizao

Source and Distribution　The cultivar originated from and mainly spreads in Xiaoyuan, Beiding, Xiying and Sanjiao in Dali County of Shaanxi Province, with an unknown history.

Botanical Characters　The medium-sized tree is spreading with a weak central leader trunk, medium-dense branches and an irregular crown. The reddish-brown 1-year-old shoots are 60.4 cm long and 1.03 cm thick, with the internodes of 6.8 cm. There are 5～7 permanent secondary branches which are 32.4 cm long with 6 nodes and developed thorns. The small lenticels are oval-shaped, protuberant and cracked. The small mother fruiting spurs can germinate 2～5 deciduous fruiting shoots which are 18.8 cm long with 16 leaves. The small thick leaves are oval-shaped and dark green. The number of flowers is small. The medium-sized blossoms have a diameter of 7.5 mm. It blooms at night.

Biological Characters　The tree has moderate vigor, high germination rate and medium branching ability. It bears early, generally in the 2nd year after planting and enters the full-bearing stage in the 10th year with high fruit set and a high yet unstable yield. In Taigu County of Shanxi Province, it enters the crisp-maturing stage in mid-September. It is an early-mid-ripening variety with serious premature fruit-dropping and serious fruit-cracking.

Fruit Characteristics　The large oblong fruit has a vertical and cross diameter of 4.13 cm and 3.25 cm, averaging 18.4 g with a regular size. It has medium-thick red skin with some protuberances on the surface. The dots are small and round, sparsely distributed. The greenish-white thick flesh is loose-textured, delicate and sweet, with less juice and a better than normal quality for fresh eating, dried fruits and candied fruits. It is mainly used for processing candied fruits. The percentage of edible part of fresh fruit is 95.8%, and SSC, TTS, TA and Vc is 28.80%, 21.47%, 0.55% and 368.98 mg per 100 g fresh fruit. The rate of fresh fruits which can be made into dried ones is 50%, and the content of TTS and TA in dried fruit is 60.71% and 0.84%. The small spindle-shaped stone weighs 0.77 g.

Evaluation　The cultivar has strong adaptability. It grows better in sandy soils. The tree has moderate vigor, bearing early with a high and stable yield. The large fruit has loose-textured flesh, suitable for making dried fruits and processing candied fruits with an excellent quality. Yet it has poor resistance to fruit-cracking. It can be developed moderately in areas with dry climate and light rainfall in maturing stage.

蜜 枣
Candied Varieties
品 种

小果算盘枣

品种来源及分布 原产和分布于湖南的溆浦、祁东、祁阳等地。

植物学性状 树体较大，树姿开张，下层主枝平展或下垂，枝叶稀疏，树冠自然半圆形。主干灰褐色，粗糙，裂纹较深，块状，树皮易片状剥落。枣头紫褐色，平均长83.0cm，粗1.04cm，节间长6.3cm。二次枝长25.6cm，7节。针刺不发达。枣股圆锥形，平均抽生枣吊4.0个。枣吊长24.9cm，着叶17片。叶片较小，椭圆形，绿色，先端渐尖，先端钝圆，叶基圆楔形，叶缘具细浅锯齿。花量多，花序平均着花5朵。花中大，昼开型。

生物学特性 适应性强，耐旱，耐瘠薄。树势中等，发枝力弱，结果较晚，结果能力强，产量较高，且稳定。定植后3年开始结果，10年生后进入盛果期。在山西太谷地区，4月中旬萌芽，5月下旬始花，9月下旬果实成熟，果实生育期约110d，为晚熟品种类型。

果实性状 果实中大，扁圆形，纵径2.15cm，横径2.10cm，单果重10.9g，大小不整齐。果肩平圆，梗洼中深、广，果顶略凹，果皮厚，紫红色，光滑，外观好。果肉白色，质地较致密，汁液中多，味酸，品质差，用于制干和加工蜜枣。鲜枣可食率91.4%，含可溶性固形物26.40%，含总糖25.77%，酸0.65%，100g果肉维生素C含量462.30mg；果皮含黄酮14.23mg/g，cAMP含量237.28μg/g。制干率32.4%。果核大，椭圆形，核重0.94g，含仁率50%左右。

评价 该品种适应性强，产量高而稳定。果实外观较好，肉质致密，味酸，可食率低，品质差，经济价值不大。

Xiaoguosuanpanzao

Source and Distribution The cultivar originated from and spreads in Xupu, Qidong and Qiyang in Hunan Province.

Botanical Characters The large tree is spreading with flat or pendulous lower-branches, sparse branches, and a semi-round crown. The grayish-brown trunk bark has rough, deep, massive fissures, easily shelled off. The purplish-brown 1-year-old shoots are 83.0 cm long and 1.04 cm thick. The internodes are 6.3 cm long with less-developed thorns. The secondary branches are 25.6 cm long with 7 nodes. The conical mother fruiting spurs can germinate 4 deciduous fruiting shoots which are 24.9 cm long with 17 leaves. The small leaves are green and oval-shaped, with a gradually-cuspate apex, a round-cuneiform base and a thin, shallow saw-tooth pattern on the margin. There are many medium-sized flowers produced, averaging 5 ones per inflorescence. It blooms in the daytime.

Biological Characters The tree has strong adaptability, strong tolerance to drought and poor soils, moderate vigor and weak branching ability. It bears late with strong fruiting ability, a high and stable yield. The tree generally bears in the 3rd year after planting, and enters the full-bearing stage in the 10th year. In Taigu County of Shanxi Province, it germinates in mid-April, begins blooming in late May and matures in late September with a fruit growth period of 110 d. It is a mid-ripening variety.

Fruit Characteristics The medium-sized oblate fruit has a vertical and cross diameter of 2.15 cm and 2.10 cm, averaging 10.9 g with irregular sizes. It has a flat-round shoulder, a medium-deep and wide stalk cavity, a slightly-sunken fruit apex, and thick, purplish-red, smooth and attractive skin. The white flesh is tight-textured and sour, with medium juice. It has a low quality, mainly used for making dried fruits and processing candied fruits. The percentage of edible part of fresh fruit is 91.4%, and SSC, TTS, TA and Vc is 26.40%, 25.77%, 0.65% and 462.30 mg per 100 g fresh fruit. The content of flavones and cAMP in mature fruit skin is 14.23 mg/g and 237.28 μg/g. The percentage of fresh fruits which can be made into dried ones is 32.4%. The large oval-shaped stone weighs 0.94 g, and the percentage of containing kernels is 50%.

Evaluation The cultivar has strong adaptability with a high and stable yield. The fruit has a good appearance and tight-textured, sour flesh with low edibility and low economic value.

蜜 枣
Candied Varieties
品 种

· 469 ·

糠 头 枣

品种来源及分布 别名糠皮枣。原产和分布于湖南的祁阳、梅溪和衡山萱洲乡等地。

植物学性状 树体中大，树姿开张，树冠乱头形。主干灰褐色或灰黑色，粗糙，皮裂条状，易剥落。枣头黄褐色，平均长78.5cm，粗0.94cm，节间长7.2cm，蜡质少。二次枝平均长30.2cm，6节，弯曲度中等。针刺发达。枣股圆柱形，平均抽生枣吊4.0个。枣吊长24.5cm，着叶18片。叶片小，椭圆形，深绿色，较光亮，先端钝尖，叶基偏斜形，叶缘锯齿锐细。花量少，花序平均着花3朵，花中大，萼片短。

生物学特性 抗逆性较强，树势中庸偏弱。定植后2～3年开始结果。成龄树一般株产40kg，大小年不明显。在山西太谷地区，4月中旬萌芽，10月上旬果实成熟，果实生育期120天左右。为极晚熟品种类型。极抗裂果，自然落果较少。

果实性状 果实较小，倒卵圆形，纵径2.67cm，横径2.12cm，单果重8.4g，最大9.0g，大小较整齐。梗洼中深、窄。果皮红色，果点小而明显。果肉浅绿色，质地致密，汁液中多，味酸，鲜食和制干品质差，可加工蜜枣。鲜枣可食率89.1%，含可溶性固形物24.00%，含总糖22.85%，酸0.42%，100g果肉维生素C含量758.06mg；果皮含黄酮5.07mg/g，cAMP含量124.69μg/g。制干率47.3%，含总糖39.40%。果核较大，椭圆形，平均重0.92g，含仁率8.3%。

评价 该品种抗逆性较强，产量高而稳定，极抗裂果，维生素C含量极高。但果实较小，且果核大，糖分低，干物质少，品质差，不宜经济栽培。

Kangtouzao

Source and Distribution The cultivar, also called Kangpizao, originated from and spreads in Qiyang, Meixi and Xuanzhou Village of Hengshan in Hunan Province.

Botanical Characters The medium-sized tree is spreading with an irregular crown. The grayish-brown or black trunk bark has rough striped fissures, easily shelled off. The yellowish-brown 1-year-old shoots are 78.5 cm long and 0.94 cm thick. The internodes are 7.2 cm long with less wax and developed thorns. The secondary branches are 30.2 cm long with 6 nodes of medium curvature. The column-shaped mother fruiting spurs can germinate 4 deciduous fruiting shoots which are 24.5 cm long with 18 leaves. The small oval-shaped leaves are dark green and glossy, with a bluntly-cuspate apex, a deflective base and a thin, sharp saw-tooth pattern on the margin. The number of flowers is small, averaging 3 ones per inflorescence. The medium-sized blossoms have short sepals.

Biological Characters The tree has strong resistance to adverse conditions, with moderate or weak vigor. It generally bears in the 2nd or 3rd year after planting. A mature tree has a yield of 40 kg with indistinct alternate bearing. In Taigu County of Shanxi Province, it germinates in mid-April and matures in early October with a fruit growth period of 120 d. It is an extremely late-ripening variety with light natural fruit-dropping and very strong resistance to fruit-cracking.

Fruit Characteristics The small obovate fruit has a vertical and cross diameter of 2.67 cm and 2.12 cm, averaging 8.4 g (maximum 9 g) with a regular size. It has a medium-deep and narrow stalk cavity, and red skin with small and distinct dots. The light-green flesh is tight-textured and sour with medium juice. It has a low quality for fresh-eating and dried fruits, so it is mainly used for processing candied fruits. The content of SSC, TTS, TA and Vc in fresh fruit is 24.00%, 22.85%, 0.42% and 758.06 mg per 100 g fresh fruit. The content of flavones and cAMP in mature fruit skin is 5.07 mg/g and 124.69 μg/g. The percentage of fresh fruits which can be made into dried ones is 47.3%, and the content of TTS in dried fruit is 39.40%. The large oval-shaped stone weighs 0.92 g, and the percentage of containing kernels is 8.3%.

Evaluation The Kangtouzao cultivar has strong resistance to adverse conditions with a high and stable yield. The fruit has very strong resistance to fruit-cracking and a very high content of Vc. Yet it has a small size, a large stone, low sugar content, fewer dry materials and a low quality and is unsuitable for commercial production.

蜜 枣
Candied Varieties
品 种

· 471 ·

涪陵鸡蛋枣

品种来源及分布 原产和分布在重庆市江北、奉节、涪陵、巴县等地，为当地的主栽品种。有100年以上栽培历史。

植物学性状 树体小，树姿较开张，干性弱，树冠偏斜形。主干灰褐色，皮裂纹浅，呈不规则纵条形。枣头较细软，红褐色，平均长73.5cm，粗1.0cm，节间长7.2cm，蜡层中多。针刺发达。枣股圆锥形，平均抽生枣吊3.0个。枣吊长23.7cm，着生叶16片。叶片较小，椭圆形，绿色，较光亮，先端急尖，叶基圆形，叶缘波状，具浅钝锯齿。花量少，花中大，每花序着花3朵。

生物学特性 对气候、土壤适应性强，抗病力较弱。树势强健，发枝力强。结果早，产量高而稳定。在山西太谷地区，4月中旬萌芽，5月下旬始花，8月下旬开始进入白熟期，白熟果生育期85d左右，为中晚熟品种类型。

果实性状 果实较大，圆形，纵径3.69cm，横径3.04cm，单果重16.4g，大小较整齐。果肩平。梗洼中深、较窄，果顶凹陷。果柄粗，长4.7mm。果面有隆起，红色。果肉浅绿色，质地疏松，汁液中多，味甜，品质中等，可加工蜜枣和鲜食。鲜枣可食率95.4%，含可溶性固形物25.60%，糖18.33%，酸0.23%，100g果肉维生素C含量317.18mg；果皮含黄酮13.29mg/g，cAMP含量134.47μg/g。果核中大，纺锤形，核重0.64g。核内不含种子。

评价 该品种适应性强，树体小，丰产性好。果实较整齐，较大，肉质疏松，味甜，可加工蜜枣和鲜食。

Fulingjidanzao

Source and Distribution The cultivar originated from and spreads in Fengjie, Fuling, Jiangbei and Baxian in Sichuan Province. It is the dominant variety there, with a cultivation history of over 100 years.

Botanical Characters The small tree is spreading with a weak central leader trunk and a deflective crown. The grayish-brown trunk bark has irregular, shallow striped fissures. The thin and soft 1-year-old shoots are reddish-brown, 73.5 cm long and 1.0 cm thick. The internodes are 7.2 cm long, with medium wax and developed thorns. The conical mother fruiting spurs can germinate 3 deciduous fruiting shoots which are 23.7 cm long with 16 leaves. The small oval-shaped leaves are green and glossy, with a sharply-cuspate apex, a round base and a shallow, blunt saw-tooth pattern on the wavy margin. The number of flowers is small, averaging 3 ones per inflorescence. The blossoms have a medium large size.

Biological Characters The tree has strong adaptability, strong vigor and strong branching ability, bearing early with a high and stable yield. It is susceptible to diseases. In Taigu County of Shanxi Province, it germinates in mid-April, begins blooming in late May and enters the white-maturing stage in late August with a fruit growth period of 85 d for the white-maturing fruit. It is a mid-late-ripening variety.

Fruit Characteristics The large globose-shaped fruit has a vertical and cross diameter of 3.69 cm and 3.04 cm, averaging 16.4 g with a regular size. It has a flat shoulder, a medium-deep and narrow stalk cavity, a sunken fruit apex, a thick stalk of 4.7 mm long, and red skin with some protuberances. The light-green flesh is loose-textured, sweet and juicy. It has medium quality for candied fruits and fresh-eating. The percentage of edible part of fresh fruit is 95.4%, and SSC, TTS, TA and Vc is 25.60%, 18.33%, 0.23% and 317.18 mg per 100 g fresh fruit. The content of flavones and cAMP in mature fruit skin is 13.29 mg/g and 134.47 μg/g. The medium-sized stone is spindle-shaped, averaging 0.64 g without kernel.

Evaluation The small plant of Fulingjidanzao cultivar has strong adaptability and high productivity. The medium-sized fruit has a regular size with juicy and sweet flesh and is suitable for processing candied fruits and for fresh-eating.

蜜 枣
Candied Varieties
品 种

· 473 ·

宣 城 尖 枣

品种来源及分布 别名长枣。原产于安徽宣城市的水东乡。主要分布于水东、孙埠、杨林等地，为当地主栽品种。栽培历史200余年。

植物学性状 树体较小，树姿开张，树冠伞形。主干深灰色，皮裂条状、较大。枣头红褐色，连续生长力强，平均长71.6cm，粗0.84cm，节间长8.7cm，着生永久性二次枝2～6个。二次枝长27.0cm，平均6节，无针刺。皮目较小，圆形，凸起，开裂。枣股大，圆柱形，多年生枣股有分歧现象，抽吊力中等，一般抽生枣吊3～4个。枣吊平均长22.3cm，着叶12片。叶片较大，叶长6.5cm，宽3.3cm，卵圆形，绿色，先端急尖，叶基圆形，叶缘锯齿稀而粗。花量多，每花序着花多达9朵。

生物学特性 树势中庸偏弱，发枝力较弱，结果早，嫁接苗当年即可结果，丰产，15～20年生树株产50kg以上。成龄树坐果率高，枣头吊果率73.3%，2～3年生枝为160.3%，4年生枝为48.1%。在山西太谷地区，9月下旬果实脆熟期，10月上旬完熟，果实生育期115d左右，为极晚熟品种。

果实性状 果实大，倒卵圆形，纵径4.80cm，横径3.70cm，单果重22.5g，大小整齐。果梗长，较粗，梗洼稍广，较深，果顶尖，柱头较大，明显。果皮红色，果面光滑。果点小而圆，分布较密。果肉厚，乳黄色，汁液少。白熟期含总糖9.90%，酸0.27%，100g果肉维生素C含量 351.10mg。鲜枣可食率97.1%，脆熟期含总糖21.02%，酸0.38%，100g果肉维生素C含量 506.55mg；成熟期果皮含黄酮12.63mg/g，cAMP含量25.23μg/g。核小，纺锤形，纵径2.90cm，横径0.70cm，核重0.65g，核纹浅，核尖中长，含仁率1.7%，种仁不饱满。

评价 该品种树体较小，高产稳产，蜜枣果形整齐，透明度高，吸糖量高，肉厚核小，品质上等，为加工蜜枣优良品种。

Xuanchengjianzao

Source and Distribution The cultivar originated from Shuidong Village of Xuancheng City in Anhui Province, and mainly spreads in Shuidong, Sunbu and Yanglin villages. It is the dominant variety there with a history of over 200 years.

Botanical Characters The small tree is spreading with an umbrella-shaped crown. The reddish-brown 1-year-old shoots have strong consecutive-growing ability, averaging 71.6 cm long and 0.84 cm thick with the internodes of 8.7 cm. There are 2～6 secondary branches which are 27.0 cm long with 6 nodes and without thorns. The small round lenticels are protuberant and cracked. The large mother fruiting spurs are column-shaped. The perennial spurs have bifurcation phenomenon. They can germinate 3～4 deciduous fruiting shoots which are 22.3 cm long with 12 leaves. The large green leaves are oval-shaped. There are many flowers produced, averaging 9 ones per inflorescence.

Biological Characters The tree has moderate or weak vigor and weak branching ability, bearing early with a high yield and high fruit set. The grafted seedlings can bear fruit in the year of planting. A 15～20-year-old tree has a yield of over 50 kg. In Taigu County of Shanxi Province, it enters the crisp-maturing stage in late September and completely matures in early October. It is an extremely late-ripening variety.

Fruit Characteristics The large column-shaped fruit has a vertical and cross diameter of 4.80 cm and 3.70 cm, averaging 22.5 g with a regular size. It has a long and thick stalk, a wide and deep stalk cavity, a cuspate fruit apex, a large and distinct stigma, red skin and a smooth surface with small, round and dense dots. The ivory-yellow flesh is thick with less juice. The content of TTS, TA and Vc in fruits of white-maturing stage is 9.90%, 0.27% and 351.10 mg per 100 g fresh fruit. The percentage of edible part of fresh fruit is 97.1%, and the content of TTS, TA and Vc is 21.02%, 0.38% and 506.55 mg per 100 g fresh fruit. The content of flavones and cAMP in mature fruit skin is 12.63 mg/g and 25.23 μg/g. The small spindle-shaped stone has a vertical and cross diameter of 2.90 cm and 0.70 cm, averaging 0.65 g with shallow veins, a medium-long apex and a shriveled kernel. The percentage of containing kernels is 1.7%.

Evaluation The plant of Xuanchengjianzao cultivar has a long life with a high and stable yield. Candied fruits have regular sizes, high transparency, high sugar absorption, thick flesh, a small stone with an excellent quality. It is a good variety for processing candied fruits.

蜜枣
Candied Varieties
品种

· 475 ·

阜阳木头枣

品种来源及分布　原产和分布于安徽省阜阳等地。

植物学性状　树体较大，树姿半开张，树冠呈圆锥形。主干皮裂条状。枣头黄褐色，平均长74.5cm，节间长7.9cm。二次枝长28.5cm，5～7节。针刺退化。枣股平均抽生枣吊4.0个。枣吊长27.5cm，着叶16片。叶片椭圆形，先端急尖，叶基圆楔形，叶缘具钝锯齿。花量较多，花序平均着花7朵。

生物学特性　树势较强，萌芽率和成枝力强。一般定植第三年结果，成龄树产量较低，平均吊果率42.6%。在山西太谷地区，9月中旬果实进入白熟期，10月上旬开始成熟采收，果实生育期118d左右，属极晚熟品种。果实抗裂果能力极强。

果实性状　果个中大，圆柱形，纵径3.59cm，横径2.52cm，单果重13.1g，大小较整齐。果皮厚，红色，果面有隆起。梗洼窄、中广。果顶平，柱头残存。肉质致密，味甜酸，汁液中多，品质中等，适宜制作蜜枣。鲜枣可食率97.0%，含可溶性固形物36.00%，总糖22.36%，酸0.39%，100g果肉维生素C含量532.76mg；果皮含黄酮2.37mg/g，cAMP含量180.77μg/g。果核较小，纺锤形，核重0.39g。种仁较饱满，含仁率96.7%。

评价　该品种成熟期极晚，抗裂果能力极强。果个中大，品质中等，适宜制作蜜枣。

Fuyangmutouzao

Source and Distribution　The cultivar originated from and spreads in Fuyang of Anhui Province.

Botanical Characters　The large tree is half-spreading with a conical crown. The trunk bark has striped fissures. The yellowish-brown 1-year-old shoots are 74.5 cm long with the internodes of 7.9 cm. The secondary branches are 28.5 cm long with 5～7 nodes and degraded thorns. The mother fruiting spurs can germinate 4 deciduous fruiting shoots which are 27.5 cm long with 16 leaves. The oval-shaped leaves have a sharply-cuspate apex, a round-cuneiform base and a blunt saw-tooth pattern on the margin. There are many flowers produced, averaging 7 ones per inflorescence.

Biological Characters　The tree has strong vigor, strong germination and branching ability. It generally bears in the 3rd year after planting with a low yield. The percentage of deciduous fruiting shoots to fruit is 42.6% on average. In Taigu County of Shanxi Province, it enters the white-maturing stage in mid-September and matures in early October with a fruit growth period of 118 d. It is an extremely late-ripening variety with very strong resistance to fruit-cracking.

Fruit Characteristics　The medium-sized oblong fruit has a vertical and cross diameter of 3.59 cm and 2.52 cm, averaging 13.1 g with a regular size. It has thick red skin, a rough surface with some protuberances, a shallow and medium-wide stalk cavity, a flat fruit apex and a remnant stigma. The tight-textured flesh is sour, sweet and juicy. It has medium quality for candied fruits. The percentage of edible part of fresh fruit is 97.0%, and SSC, TTS, TA and Vc is 36.00%, 22.36%, 0.39% and 532.76 mg per 100 g fresh fruit. The content of flavones and cAMP in mature fruit skin is 2.37 mg/g and 180.77 μg/g. The small stone is spindle-shaped, averaging 0.39 g with a well-developed kernel. The percentage of containing kernels is 96.7%.

Evaluation　The cultivar matures very late. The fruit has very strong resistance to fruit-cracking. It has a medium-large fruit size and medium quality ans is suitable for processing candied fruits.

宁阳暄铃枣

品种来源及分布 别名酸铃、暄铃、大个子枣。分布于山东的宁阳、泗水、汶上、曲阜、兖州等地，其中宁阳的葛石、堽城、蒋集一带为其集中产区，有大面积生产栽培，数量占当地枣树的90%以上。为当地原产的乡土品种。

植物学性状 树体较高大，树姿半开张，干性较弱，枝叶密度中等，树冠呈自然圆头形。主干浅灰褐色，条状浅裂，易剥落。枣头红褐色，平均长65.5cm，节间长6.7cm，蜡质多。二次枝平均长27.8cm，6节，弯曲度小，无针刺。枣股圆锥形，平均抽生枣吊4.0个。枣吊长16.6cm，着叶9～11片。叶片中大，卵状披针形，中厚，深绿色，有光泽，先端渐尖，叶基圆形。叶缘平或略呈波状，齿尖较钝。花量大，花序平均着花8朵，花较大，昼开型。

生物学特性 适应性较强，耐干旱、瘠薄。树势强，发枝力中等，坐果要求温度较高，花期气温不稳定时坐果率低。定植后3、4年开始结果，盛果期产量较高，一般株产50kg。在山西太谷地区，4月中旬萌芽，5月下旬始花，10月上旬进入完熟期，果实生育期115d左右，为极晚熟品种类型。果实着色期遇雨裂果严重。

果实性状 果实较大，圆柱形，纵径3.63cm，横径2.52cm，单果重14.7g，大小较整齐。果肩平。梗洼中深而窄。果柄粗，较短，长6.1mm。果顶平，柱头残存。果面较平滑光亮。果皮厚，浅色，富光泽。果肉厚，白色，质地致密，汁液少，适宜制作蜜枣，鲜食品质中等，制干率较低，品质较差。鲜枣可食率95.4%，白熟期含可溶性固形物20.40%，100g果肉维生素C含量461.91mg；果皮含黄酮4.62mg/g，cAMP含量201.83μg/g。干枣含总糖66.70%。果核较大，纺锤形，核重0.68g，含仁率仅6.7%。

评价 该品种耐旱，耐瘠，生长势强，产量较高。果实较大，肉质疏松，少汁，味淡，可制蜜枣等加工品。

Ningyangxuanlingzao

Source and Distribution The cultivar originated from and spreads in Ningyang, Sishui, Hanshang and Qubu in Shandong Province. Geshi and Jiangji in Ningyang are the central producing areas of this variety, with a large-scale cultivation, which occupies over 90% of the total jujube area there.

Botanical Characters The large tree is half-spreading with a weak central leader trunk, medium-dense branches and a natural-round crown. The reddish-brown 1-year-old shoots are 65.5 cm long. The internodes are 6.7 cm long with much wax and no thorns. The secondary branches are 27.8 cm long with 6 nodes of small curvature. The conical mother fruiting spurs can germinate 4 deciduous fruiting shoots which are 16.6 cm long with 9～11 leaves. The medium-sized thick leaves are ovate-lanceolate, dark green and glossy. There are many large flowers produced, averaging 8 ones per inflorescence.

Biological Characters The tree has strong adaptability, strong tolerance to drought and poor soils, strong vigor and medium branching ability. It requires higher temperature for the flowers to set fruit. It generally bears in the 3rd or 4th year after planting. In Taigu County of Shanxi Province, it begins blooming in late May and matures in early October. It is an extremely late-ripening variety with serious fruit-cracking if it rains in the coloring stage.

Fruit Characteristics The large column-shaped fruit has a vertical and cross diameter of 3.63 cm and 2.52 cm, averaging 14.7 g with a regular size. It has a smooth and bright surface, and thick, light-red and glossy skin. The white flesh is thick and tight-textured with less juice. It is suitable for processing candied fruits. It has medium quality for fresh-eating and a low percentage of fresh fruits which can be made into dried ones. Dried fruits have a low quality. The percentage of edible part of fresh fruit is 95.4%, and the content of SSC and Vc is 20.40% and 461.91 mg per 100 g fresh fruit. The large spindle-shaped stone weighs 0.68 g.

Evaluation The cultivar has strong tolerance to drought and poor soils, strong vigor, with a long life and a high yield. The large fruit has loose-textured flesh with a light flavor and less juice. It can be used for processing candied fruits.

蜜 枣
Candied Varieties
品 种

观赏品种 Ornamental Varieties

枣树公园（宁夏 灵武）
Jujube Tree Park (Lingwu, Ningxia)

磨 盘 枣

品种来源及分布 又名磴磴枣、磨子枣、葫芦枣、药葫芦枣。陕西大荔，甘肃庆阳，山东乐陵、无棣、夏津，河北交河、青县、献县、曲阳、大名等地均有分布栽培，但数量很少，多为四旁零星栽植。栽培历史悠久，可能起源于陕西关中一带。

植物学性状 树体较大，树姿开张，干性中强，枝条中密，粗壮，树冠呈自然圆头形。主干皮裂深、块状。枣头红褐色，生长势较强，木质较软，平均生长量83.8cm，节间长9.5cm。二次枝长33.3cm，5～7节。针刺不发达。枣股较大，圆柱形，抽吊力较强，一般抽生枣吊3～5个。枣吊平均长24.67cm，着叶16片，少数枣吊有副吊生长现象。叶片中大，卵状披针形，浓绿色，先端尖凹，叶基圆形，叶缘锯齿浅钝。花量多，每花序着花7～9朵。花大，花径8mm左右。

生物学特性 树势强旺，成枝力较强，开花结果较早，一般定植后2～3年开始结果，结实力中等，盛果期产量较低，枣头吊果率45.9%，2～3年生枝为30.0%，4年生枝为0.5%。在山西太谷地区，9月下旬果实脆熟，果实生育期110d左右，为晚熟品种类型。果实抗裂果、抗病。

果实性状 果实较小，单果重7.2g，大小较整齐。果形石磨状，果实中部有一条缢痕，深宽约2～3mm，缢痕上部大、下部小，极具观赏价值。梗洼较广，中深，果顶凹，柱头遗存，不明显。果肉较厚，绿白色，肉质粗松，甜味较淡，汁液少，品质较差。鲜枣可食率90.1%，含可溶性固形物29.40%，总糖23.9%，酸0.51%，100g果肉维生素C含量448.27mg；果皮含黄酮3.19mg/g，cAMP含量73.53μg/g。制干率50.5%，干枣含总糖58.86%，酸1.23%。果核中大，短纺锤形或卵圆形，核重0.71g，含仁率63.3%。

评价 该品种适应性较强，树体健壮，产量较低，抗裂果，可制干，但品质较差。可作为观赏树木于庭院栽培，也可用作培育抗裂品种的育种材料。

Mopanzao

Source and Distribution The cultivar mainly spreads in Dali County of Shaanxi Province, Qingyang of Gansu Province, Leling, Wudi, Xiajin of Shandong Province and Jiaohe, Qingxian, Xianxian of Hebei Province. Yet it mainly dispersedly planted. It has a long history, originating from Guanzhong in Shaanxi Province.

Botanical Characters The large tree is spreading with a medium-strong central leader trunk, medium-dense and strong branches and a natural-round crown. The reddish-brown 1-year-old shoots have strong growth vigor with soft woody part. The secondary branches are 33.3 cm long with less-developed thorns. The large column-shaped mother fruiting spurs can germinate 3～5 deciduous fruiting shoots. The medium-sized leaves are ovate-lanceolate and dark green. There are many large flowers with a diameter of 8 mm produced, averaging 7～9 ones per inflorescence.

Biological Characters The tree has strong vigor and strong branching ability, blooming and bearing early and a low yield in full-bearing stage. It is a late-ripening variety with strong resistance to fruit-cracking and diseases.

Fruit Characteristics The small fruit is stone-mill-shaped, with a constriction (2～3 mm deep and wide) in the middle part of the fruit which is larger on the upper side and smaller on the lower side. The fruit has a vertical and cross diameter of 3.13 cm and 3.04 cm, averaging 7.2 g (maximum 13.0 g) with a regular size. It has a medium-long and medium-thick stalk, a wide and medium-deep stalk cavity, a sunken fruit apex and a remnant yet indistinct stigma. The greenish-white thick flesh is loose-textured and slightly sweet with less juice. It has a very poor quality for fresh eating and a lower than normal quality for dried fruits. The percentage of edible part of fresh fruit is 90.1%, and SSC, TTS, TA and Vc is 29.40%, 23.9%, 0.51% and 448.27 mg per 100 g fresh fruit. The content of flavones and cAMP in mature fruit skin is 3.19 mg/g and 73.53 μg/g. The rate of fresh fruits which can be made into dried ones is 50.5%, and the content of TTS and TA in dried fruit is 58.86% and 1.23%. The medium-sized stone is short spindle-shaped or oval-shaped, averaging 0.71 g.

Evaluation The cultivar has strong adaptability and strong vigor with a low yield and strong resistance to fruit-cracking. It can be planted in family gardens as an ornamental plant, and can also be considered as a breeding material for varieties with strong resistance to fruit-cracking.

茶壶枣

品种来源及分布 原产山东夏津、临清等地。数量极少,多为庭院零散栽植,用于观赏。栽培历史不详。临清县现有百年以上大树生长。目前北方各地都引种栽植观赏。

植物学性状 树体中大,树姿开张,干性较强,枝条中密,粗壮,树冠自然半圆形或伞形。主干皮裂浅、条状,不易剥落。枣头红褐色,生长势强,木质较松,髓部大,平均长81.9cm,节间长9.4cm。二次枝长36.4cm,平均5节。针刺发达。皮目小而圆,分布较稀,凸起,不开裂,灰白色。枣股中等大,圆锥形,抽吊力较强,一般抽生枣吊3~4个。枣吊粗,平均长18.6,着叶10片,部分枣吊有副吊生长现象。叶片中厚,宽大,叶长7.7cm,叶宽3.9cm,卵圆形,浓绿色,先端急尖,叶基圆形或心形,叶缘锯齿中密。花量特多,每花序着花15朵左右,昼开型。

生物学特性 树势较强,萌芽率高,成枝力强。结果较早,一般栽后第二年开始结果,10年后进入盛果期。坐果率高,结实力强,主要坐果部位在枣吊的5~9节,占坐果总数的79.6%。较丰产,产量稳定,枣头吊果率2.0%,2~3年生枝为130.8%,4年生枝为51%。在山西太谷地区,9月下旬果实成熟。果实生育期115d左右,为极晚熟品种类型。果实抗裂果和抗病性能极强。

果实性状 果实较小,果形奇特,果肩到果顶有1~5条长短不等的肉质状突出物,有的果实在肩部两端各有1个肉质突出物,与果实连成一体,形似茶壶的壶嘴和壶把,故名茶壶枣。单果重8.4g,大小不整齐。果梗中长、粗,梗洼中广、深,果顶凹,柱头不明显。果皮较薄,红色。果点中大,分布密,圆形,浅黄色,不明显。果肉较厚,绿白色,肉质较粗松,味甜略酸,汁液中多,品质中等,适宜观赏和制干。鲜枣可食率92.0%,含可溶性固形物30.40%;果皮含黄酮19.22mg/g,cAMP含量208.19μg/g。核较小,短纺锤形,核重0.67g,核纹浅,核尖短,含仁率26.7%。

评价 该品种适应性强,结果较早,坐果稳定,产量高。果实形状奇特艳丽美观,有极高的观赏价值,适于庭院栽培。

Chahuzao

Source and Distribution The cultivar originated from Xiajin and Linqing County in Shandong Province with a very small quantity and an unknown history. It is mainly planted in family gardens for ornament. At present, many northern places have introduced and planted this cultivar for ornament.

Botanical Characters The medium-sized tree is spreading with a strong central leader trunk, medium-dense and strong branches and a semi-round or umbrella-shaped crown. The reddish-brown 1-year-old shoots have strong growth vigor with loose woody part and large medulla. The secondary branches are 36.4 cm long with developed thorns. The conical-shaped mother fruiting spurs can germinate 3~4 thick deciduous fruiting shoots which are 18.6 cm long with 10 leaves. Some of the deciduous shoots have accessory ones growing. The leaves are medium-thick and wide, oval-shaped and dark green. The number of flowers is extremely large, averaging 15 ones per inflorescence.

Biological Characters The tree has strong vigor, high germination rate and strong branching ability. It bears early and enters the full-bearing stage in the 10th year with high fruit set, strong fruiting ability and a high and stable yield. It is an extremely late-ripening variety with strong resistance to fruit-cracking and diseases.

Fruit Characteristics The small fruit has a peculiar shape. From shoulder to the top, there may be 1~5 fleshy projections of different lengths. Some fruits have a fleshy projection on both shoulders, united with the fruit as one body, just like the mouth and handle of a tea pot. That is why it is called Chahuzao (Chahu means tea pot in Chinese). The fruit has an average weight of 8.4 g with irregular sizes. It has thin red skin. The greenish-white thick flesh is loose-textured, sweet and a little sour, with medium juice and a medium quality. It is suitable for ornament or making dried fruits. The percentage of edible part of fresh fruit is 92.0%, and SSC is 30.40%. The small stone is short spindle-shaped, averaging 0.67 g.

Evaluation The cultivar has strong adaptability with stable fruit set and a high yield. The fruit has a peculiar and attractive shape with a very high value for ornament. It can be planted in family gardens.

三 变 红

品种来源及分布 别名三变色、三变丑。原产分布于河南省永城市的十八里、鄼阳、城关、黄口、演集等地，为当地主栽品种之一。山东省兖州市道沟、陵城等地也有零星栽培。

植物学性状 树体较大，树姿直立，干性中强，枝条较稀，树冠圆锥形。主干皮裂条状、中深，易剥落。枝系嫩梢及幼叶呈淡棕绿色或紫色。枣头红褐色，平均生长量70.9cm，节间长8.4cm。二次枝长37.5cm，平均6节，无针刺。皮目中大，圆形或椭圆，凸起，开裂，灰白色。枣股中大，圆锥形，抽吊力较强，一般抽生枣吊3～5个。枣吊平均长20.5cm，着叶12片。叶较大，叶长6.8cm，叶宽3.1cm，卵圆形，浓绿色，先端锐尖，叶基圆楔形，叶缘锯齿细，较密。花量多，每花序着花8朵左右。花中大，花径7mm，特点是柱头呈紫色，昼开型。

生物学特性 树势中等，萌发力较弱。结果早，定植后2年左右开始结果，10年左右进入盛果期，较丰产，坐果率较高，枣头吊果率28.5%，2～3年生枝为66.4%，4年生枝为21.5%。主要坐果部位在枣吊的3～7节，占坐果总数的81%左右。在山西太谷地区，9月下旬果实脆熟，果实生育期110d左右，为中晚熟品种类型。成熟期遇雨易裂果。

果实性状 果实中大，长卵形或圆柱形，纵径4.51cm，横径2.71cm，单果重14.6g，大小整齐。果梗中长，较细，梗洼浅而较窄，果顶平，柱头遗存。果皮中厚，盛花末期落花后子房为紫色，幼果尖部呈淡紫色，随后全部变成紫色，7月上旬开始紫色渐变为条状绿色，成熟时变为深红色，色泽变化与胎里红品种完全不同。因果皮颜色从坐果到成熟变化三次，由此而得名三变红、三变色。果点中大，分布密，浅黄色，圆形，较明显。果肉厚，绿白色肉质致密，较酥脆，味甜，汁液中多，品质中上，可鲜食、制干和观赏。鲜枣可食率94.7%，含可溶性固形物29.40%，总糖28.25%，酸0.54%，100g果肉维生素C含量340.06mg；果皮含黄酮5.52mg/g，cAMP含量67.01μg/g。核小，长纺锤形，纵径2.98cm，横径0.81cm，核重0.78g，核尖长，核纹较浅，含仁率55.0%，种仁不饱满。

评价 该品种果实发育期皮色变化三次，亮丽美观，主要用作观赏，还可鲜食和制干利用。

Sanbianhong

Source and Distribution The cultivar originated from and spreads in Shibali, Chengguan, Huangkou and Yanji in Yongcheng City of Henan Province. It is one of the dominant varieties there.

Botanical Characters The large tree is vertical with a medium-strong central leader trunk, sparse branches and a conical-shaped crown. The reddish-brown 1-year-old shoots are 70.9 cm long. The secondary branches are 37.5 cm long without thorns. The medium-sized lenticels are grayish white, round or oval-shaped, protuberant and cracked. The conical-shaped mother fruiting spurs can germinate 3～5 deciduous fruiting shoots. The large leaves are oval-shaped and dark green. There are many flowers with a diameter of 7 mm, averaging 8 ones per inflorescence.

Biological Characters The tree has moderate vigor and weak germination ability. It bears early, generally in the 2nd year after planting and enters the full-bearing stage in the 10th year with a high yield and high fruit set. It is a mid-late-ripening variety with serious fruit-cracking if it rains in the maturing stage.

Fruit Characteristics The medium-sized fruit is long oval-shaped or column-shaped averaging 14.6 g with a regular size. It has medium-thick skin. The ovary is purple at the end of blooming. The skin of young fruits gradually becomes striped green from purple and then turns dark red when completely maturing. That is why it is called Sanbianhong or Sanbianse (san means three, bian means change and hong means red). It is totally different from the cultivar of Tailihong. The greenish-white thick flesh is tight-textured, crisp and sweet, with medium juice and a better than normal quality. It can be used for fresh eating, dried fruits and ornament. The percentage of edible part of fresh fruit is 94.7%, and SSC, TTS, TA and Vc is 29.40%, 28.25%, 0.54% and 340.06 mg per 100 g fresh fruit. The small stone is long spindle-shaped, averaging 0.78 g.

Evaluation The fruit of Sanbianhong cultivar changes color for 3 times in the growing stage. It is bright and attractive, mainly used for ornament. It can also be used for fresh eating and dried fruits.

观 赏
Ornamental Varieties
品 种

葫芦长红

品种来源及分布 别名亚腰葫芦枣。由长红枣品种群的大马牙枣演变而来。原产于山东枣庄市山亭区店子乡东剪子山村。栽培数量少。

植物学性状 树体较大，树姿半开张，树冠呈乱头形。主干皮色灰褐色，裂纹浅，宽条状，易剥落。枣头紫褐色，长63.7cm，粗0.90cm，节间长8.2cm，蜡质少。二次枝平均长27.6cm，6节左右，弯曲度小。针刺发达。枣股圆柱形，平均抽生枣吊4.0个。枣吊长23.6cm，着叶16片。叶片较大，卵状披针形，深绿色，叶厚，叶面平滑光亮，先端长，渐尖。叶基圆楔形。叶缘具钝锯齿。花量中等，花序平均着花5朵，花中大，花径6.0～7.0mm，为昼开型。

生物学特性 耐旱耐瘠，能适应贫瘠的山地土壤条件，树势强旺，发枝力中等。开花结果早，嫁接后1、2年开始结果，盛果期坐果率极高，枣头、2～3年和3年以上枝的吊果率分别为26.7%、279.1%和193.4%。丰产性强，并且产量稳定。在山西太谷地区，9月下旬果实成熟采收，果实生育期110d左右，为晚熟品种类型。果实成熟期遇雨极少有裂果。

果实性状 果实较小，长倒卵形或倒卵形，大小较整齐。纵径2.70cm，横径2.20cm，单果重7.7g，最大18.2g。果肩圆形耸起，较窄小，有数条深浅不等的沟纹。梗洼小、中深，环洼中深、广。从果顶与胴部连接处开始向下突然收缩变细呈乳头状，顶端圆形或平圆，顶点略凹陷，成一字纹，果形酷似葫芦，奇特美观，极具观赏价值。柱头宿存，稍突起。果柄粗，长3.5mm。果面光洁，平滑。果皮赭红色，富光泽。果肉浅绿色或乳白色，质地稍粗松，汁液少，鲜食风味不佳，干制红枣品质中等。鲜枣可食率92.0%，含可溶性固形物35.00%，100g果肉维生素C含量415.38mg；果皮含黄酮13.00mg/g，cAMP含量134.02μg/g。制干率48.0%左右。果核较大，梭形，两端尖锐，核重0.62g，核纹粗深短斜，含仁率11.7%。

评价 该品种树体较大，树势强健，耐旱，耐瘠，产量高而稳定。鲜食风味不佳，干制红枣品质中等，由于果形奇特美观，也可作为观赏品种发展。

Huluchanghong

Source and Distribution The cultivar, also called Yayaohuluzao, originated from Dongjianzishancun of Dianzi Village in Shanting District of Zaozhuang City, Shandong Province with a small quantity.

Botanical Characters The large tree is half-spreading with an irregular crown. The purplish-brown 1-year-old shoots are 63.7 cm long. The internodes are 8.2 cm long, with less wax and developed thorns. The secondary branches are 27.6 cm long with 6 nodes of small curvature. The column-shaped mother fruiting spurs can germinate 4 deciduous fruiting shoots. The large dark-green leaves are ovate-lanceolate, thick and glossy. The number of flowers is medium large. The blossoms are of medium size of a diameter of 6.0～7.0 mm.

Biological Characters The tree has strong tolerance to drought and poor soils, with strong vigor and medium branching ability, blooming and bearing early. It generally bears in the 1st or 2nd year after grafting with very high fruit set, a high and stable yield. It is a late-ripening variety with light fruit-cracking even if it rains in the maturing stage.

Fruit Characteristics The small fruit is long obovate or obovate with a vertical and cross diameter of 2.70 cm and 2.20 cm, averaging 7.7 g (maximum 18.2 g) with a regular size. It has a smooth surface and brownish-red glossy skin. Seen from the top to the bottom, it looks like a gourd which gives it a special appearance, high in ornamental value. The flesh is light green or ivory white, loose-textured, with less juice. It has medium quality for dried fruits. The percentage of edible part of fresh fruit is 92.0%, and SSC and Vc is 35.00% and 415.38 mg per 100 g fresh fruit. The large shuttle-shaped stone has weighs 0.62 g.

Evaluation The large plant of Huluchanghong cultivar is tolerant to drought and poor soils, with strong tree vigor, a high and stable yield. It has a poor taste for fresh-eating and medium quality for dried fruits. Because of its unique and attractive fruit shape, it can be developed as an ornamental variety.

胎 里 红

品种来源及分布 别名老来变。原产河南省镇平县的官寺、侯集、八里庙一带。数量极少，栽培历史不详。

植物学性状 树体中大，树姿开张，枝条中密，树冠自然半圆形。主干树皮粗糙，皮裂呈条状，不易剥落。枣头紫红色，平均长71.7cm，节间长7.6cm，二次枝长28.2cm，平均6节。无针刺。皮目小，卵圆形，凸起，不开裂，灰白色。枣股中大，抽吊力较强，一般抽生枣吊3~4个，多者达8个。枣吊粗而较长，平均吊长24.5cm，着叶14片。叶片中大，叶长4.4cm，叶宽2.6cm，卵状披针形，幼叶紫红色，成熟叶片绿色，先端急尖，叶基圆形或楔形，叶缘锯齿粗，较密。花量多，幼蕾为紫色，至开花时逐渐变浅。花中大，花径7mm左右，昼开型。

生物学特性 树势较强，萌芽率高，成枝力强，枣头枝较细弱。开花结果较早，盛果期产量中等而稳定，但坐果率较低，枣头吊果率2.0%，2~3年生枝为20.3%，4年生枝为15%，主要坐果部位在枣吊的4~10节，占坐果总数的76.1%。在山西太谷地区，9月下旬果实脆熟，果实生育期110d左右，为晚熟品种类型。果实成熟不一致，较抗裂果和果实病害。

果实性状 果实较小，柱形，单果重11.0g，大小整齐。落花后幼果为紫色，至果实接近成熟时变为水红或粉红色，成熟时变为鲜红色，十分美观。果梗细，较短，梗洼窄而深。果顶平，柱头遗存，较明显。果皮薄，鲜红色，果面光滑。果点小而圆，分布密，浅黄色，较明显。果肉厚，绿白色，肉质细，较酥脆，味酸甜，汁液中多，品质中上，适宜观赏利用。鲜枣可食率95.8%，含可溶性固形物32.50%，100g果肉维生素C含量354.33mg；果皮含黄酮2.76mg/g，cAMP含量77.08μg/g。果核较小，纺锤形，核重0.46g，核尖细长，核纹深，含仁率96.7%。

评价 该品种花朵和幼果均呈红色，且随果实发育果皮色泽由紫色变为粉红再变为鲜红色，极富观赏价值。适应性强，较丰产，果实中大，品质中等，也可作为蜜枣加工品种利用。

Tailihong

Source and Distribution The cultivar, also called Laolaibian, originated from Guansi, Houji and Balimiao in Zhenping County of Henan Province with a very small quantity and an unknown history.

Botanical Characters The medium-sized tree is spreading with medium-dense branches and a semi-round crown. The purplish-red 1-year-old shoots are 71.7 cm long with the internodes of 7.6 cm. The secondary branches are 28.2 cm long with 6 nodes and without thorns. The medium-sized mother fruiting spurs can germinate 3~4 thick deciduous fruiting shoots which are 24.5 cm long with 14 leaves. The medium-sized leaves are ovate-lanceolate. Young leaves are purplish red, while the mature ones turn green. There are many medium-sized flowers with a diameter of 7 mm. Young blossoms are purple and will fade gradually till blooming.

Biological Characters The tree has strong vigor, high germination rate and strong branching ability, with thin and weak 1-year-old shoots. It blooms and bears early with low fruit set. It has a medium and stable yield in full-bearing stage. It is a late-ripening variety with strong resistance to fruit-cracking and diseases.

Fruit Characteristics The small fruit is sharp column-shaped, averaging 11.0 g with a regular size. Young fruit is purple after blossom-dropping. Yet it turns pink when nearly maturing and after completely maturing, it becomes brightly red, which is rather attractive. It has brightly-red thin skin and a smooth surface. The greenish-white thick flesh is delicate, crisp, sweet and sour, with medium juice. It has a better quality, suitable for ornament. The percentage of edible part of fresh fruit is 95.8%, and SSC and Vc is 32.50% and 354.33 mg per 100 g fresh fruit. The content of flavones and cAMP in mature fruit skin is 2.76 mg/g and 77.08 μg/g. The small spindle-shaped stone has a thin long apex and deep veins. The percentage of containing kernels is 96.7%.

Evaluation The cultivar has red flowers and red fruits. The purple fruit skin will become pink and then brightly red with the growth of fruit, valuable for ornament. It has strong adaptability, high productivity, medium-large fruit size with a medium or poor quality. It can be used for processing candied fruits.

观 赏
Ornamental Varieties
品 种

大 荔 龙 枣

品种来源及分布 别名陕西龙枣、龙爪枣、曲枝枣。原产和分布于陕西大荔县石槽、八渔、苏村、西漠一带。现多用作制干和观赏品种用途引种栽培。

植物学性状 树体中大，树姿开张，干性弱，枝条中密，树冠自然圆头形或半圆形。主干皮裂条状、较深，易剥落。枣头黄褐色，平均长72.2cm，节间长8.0cm，二次枝长24.1cm，平均6节。针刺不发达。皮目小，椭圆形，凸起，分布中密。枣股较小，抽吊力中等，一般抽生枣吊3~4个。枣吊平均长19.1cm，着叶14片。枣头、二次枝、枣吊都弯曲生长，故又名龙爪枣和曲枝枣。叶片较小，叶长6.2cm，叶宽2.5cm，长卵形或卵状披针形，先端渐尖，叶基圆楔形，叶缘锯齿细而浅。花量较少，花中大，花径7mm，昼开型。

生物学特性 树势中等，枣头生长势较旺，成枝力强。开花结果早，丰产，产量较稳定，枣头吊果率98.7%，2~3年生枝为110.3%，4年生枝为47.0%。在山西太谷地区，9月下旬果实成熟，果实生育期100d左右。成熟期遇雨易裂果。

果实性状 果实中等大小，倒卵圆形，纵径3.73cm，横径2.60cm，单果重11.6g，最大14.6g，大小较整齐。果梗中长，较粗，梗洼中广，较深。果顶平或微凹，柱头遗存。果皮厚，紫红色，果面较平滑。果点中大，分布中密，圆形，较明显。果肉厚，绿白色，肉质较粗，味甜，汁液少，品质中等，可制干和加工蜜枣。鲜枣可食率95.3%，含可溶性固形物28.80%，总糖21.30%，酸0.59%，100 g果肉维生素C含量367.88mg；果皮含黄酮38.19mg/g，cAMP含量127.40μg/g。制干率50%左右，干枣含总糖67.41%，酸2.05%。核小，长纺锤形，纵径2.20~2.50cm，横径0.70~1.00cm，核重0.55g，核尖较短，核纹中深，核面较粗糙，含仁率6.7%。

评价 该品种枝条扭曲生长，枝形奇特，观赏价值高，可作庭院观赏树木和盆景栽培，但弯曲度不如龙枣品种大。适应性较强，产量高，但枣果鲜食品质较差，可作为制干品种适度发展。

Dalilongzao

Source and Distribution The cultivar originated from and spreads in Shicao, Bayu, Sucun and Ximo in Dali County of Shaanxi Province. It is mainly used for making dried fruits or used as an ornamental plant.

Botanical Characters The medium-sized tree is spreading with a weak central leader trunk, medium-dense branches and a natural-round or semi-round crown. The yellowish-brown 1-year-old shoots are 72.2 cm long with the internodes of 8.0 cm. The secondary branches are 24.1 cm long with less-developed thorns. The small mother fruiting spurs can germinate 3~4 deciduous fruiting shoots which are 19.1 cm long with 14 leaves. The 1-year-old shoots, secondary branches and the deciduous fruiting shoots all grow crookedly, so it is also called Longzhuazao (Dragon-Foot Jujube) or Quzhizao (Curly-Branch Jujube). The small leaves are long oval-shaped or ovate-lanceolate. The number of flowers is small. The medium-sized blossoms have a diameter of 7 mm.

Biological Characters The tree has moderate vigor with strong growth vigor for the 1-year-old shoots and strong branching ability. It blooms and bears early with a high and stable yield. In Taigu County of Shanxi Province, it matures in late September. Fruit-crack easily occurs if it rains in maturing stage.

Fruit Characteristics The medium-Sized obovate fruit has a vertical and cross diameter of 3.73 cm and 2.60 cm, averaging 11.6 g with a regular size. It has medium-thick and purplish-red skin and a smooth surface. The greenish-white thick flesh is rough and sweet with less juice and a medium quality for dried fruits and candied fruits. The percentage of edible part of fresh fruit is 95.3%, and SSC, TTS, TA and Vc is 28.80%, 21.30%, 0.59% and 367.88 mg per 100 g fresh fruit. The rate of fresh fruits which can be made into dried ones is 50%. The small stone is long spindle-shaped, averaging 0.55 g.

Evaluation The branches of Dalilongzao cultivar grow crookedly with very special shapes, which make it valuable for ornament. It can be planted in family gardens for ornament or used as bonsai. It has a smaller crookedness than Longzao. The cultivar has strong adaptability with a high yield.

龙 枣

品种来源及分布 又名龙须枣、曲枝枣、蟠龙枣、龙爪枣。广泛分布于北京、河北献县、河南淇县、山东乐陵、庆云、夏津、泰安等地。多作为观赏树于庭院和四旁零星栽植。栽培历史不详。

植物学性状 树体较小，树姿开张，干性弱，枝条密，树冠呈自然圆头形。主干皮裂浅、条状、较易剥落。枣头紫红色或紫褐色，平均生长量52.2cm，节间长6.8cm，枝条弯曲或盘圈生长。二次枝、枣吊和果实也弯曲生长，二次枝长16.3cm，自然生长3～5节。针刺不发达。皮目小而圆，凸起，不开裂，分布中密。枣股小，圆柱形，抽吊力中等，一般抽生枣吊3～4个。枣吊细，较长，长23.7cm，着叶13片。叶片小，叶长5.0cm，叶宽2.2cm，卵状披针形，浓绿色，较厚，先端渐尖，叶基楔形，叶缘锯齿细。花量少，枣吊着花25～35朵，每花序着花5朵左右。花较大，花径7mm左右，昼开型。

生物学特性 树势较弱，成枝力中等，枣头生长细弱，生长量小，嫁接苗开花结果较迟，一般第三年开始结果。坐果率低，结实力差，产量极低。在山西太谷地区，4月中旬萌芽，6月初始花，9月下旬果实成熟。果实生育期110d左右，为晚熟品种类型。果实抗病、抗裂果。

果实性状 果实小，细腰扁柱形，多偏斜，纵径2.73cm，横径1.99cm，单果重4.7g，大小较整齐。果梗细长，梗洼窄，较深。果顶凹，柱头遗存，不明显。果皮厚，深红色，果面不平滑，果实中部细腰状。果点小而密，浅黄色，不明显。果肉厚，绿白色，肉质较硬，味较甜，汁液少，可制干，但干枣品质差。鲜枣可食率91.0%，含可溶性固形物34.50%，总糖27.68%。核中大，长纺锤形，核重0.42g，核尖中长，核纹中深，不含种仁。

评价 该品种丰产性差，果实小，品质较差，食用价值不高。树体矮小，枝形奇特，有很高的观赏价值。可庭院栽培或制作盆景。

Longzao

Source and Distribution The cultivar widely spreads in the Forbidden City of Beijing, Xianxian of Hebei Province, Qixian of Henan Province, Leling, Qingyun, Xiajin and Taian in Shandong Province. It is mostly Planted in family gardens or surroundings as an ornamental plant.

Botanical Characters The small tree is spreading with a weak central leader trunk, dense branches and a natural-round crown. The purplish-red or purplish-brown 1-year-old shoots are 52.2 cm long with the internodes of 6.8 cm. The branches grow crookedly or coiled. The secondary branches also grow crookedly, averaging 16.3 cm long with 3～5 nodes and less-developed thorns. The small lenticels are round, protuberant, medium-dense and not cracked. The small column-shaped mother fruiting spurs can germinate 3～4 thin and long deciduous fruiting shoots, which also grow crookedly, averaging 23.7 cm long with 13 leaves. The small thick leaves are ovate-lanceolate and dark green. The number of flowers is small, averaging 25～35 ones per deciduous fruiting shoot and 5 ones per inflorescence. The daytime-bloomed large blossoms have a diameter of 7 mm.

Biological Characters The tree has weak vigor and medium branching ability. The 1-year-old shoots have weak growth vigor and small growth accretion. The grafted seedlings bear late with low fruit set, low fruiting ability and a very low yield. It is a late-ripening variety with strong resistance to diseases and fruit-cracking.

Fruit Characteristics The small deflective fruit is flat column-shaped with a thin waist, averaging 4.68 g with a regular size. It has dark-red thick skin and an unsmooth surface. The middle part of the fruit is thinner. The greenish-white thick flesh is hard and sweet with less juice. It can be used for making dried fruits, yet with a poor quality. The percentage of edible part of fresh fruit is 91.0%, and SSC and TTS is 34.50% and 27.68%. The medium-sized stone is long spindle-shaped, averaging 0.42 g.

Evaluation The cultivar has a small fruit size with a poor quality and low edible value. The tree has a small size and peculiar branches, which makes it valuable for ornament. It can be planted in family gardens or making bonsai.

观 赏

Ornamental Varieties

品 种

大 柿 饼 枣

品种来源及分布 原产和分布于山东省宁阳县、肥城市、兖州市等地，数量极少。

植物学性状 树体较小，树姿半开张，干性强，枝叶较密，树冠呈圆锥形。主干深灰褐色，树皮裂纹深，窄条状，不易剥落。枣头黄褐色，稍有光泽，平均长81.2cm，节间长8.7cm，很少被覆蜡质。二次枝生长较差，平均长25.4cm，节数5节。无针刺。枣股圆柱形，平均抽生枣吊4.0个。枣吊长21.0cm，着叶11片，常有二次生长。叶片椭圆形，中大，质厚，深绿色，富光泽，先端较宽，钝尖，叶基圆形，叶缘锯齿钝，细疏，齿角圆，裂刻较深。花量多，花序平均着花11朵，花中大，昼开型。花具有特殊性，70%左右花的柱头为3裂，子房分室不明显，中间隔膜消失。

生物学特性 适应性较强，树势较弱，要求较深厚的土壤条件。树势中庸，发枝力较强。结果能力较强，结果早，嫁接苗定植后2年开始结果。成龄树枣吊一般结果1～2个。在山西太谷地区，4月中旬发芽，5月下旬始花，9月下旬果实成熟，果实生育期110d左右，为晚熟品种类型。抗裂果能力极强。

果实性状 果中大，扁圆形，如柿饼状或蟠桃，纵径2.13cm，横径3.30cm，单果重9.6g，大小极不整齐，小果仅5g左右，大果可达20g以上。果肩平，宽圆。梗洼浅、广。果柄短细，长约2mm，粗不足1mm。果顶凹陷，较深。果面不平，有隆起和8～10条纵行沟纹。果皮厚，红色，光泽较差。果肉绿色，质地致密，汁液少，味甜酸，鲜食品质差。鲜枣100g果肉维生素C含量710.72mg。小果基本无核或残核，部分大果具有果核，核短小，陀螺状，核壳软，易破碎。核内无仁。

评价 该品种适应性较强，树势较弱，易坐果，产量中等较稳定。品质较差，维生素C含量极高。果实大小、成熟期均极不一致，不适于商品栽培。该品种果形奇特，形似蟠桃，可用于观赏果实栽培。

Dashibingzao

Source and Distribution The cultivar originated from and spreads in Ningyang, Feicheng and Yanzhou in Shandong Province, with a very small quantity.

Botanical Characters The small tree is half-spreading with a strong central leader trunk, dense branches and a conical crown. The yellowish-brown 1-year-old shoots are a little glossy, averaging 81.2 cm long. The internodes are 8.7 cm long, almost without wax and thorns. The secondary branches are 25.4 cm long with 5 nodes. The column-shaped mother fruiting spurs can germinate 4 deciduous fruiting shoots which are 21 cm long with 11 leaves. The deciduous fruiting shoots often have secondary growth. The oval-shaped leaves are medium large, thick, dark green and glossy, with a wide blunt-cuspate apex, a round base and a blunt, thin and deep saw-tooth pattern on the margin. There are many medium-sized special flowers produced, averaging 11 ones per inflorescence. The stigmas of 70% of the total flowers crack into 3 pieces with indistinct chambers in the ovary and disappeared membrane in the middle part.

Biological Characters The tree has strong adaptability, weak branching ability and weak vigor. It requires deep soils. The tree bears early, with strong productivity. The grafting seedlings begin fruiting in the 2nd year after planting. In Taigu County of Shanxi Province it germinates in mid-April, begins blooming in late May and matures in late September. It is a late-ripening variety with very strong resistance to fruit-cracking.

Fruit Characteristics The medium-sized oblate fruit looks like a dried persimmon, averaging 9.6 g (minimum 5 g and maximum over 20 g) with irregular sizes. It has a flat shoulder, a shallow and wide stalk cavity, a stalk of 2 mm long and less than 1 mm thick, an unsmooth surface with 8 or 10 vertical rills, and thick red skin with poor gloss. The light-green flesh is tight-textured, sour and sweet, with less juice. It has a low quality for fresh-eating. The content of Vc in fresh fruit is 710.72 mg per 100 g fresh fruit. The short and small stone is top-shaped, with a soft shell, easily broken. Some small fruits even have no stones. There are no kernels inside the stones.

Evaluation The cultivar has strong adaptability and weak vigor, with strong fruit-set ability, a medium and stable yield. The fruit has a low quality and irregular sizes and is unsuitable for commercial cultivation. Yet it has a special fruit shape, just like flat peaches. So it can be cultivated as an ornamental plant.

观 赏
Ornamental Varieties
品 种

葫芦枣

品种来源及分布　主要分布于山西襄汾、稷山、闻喜和河南内黄、淇县等地。栽培数量不多，栽培历史和起源不详。据说20世纪70年代资源调查时发现山西省襄汾县城关新城庄有几百年该品种的老枣树，当地群众也证实了这一说法，但已无现存古树。

植物学性状　树体较小，树姿半开张，树冠呈自然半圆形。主干灰褐色，皮呈宽条形纵裂，容易剥落。枣头红褐色，平均长96.6cm，粗1.0cm，平均节间长7.9cm，蜡质少。二次枝平均长29.0cm，6节，弯曲度中等。针刺细短，不发达。枣吊平均长32.8cm，着叶20片。叶片中大，卵状披针形，绿色，先端渐尖，较短，先端尖圆，叶基圆楔形，叶缘具锐齿。花序平均着花10.7朵。

生物学特性　风土适应性较强，树势中等，发枝力较强，枣头生长势中庸。定植后第二年开始结果，成龄树产量中等而稳定，50年生树可产鲜枣50kg。在山西太谷地区，4月中旬萌芽，5月下旬始花，9月中旬进入成熟期，果实生育期105d左右，为中熟品种类型。

果实性状　果实中大，绝大多数枣果中部有明显的缢痕，上部粗而下部细，形状似葫芦而得名，也有少数为圆锥形，似纺锤辣椒形，最大纵径3.52cm，横径2.74cm，单果重10.3g，大小较整齐。果肩较小，平圆。梗洼中深，狭窄。果顶尖。果皮红色，较薄。果点小，中密。果肉白色，肉质细腻，酥脆多汁，味酸甜，适宜鲜食和观赏，鲜食品质上等。鲜枣可食率93.6%，含可溶性固形物30.00%，总糖28.48%，酸0.43%。果核椭圆形，核重0.66g。核内多数含种子，含仁率96.7%。

评价　该品种树体较小，树势中庸，产量一般。果个中大，形似葫芦状，且极具观赏价值，品质上等，适宜鲜食，在交通便利地区和城郊可适量发展，是鲜食和观赏兼用的多用途优良品种。

Huluzao

Source and Distribution　The cultivar mainly spreads in Xiangfen, Jishan, Wenxi of Shanxi Province and Neihuang, Qixian of Henan Province, with a small quantity and unknown origin and history. It is said that a tree of several hundreds of years old for this cultivar was found in Xiangfen County of Shanxi Province when investigation on jujube resource was being done in 1970s.

Botanical Characters　The small tree is half-spreading with a semi-round crown. The grayish-brown trunk bark has vertical wide-striped fissures, easily shelled off. The reddish-brown 1-year-old shoots are 96.6 cm long and 1.0 cm thick. The internodes are 7.9 cm long with less wax and thin, short and less-developed thorns. The secondary branches are 29.0 cm long with 6 nodes of medium curvature. The deciduous fruiting shoots are 32.8 cm long with 20 leaves. The medium-sized green leaves are ovate-lanceolate with a gradually-cuspate short apex, a round-cuneiform base and a sharp saw-tooth pattern on the margin. Each inflorescence has 10.7 flowers.

Biological Characters　The tree has strong adaptability, moderate vigor and strong branching ability with moderate growth potential for the 1-year-old shoots. It generally bears in the 2nd year after planting with a medium and stable yield. A 50-year-old tree has a yield of 50 kg on average. In Taigu County of Shanxi Province, it enters the maturing stage in mid-September. It is a mid-ripening variety.

Fruit Characteristics　The medium-sized fruit has a vertical and cross diameter of 3.52 cm and 2.74 cm, averaging 10.3 g with a regular size. Most fruits have an obvious constriction in the middle part, thicker in the upper part and thinner in the lower part, which makes it like a gourd. That is why it is called Huluzao, for Hulu in Chinese means gourd. It has a small and flat-round shoulder, a medium-deep and narrow stalk cavity, a pointed fruit apex and thin red skin with small medium-dense dots. The white flesh is crisp, juicy, sour and sweet. It has a very good quality for fresh eating, also used for ornament. The percentage of edible part of fresh fruit is 93.6%, and SSC, TTS and TA is 30.00%, 28.48% and 0.43%. The oval-shaped stone weighs 0.66 g. Most stones contain kernels, the percentage of which is 96.7%.

Evaluation　The small plant of Huluzao cultivar has moderate tree vigor and medium productivity. The medium-sized fruit is gourd-shaped with a very good quality for fresh-eating and a high value for ornament. It can be developed in areas with convenient transportation or in suburban areas.

观赏
Ornamental Varieties

羊 奶 枣

品种来源及分布 别名牛奶枣、狗巴枣。原产和分布于陕西西安近郊和大荔县石槽、八渔、苏村等地，多零星栽植。栽培历史不详。

植物学性状 树体中大，树姿开张，枝条较密，树冠呈乱头形，主干皮裂条状，不易剥落。枣头黄褐色，平均生长量66.0cm，节间长7.7cm。二次枝长27.8cm，平均6节。针刺不发达。皮目小，椭圆形，灰白色。枣股中大，圆柱形，抽吊力中等，一般抽生枣吊2～5个。枣吊平均长22.2cm，着叶12片。叶片中大，叶长7.4cm，叶宽2.9cm，卵状披针形，先端锐尖，叶基偏斜，叶缘锯齿粗。花量特多，每花序着花13朵以上。花中大，花径7mm左右，昼开型。

生物学特性 树势中等，枝条生长健壮，萌芽率高，成枝力较强。结果较早，一般栽后第二年开始结果，15年后进入盛果期，产量低，不稳产。花期需要较高温度，低温易造成坐果率下降。在山西太谷地区，9月中旬果实成熟。果实生育期100d左右，为中熟品种类型。成熟期遇雨易裂果。

果实性状 果实中大，长葫芦形，果顶1/4左右处有不明显的缢痕，纵径4.20～4.50cm，横径2.00～2.40cm，单果重13.4g，大小较整齐。果梗细，中长，梗洼浅而窄。果顶尖，柱头明显。果皮薄，红色，果面较光滑。果点小而密，圆形，浅黄色，较明显。果肉厚，绿白色，质细，肉脆，味甜，汁液多，鲜食品质上等，适宜鲜食和观赏。鲜枣可食率96.9%，100g果肉维生素C含量361.90mg。果核小，长纺锤形，纵径2.10～2.20cm，横径0.60～0.70cm，核重0.42g。核纹浅，核尖特长，核内无种仁。

评价 该品种开花结果早，果实较大，果形细长，鲜食品质优异。但结果少，产量低，成熟期易裂果，不耐贮运，不宜用作生产栽培。但枣果形状奇特美观，可供观赏利用。

Yangnaizao

Source and Distribution The cultivar originated from and spreads in the suburbs of Xian City and Shicao, Bayu, Sucun in Dali County of Shaanxi Province. It is mainly dispersedly planted with an unknown history.

Botanical Characters The medium-sized tree is spreading with dense branches and an irregular crown. The trunk bark has striped fissures, uneasily shelled off. The yellowish-brown 1-year-old shoots are 66.0 cm long with the internodes of 7.7 cm. The secondary branches are 27.8 cm long with 6 nodes and less-developed thorns. The small lenticels are grayish white and oval-shaped. The medium-sized column-shaped mother fruiting spurs can germinate 2～5 deciduous fruiting shoots which are 22.2 cm long with 12 leaves. The medium-sized leaves are ovate-lanceolate, 7.4 cm long and 2.9 cm wide, with a sharply-cuspate apex, a deflective base and a thick saw-tooth pattern on the margin. The number of flowers is extremely large, averaging 13 ones per inflorescence. The medium-sized blossoms have a diameter of 7 mm. It blooms in the daytime.

Biological Characters The tree has moderate vigor, strong branches, high germination rate and strong branching ability, bearing early (generally in the 2nd year after planting) with a very low yield for a mature tree. In Taigu County of Shanxi Province, it matures in mid-September with a fruit growth period of 100 days. It is a mid-late-ripening variety with serious fruit-cracking if it rains in the maturing stage.

Fruit Characteristics The medium-sized fruit is long calabash-shaped with a constriction at 1/4 of the fruit apex. It has a vertical and cross diameter of 4.20～4.50 cm and 2.00～2.40 cm, averaging 13.4 g with a regular size. It has a medium-long and thin stalk, a shallow and narrow stalk cavity, a cuspate fruit apex, a distinct stigma, thin red skin and a smooth surface with small, dense, round, light-yellow and distinct dots. The greenish-white thick flesh is delicate, crisp, sweet and juicy. It has a good quality for fresh eating and is suitable for fresh eating and ornament. The percentage of edible part of fresh fruit is 96.9%, and the content of Vc is 361.90 mg per 100 g fresh fruit. The small stone is long spindle-shaped, averaging 0.42 g.

Evaluation The cultivar blooms and bears early with a large fruit size, thin and long fruit shape and an excellent quality for fresh eating. Yet it has poor fruiting ability with a low yield, serious fruit-cracking in the maturing stage and poor tolerance to storage and transport. It is not suitable for commercial production. Yet the fruit has a peculiar and attractive shape, which makes it valuable for ornament.

观 赏
Ornamental Varieties
品 种

· 501 ·

柿 顶 枣

品种来源及分布 又名柿蒂枣、柿萼枣、柿花枣、柿把枣。分布于陕西省大荔县石槽乡三教、王马、马二等村。数量不多，栽培和起源历史不详。

植物学性状 树体中大，树姿开张，树冠自然半圆形。主干皮裂中深，不易剥落。枣头红褐色，皮目中大，椭圆形，分布较密，凸起，不开裂。二次枝生长3～9节。针刺不发达。枣股中大，圆锥形，抽吊力中等，一般抽生枣吊2～4个，多者达6个。枣吊长11～20cm。叶片中大，叶长4.5～5.2cm，叶宽2.3～2.6cm，长卵形，有皱褶现象，先端渐尖，叶基圆形，叶缘锯齿细，中密。花朵开放状态特殊，初花期普遍存在雄蕊优先伸出花蕾生长的现象。

生物学特性 树势中等，萌发力较强，坐果率较低，平均吊果率仅37.0%，产量较低。在山西太谷地区，9月中旬果实成熟。果实生育期100～110d，为中熟品种类型。果实抗裂果和抗病性能强。

果实性状 果实中大，柱形，纵径3.50cm，横径2.90cm，单果重12.0g，最大14.7g，大小不整齐。果梗中长，萼片宿存，随果实发育而逐渐肉质化，呈五角形，盖住梗洼和果肩，形如柿萼，故又名柿萼枣。果顶凹，不平整。果皮厚，深红色，果面平滑。果点小而圆，分布稀，不明显。果肉较厚，乳白色，肉质较脆，味甜，汁液少，品质中等，可制干和观赏。鲜枣可食率94.8%，100g果肉维生素C含量473.29mg。果核较小，短纺锤形，略弯曲，纵径1.80cm，横径0.90cm，核重0.63g，核尖短，核纹粗细不一，含仁率高。

评价 该品种为枣树的特殊类型，部分花萼肥大宿存，极适宜观赏栽培。作为栽培品种利用的经济价值不大，为性状特异的种质资源。

Shidingzao

Source and Distribution The cultivar, also called Shidingzao, Shiezao or Shihuazao, spreads in Sanjiao, Wangma and Maer in Shicao Village of Dali County, Shaanxi Province. It has a small quantity with an unknown history and orgin.

Botanical Characters The medium-sized tree is spreading with a semi-round crown. The trunk bark has medium-deep fissures, uneasily shelled off. The reddish-brown 1-year-old shoots have medium-large and oval-shaped, densely-distributed, protuberant and uncracked lenticels. There are 3～9 nodes on the secondary branches with less-developed thorns. The medium-sized conical-shaped mother fruiting spurs can germinate 2～4 (maximum 6) deciduous fruiting shoots which are 11～20 cm long. The medium-sized leaves are long oval-shaped with some wrinkling. The flowers have a special blooming situation. At the beginning of blooming, the stamen will emerge from the unopened blossom when growing.

Biological Characters The tree has moderate vigor and strong germination ability, with low fruit set and a low yield. The percentage of fruits to deciduous fruiting shoots is only 37.0% on average. In Taigu County of Shanxi Province, it matures in mid-September with a fruit growth period of 100～110 d. It is a mid-ripening variety with strong resistance to fruit-cracking and diseases.

Fruit Characteristics The medium-sized fruit is column-shaped, with a vertical and cross diameter of 3.50 cm and 2.90 cm, averaging 12.0 g (maximum 14.7 g) with irregular sizes. It has a medium-long stalk and a persistent calyx, which will become fleshy and five-star-shaped with the growth of the fruit, covering the stalk cavity and fruit shoulder, just like the calyx of persimmon. That is why it is also called Shiezao (shi means persimmon, e means calyx). The fruit has a sunken and unsmooth fruit apex, dark-red thick skin and a smooth surface with small, round, sparse and indistinct dots. The ivory-white thick flesh is crisp and sweet, with less juice and a medium quality. It can be used for dried fruits or ornament. The percentage of edible part of fresh fruit is 94.8%, and the content of Vc is 473.29 mg per 100 g fresh fruit. The small stone is short spindle-shaped, slightly curly with a vertical and cross diameter of 1.80 cm and 0.90 cm, averaging 0.63 g with a short apex and thin or thick veins.

Evaluation The cultivar is a special type of jujube with some large and persistent calyx and is valuable for ornament. The economic value as a cultivation variety is low, yet it can be collected and preserved as a special germplasm.

观 赏
Ornamental Varieties
品 种

·503·

临猗辣椒枣

品种来源及分布 原产山西临猗等地。多为零星栽植。栽培历史不详。

植物学性状 树体中大，树姿开张，树冠自然半圆形。主干皮裂深、条状，易剥落。枣头红褐色，平均生长量66.2cm，节间长7.3cm左右。二次枝长27.8cm，平均6节，无针刺。皮目小，圆形或椭圆形，凸起，开裂，灰白色，分布密。枣股较大，圆锥形，抽吊力中等，一般抽生枣吊3～4个。枣吊长24.2cm左右，着叶20片。叶片中大，叶长6.4cm，叶宽3.0cm，椭圆形，深绿色，叶片合抱，先端钝尖，叶基圆形或楔形，叶缘锯齿钝，较密。花量多，花径6.3mm，昼开型。

生物学特性 树势强健，枣头枝生长旺盛，发枝力较弱。结果较迟，产量较高而稳定。成龄树最高株产可达100kg。在山西太谷地区，9月底至10月初果实成熟。果实生育期110d左右，为晚熟品种类型。果实抗裂果和抗病性较强。

果实性状 果实较大，长锥形，似辣椒形，纵径4.10cm，横径2.60cm，单果重20.8g，大小较整齐。果梗中长，较粗，梗洼窄而较深，果顶尖。果皮厚，紫红色，果面凹凸不平。果点小而圆，分布密，浅黄色，较明显。果肉厚，白绿色，肉质致密、细脆，味甜，汁液中多，品质较差，主要用于观赏。鲜枣可食率93.3%，含可溶性固形物28.00%，总糖19.25%，酸0.36%，100g果肉维生素C含量288.68mg；果皮含黄酮5.59mg/g，cAMP含量296.22μg/g。果核大，长纺锤形，纵径2.2cm，横径0.57cm，核重1.40g，核纹浅，核尖长，含仁率12.2%。

评价 该品种适应较强，树体中大，树势强健，产量高而稳定。果实形状奇特，亮丽美观，可作为观赏品种利用。

Linyilajiaozao

Source and Distribution The cultivar originated from Linyi County of Shanxi Province with an unknown history. It is dispersedly planted.

Botanical Characters The medium-sized tree is spreading with a semi-round crown. The trunk bark has deep striped fissures, easily shelled off. The reddish-brown 1-year-old shoots are 66.2 cm long with the internodes of 7.3 cm. The secondary branches are 27.8 cm long with 6 nodes and without thorns. The small lenticels are grayish white, round or oval-shaped, protuberant and cracked, densely-distributed. The large conical-shaped mother fruiting spurs can germinate 3～4 deciduous fruiting shoots, which are 24.2 cm long with 20 leaves. The medium-sized leaves are oval-shaped and dark green, curling toward the center, 6.4 cm long and 3.0 cm wide, with a bluntly-cuspate apex, a round or cuneiform base and a blunt and dense saw-tooth pattern on the margin. There are many flowers with a diameter of 6.3 mm produced. It blooms in the daytime.

Biological Characters The tree has strong vigor and weak branching ability with strong growth vigor for the 1-year-old shoots. It bears late with a high and stable yield. A mature tree has a yield of 100 kg at most. In Taigu County of Shanxi Province, it matures at the end of September or at the beginning of October with a fruit growth period of 110 d. It is a late-ripening variety with strong resistance to fruit-cracking and diseases.

Fruit Characteristics The large fruit is long conical-shaped, just like pepper, with a vertical and cross diameter of 4.10 cm and 2.60 cm, averaging 20.8 g with a regular size. It has a medium-long and thick stalk, a narrow and deep stalk cavity, a cuspate fruit apex, purplish-red thick skin and a jagged surface with small, round, dense, light-yellow and distinct dots. The greenish-white thick flesh is tight-textured, crisp and sweet with medium juice and a poor quality, mainly used for ornament. The percentage of edible part of fresh fruit is 93.3%, and SSC, TTS, TA and Vc is 28.00%, 19.25%, 0.36% and 288.68 mg per 100 g fresh fruit. The content of flavones and cAMP in mature fruit skin is 5.59 mg/g and 296.22 μg/g. The small stone is long spindle-shaped, with a vertical and cross diameter of 2.2 cm and 0.57 cm, averaging 1.40 g with shallow veins and a long apex.

Evaluation The medium-sized plant of Linyilajiaozao cultivar has strong adaptability and strong vigor with a high and stable yield. The fruit has a peculiar shape and a bright and attractive appearance, which can be used as an ornamental plant.

观 赏
Ornamental Varieties
品 种

大果算盘枣

品种来源及分布　原产于湖南溆浦的低庄镇、花桥乡一带，数量较多，各村都有栽培。

植物学性状　树体较大，树姿开张，枝较稀，树冠呈乱头形。主干深灰褐色，树皮宽条纵裂，易剥落。枣头红褐色，平均长80.3cm，粗1.10cm，节间长6.6cm，蜡质多。二次枝长22.0cm，6节，弯曲度中等。枣股圆锥形，平均抽生枣吊4.0个。枣吊长20.3cm，着叶18片。叶片小，卵圆形，深绿色，先端渐尖，叶基偏圆形，叶缘具钝锯齿。花量较大，花序平均着花5朵。花中大，昼开型。

生物学特性　适应性中等，抗病力较强。树势和发枝力均弱。较丰产稳定。定植后3年开始结果，15年左右进入盛果期，成龄树一般株产35kg。在山西太谷地区，4月下旬萌芽，5月底始花，10月上旬果实成熟，果实生育期120d左右，为极晚熟品种类型。抗裂果。

果实性状　果实大，扁圆形，纵径2.30cm，横径2.93cm，单果重17.1g，大小较整齐。果肩平，梗洼浅广。果顶圆，顶点凹陷。果皮厚，红色，光滑，果实外观好。果肉浅绿色，质地致密，汁液较少，味甜，鲜食品质差，可用于加工和观赏。鲜枣可食率90.1%，含总糖24.14%，酸0.43%，100g果肉维生素C含量451.78mg；果皮含黄酮25.85mg/g，cAMP含量77.04μg/g。制干率37.0%。果核大，圆形，核重1.69g，含仁率18.3%。

评价　该品种适应性中等，抗病力强，较丰产稳产。果实外观奇特，抗裂果，肉质致密少汁，核大，可食率低，鲜食和制干品质差，不宜生产栽培，可观赏利用。

Daguosuanpanzao

Source and Distribution　The cultivar originated from and spreads in Dizhuang and Huaqiao villages of Xupu County in Hunan Province, with a large quantity in different areas of the villages.

Botanical Characters　The large tree is spreading with sparse branches and an irregular crown. The dark grayish-brown trunk bark has vertical, wide-striped fissures, easily shelled off. The reddish-brown 1-year-old shoots are 80.3 cm long and 1.10 cm thick. The internodes are 6.6 cm long with much wax. The secondary branches are 22.0 cm long with 6 nodes of medium curvature. The conical mother fruiting spurs can germinate 4 deciduous fruiting shoots which are 20.3 cm long with 18 leaves. The small oval-shaped leaves are dark green, with a gradually-cuspate apex, a deflective-round base and a blunt saw-tooth pattern on the margin. There are many medium-sized flowers produced, averaging 5 ones per inflorescence. It blooms in the daytime.

Biological Characters　The tree has moderate adaptability, strong resistance to diseases, weak vigor and weak branching ability, with a high and stable yield. It generally bears in the 3rd year after planting, and enters the full-bearing stage in the 15th year with an average yield for a mature tree of 35 kg. In Taigu County of Shanxi Province, it germinates in late April, begins blooming in late May and matures in early October with a fruit growth period of 120 d. It is an extremely late-ripening variety with strong resistance to fruit-cracking.

Fruit Characteristics　The large oblate fruit has a vertical and cross diameter of 2.30 cm and 2.93 cm, averaging 17.1 g with a regular size. It has a flat shoulder, a shallow and wide stalk cavity, a round fruit apex, a smooth surface, thick red skin and an attractive appearance. The light-green flesh is tight-textured and sweet, with less juice. It has a low quality for fresh-eating, yet it can be used for making dried fruits and for ornament. The percentage of edible part of fresh fruit is 90.1%, and the content of TTS, TA and Vc is 24.14%, 0.43% and 451.78 mg per 100 g fresh fruit. The content of flavones and cAMP in mature fruit skin is 25.85 mg/g and 77.04 μg/g. The percentage of fresh fruits which can be made into dried ones is 37.0%. The large round stone weighs 1.69 g, and the percentage of containing kernels is 18.3%.

Evaluation　The cultivar has moderate adaptability and strong resistance to diseases with a high and stable yield. The fruit has an attractive appearance, strong resistance to fruit-cracking, tight-textured and less juicy flesh, a large stone and low edibility. It has a low quality for fresh-eating and dried fruits and is unsuitable for commercial production. Yet it can be used as an ornamental plant.

观 赏
Ornamental Varieties
品 种

大叶无核枣

品种来源及分布 原产于河南省内黄县、浚县，数量极少。

植物学性状 树体中大，树姿开张，干性较弱，树冠自然圆头形。主干灰褐色，树皮裂纹浅，条片状，较平滑。枣头红褐色，较粗壮，平均长65.0cm，粗1.00cm，节间长8.0cm，顶芽延续生长能力强。二次枝节数6节。针刺细短，直刺长0.8cm。枣股圆柱形，一般抽生枣吊5个左右。枣吊长22.8cm，着生叶片14片，少数枣吊花后有二次生长，吊长可达42.0cm，全枝着叶18片。叶片特大，长10.1cm，叶宽7.6cm，卵圆形，厚，深绿色，光亮，不平展，先端尖圆，叶基心形，叶缘波状形，锯齿粗大。花为多花型。

生物学特性 适应性较弱，不耐瘠薄，要求土层深厚、肥沃。树势较弱，发枝力差。坐果性能极差，产量低。在山西太谷地区，4月中旬萌芽，5月下旬始花，10月上旬成熟采收，果实生育期120d左右，为极晚熟品种类型。

果实性状 果实中大，扁圆形，纵径3.03cm，横径3.19cm，单果重13.2g，最大23.1g，大小不整齐。果肩凸。梗洼深、窄。果顶广圆，顶点凹陷，浅沟状。果皮浅红色，光亮美观。果点小，中密。果肉厚，白色，质地疏松，汁液中多，味甜，鲜食品质中等。鲜枣含可溶性固形物32.40%，总糖24.70%，酸0.33%。果核退化，核壳薄膜状，不易和果肉分离，可随果肉一起食用。核内无种子。

评价 该品种适应性较差，树势弱，结果晚，产量低，不宜生产栽培。但叶片特大，浓绿，果核退化，是具有一定的观赏价值的稀有无核品种资源。

Dayewuhezao

Source and Distribution The cultivar originated from Neihuang and Junxian in Henan Province.

Botanical Characters The medium-sized tree is spreading with a weak central leader trunk and a natural-round crown. The grayish-brown trunk bark has shallow, smooth, striped fissures. The strong reddish-brown 1-year-old shoots are 65.0 cm long and 1.0 cm thick with the internodes of 8.0 cm long. The terminal bud has strong continuous-growing potential. The thorns are thin and short, 0.8 cm long. There are 6 nodes on each secondary branch. The column-shaped mother fruiting spurs can germinate 5 deciduous fruiting shoots, which are 22.8 cm long with 14 leaves. Some of the deciduous fruiting shoots have secondary growth after flowering, and their length reaches as high as 42.0 cm with 18 leaves. The extremely large leaves are 10.1 cm long and 7.6 cm wide, oval-shaped, thick, uneven, dark green and glossy, with a sharply-round apex, a sub-heart-shaped base and a thick saw-tooth pattern on the wavy margin. The number of flowers is small, averaging 3 ones per inflorescence. The medium-sized blossoms have short sepals, with a multi-florous character.

Biological Characters The tree has weak vigor, weak branching ability, poor adaptability and poor tolerance to infertile soils. It requires deep and fertile soils. It has very low fruit set and a low yield. In Taigu County of Shanxi Province, it germinates in mid-April, begins blooming in late May and matures in early October with a fruit growth period of 120 d. It is an extremely late-ripening variety.

Fruit Characteristics The medium-sized oblate fruit has a vertical and cross diameter of 3.03 cm and 3.19 cm, averaging 13.2 g (maximum 23.1 g) with irregular sizes. It has a protuberant shoulder, a deep and narrow stalk cavity, a flat-round fruit apex, and attractive light-red skin with small and medium-dense dots. The white flesh is thick, loose-textured and sweet, with medium juice. It has medium quality for fresh-eating. The content of SSC, TTS and TA in fresh fruit is 32.40%, 24.70% and 0.33%. The degraded stone has a membrane-shaped shell, uneasily separated from the flesh. It can be eaten together with the flesh. The stone contains no kernel.

Evaluation The cultivar has poor adaptability and weak vigor, bearing late with a low yield. It is unsuitable for commercial cultivation. Yet it has very large and dark-green leaves, and a degraded fruit stone, which make it valuable in ornament and research.

其他品种 Other Varieties

设施枣树（陕西　大荔）
Jujube Trees in the Greenhouse (Dali, Shanxi)

直社枣
Zhishezao

枣强骨头小枣
Zaoqianggutouxiaozao

榆次面枣
Yucimianzao

榆次毛猴枣
Yucimaohouzao

义乌甜酸枣
Yiwutiansuanzao

延川奶枣
Yanchuannaizao

溆浦圆枣
Xupuyuanzao

溆浦尖枣
Xupujianzao

襄汾木枣
Xiangfenmuzao

枣庄贡枣
Zaozhuanggongzao

榆次奶头枣
Yucinaitouzao

榆次大馍枣
Yucidamozao

榆次长木枣
Yucichangmuzao

永城圆红枣
Yongchengyuanhongzao

兖洲三变红
Yanzhousanbianhong

延川牛奶脆
Yanchuanniunaicui

溆浦长枣
Xupuchangzao

盱眙雁来红
Xuyiyanlaihong

新蔡大圆丰
Xincaidayuanfeng

武乡甜枣
Wuxiangtianzao

武乡碳枣
Wuxiangtanzao

历城串铃枣
Lichengchuanlingzao

濮阳三变丑
Puyangsanbianchou

韶关大枣
Shaoguandazao

泾渭鲜枣
Jingweixianzao

天津大马牙
Tianjindamaya

深县串杆枣
Shenxianchuanganzao

太谷大酸枣
Taigudasuanzao

束鹿糖枣
Shulutangzao

乾陵大枣
Qianlingdazao

五台木枣
Wutaimuzao

太原秋团枣
Taiyuanqiutuanzao

万荣玻璃脆
Wanrongbolicui

五台绵枣
Wutaimianzao

万荣楼疙瘩
Wanronglougeda

泰安灵枣
Taianlingzao

太谷龙壶枣
Taigulonghuzao

乐陵蚂蛉枣
Lelingmalingzao

韶关白枣
Shaoguanbaizao

山西大令枣
Shanxidalingzao

南京大木枣
Nanjingdamuzao

灵宝灵1号
lingbaoling 1

临泽吊吊婆 Linzediaodiaopo	临猗圆铃枣 Linyiyuanlingzao	临猗甜酸枣 Linyitiansuanzao
临猗马铃枣 Linyimalingzao	库尔勒小枣 Kuerlexiaozao	湖北铃铛枣 Hubeilingdangzao
江苏木枣 Jiangsumuzao	甘肃冬枣 Gansudongzao	朝阳瓶子枣 Chaoyangpingzizao
朝阳麻枣 Chaoyangmazao	朝阳凌枣 Chaoyanglingzao	涿鹿悠悠枣 Zhuoluyouyouzao

其他 品种
Other Varieties

广东珍珠枣
Guangdongzhenzhuzao

哈密大枣
Hamidazao

河南龙枣
Henanlongzao

大荔铃铃枣
Dalilinglingzao

湖北鸡心枣
Hubeijixinzao

阜阳晒枣
Fuyangshaizao

运城蛤蟆枣
Yunchenghamazao

榆次面美枣
Yucimianmeizao

新郑小圆枣
Xinzhengxiaoyuanzao

西营笨枣
Xiyingbenzao

五台面枣
Wutaimianzao

五台醋枣
Wutaicuzao

太谷葫芦枣
Taiguhuluzao

运城脆枣
Yunchengcuizao

南谷丰葫芦枣
Nangufenghuluzao

月初（韩国）
Yuechu (Korea)

无等（韩国）
Wudeng (Korea)

桐柏大枣
Tong baidazao

义县木枣 Yixianmuzao	安宁小枣 Anningxiaozao	北京老虎眼 Beijinglaohuyan
新疆长圆枣 Xinjiangchangyuanzao	太原十月红 Taiyuanshiyuehong	嵊县白蒲枣 Shengxianbaipuzao
赞皇紫铃蛋 Zanhuangzilingdan	山西寿星枣 Shanxishouxingzao	景泰大枣 Jingtaidazao
清徐葫芦枣 Qingxuhuluzao	临猗胜利枣 Linyishenglizao	泗洪大枣 Sihongdazao

朝阳金丝蜜
Chaoyangjinsimi

佳县白枣
Jiaxianbaizao

阿拉尔圆脆枣
Alaeryuancuizao

昆明枣
Kunmingzao

朝阳软核枣
Chaoyangruanhezao

佳县细腰腰枣
Jiaxianxiyaoyaozao

怀柔脆枣
Huairoucuizao

阜阳蚂蚁枣
Fuyangmayizao

京南大白枣
Jingnandabaizao

根德大枣
Gendedazao

朝阳小圆铃
Chaoyangxiaoyuanling

朝阳大尖顶
Chaoyangdajianding

运城绵枣 Yunchengmianzao　　苏子峪大枣 Suziyudazao　　临猗牙枣 Linyiyazao

献县辣椒枣 Xianxianlajiaozao　　北京大红袍枣 Beijingdahongpaozao　　佳县牙枣 Jiaxianyazao

朝阳小尖枣 Chaoyangxiaojianzao　　朝阳小平顶 Chaoyangxiaopingding　　朝阳牛心枣 Chaoyangniuxinzao

鲍庄大铃枣
Baozhuangdalingzao

北京大酸枣
Beijingdasuanzao

北京嘎嘎枣
Beijinggagazao

北京黑腰枣
Beijingheiyaozao

北京花生枣
Beijinghuashengzao

朝阳秤砣枣
Chaoyangchengtuozao

朝阳晚枣
Chaoyangwanzao

福枣(韩国)
Fuzao (Korea)

绵城(韩国)
Miancheng (Korea)

品种索引 Varieties Index

A

阿克苏小枣（Akesuxiaozao）/244
安阳团枣（Anyangtuanzao）/302

B

八升胡（Bashenghu）/342
斑枣（Banzao）/382
板枣（Banzao）/292
保德小枣（Baodexiaozao）/358
保德油枣（Baodeyouzao）/122
北碚小枣（Beibeixiaozao）/274
北京白枣（Beijingbaizao）/6
北京笨枣（Beijingbenzao）/384
北京鸡蛋枣（Beijingjidanzao）/16
北京坠子白（Beijingzhuizibai）/94
扁核酸（Bianhesuan）/136
彬县黑疙瘩（Binxianheigeda）/184
彬县耙齿枣（Binxianpachizao）/324
彬县酸疙瘩（Binxiansuangeda）/364
彬县圆枣（Binxianyuanzao）/312
不落酥（Buluosu）/18

C

沧县傻枣（Cangxianshazao）/380
沧县屯子枣（Cangxiantunzizao）/406
沧县小枣（Cangxianxiaozao）/400
沧县长小枣（Cangxianchangxiaozao）/416
茶壶枣（Chahuzao）/484
成武冬枣（Chengwudongzao）/36

D

大白铃（Dabailing）/10
大瓜枣（Daguazao）/60
大果算盘枣（Daguosuanpanzao）/506
大荔墩墩枣（Dalidundunzao）/376
大荔干尾巴（Daliganweiba）/164
大荔疙瘩枣（Daligedazao）/180
大荔鸡蛋枣（Dalijidanzao）/58
大荔林檎枣（Dalilinqinzao）/362
大荔龙枣（Dalilongzao）/492
大荔马牙枣（Dalimayazao）/68
大荔面枣（Dalimianzao）/356
大荔水枣（Dalishuizao）/466
大荔小墩墩枣（Dalixiaodundunzao）/174
大荔小圆枣（Dalixiaoyuanzao）/428
大荔圆枣（Daliyuanzao）/160
大荔知枣（Dalizhizao）/410
大柿饼枣（Dashibingzao）/496
大叶无核枣（Dayewuhezao）/508
定襄山枣（Dingxiangshanzao）/464
定襄小枣（Dingxiangxiaozao）/420
定襄星星枣（Dingxiangxingxingzao）/308
定襄油荷枣（Dingxiangyouhezao）/340
冬枣（Dongzao）/2
敦煌大枣（Dunhuangdazao）/306

E

鹅子枣（Ezizao）/452

F

蜂蜜罐（Fengmiguan）/38
涪陵鸡蛋枣（Fulingjidanzao）/472
阜阳木头枣（Fuyangmutouzao）/476

G

疙瘩脆（Gedacui）/ 14
姑苏小枣（Gusuxiaozao）/ 432
官滩枣（Guantanzao）/ 130
灌阳短枣（Guanyangduanzao）/ 408
灌阳长枣（Guanyangchangzao）/ 446
广洋枣（Guangyangzao）/ 352

H

合阳铃铃枣（Heyanglinglingzao）/ 74
河津水枣（Hejinshuizao）/ 394
河津条枣（Hejintiaozao）/ 230
核桃纹（Hetaowen）/ 318
衡山长大枣（Hengshanchangdazao）/ 222
衡阳珍珠枣（Hengyangzhenzhuzao）/ 386
洪赵脆枣（Hongzhaocuizao）/ 392
洪赵葫芦枣（Hongzhaohuluzao）/ 346
洪赵十月红（Hongzhaoshiyuehong）/ 178
洪赵小枣（Hongzhaoxiaozao）/ 310
壶瓶枣（Hupingzao）/ 288
葫芦枣（Huluzao）/ 498
葫芦长红（Huluchanghong）/ 488
湖南鸡蛋枣（Hunanjidanzao）/ 8
灰枣（Huizao）/ 290

J

鸡心枣（Jixinzao）/ 146
稷山柳罐枣（Jishanliuguanzao）/ 158
稷山圆枣（Jishanyuanzao）/ 152
稷山长枣（Jishanchangzao）/ 172
佳县密点脆木枣（Jiaxianmidiancuimuzao）/ 200
佳县牙枣（Jiaxianyazao）/ 350
交城端枣（Jiaochengduanzao）/ 328
交城甜酸枣（Jiaochengtiansuanzao）/ 92
金丝小枣（Jinsixiaozao）/ 282
晋枣（Jinzao）/ 294
俊枣（Junzao）/ 166
骏枣（Junzao）/ 286

K

喀什噶尔小枣（Kashigeerxiaozao）/ 436
糠头枣（Kangtouzao）/ 470
孔府酥脆枣（Kongfusucuizao）/ 22

L

辣椒枣（Lajiaozao）/ 44
兰溪马枣（Lanximazao）/ 456
郎家园枣（Langjiayuanzao）/ 32
郎溪牛奶枣（Langxiniunaizao）/ 454
郎枣（Langzao）/ 138
乐陵长木枣（Lelingchangmuzao）/ 170
离石合钵枣（Lishihebozao）/ 264
黎城大马枣（Lichengdamazao）/ 334
黎城小枣（Lichengxiaozao）/ 332
连县苦楝枣（Lianxiankulianzao）/ 238
连县木枣（Lianxianmuzao）/ 448
连县糖枣（Lianxiantangzao）/ 78
林县无头枣（Linxianwutouzao）/ 54
临汾蜜枣（Linfenmizao）/ 28
临汾木疙瘩（Linfenmugeda）/ 232
临汾团枣（Linfentuanzao）/ 298
临汾针葫芦（Linfenzhenhulu）/ 320
临潼轱辘枣（Lintongguluzao）/ 62
临猗笨枣（Linyibenzao）/ 240
临猗脖脖枣（Linyibobozao）/ 254
临猗鸡蛋枣（Linyijidanzao）/ 242
临猗辣椒枣（Linyilajiaozao）/ 504
临猗梨枣（Linyilizao）/ 4
临泽大枣（Linzedazao）/ 150
临泽小枣（Linzexiaozao）/ 148
灵宝大枣（Lingbaodazao）/ 124
灵武长枣（Lingwuchangzao）/ 40
龙枣（Longzao）/ 494
吕梁木枣（Lvliangmuzao）/ 120

M

马连小枣（Malianxiaozao）/ 338
马牙白枣（Mayabaizao）/ 42
密云小枣（Miyunxiaozao）/ 156
民勤小枣（Minqinxiaozao）/ 348

旻枣（Minzao）/ 82
鸣山大枣（Mingshandazao）/ 314
磨盘枣（Mopanzao）/ 482

N

南京冷枣（Nanjinglengzao）/ 100
南京鸭枣（Nanjingyazao）/ 422
南京枣（Nanjingzao）/ 458

内黄苹果枣（Neihuangpingguozao）/ 70
宁阳六月鲜（Ningyangliuyuexian）/ 20
宁阳暄铃枣（Ningyangxuanlingzao）/ 478

P

泡泡红（Paopaohong）/ 256
平陆棒槌枣（Pinglubangchuizao）/ 246
平陆尖枣（Pinglujianzao）/ 34
平顺笨枣（Pingshunbenzao）/ 192
平遥大枣（Pingyaodazao）/ 194
平遥苦端枣（Pingyaokuduanzao）/ 212
婆婆枣（Popozao）/ 226

婆枣（Pozao）/ 126
婆枣枝变1号（Pozaozhibian 1）/ 248
蒲城晋枣（Puchengjinzao）/ 168
蒲城绵枣（Puchengmianzao）/ 398
蒲城圆梨枣（Puchengyuanlizao）/ 402
濮阳糖枣（Puyangtangzao）/ 52
濮阳小枣（Puyangxiaozao）/ 116

Q

清徐圆枣（Qingxuyuanzao）/ 374
清苑大丹枣（Qingyuandadanzao）/ 112

庆云小梨枣（Qingyunxiaolizao）/ 72
曲阜猴头枣（Qufuhoutouzao）/ 442

R

汝城枣（Ruchengzao）/ 262

S

三变红（Sanbianhong）/ 486
山东梨枣（Shandonglizao）/ 26
歙县马枣（Shexianmazao）/ 460

柿顶枣（Shidingzao）/ 502
嵩县大枣（Songxiandazao）/ 344
苏南白蒲枣（Sunanbaipuzao）/ 462

T

胎里红（Tailihong）/ 490
太谷端子枣（Taiguduanzizao）/ 370
太谷墩墩枣（Taigudundunzao）/ 322
太谷黑叶枣（Taiguheiyezao）/ 440
太谷壶瓶酸（Taiguhupingsuan）/ 208
太谷鸡心蜜（Taigujixinmi）/ 30
太谷铃铃枣（Taiguglinglingzao）/ 48
太谷美蜜枣（Taigumeimizao）/ 46
太原驴粪蛋（Taiyuanlvfendan）/ 276
太原圆枣（Taiyuanyuanzao）/ 268

太原长枣（Taiyuanchangzao）/ 278
泰安马铃脆（Taianmalingcui）/ 66
泰安酥圆铃（Taiansuyuanling）/ 368
糖枣（Tangzao）/ 162
滕州大马牙（Tengzhoudamaya）/ 140
滕州大马枣（Tengzhoudamazao）/ 80
滕州落地红（Tengzhouluodihong）/ 252
天津二秋枣（Tianjinerqiuzao）/ 102
天津夵夵枣（Tianjingagazao）/ 56
天津快枣（Tianjinkuaizao）/ 88

W

万荣翠枣（Wanrongcuizao）/ 234
万荣福枣（Wanrongfuzao）/ 250
文水沙枣（Wenshuishazao）/ 330

无核小枣（Wuhexiaozao）/ 142
吴县水团枣（Wuxianshuituanzao）/ 90
武乡牙枣（Wuxiangyazao）/ 218

X

西双版纳小枣（Xishuangbannaxiaozao）/ 266
夏津妈妈枣（Xiajinmamazao）/ 50
夏县圆脆枣（Xiaxianyuancuizao）/ 336
夏县紫圆枣（Xiaxianziyuanzao）/ 154
献县绵枣（Xianxianmianzao）/ 110
献县木枣（Xianxianmuzao）/ 182
献县酸枣（Xianxiansuanzao）/ 424
献县小大枣（Xianxianxiaodazao）/ 196
献县小小枣（Xianxianxiaoxiaozao）/ 404
献县圆小枣（Xianxianyuanxiaozao）/ 378
相枣（Xiangzao）/ 132
香山小白枣（Xiangshanxiaobaizao）/ 412
襄汾崖枣（Xiangfenyazao）/ 210
襄汾圆枣（Xiangfenyuanzao）/ 64
小果算盘枣（Xiaoguosuanpanzao）/ 468
新疆小圆枣（Xinjiangxiaoyuanzao）/ 220
新乐大枣（Xinledazao）/ 128
新郑大马牙（Xinzhengdamaya）/ 438
新郑大枣（Xinzhengdazao）/ 430
新郑鸡蛋枣（Xinzhengjidanzao）/ 372

新郑尖头灰枣（Xinzhengjiantouhuizao）/ 190
新郑九月青（Xinzhengjiuyueqing）/ 106
新郑齐头白（Xinzhengqitoubai）/ 418
新郑酥枣（Xinzhengsuzao）/ 104
新郑长鸡心（Xinzhengchangjixin）/ 202
溆浦薄皮枣（Xupubopizao）/ 186
溆浦秤锤枣（Xupuchengchuizao）/ 236
溆浦秤砣枣（Xupuchengtuozao）/ 188
溆浦观音枣（Xupuguanyinzao）/ 354
溆浦葫芦枣（Xupuhuluzao）/ 96
溆浦米枣（Xupumizao）/ 224
溆浦蜜蜂枣（Xupumifengzao）/ 76
溆浦木枣（Xupumuzao）/ 228
溆浦沙糖枣（Xupushatangzao）/ 390
溆浦柿饼枣（Xupushibingzao）/ 198
溆浦甜酸枣（Xuputiansuanzao）/ 214
溆浦香枣（Xupuxiangzao）/ 360
溆浦岩枣（Xupuyanzao）/ 204
宣城尖枣（Xuanchengjianzao）/ 474
薛城冬枣（Xuechengdongzao）/ 86

Y

亚腰长红（Yayaochanghong）/ 366
延川白枣（Yanchuanbaizao）/ 270
延川跌牙枣（Yanchuandieyazao）/ 434
延川狗头枣（Yanchuangoutouzao）/ 304
延川条枣（Yanchuantiaozao）/ 272
羊奶枣（Yangnaizao）/ 500
义乌大枣（Yiwudazao）/ 450
义乌棉絮枣（Yiwumianxuzao）/ 258
缨络枣（Yingluozao）/ 24

永城长红（Yongchengchanghong）/ 144
永济蛤蟆枣（Yongjihamazao）/ 12
永济鸡蛋枣（Yongjijidanzao）/ 108
榆次九月青（Yucijiuyueqing）/ 114
榆次团枣（Yucituanzao）/ 300
榆次牙枣（Yuciyazao）/ 98
玉田小枣（Yutianxiaozao）/ 326
垣曲枣（Yuanquzao）/ 176
圆铃枣（Yuanlingzao）/ 134

Z

赞皇大枣（Zanhuangdazao）/ 284
赞皇长枣（Zanhuangchangzao）/ 426
赞新大枣（Zanxindazao）/ 316
枣强脆枣（Zaoqiangcuizao）/ 84
镇平九月寒（Zhenpingjiuyuehan）/ 388
直社疙瘩枣（Zhishegedazao）/ 206

中草笨枣（Zhongcaobenzao）/ 414
中宁大红枣（Zhongningdahongzao）/ 396
中宁小圆枣（Zhongningxiaoyuanzao）/ 260
中阳团枣（Zhongyangtuanzao）/ 296
遵义甜枣（Zunyitianzao）/ 216

主要参考文献

陈贻金. 1991. 中国枣树学概论[M]. 北京：中国科学技术出版社.
李登科. 2006. 枣种质资源描述规范和数据标准[M]. 北京：中国农业出版社.
刘孟军. 2004. 枣优质生产技术手册[M]. 北京：中国农业出版社.
刘孟军，汪民. 2009. 中国枣种质资源[M]. 北京：中国林业出版社.
曲泽州，王永蕙. 1993. 中国果树志:枣卷[M]. 北京：中国林业出版社.
张毅，孙岩. 2002. 枣推广新品种图谱[M]. 济南：山东科学技术出版社.
中国农业科学院果树研究所. 1993. 果树种质资源目录:第一集[G]. 北京：中国农业出版社.
中国农业科学院果树研究所. 1998. 果树种质资源目录:第二集[G]. 北京：中国农业出版社.

Main References

CHEN Y J. 1991. An Outline of Chinese Jujube[M]. Beijing: Science and Technology Press of China.
LI D K. 2006. Description Criterion and Data Standardization of Jujube Genetic Resources[M]. Beijing: China Agriculture Press.
LIU M J. 2004. Technical Manual for High-Quality Production of Jujube[M]. Beijing: China Agricultural Press.
LIU M J, WANG M. 2009. Chinese jujube Germplasm Resources [M]. Beijing: Chinese Forestry Press.
QU Z Z, WANG Y H. 1993. Records of Chinese Fruits·Jujube[M]. Beijing: Chinese Forestry Press.
ZHANG Y, SUN Y. 2002. Mapping on Popularized New Cultivars of Jujube[M]. Jinan: Shandong Science and Technology Press.
Pomology Institute, Chinese Academy of Agricultural Sciences. 1993. List of Genetic Resources for Fruits Vol. 1[G]. Beijing: China Agriculture Press.
Pomology Institute, Chinese Academy of Agricultural Sciences. 1998. List of Genetic Resources for Fruits Vol. 2[G]. Beijing: China Agriculture Press.